Solid and Hazardous
Waste
Management

Solid and Hazardous
Waste
Management

PM CHERRY

CBS

CBS Publishers & Distributors Pvt Ltd

New Delhi • Bengaluru • Chennai • Kochi • Kolkata • Mumbai • Pune
Hyderabad • Nagpur • Patna • Vijayawada

Solid and Hazardous
Waste Management

ISBN: 978-81-239-2830-2

First Edition: 2016
Reprint: 2017, 2021, 2023

Published by **Satish Kumar Jain** and produced by **Varun Jain** for

CBS Publishers & Distributors Pvt Ltd

4819/XI Prahlad Street, 24 Ansari Road, Daryaganj, New Delhi 110 002, India.
Ph: 011-23289259, 23266861

Website: www.cbspd.com
e-mail: delhi@cbspd.com

Corporate Office: 204 FIE, Industrial Area, Patparganj, Delhi 110 092

Ph: 011-4934 4934   Fax: 011-4934 4935

e-mail: publishing@cbspd.com; publicity@cbspd.com

Branches

- **Bengaluru:** Seema House 2975, 17th Cross, KR Road, Banasankari 2nd Stage, Bengaluru 560 070, Karnataka, India
  Ph: +91-80-26771678/79   Fax: +91-80-26771680   e-mail: bangalore@cbspd.com
- **Chennai:** 7, Subbaraya Street, Shenoy Nagar, Chennai 600 030, Tamil Nadu, India
  Ph: +91-44-26680620, 26681266   Fax: +91-44-42032115   e-mail: chennai@cbspd.com
- **Kochi:** 42/1325, 1326, Power House Road, Opp KSEB, Power House, Ernakulum Kochi 682 018, Kerala, India
  Ph: +91-484-4059061-65,67   Fax: +91-484-4059065   e-mail: kochi@cbspd.com
- **Kolkata:** 147, Hind Ceramics Compound, 1st Floor, Nilgunj Road, Belghoria, Kolkata-700056, West Bengal, India
  Ph: +033-25633055, 033-25633056   e-mail: kolkata@cbspd.com
- **Lucknow:** Basement, Khushnuma Complex, 7 Meerabai Marg (Behind Jawahar Bhawan),Lucknow-226001, UP, India
  Ph: +0522-4000032   e-mail: tiwari.lucknow@cbspd.com
- **Mumbai:** PWD Shed, Gala no 25/26, Ramchandra Bhatt Marg, Next to JJ Hospital Gate no. 2, Opp. Union Bank of India,
  Noorbaug, Mumbai-400009, Maharashtra, India
  Ph: 022-66661880/89   e-mail: mumbai@cbspd.com

Representatives

- Hyderabad   0-9885175004
- Patna   0-9334159340
- Jharkhand   0-9811541605
- Pune   0-9923910676
- Nagpur   0-9421945513
- Uttarakhand   0-9716462459

Printed at SRK Graphics, Shahdara,Delhi, India

Preface

Waste management is the collection, transport, processing, recycling or disposal, and monitoring of waste materials. The term usually relates to materials produced by human activity, and is generally undertaken to reduce their effect on health, the environment or aesthetics. Waste management is also carried out to recover resources from it. Waste management can involve solid, liquid, gaseous or radioactive substances, with different methods and fields of expertise for each. Waste management practices differ for developed and developing nations, for urban and rural areas, and for residential and industrial producers. Management for nonhazardous waste residential and institutional waste in metropolitan areas is usually the responsibility of local government authorities, while management for nonhazardous commercial and industrial waste is usually the responsibility of the generator.

This reference textbook summarises the important aspects from various areas of solid waste. The purpose of this book is twofold: first, to bring into focus those aspects of solid waste which are particularly valuable to solid waste management practice and, secondly, to lay the groundwork for understanding in the area of specialised analysis that will serve the reader as a basis in all the common phases of solid waste management practice and research. In addition, emphasis is laid more on advanced methods of management, which are required to help solve many of the complex modern problems facing the sanitary/ public health department officials.

This reference textbook on solid and hazardous waste management is divided into ten sections and contains 33 chapters. Each chapter covers an important aspects of solid and hazardous waste. Section I is devoted to general considerations. Chapter 1 is focused on solid waste management: a review. Chapter 2 deals with municipal solid waste management in India and aims to outline the existing situation of solid waste management problems and their remedies.

Section II concentrates on risks and hazardous management. Hazardous waste management is a multidimensional issues and water pollution, air pollution, solid waste and groundwater hydraulics are all to be taken into account simultaneously. Hazardous waste management is truly a multi-disciplinary field. The participants in programs at problem sites include owners, operators regulators, consultants, contractors and suppliers. However, the key role is played by the general public and the political system. Chapter 3 is devoted to hazardous waste management and discusses the sources, classification, reduction and their treatment. Chapter 4 focuses on abiotic treatment and highlights liquids–solid separations and their biological and chemical treatment. Chapter 5 acquaints the readers with treatment of sewage sludge. Chapter 6 brings to light managing risks and hazards which are not dictated by the method but by the performing teams having depth of experience in the nature of process, materials, construction, operations and maintenance of the plant.

Section III is devoted to recycling of wastes. Chapter 7 focuses on recycling of solid wastes. Chapter 8 concentrates on life cycle analysis (LCA) of material recycling. LCA enable a manufacturer to quantify, how much energy and raw materials are used, and how much solid, liquid and gaseous waste is generated at each stage of product life cycle. Chapter 9 acquaints the readers with paper recycling which is the process of recovering waste paper and remarking it into new paper products. Chapter 10 concentrates on glass recycling which is the process of turning waste glass into usable products. Chapter 11 is devoted to recycling of metals. Metal plays an important role with industrial developments and improved living standards. Chapter 12 deals with plastic recycling which is the process of recovering scrap or waste plastics and reprocessing the materials into useful products, sometimes completely in different in form from their original state. Chapter 13 discusses mechanical recycling of PVC waste. Chapter 14 brings to light tyre recycling which is the process of recycling tyres that are no longer suitable for use on vehicle due to wear or irreparable damage.

Section IV concentrates on E-waste management. Chapter 15 is devoted to electronic wastes and discusses treatment and disposal methods of these products. Chapter 16 is devoted to environmentally sound options for E-waste management. Chapter 17 deals with electronic waste management in India. The impact, status and electronic waste management strategies are discussed.

Section V acquaints the readers from waste to energy. Chapter 18 focuses on pyrolysis, gasification and combined pyrolysis gasification systems. Chapter 19 concentrates on incineration which is waste treatment process that involves the combustion of organic substances contained in waste materials.

Section VI discusses composting. Compost is a plant matter that has been decomposed and recycled as a fertiliser and soil amendment. Thus, composting is a biological process in which micro-organisms, mainly fungi and bacteria convert degradable organic waste into humus like substance. Chapter 20 concentrates on pretreatment of municipal solid waste by windrow composting. Chapter 21 is devoted to composting toilet and aerated static pile composting. A composting toilet is an aerobic processing system that treats excreta with no water or small volumes of flush water, via composting or managed aerobic decomposition.

Section VII focuses on waste landfill. Sanitary landfill involves well-designed engineering methods to protect the environment from contamination of solid or liquid wastes. Chapter 22 deals with landfill—sanitary landfill: a review. Chapter 23 discusses landfilling methods and operations. Chapter 24 acquaints the readers with landfill gas emission to the atmosphere. The gas generated from landfills is resulted from the process of waste decomposition and related to the waste landfilled and landfilled technologies used. Chapter 25 is devoted to leachate which is any liquid that, in passing through matter, extracts solutes, suspended solids or any other component of the material through which it has passed. Chapter 26 aims at providing electrochemical oxidation of landfill leachate treatment. Chapter 27 concentrates on pretreatment of industrial landfill leachate by Fenton's oxidation. Chapter 28 focuses on combined landfill gas and leachate extraction systems. One of the main impetuses for the installation of landfill gas collection and disposal systems is the already considerable weight of environmental legislation, driven by an increasingly aware public opinion.

Section VIII is devoted to hospital and biomedical wastes. Chapter 29 acquaints the readers with hospital waste and its management and discusses classification and treatment of wastes. Chapter 30 deals with biomedical waste management.

Section IX focuses on special topics. Chapter 31 brings to light environmental audit of municipal solid waste management. Chapter 32 concentrated on integrated waste management which can be defined as the integration of waste streams, collection and treatment methods, environmental benefits, economic optimisation and societal acceptability into a practical system for any region.

Section X focuses on various case studies. Chapter 33 deals with case studies related to sustainable waste management in Nagpur, solid waste management in Jalandhar, environmental impact assessment of municipal solid waste landfills (Jordan), solid waste management in Mangalore.

Glossary and index have been provided at the end for quick reference. Diagrams, figures and tables supplement the text. All the topics have been covered in a cogent and lucid style to help the reader grasp the information quickly and easily.

It may not be wrong to hold that the present reference textbook on *Solid and Hazardous Waste Management* is a complete treatise on this subject. It is essential reading for all students for B.Tech/ M.Tech (Chemical/Civil/Environmental Engineering). Besides students, the book will prove useful to industrialists and consultants in solid and hazardous waste management and decision-makers.

The reference textbook also caters to the requirement of the syllabus prescribed by various Indian universities for undergraduate students pursuing engineering, life sciences, environment and allied courses. It has been prepared with meticulous care, aiming at making the book error-free. Constructive suggestions are always welcome from users of this book.

PM Cherry

Contents at a Glance

Contents

SECTION II

SECTION III

SECTION IV

SECTION V

SECTION VII

SECTION VIII

SECTION I

General Considerations

Chapter 1

Solid Waste Management: A Review

INTRODUCTION

Solid waste refers here to all non-liquid wastes. In general this does not include excreta, although sometimes nappies and the faeces of young children may be mixed with solid waste. Solid waste can create significant health problems and a very unpleasant living environment if not disposed of safely and appropriately. If not correctly disposed of, waste may provide breeding sites for insect-vectors, pests, snakes and vermin (rats) that increase the likelihood of disease transmission. It may also pollute water sources and the environment.

ASSOCIATED RISKS

Disease Transmission

Decomposing organic waste attracts animals, vermin and flies. Flies may play a major role in the transmission of faecal-oral diseases, particularly where domestic waste contains faeces (often those of children). Rodents may increase the transmission of diseases such as leptospirosis and salmonella, and attract snakes to waste heaps.

Solid waste may also provide breeding sites for mosquitoes. Mosquitoes of the Aedes genus lay eggs in water stored in discarded items such as tins and drums; these are responsible for the spread of dengue and yellow fevers. Such conditions may also attract mosquitoes of the Anopheles genus, which transmit malaria. Mosquitoes of the Culex genus breed in stagnant water with high organic content and transmit microfilariases, appropriate conditions are likely to arise where leachate from waste enters pooling water. In times of famine or food scarcity, members of the affected population may be attracted to waste heaps to scavenge for food; this is likely to increase the risk of gastro-enteritis, dysentery and other illnesses.

Pollution

Poor management of the collection and disposal of solid waste may lead to leachate pollution of surface water or groundwater. This may cause significant problems if the waste contains toxic substances, or if nearby water sources are used for water supplies. Where large quantities of dry waste are stored in hot climates this may create a fire hazard. Related hazards include smoke pollution and fire threat to buildings and people.

Effect on Morale

The effect of living in an unhygienic and untidy environment may lead people to become demoralised and less motivated to improve conditions around them. Waste attracts more waste and leads to less hygienic behaviour in general.

SOURCES AND TYPES OF SOLID WASTE

Sources of Solid Waste

In most emergency situations the main sources of solid waste are:
1. Medical centres.
2. Food stores.
3. Feeding centres.
4. Food distribution points.
5. Slaughter areas.
6. Warehouses.
7. Agency premises.
8. Markets.
9. Domestic areas.

Appropriate solid waste management strategies may vary for institutional, communal and domestic sources, depending on types and volumes of waste. Waste from medical centres poses specific health hazards and for this reason is considered separately.

Type and Quantity of Waste

The type and quantity of waste generated in emergency situations varies greatly. The main factors affecting these are:
1. The geographical region (developed or less-developed country or region).
2. Socio-cultural practices and material levels among affected population.
3. Seasonal variations (affecting types of food available).
4. The stage of emergency (volume and composition of waste may change over time).
5. The packaging of food rations.

In general, the volume of waste generated is likely to be small and largely degradable where the population is of rural origin and the food rations supplied are unpackaged dry foodstuffs. Displaced urban populations are more likely to generate larger volumes of non-degradable waste, especially where packaged food rations are provided.

Guideline values suggest that each person is likely to produce 0.5–1.0 litre of refuse per day with an organic content of 25 to 35 per cent and a moisture content between 10 and 60 per cent. However, this is likely to vary greatly and estimates should be made locally.

Different categories of solid waste include:

Organic waste	Waste from preparation of food, market places, etc.
Combustibles	Paper, wood, dried leaves, packaging for relief items, etc. (high organic and low moisture content).
Non-combustibles	Metal, tin cans, bottles, stones, etc.
Ashes/dust	Residue from fires used for cooking.

Bulky waste	Tree branches, tyres, etc.
Dead animals	Carcasses of domestic animals and livestock.
Hazardous waste	Oil, battery acid, medical waste.
Construction waste	Roofing, rubble, broken concrete, etc.

INITIAL STEPS

In order to establish effective solid waste management in the affected area the process is as shown in Fig. 1.1.

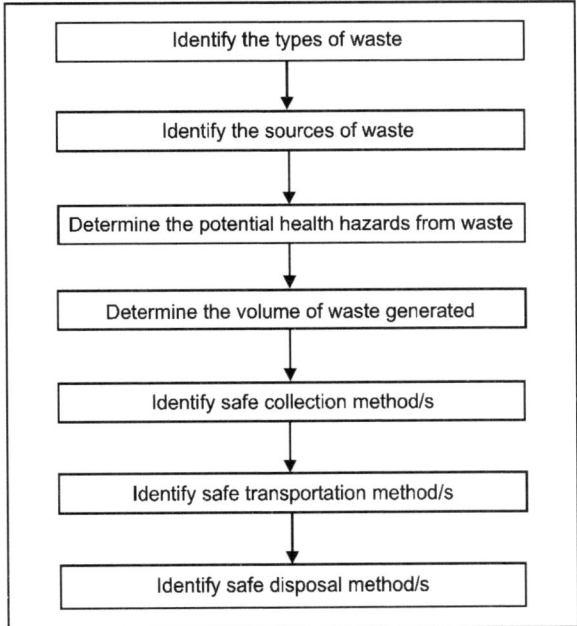

Fig. 1.1 Initial steps in solid waste management.

Key Components of Solid Waste Management

Solid waste management can be divided into five key components:

1. Generation.
2. Storage.
3. Collection
4. Transportation.
5. Disposal.

Generation

Generation of solid waste is the stage at which materials become valueless to the owner and since they have no use for them and require them no longer, they wish to get rid of them. Items which may be valueless to one individual may not necessarily be valueless to another. For example, waste items such as tins and cans may be highly sought after by young children.

Storage

Storage is a system for keeping materials after they have been discarded and prior to collection and final disposal. Where on-site disposal systems are implemented, such as where people discard items directly into family pits, storage may not be necessary. In emergency situations, especially in the early stages, it is likely that the affected population will discard domestic waste in poorly defined heaps close to dwelling areas. If this is the case, improved disposal or storage facilities should be provided fairly quickly and these should be located where people are able to use them easily. Improved storage facilities include:

1. Small containers: Household containers, plastic bins, etc.
2. Large containers: Communal bins, oil drums, etc.
3. Shallow pits.
4. Communal depots: Walled or fenced-in areas.

In determining the size, quantity and distribution of storage facilities the number of users, type of waste and maximum walking distance must be considered. The frequency of emptying must also be determined, and it should be ensured that all facilities are reasonably safe from theft or vandalism.

Collection

Collection simply refers to how waste is collected for transportation to the final disposal site. Any collection system should be carefully planned to ensure that storage facilities do not become overloaded. Collection intervals and volumes of collected waste must be estimated carefully.

Transportation

This is the stage when solid waste is transported to the final disposal site. There are various modes of transport which may be adopted and the chosen method depends upon local availability and the volume of waste to be transported. Types of transportation can be divided into three categories:

1. Human-powered: Open hand-cart, hand-cart with bins, wheelbarrow, tricycle.
2. Animal-powered: Donkey-drawn cart.
3. Motorised: Tractor and trailer, standard truck, tipper-truck.

Disposal

The final stage of solid waste management is safe disposal where associated risks are minimised. There are four main methods for the disposal of solid waste:

1. Land application: Burial or landfilling.
2. Composting.
3. Burning or incineration.
4. Recycling (resource recovery).

The most common of these is undoubtedly land application, although all four are commonly applied in emergency situations.

ON-SITE DISPOSAL OPTIONS

The technology choices outlined below are general guidelines for disposal and storage of waste on-site, these may be adapted for the particular site and situation in question.

Communal Pit Disposal

Perhaps the simplest solid waste management system is where consumers dispose of waste directly into a communal pit. The size of this pit will depend on the number of people it serves. The long-term recommended objective is six cubic metres per fifty people. The pit should be fenced off to prevent small children falling in and should generally not be more than 100 m from the dwellings to be served. Ideally, waste should be covered at least weekly with a thin layer of soil to minimise flies and other pests.

Advantages: It is rapid to implement, and requires little operation and maintenance.

Constraints: The distance to communal pit may cause indiscriminate disposal, and waste workers required to manage pits.

Family Pit Disposal

Family pits may provide a better long-term option where there is adequate space. These should be fairly shallow (up to 1 m deep) and families should be encouraged to regularly cover waste with soil from sweeping or ash from fires used for cooking. This method is best suited where families have large plots and where organic food wastes are the main component of domestic refuse.

Advantages: Families are responsible for managing their own waste; no external waste workers are required; and community mobilisation can be incorporated into hygiene promotion program.

Constraints: Involves considerable community mobilisation for construction, operation and maintenance of pits; and considerable space is needed.

Communal Bins

Communal bins or containers are designed to collect waste where it will not be dispersed by wind or animals, and where it can easily be removed for transportation and disposal. Plastic containers are generally inappropriate since these may be blown over by the wind, can easily be removed and may be desirable for alternative uses. A popular solution is to provide oil drums cut in half. The bases of these should be perforated to allow liquid to pass out and to prevent their use for other purposes. A lid and handles can be provided if necessary.

In general, a single 100 litre bin should be provided for every fifty people in domestic areas, everyone hundred people at feeding centres and every ten market stalls. In general, bins should be emptied daily.

Advantages: Bins are potentially a highly hygienic and sanitary management method; and final disposal of waste well away from dwelling areas.

Constraints: Significant collection, transportation and human resources are required; system takes time to implement; and efficient management is essential.

Family Bins

Family bins are rarely used in emergency situations since they require an intensive collection and transportation system and the number of containers or bins required is likely to be huge. In the later stages of an emergency, however, community members can be encouraged to make their own refuse baskets or pots and to take responsibility to empty these at communal pits or depots.

Advantages: Families are responsible for maintaining collection containers; and potentially a highly sanitary management method.

Constraints: In general, the number of bins required is too large; significant collection, transportation and human resources are required; takes time to implement; and efficient management essential.

Communal Disposal Without Bins

For some public institutions, such as markets or distribution centres, solid waste management systems without bins can be implemented, whereby users dispose of waste directly onto the ground. This can only work if cleaners are employed to regularly sweep around market stalls, gather waste together and transport it to a designated off-site disposal site. This is likely to be appropriate for vegetable waste but slaughterhouse waste should be disposed of in liquid-tight containers and buried separately.

Advantages: System rapid to implement; there is minimal reliance on actions of users; and it may be in line with traditional/usual practice.

Constraints: Requires efficient and effective management; and full-time waste workers must be employed.

TRANSPORTATION OPTIONS

Where bins or collection containers require emptying, transportation to the final disposal point is required. As described, waste transportation methods may be human-powered, animal-powered or motorised.

Human-Powered

Wheelbarrows are ideal for the transportation of waste around small sites such as markets but are rarely appropriate where waste must be transported considerable distances off-site. Handcarts provide a better solution for longer distances since these can carry significantly more waste and can be pushed by more than one person. Carts may be open or can be fitted with several containers or bins.

Animal-Powered

Animal-powered transportation means such as a horse or donkey with cart are likely to be appropriate where they are commonly used locally. This may be ideal for transportation to middle distance sites.

Motorised

Where the distance to the final disposal site is great, or where the volume of waste to be transported is high, the use of a motorised vehicle may be the only appropriate option. Options include tractor and trailer, a standard truck, or a tipper-truck, the final choice depending largely on availability and speed of procurement. For large volumes of waste it may sometimes be appropriate to have a two-stage transportation system requiring a transfer station. For example, waste is transported by handcart to a transfer station where it is loaded into a truck to be taken to an off-site disposal site several kilo metres away.

OFF-SITE DISPOSAL OPTIONS

The technology choices outlined below are general options for the final disposal of waste off-site.

Landfilling

Once solid waste is transported off-site it is normally taken to a landfill site. Here the waste is placed in a large excavation (pit or trench) in the ground, which is back-filled with excavated soil each day waste is tipped. Ideally, about 0.5 m of soil should cover the deposited refuse at the end of each day to prevent animals from digging up the waste and flies from breeding.

The location of landfill sites should be decided upon through consultation with the local authorities and the affected population. Sites should preferably be fenced, and at least one kilometre downwind of the nearest dwellings.

Advantages: A sanitary disposal method if managed effectively.

Constraints: A reasonably large area is required.

Incineration

Although burning or incineration is often used for the disposal of combustible waste, this should generally only take place off-site or a considerable distance downwind of dwellings. Burning refuse within dwelling areas may create a significant smoke or fire hazard, especially if several fires are lit simultaneously. Burning may be used to reduce the volume of waste and may be appropriate where there is limited space for burial or landfill. Waste should be ignited within pits and covered with soil once incinerated, in the same manner as landfilling. The same constraints for siting landfill sites should be applied here also.

Advantages: Burning reduces volume of combustible waste considerably; and it is appropriate in off-site pits to reduce scavenging.

Constraints: There can be smoke or fire hazards.

Composting

Simple composting of vegetables and other organic waste can be applied in many situations. Where people have their own gardens or vegetable plots, organic waste can be dug into the soil to add humus and fibre. This makes the waste perfectly safe and also assists the growing process. This should be encouraged wherever possible, particularly in the later stages of an emergency program.

Properly managed composting requires careful monitoring of decomposing waste to control moisture and chemical levels and promote microbial activity. This is designed to produce compost which is safe to handle and which acts as a good fertiliser. Such systems require considerable knowledge and experience and are best managed centrally. In general, they are unlikely to be appropriate in emergencies.

Advantages: Composting is environmentally friendly; and beneficial for crops.

Constraints: Intensive management and experienced personnel are required for large-scale operations.

Recycling

Complex recycling systems are unlikely to be appropriate but the recycling of some waste items may be possible on occasions. Plastic bags, containers, tins and glass will often be automatically recycled since they are likely to be scarce commodities in many situations. In most developing country contexts there exists a strong tradition of recycling leading to lower volumes of waste than in many more developed societies.

Advantages: Recycling is environmentally friendly.

Constraints: There is limited potential in most emergency situations, and it is expensive to set up.

PROTECTIVE MEASURES

In order to minimise disease transmission there are several protective measures that can be undertaken. These concern equipment for staff and the siting and management of disposal sites.

Staff

It is important that workers employed to collect and transport solid waste are provided with appropriate clothing and equipment. Gloves, boots and overalls should be provided wherever possible. Where waste is burned, or is very dusty, workers should have protective masks. Water and soap should be available for hand and face washing, and changing facilities should be provided where appropriate.

Siting of Disposal Sites

The location of all disposal sites should be determined through consultation with key stakeholders · including local government officials, representatives of local and displaced populations, and other agencies working in the area. Appropriate siting should minimise the effects of odour, smoke, water pollution, insect vectors and animals.

On-site disposal is generally preferred since this requires no transportation and staff needs are low. This is appropriate where volumes of waste are relatively small, plenty of space is available and waste is largely organic or recyclable.

If the volumes of waste generated are large, or space within the site is severely limited, it may be necessary to dispose of waste off-site. Where off-site disposal is to be used the following measures should be taken in selecting and developing an appropriate site:

1. Locate sites at least 500 m (ideally 1 km) downwind of nearest settlement.
2. Locate sites downhill from groundwater sources.
3. Locate sites at least 50 m from surface water sources.
4. Provide a drainage ditch downhill of landfill site on sloping land.
5. Fence and secure access to site.

Careful assessment should be made to determine who owns the proposed site and to ensure that apparently unused areas are not in fact someone's farm or back yard.

Chapter 2

Municipal Solid Waste Management in India

INTRODUCTION

Urbanisation is now becoming a global phenomenon, but its ramifications are more pronounced in developing countries. Natural growth of population, reclassifications of habitation and migration trends are important in urban population in India. Due to rapid urbanisation and uncontrolled growth rate of population, solid waste management (SWM) has become acute in India. Municipal bodies in India render SWM services. Though, it is an essential service, it is not attaining proper priority, which it deserves and services are poor. NEERI has provided extensive services to municipal bodies in India to improve their MSWM system.

The present chapter aims to outline the existing situation of SWMS, problems associated with the system and also highlights some best practices and lessons learnt by NEERI's experience along with EXNORA's Zero Waste Management in two South Indian cities. (EXNORA, which is an acronym for 'Excellent Novel Redical'). An approach for design of sustainable SWMS compatible to Indian situations is also detailed.

SITUATION ANALYSIS

Municipal solid waste management (MSWM) is a part of public health and sanitation, and is entrusted to the municipal government for execution. Presently, the systems are assuming larger importance due to population explosion in municipal areas, legal intervention, emergence of newer technologies and rising public awareness towards cleanliness.

Except in the metropolitan cities, SWM is the responsibility of a health officer who is assisted by the engineering department in the transportation work. The activity is mostly labour intensive, and 2–3 workers are provided per 1000 residents served. The municipal agencies spend 5–25 per cent of their budget on SWM, which is Rs. 75–250 per capita per year. Normally a city of 1 million populations spends around Rs. 10 crores for this activity. In spite of this huge expenditure, services are not provided to the desired level.

Quantity and characteristics are two major factors, which are considered as the basis for the design of efficient, cost effective and environmentally compatible waste management system. The municipal corporation often depends on the vehicle trips record to estimate the waste quantity.

This does not give the actual picture of waste generation. NEERI has conducted extensive studies on quantum of waste generation in various cities. Studies have revealed that quantum of waste generation varies between 0.2–0.4 kg/capita/day in the urban centres and it goes up to 0.5 kg/capita/day in metropolitan cities. Per capita waste quantity for various cities with different population is presented in Table 2.1.

Table 2.1. Per capita quantity of municipal solid waste in Indian cities (NEERI, 2008).

Population range (in million)	Average per capita value kg/capita/day
0.1–0.5	0.21
0.5–1.0	0.25
1.0–2.0	0.27
2.0–5.0	0.35
>5.0	0.50

Characterisation studies carried out by NEERI indicate that MSW contains large organic fraction (30–40 per cent), ash and fine earth (30–40 per cent), paper (3–6 per cent) along with plastic glass and metal (each less than 1 per cent), calorific value of refuse ranges between 800–1000 kcal/kg and C/N ratio ranges between 20 and 30. Presently, NEERI has again been retained by Central Pollution Control Board (CPCB), New Delhi in quantification and characterisation of MSW in metro, class I and class II cities and towns to know the actual quantities as well as characteristics of solid wastes in designing MSWM system.

The collection bin and implements used in various cities are not properly designed. It has been observed that community bins have not been installed at proper location. This has resulted in poor collection efficiency. Lack of public awareness has made the situation worse. Various types of vehicles are used for transportation of waste to the disposal site. However, these vehicles are not designed as per requirement. In many urban centers, proper garages are not provided for the vehicles for protection from heat and rain. Preventive maintenance system is not adopted and as a result the life of the vehicle is reduced. Many of the vehicles used for transportation of waste have outlived their normal life.

Manual composting is carried out in smaller urban centres. Although in 1980's mechanical composting plants were set up in 10 cities, presently, only one plant out of them continues to be in operation. Over the years, a few more plants have been set up. Incineration has not been successful due to the low calorific value of the solid waste. Waste is disposed of in low-lying areas without taking any precautions and without any operational control. Solid waste workers handle the waste without any protective equipment and are prone to infection.

FUTURE SCENARIO

The waste quantities are estimated to increase from 150 million tons in 2007 to 250 million tons in 2015. The waste characteristics are expected to change due to urbanisation, increased commercialisation and standard of living. The present trend indicates that the paper and plastics content will increase while the organic content will decrease. The ash and earth content is also expected to decrease mainly due to an increase in the paved surface. Although, the organic content is expected to decrease, the material will still be amenable to biodegradation and the calorific value will continue to be unsuitable for incineration.

In keeping with the present practices and estimates of waste generation, around 90 per cent of the generated wastes are landfilled requiring around 1200 hectare of land every year with an average depth of 3 m. Due to rapid urbanisation, prevailing land use regulation and completing demands for available land, it is desirable that adequate land be earmarked at the planning stage itself for solid waste disposal. The larger quantities of solid waste and higher degree of urbanisation will necessitate better management involving a higher level of expenditure on manpower and equipment.

PROBLEMS ASSOCIATED WITH THE SYSTEM

SWM systems exist in most of the urban centres since last few decades. However, these systems have yet to emerge as a well-organised practice. Although, the solid waste characteristics in different urban centers vary significantly, there is a meager effort to tailor the system configuration to the waste characteristics.

Major Deficiencies Associated with System

The major deficiencies associated with the system are discussed below.

Rapidly increasing areas to be served and quantity of waste

The solid waste quantities generated in urban centres are increasing due to rise in the population and increase in the per capita waste generation rate. The increasing solid waste quantities and the areas to be served strain the existing SWM system.

Inadequate resources

While allocating resources including finance, SWM is assigned with a low priority resulting in inadequate provision of funds. Often there is a common budget for collection and treatment of sewage and SWM and the later receives a minor share of the funds. The inadequacy of human resource is mainly due to the absence of suitably trained staff.

Inappropriate technology

The equipment and machinery presently used in the system are usually that which have been developed for general purpose or that which have been adopted from other industry. This results in underutilisation of existing resources and lowering of the efficiency. A few attempts have been made to borrow the technology developed in other countries like highly mechanised compost plants, incinerator-cum-power plants, compactor vehicles, etc. However, these attempts have met with little success, since, the solid waste characteristics and local conditions in India are much different from those for which the technology is developed.

Disproportionately high cost of manpower

Mostly out of the total expenditure, around 90 per cent is accounted for manpower of which major portion is utilised for collection. Since citizens tend to throw the waste on the adjoining road and outside the bin, the work of the collection staff is increased. Hence, the cost of collection increases considerably.

Societal and management apathy

The operational efficiency of SWM depends on the active participation of both the municipal agency and the citizens. Since the social status of SWM is low, there is a strong apathy towards it, which can be seen from the uncollected waste in many areas and the deterioration of aesthetic and environmental quality at the uncontrolled disposal sites.

Low efficiency of the system

The SWM system is unplanned and is operated in an unscientific way. Neither the work norms are specified nor the work of collection staff appropriately supervised. The vehicles are poorly maintained and no schedule is observed for preventive maintenance. Due to shortage of financial resources, the vehicles are often used beyond their economical life resulting in inefficient operation. Further, there is no co-ordination of activities between different components of the system. The cumulative effect of all these factors is an inefficient SWM system.

NEERI'S CONTRIBUTION TOWARDS MSWM IN INDIA

Since last three decades, at NEERI, SWM Division has been carrying research, development and expertise extension program to improve the status of waste management in the country. In order to perform practice-driven research, NEERI has been consistently working with the partnership of related organisations in the country.

The prominent best practices evolved during the last five years, are described below:
1. Preparation of strategy paper on SWM in India.
2. Long-term planning of SWM.
3. Biomethanation of vegetable market wastes.
4. Greenhouse gas inventory estimation for waste sector, its uncertainty analysis and formulated measures to mitigate the same.
5. Utilisation of landfill site for construction of Rail Car Depot.
6. Site selection criteria for sanitary landfills.
7. Utilisation of residue from destruction of soiled currency notes.

NEERI has successfully developed strategic long-term plans for a number of cities including metropolitan cities like Mumbai, Delhi and Islands like Lakshadweep and Port Blair, among which many municipalities adopted the plan. NEERI is also assisting in implementation of the long-term plans. NEERI has been actively engaged in various ecosystems like Island areas such as Lakshadweep, coastal areas such as Mumbai and others metropolitan cities and towns having different geographical, climatological and social environment in the country.

LESSONS LEARNT FROM NEERI'S CONTRIBUTION AND EXNORA'S ZEROS WASTE MANAGEMENT

The involvement of NEERI in SWM activity emphasises that the improvement of system needs to be developed addressing the following issues:
1. Financial weakness of managing agency.
2. Difficulties in changing the prevailing nature of infrastructure service.
3. Low recycling potential of waste material.
4. Non-availability of skilled labours.
5. Societal and managerial apathy.

EXNORA, a local non-governmental organisation (NGO) based on a 'zero waste management scheme' set up, run and financed by the residents themselves. EXNORA, which is an acronym for 'Excellent Novel Radical', has been driving the environmentalist movement for sustainable urban development in Chennai (India) since 1989. It has been promoting community-based projects in areas where the local government is unable to provide sufficient service. EXNORA believes that areabased project led by the local community is an ideal way to spread environmental, social and civic messages resulting in a more equal and responsible society.

EXNORA identified SWM as a medium to promote a new model of society based on participative democracy. It places SWM at the core of society: its moral values, its social structure, its lifestyle, its economics and politics. The model itself raises the fundamental issue of governance. It is an idealistic model because if it were to work as originally envisaged it suggests that citizens would no longer need local government for the provision of basic needs such as a clean environment. It stresses the need for local bodies to be more aware of people's concerns and for citizens to be more involved in civic affairs. This model has been implemented in two communities viz. Chennai and Hyderabad. Results from two Indian cities, indicate limited success of the schemes both in saving a significant fraction of the generated waste from dumping, and in rehabilitating the local poor. However, they show that motivated individuals can successfully set up and manage waste collection systems that lead to overall environmental improvements. The system advocated by EXNORA seems to require significant local resources, and political and technical support which are hard to find and sustain without strong local leaders. This is based on triangular contracts between the municipality, the residents and microenterprises and may provide a good solution in dealing with the technical and commercial aspects which communities find difficult.

DESIGN OF APPROPRIATE WASTE MANAGEMENT SYSTEM

Approach to Design Waste Management System

An approach to design sustainable waste management system and operating guidelines is outlined below.

Quantity and characteristics

Quantity and characteristics of the waste are the major factors, which decide magnitude of waste management problem. It is necessary to carry out weighment exercise regularly to assess the quantity of waste. Future per capita quantity can be estimated with the help of projected population and annual increase of per capita quantity. On the basis of the waste quantity, infrastructure requirement can be estimated. It is also necessary to carry out characterisation studies frequently in order to assess the changes in waste characteristics due to ever-changing scenario. This data will also serve as a basis for selection of disposal/treatment option.

Collection of waste

Properly designed collection bins and implements should be used for collection and storage of waste. Wastes should be collected frequently in order to avoid accumulation, which leads to degradation of environmental and aesthetic quality. Suggested collection frequencies for cities with different population ranges have been presented in Table 2.2. Suggestion from the citizen as well as workers for improvement in the design of bins and implements will be useful. Spacing and location of the bins should be fixed on the basis of the waste load and public opinion. House to house collection system can be introduced gradually to ensure environment friendly collection practices.

Transportation of waste

Selection of properly designed vehicles is important. Various factors like width of the road, transport volume, road conditions, etc. play important role in selection of vehicles. Proper garage should be provided to save the vehicles from wear and tear due to heat and rain. Preventing maintenance system should be introduced which is useful for longer life of the vehicles. Vehicle route should be properly planned for proper utilisation of manpower, saving of fuel and reduction of time. Time and motion study should be conducted to reduce the non-productive idle time of the vehicles and increase productivity.

Table 2.2. Collection frequency of solid waste (NEERI, 2008).

Types of locality	Frequency	
	Class I cities (> 1 lakh population)	Class II cities (< 1 lakh population)
Residential areas		
Area with high population density	Once or twice a day	Once a day
Area with medium population density	Once or twice a day	Once in two days
High income and VIP area	Once or twice a day	Once a day
Area with low population density	Once in two days	Once in three days
Markets	Once or twice a day	Once or twice a day
Commercial areas	Twice a day	Once a day
Industrial areas	Once a day	Once a day

Disposal of waste

Sanitary landfill technique should be adopted for disposal for waste. Compaction of waste should be carried out regularly preferably with bulldozer. A daily earth cover of 15 cm thickness and final cover of 60 cm thickness should be applied over the compacted waste. These practices will minimise migration of leachate through soil strata, suppress the foul odour and improve the aesthetic quality. Impervious clay liner/synthetic liner should be provided at the bottom of the landfill for protection of groundwater from environmental pollution. Perforate polyvinyl chloride (PVC) pipe can be provided for leachate collection. It is also desirable to install gas collection and flaring system to prevent continuous escape of methane in the surrounding atmosphere.

Treatment/recycling of waste

Composting is the process of decomposition and stabilisation of organic matter under controlled condition. Since India is an agriculture-based country, there is a need for popularisation of the product among the farmers and to exploit the manure value of the product. Waste minimisation, through segregation of recyclable materials like plastics, glass, metals, etc. is another aspect, which needs special attention. NGOs may come forward to promote the activity. Waste pickers may be trained so that the segregation of recyclable items can be done in a more systematic and organised way.

Financial structure

A new tax scheme can be introduced to meet the expenditure for modernisation of SWM system and to improve the financial status of municipal corporation. Additional charges can be collected from the individuals availing house-to-house collection facility.

Community participation

Community participation is essential for smooth and efficient operation of SWM system. In every area, citizen forums should be formed. These forums should comprise citizen's representatives, social workers and municipal officers. Immediate action based on feedback from such forum will go a long way in improving the situation. Various programs should be conducted for increasing public awareness.

Thus SWM is a vital, ongoing and large public service system, which needs to be efficiently provided to the community to maintain aesthetic and public health standards. Municipal agencies will have to plan and execute the system in keeping with the increasing urban areas and population. There has to be a systematic effort in the improvement in various factors like institutional arrangement, financial provisions, appropriate technology, operations management, human resource development, public participation and awareness, and policy and legal framework for an integrated SWM system. To achieve cleanliness, which is next to Godliness, it is necessary to design and operate an efficient SWM system. Public co-operation is essential for successful operation of such a system. Finally, there is also a need to develop a methodology of research for developing interactive techniques for system's design and operational control as indicated in Fig. 2.1.

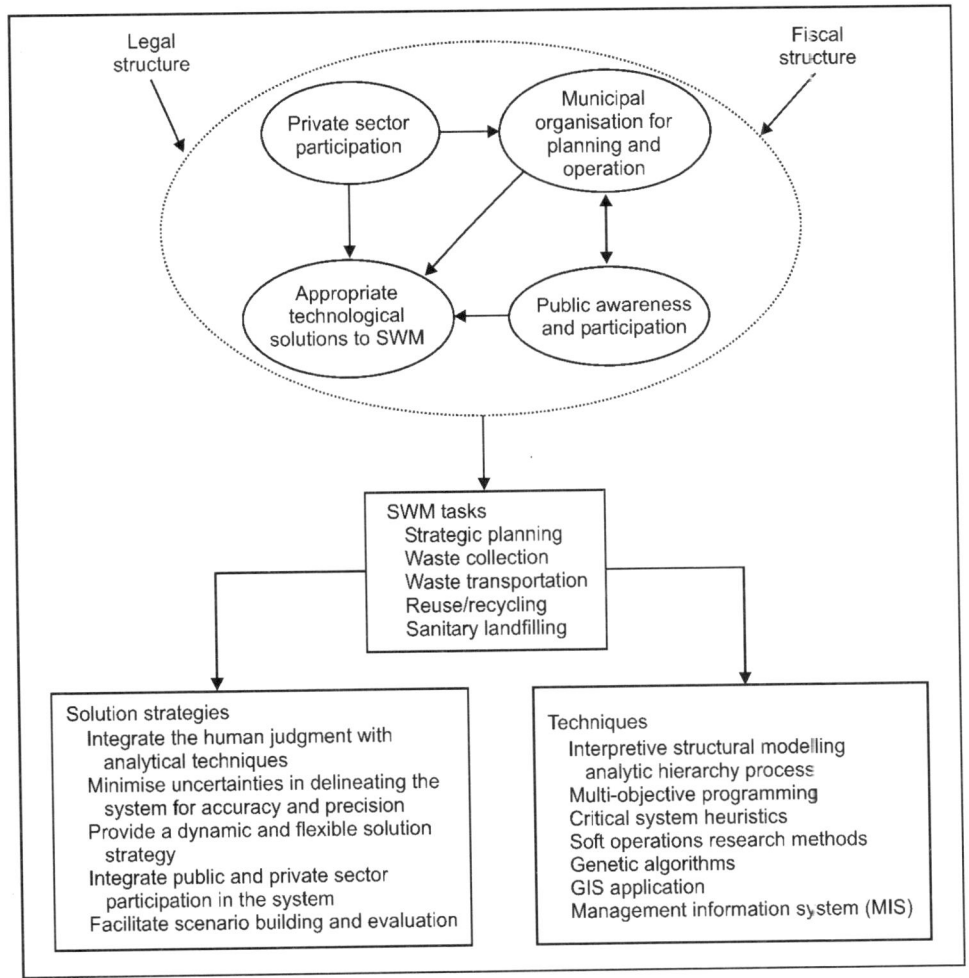

Fig. 2.1. Methodology of research for developing interactive techniques for system's design and operational control.

SECTION II

Risks and Hazardous Management

Chapter 3

Hazardous Waste Management

INTRODUCTION

A hazardous waste is waste that poses substantial or potential threats to public health or the environment. A hazardous substance is a material that may pose a danger to living organisms, materials, structures or the environment by explosion or fire hazards, corrosion, toxicity to organisms or other detrimental effects. A simple definition of a hazardous waste is that it is a hazardous substance that has been discarded, abandoned, neglected, released or designated as a waste material, or one that may interact with other substances to be hazardous.

Listed hazardous waste are materials specifically listed by the EPA or State as a hazardous waste. Hazardous wastes listed by EPA fall into two major categories:

1. Process wastes from general activities (F-listed) and from specific industrial processes (K-listed).
2. Unused or off-specification chemicals, container residues and spill cleanup residues of acute hazardous waste chemicals (P-listed) and other chemicals (U-listed).

These wastes may be found in different physical states such as gaseous, liquids or solids. Furthermore, a hazardous waste is a special type of waste because it cannot be disposed of by common means like other by-products of our everyday lives. Depending on the physical state of the waste, treatment and solidification processes might be available. In other cases, however, there is not much that can be done to prevent harm.

NATURE AND SOURCES OF HAZARDOUS WASTE

Classification of Hazardous Substances and Wastes

Many specific chemicals in widespread use are hazardous because of their chemical reactivates, fire hazards, toxicities, and other properties. There are numerous kinds of hazardous substances, usually consisting of mixtures of specific chemicals. These include the following:

1. Explosives, such as dynamite, or ammunition.
2. Compressed gases, such as hydrogen and sulphur dioxide.
3. Flammable liquids, such as gasoline and aluminium alkyls.
4. Flammable solids, such as magnesium metal, sodium hydride, and calcium carbide that burn readily, are water-reactive or spontaneously combustible.
5. Oxidising materials, such as lithium peroxide, that supply oxygen for the combustion of normally non-flammable materials.

21

6. Corrosive materials, including oleum, sulphuric acid and caustic soda, which may wound exposed flesh or cause disintegration of metal containers.
7. Poisonous materials, such as hydrocyanic acid or aniline.
8. Etiologic agents, including causative agents of anthrax, botulism or tetanus.
9. Radioactive materials, including plutonium, cobalt-60, and uranium hexafluoride.

Characteristics of listed wastes

The characteristics of listed wastes are:

1. Ignitability, characteristic of substances that are liquids whose vapours are likely to ignite in the presence of ignition sources, non-liquids that may catch fire from friction or contact with water and which burn vigorously or persistently, ignitable compressed gases, and oxidisers.
2. Corrosivity, characteristic of substances that exhibit extremes of acidity or basicity or a tendency to corrode steel.
3. Reactivity, characteristic of substances that have a tendency to undergo violent chemical change (examples are explosives, pyrophoric materials, water-reactive substances, or cyanide sulphide-bearing wastes).
4. Toxicity, defined in terms of a standard extraction procedure followed by chemical analysis for specific substances.

In addition to classification by characteristics, EPA designates more that 450 listed wastes which are specific substances or classes of substances known to be hazardous. Each such substance is assigned an EPA hazardous waste number in the format of a letter followed by 3 numerals, where a different letter is assigned to substances from each of the following lists:

1. *F-type wastes from non-specific sources:* For example, quenching waste-water treatment sludges from metal heat treating operations where cyanides are used in the process.
2. *K-type wastes from specific sources:* For example, heavy ends from the distillation of ethylene dichloride in ethylene dichloride production.
3. *P-type acute hazardous wastes:* Wastes that have been found to be fatal to humans in low doses, or capable of causing or significantly contributing to an increase in serious irreversible or incapacitating reversible illness. These are mostly specific chemical species such as fluorine or 3-chloropropane nitrile.
4. *U-type miscellaneous hazardous wastes:* These are predominantly specific compounds such as calcium chromate or phthalic anhydride.

Hazardous wastes

Three basic approaches to defining hazardous wastes are: (i) a qualitative description by origin, type, and constituents, (ii) classification by characteristics largely based upon testing procedures, and (iii) by means of concentrations of specific hazardous substances. Wastes may be classified by general type such as 'spent halogenated solvents' or by industrial sources such as 'pickling liquor from steel manufacturing'.

Various countries have different definitions of hazardous wastes. Radioactive wastes are a problem for any country with a significant nuclear power or weapons industry. Special problems are posed by mixed waste containing both radioactive and chemical wastes.

Hazardous wastes and air and water pollution control

Somewhat paradoxically, measures taken to reduce air and water pollution (Fig. 3.1) have had a tendency to increase production of hazardous wastes. Most water treatment processes yield sludges or concentrated liquors that require stabilisation and disposal. Air scrubbing process likewise produce sludges. Baghouses and precipitators used to control air pollution all yield significant quantities of solids, some of which are hazardous.

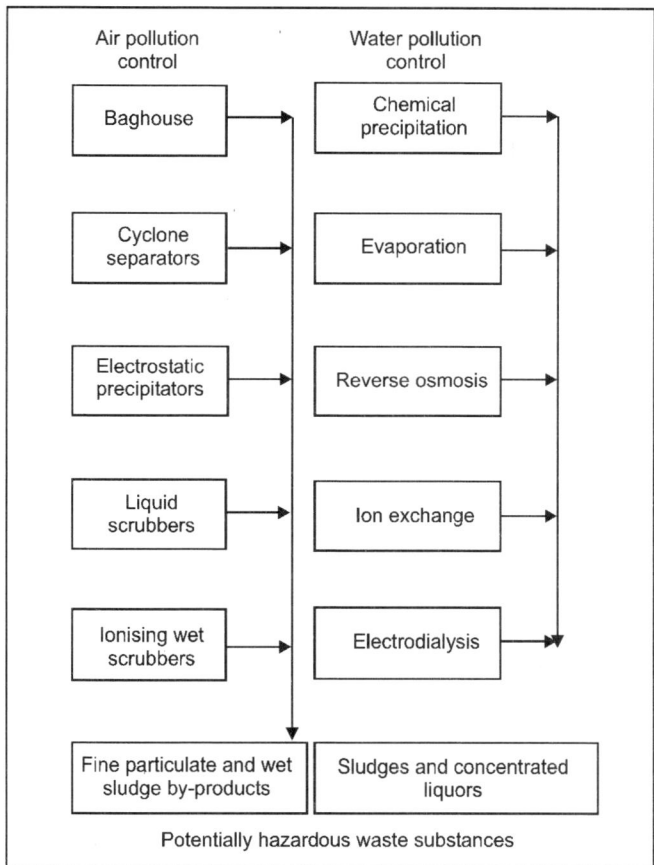

Fig. 3.1. Potential contributions of air and water pollution control measures to hazardous wastes production.

Origin and Amounts of Wastes

In a non-regulatory sense there is no sharp demarcation between hazardous and nonhazardous wastes. Some wastes, such as soluble toxic heavy metal salt wastes, are obviously hazardous. By comparison, discarded leaves and tree trimmings would be regarded as posing no danger. But, if properly treated and immobilised, the heavy metal wastes are of little danger, whereas discarded tree limbs pose a fire hazard under certain circumstances. Materials that by themselves are non-hazardous may interact with hazardous substances to increase the dangers from the latter. For example, soluble humic substances from the decay of tree leaves may solubilise and transport heavy metal ions.

Staggering amounts of wastes of all kinds are produced by human activities. Such wastes include municipal refuse, sewage sludge, agricultural residues, and toxic, chemically reactive by-products of manufacturing processes. An idea of quantities of solid wastes generated can be obtained by considering mining and milling wastes. The quantities of such wastes are enormous because large quantities of rock must be removed to get to the ore and because the metal or other economically valuable constituent is usually a small percentage of the ore. Therefore, by-products such as overburden and beneficiation wastes accumulate in vast amounts.

Non-hazardous solid wastes

It is appropriate to consider 'non-hazardous' waste (solid waste, the municipal refuse and garbage produced by human activities) along with hazardous waste because it may not be non-hazardous in all cases and situations, and it may interact with hazardous wastes. Furthermore, the amounts of solid waste produced each year are so enormous that capacity to deal with the problem is under severe strain. Disposal of about 92 per cent of municipal refuse is in landfills. However, as total quantities of solid waste have increased, the landfill capacity to handle waste has decreased.

The potential of incineration for handling municipal refuse is very high because it can reduce waste mass by 75 per cent and volume by 90 per cent. However, environmental concern about organic pollutants (particularly dioxins) in stack emissions and heavy metals in incinerator ash have slowed municipal incinerator development. Recycling can certainly reduce quantities of sold waste, perhaps as much as 50 per cent, but it is not the panacea claimed by its most avid advocates. The overall solution to the solid waste problem must involve several kinds of measures, particularly: (i) reduction of wastes at the source, (ii) recycling as much waste as is practical, (iii) reducing the volume of remaining wastes by measures such as incineration, (iv) treating residual material as much as possible to render it non-leachable and innocuous, and (v) placing the residual material in landfills, properly protected from leaching or release by other pathways.

Some of the hazardous wastes are those from specific sources produced by industries such as the manufacture of inorganic pigments, organic chemicals, pesticides, explosives, iron and steel, and nonferrous metals, and from processes such as petroleum refining or wood preservation; some examples are given below:

1. Bottoms sediment sludge from the treatment of waste-waters from wood-preserving processes that use creosote and/or pentachlorophenol.
2. Waste-water treatment sludge from the production of chrome yellow and orange pigments.
3. Heavy ends (residue) from the distillation of vinyl chloride in vinyl chloride monomer production.
4. 2,6-Dichlorophenol waste from the production of 2,4-D.
5. Pink/red water from TNT operations.
6. Slop oil emulsion solids from the petroleum refining industry.
7. Ammonia lime still sludge from coking operations.
8. Electrolytic anode slimes/sludges from primary zinc production.

The second largest category of wastes generated are reactive wastes, followed by corrosive wastes and toxic wastes.

Corrosive Substances

Conventionally, corrosive substances are regarded as those that dissolve metals or cause oxidised material, such as rust from iron, to form on the surface of metals. In a broader sense, corrosives cause deterioration of materials, including living tissue, that they contact. Most corrosives belong to at least one of the four following chemical classes: (i) strong acids, (ii) strong bases, (iii) oxidants, and (iv) dehydrating agents. Table 3.1 lists some of the major corrosive substances and their effects.

Sulphuric acid

Sulphuric acid is a prime example of a corrosive substance. As well as being a strong acid, concentrated sulphuric acid is also a dehydrating agent and oxidant. The tremendous affinity of H_2SO_4 for water is illustrated by the heat generated when water and concentrated sulphuric acid are mixed. If this is done incorrectly by adding water to the acid, localised boiling and spattering can occur that result in personal injury. The major destructive effect of sulphuric acid on skin tissue is removal of water with accompanying release of heat. Sulphuric acid decomposes carbohydrates by removal of water. In contact with sugar, for example, concentrated sulphuric acid reacts to leave a charred mass. The reaction is

$$C_{12}H_{22}O_{11} \xrightarrow{\text{H}_2\text{SO}_4} 11H_2O(H_2SO_4) + 12C + \text{heat} \qquad \dots (3.1)$$

Some dehydration reactions of sulphuric acid can be very vigorous. For example, the reaction with perchloric acid produces unstable Cl_2O_7, and a violent explosion can result. Concentrated sulphuric acid produces dangerous or toxic products with a number of other substances, such as toxic carbon monoxide (CO) from reaction with oxalic acid, $H_2C_2O_4$; toxic bromine and sulphur dioxide (Br_2, SO_2) from reaction with sodium bromide, NaBr; and toxic, unstable chlorine dioxide (ClO_2) from reaction with sodium chlorate, $NaClO_3$.

Table 3.1. Examples of some corrosive substances.

Name and formula	Properties and effects
Nitric acid, HNO_3	Strong acid and strong oxidiser, corrodes metal, reacts with protein in tissue to form yellow xanthoproteic acid, lesions are slow to heal.
Hydrochloric acid, HCl	Strong acid, corrodes metals, gives off HCl gas vapour, which can damage respiratory tract tissue.
Hydrofluoric acid, HF	Corrodes metals, dissolves glass, causes particularly bad burns to flesh.
Alkali metal hydroxides, NaOH and KOH	Strong bases, corrode zinc, lead, and aluminium, substances that dissolve tissue and cause severe burns.
Hydrogen peroxide, H_2O_2	Oxidiser, all but very dilute solutions cause severe burns.
Interhalogen compounds such as ClF, BrF_3	Powerful corrosive irritants that acidify, oxidise, and dehydrate tissue.
Halogen oxides such as OF_2, Cl_2O, Cl_2O_7	Powerful corrosive irritants that acidify, oxidise, and dehydrate tissue.
Elemental fluorine, chlorine, bromine (F_2, Cl_2, Br_2)	Very corrosive to mucous membranes and moist tissue, strong irritants.

Contact with sulphuric acid causes severe tissue destruction resulting in severe burns, which may be difficult to heal. Inhalation of sulphuric acid fumes or mists damages tissues in the upper respiratory tract and eyes. Long-term exposure to sulphuric acid fumes or mists has caused erosion of teeth.

Toxic substances

Toxicity is of the utmost concern in dealing with hazardous substances. This includes both long-term chronic effects from continual or periodic exposures to low levels of toxicants, and acute effects from a single large exposure. For regulatory and remediation purposes a standard test is needed to measure the likelihood of toxic substances getting into the environment and causing harm to organisms.

Chemical Classes of Hazardous Substances

Another way of viewing hazardous substances in the context of their chemical properties is to divide them into classes of chemicals. A number of elements are used industrially in their elemental forms, in many cases for chemical synthesis. Some of these elements pose hazards of flammability, corrosivity, reactivity, or toxicity. Elemental hydrogen, H_2, is extremely flammable and forms explosive mixtures with air. Three of the halogens — fluorine, chlorine, and bromine — are widely produced as elemental F_2, Cl_2, and Br_2, respectively. Fluorine is the strongest elemental oxidant and extremely reactive. It is very corrosive to the skin and inhalation of F_2 can cause severe lung damage. Chlorine, one of the most widely produced industrial chemicals, is a reactive oxidant that forms acid in water and is a corrosive poison to tissue, especially in the respiratory tract. Bromine is a volatile brown liquid which is corrosive to skin in both the liquid and vapour form. Elemental white phosphorus is a reactive substance that may catch fire spontaneously in air. It is a systemic poison. Elemental lithium, sodium, and potassium react with a large number of chemicals and burn readily to give off caustic oxide and hydroxide fumes. Elemental mercury vapour is especially toxic by inhalation. Some metals, commonly known as heavy metals, are particularly toxic in their chemically combined forms. These include lead, cadmium, mercury, beryllium, and arsenic.

Many inorganic compounds are hazardous because of reactivity (NH_4ClO_4), corrosivity (HNO_3) and toxicity (KCN). Many organometallic compounds, which have a metal atom or metalloid atom (such as silicon or arsenic) bonded directly to carbon in a hydrocarbon group or in carbon monoxide, CO, are volatile, reactive, and toxic.

Organic compounds

There are millions of known organic compounds, most of which can be hazardous in some way and to some degree. Most organic compounds can be divided among hydrocarbons, oxygen-containing compounds, nitrogen-containing compounds, organohalides, sulphur-containing compounds, phsophorus-containing compounds, or combinations thereof.

Physical Forms and Segregation of Wastes

Three major categories of wastes based upon their physical forms are organic materials, aqueous wastes, and sludges. These forms largely determine the course of action taken in treating and disposing of the wastes. The level of segregation, a concept illustrated in Fig. 3.2 is very important in treating, storing, and disposing of different kinds of wastes. It is relatively easy to deal with wastes that are not mixed with other kinds of wastes; that is, those that are highly segregated. For example, spent hydrocarbon solvents can be used as fuel in boilers. However, if these solvents are mixed with spent organochlorine solvents, the production of contaminant hydrogen chloride during combustion may prevent fuel use and require disposal in special hazardous waste incinerators. Further mixing with inorganic sludges adds mineral matter and water. These impurities complicate the treatment processes required by producing mineral ash in incineration or lowering the heating values of the material incinerated because of the presence of water. Among the most difficult types of wastes to handle and treat are those with the least segregation, of which a 'worst case scenario' would be 'dilute sludge consisting of mixed organic and inorganic wastes', as shown in Fig. 3.2.

Concentration of wastes is an important factor in their management. A waste that has been concentrated or preferably never diluted is generally much easier and more economical to handle than one that is dispersed in a large quantity of water or soil. Dealing with hazardous wastes is greatly facilitated when the original quantities of wastes are minimised and the wastes remain separated and concentrated insofar as possible.

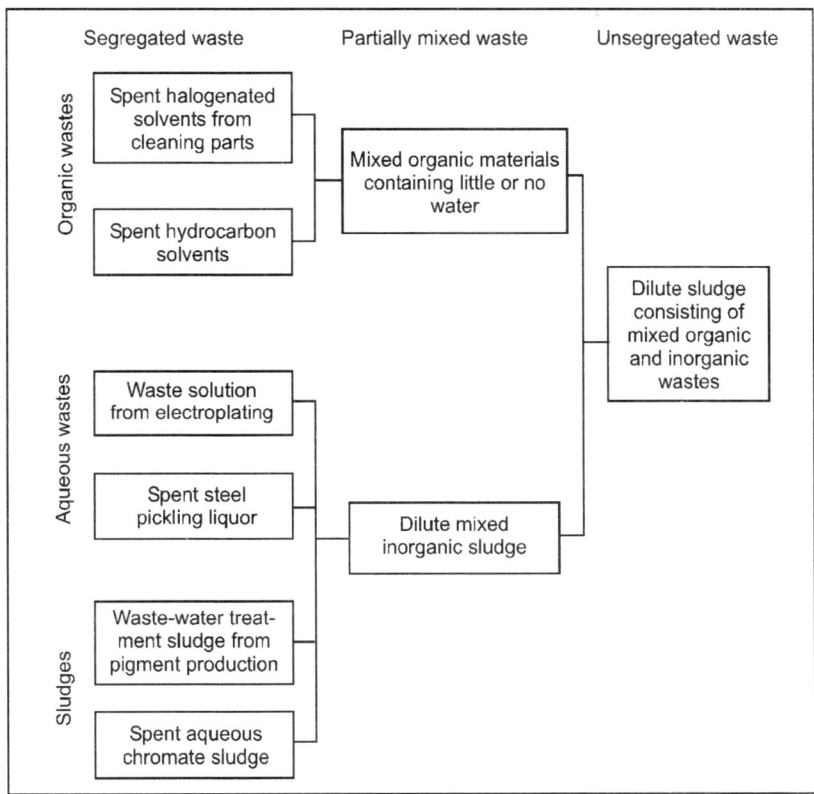

Fig. 3.2. Illustration of waste segregation.

Generation, Treatment and Disposal

Hazardous waste management refers to a carefully organised system in which wastes go through appropriate pathways to their ultimate elimination or disposal in ways that protect human health and the environment. The management of hazards posed by hazardous substances and wastes is a crucial part of the operation of any modern chemical industry. It is a significant and increasing part of the cost of any business dealing with chemical products and processes. Personnel working with such products and processes must have a good understanding of hazardous substances and hazardous wastes. Three main aspects of hazardous waste management involve generation, treatment, and disposal as illustrated in Fig. 3.3. The effectiveness of a hazardous waste system is a measure of how well it reduces the quantities and hazards of wastes, ideally approaching zero for both. In decreasing order of effectiveness, the options for handling hazardous wastes are the following:

1. Measures that prevent generation of wastes.
2. Recovery and recycle of waste constituents.
3. Destruction and treatment, conversion to nonhazardous waste forms.
4. Disposal (storage, landfill).

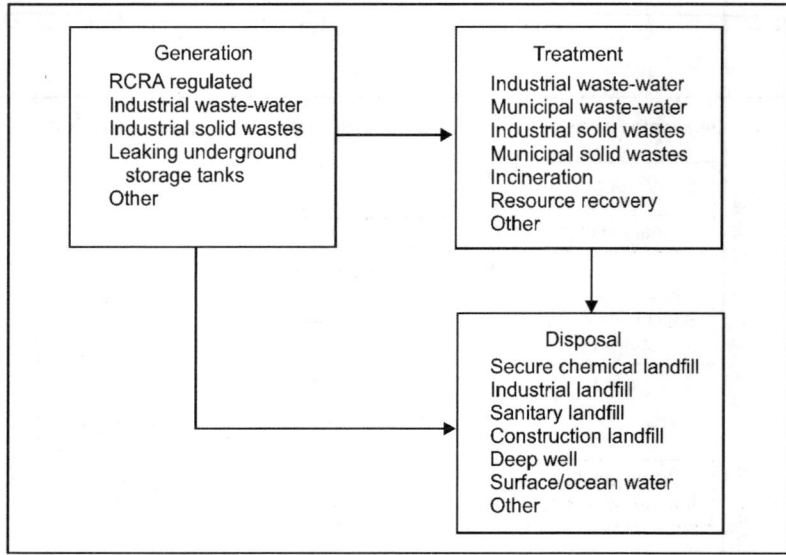

Fig. 3.3. System of generation, treatment, and disposal of hazardous wastes.

Treatment, storage, and disposal facilities

A crucial part of the regulation of hazardous wastes in India pertains to treatment, storage and disposal facilities (TSDF). Treatment alters the physical, chemical or biological character or composition of a waste to make it safer. Storage refers to the holding of hazardous wastes for a temporary period pending treatment or disposal. Disposal refers to the ultimate fate of hazardous substances or their treatment products.

Waste reduction and waste minimisation

Many hazardous waste problems can be avoided at early stages by waste reduction and waste minimisation. As these terms are most commonly used, waste reduction refers to source reduction–less waste–producing materials in, less waste out. Waste minimisation can include treatment processes, such as incineration which reduce the quantities of wastes for which ultimate disposal is required.

Waste treatment

An overall scheme for the treatment of hazardous wastes is shown in Fig. 3.4. Under the category of treatment it is necessary to consider both municipal waste-water and municipal solid wastes along with hazardous wastes. The goal of many industrial waste-water and sludge treatment processes is to produce an effluent that meets standards for release to a municipal waste-water treatment plant and in some cases to produce solids that can be co-disposed with municipal solid wastes. Incineration of municipal solid wastes may produce some solids, particularly fly ash, that have to be treated as hazardous.

The scheme outlined in Fig. 3.4 may serve as a frame of reference for subsequent discussions of waste treatment. The ideal treatment process reduces the quantity of hazardous waste material to a small fraction of the original amount and converts it to a non-hazardous form. However, most treatment processes yield material, such as sludge from waste-water treatment or incinerator ash, which requires disposal and which may be hazardous to some extent.

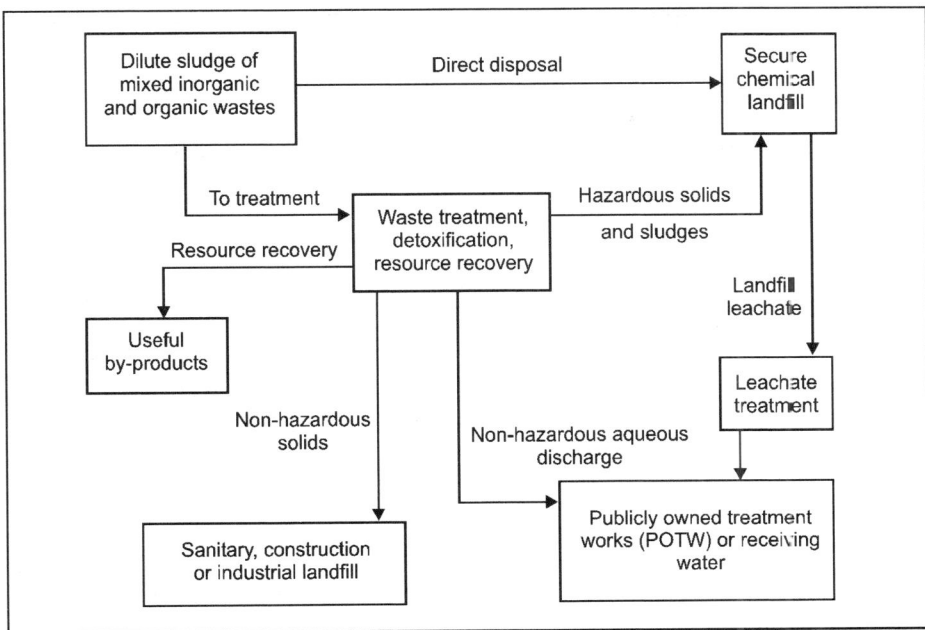

Fig. 3.4. Treatment options for mixed hazardous wastes.

Direct disposal of minimally treated hazardous wastes is becoming more severely limited with new regulations coming from the Hazardous and Solid Waste Amendments. Under its 'land-ban' rules, this Act prohibits the land disposal of more than 400 chemicals and waste streams unless they are treated or can be shown not to migrate during the time that they remain hazardous. The ultimate objective of these rules is to reduce the amounts of hazardous wastes generated, although quantities are expected to increase during the next decade. More emphasis in treatment is being placed on recovery of recyclable materials and production of innocuous by-products. There are strong regulatory and economic incentives to generate fewer wastes in manufacturing by modification of processes, product substitution, recycling, and careful control throughout the manufacturing system.

Hazardous Substances and Health

In recent years, the health aspects of hazardous substances have received increased attention by the public and by legislative bodies. A basic question is the linkage between the health of people living near Superfund sites and the chemicals found in the sites. This concern gained increased recognition in the US with passage of the 1986 SARA Act, greatly expanding the health authorities sections of the 1980 CERCLA act. The Agency for Toxic Substances and Disease Registry (ATSDR), authorised by the 1980 CERCLA act and administered by the Public Health Service of the Department of Health and Human Services is responsible for the health aspects of toxic substances release. It is charged with maintaining files of information and data on the health effects and diseases potentially caused by toxic substances, keeping records of exposure to toxic substances, and listing areas where public access has been restricted because of contamination by toxic substances. In addition, ATSDR is the major conduit of information on the health effects of hazardous substances and plays an active role in the response and remediation activities at Superfund waste sites. The agency has prepared extensive Toxicological Profile documents pertaining to specific hazardous substances encountered at Superfund sites. The materials that are the subjects of these profiles are those both commonly encountered at hazardous waste sites and likely to pose substantial health hazards.

REDUCTION, TREATMENT, AND DISPOSAL OF HAZARDOUS WASTE

Waste Reduction and Minimisation

Many hazardous waste problems can be avoided at early stage by waste reduction (cutting down quantities of wastes from their sources) and waste minimisation (utilisation of treatment processes which reduce the quantities of wastes requiring ultimate disposal).

There are several ways in which quantities of wastes can be reduced, including source reduction, waste separation and concentration, resource recovery, and waste recycling. The most effective approaches to minimising wastes centre around careful control of manufacturing processes, taking into consideration discharges and the potential for waste minimisation at every step of manufacturing. Viewing the process as a whole (as outlined for a generalised chemical manufacturing process in Fig. 3.5) often enables crucial identification of the source of a waste, such as a raw material impurity, catalyst or process solvent. Once a source is identified, it is much easier to take measure to eliminate or reduce the waste.

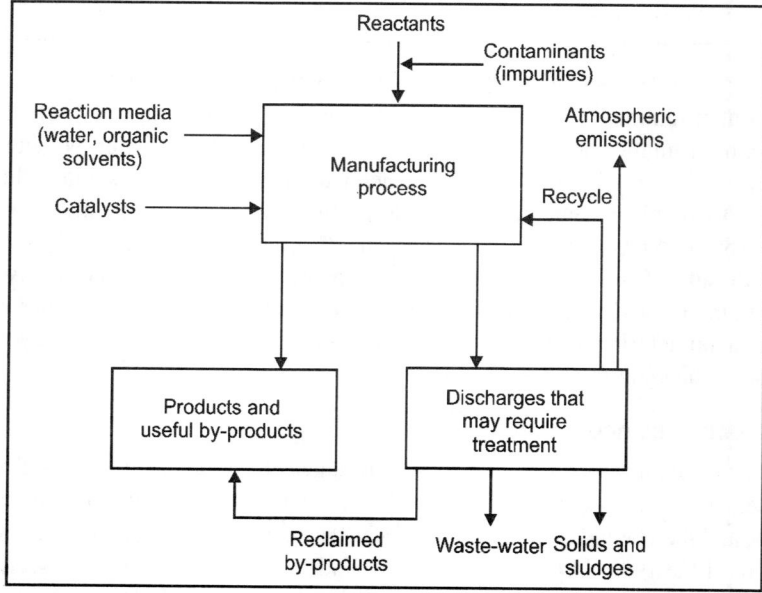

Fig. 3.5. Chemical manufacturing process from the viewpoint of discharges and waste minimisation.

Modifications of the manufacturing process can yield substantial waste reduction. Some such modifications are of a chemical nature. Changes in chemical reaction conditions can minimise production of by-product hazardous substances. In some cases, potentially hazardous catalysts, such as those formulated from toxic substances, can be replaced by catalysts that are non-hazardous or that can be recycled rather than discarded. Wastes can be minimised by volume reduction, for example, through dewatering and drying sludge.

Recycling

Wherever possible, recycling and reuse should be accomplished onsite because it avoids having to move wastes and because a process that produces recyclable materials is often the most likely to have use for them. The four broad areas in which something of value may be obtained from wastes are the following:

1. Direct recycle as raw material to the generator, as with the return to feedstock of raw materials not completely consumed in a synthesis process.
2. Transfer as a raw material to another process; a substance that is a waste product from one process may serve as a raw material for another, sometimes in an entirely different industry.
3. Utilisation for pollution control or waste treatment, such as use of waste alkali to neutralise waste acid.
4. Recovery of energy; for example, from the incineration of combustible hazardous wastes.

Examples of recycling

Recycling of scrap industrial impurities and products occurs on a large scale with a number of different materials. Most of these materials are not hazardous, but, as with most large-scale industrial operations, their recycle may involve the use or production of hazardous substances. Some of the more important examples are the following:

1. Ferrous metals composed primarily of iron and used largely as feedstock for electric-arc furnaces.
2. Nonferrous metals, including aluminium (which ranks next to iron in terms of quantities recycled), copper and copper alloys, zinc, lead, cadmium, tin, silver, and mercury.
3. Metal compounds, such as metal salts.
4. Inorganic substances including alkaline compounds (such as sodium hydroxide used to remove sulphur compounds from petroleum products), acids (steel pickling liquor where impurities permit reuse), and salts (for example, ammonium sulphate form coal coking used as fertiliser).
5. Glass, which makes up about 10 per cent of municipal refuse.
6. Paper, commonly recycled from municipal refuse.
7. Plastic, consisting of a variety of mouldable polymeric materials and composing a major constituent of municipal wastes.
8. Rubber.
9. Organic substances, especially solvents and oils, such as hydraulic and lubricating oils.
10. Catalysts from chemical synthesis or petroleum processing.
11. Materials with agricultural uses, such as waste lime or phosphate containing sludges used to treat and fertilise acidic soils.

Waste oil utilisation and recovery

Waste oil generated from lubricants and hydraulic fluids is one of the more commonly recycled materials. The collection, recycling, treatment, and disposal of waste oil are all complicated by the fact that it comes from diverse, widely dispersed sources and contains several classes of potentially hazardous contaminants. These are divided between organic constituents (polycyclic aromatic hydrocarbons, chlorinated hydrocarbons) and inorganic constituents (aluminium, chromium, and iron from wear of metal parts; barium and zinc from oil additives; lead from leaded gasoline).

Recycling waste oil

The processes used to convert waste oil to a feedstock hydrocarbon liquid for lubricant formulation are illustrated in Fig. 3.6. The first of these uses distillation to remove water and light ends that have come from condensation and contaminant fuel. The second, or processing, step may be a vacuum distillation in which the three products are oil for further processing, a fuel oil cut, and a heavy residue. The processing step may also employ treatment with a mixture of solvents including isopropyl and butyl alcohols and methylethyl ketone to dissolve the oil and leave contaminants as a sludge; or contact with sulphuric acid to remove inorganic contaminants followed by treatment with clay to take out acid and contaminants that cause odour and colour. The third step shown in Fig. 3.6 employs vacuum distillation to separate lubricating oil stocks from a fuel fraction and heavy residue. This phase of treatment may also involve hydrofinishing, treatment with clay, and filtration.

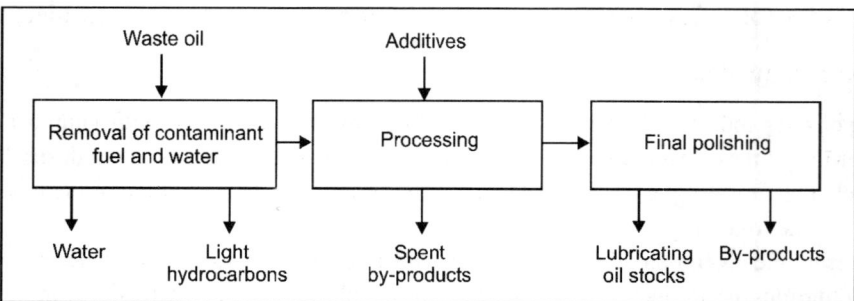

Fig. 3.6. Major steps in reprocessing waste oil.

Waste oil fuel

For economic reasons, waste oil that is to be used for fuel is given minimal treatment of a physical nature, including settling, removal of water, and filtration. Metals in waste fuel oil become highly concentrated in its fly ash, which may be hazardous.

Waste solvent recovery and recycle

The recovery and recycling of waste solvents has some similarities to the recycling of waste oil and is also an important enterprise. Among the many solvents listed as hazardous wastes and recoverable from wastes are dichloromethane, tetrachloroethylene, trichloroethylene, 1,1,1-trichloroethane, benzene, liquid alkanes, 2-nitropropane, methylisobutyl ketone, and cyclohexanone. For reasons of both economics and pollution control, many industrial processes that use solvents are equipped for solvent recycle. The basic scheme for solvent reclamation and reuse is shown in Fig. 3.7. A number of operations are used in solvent purification. Entrained solids are removed by settling, filtration or centrifugation. Drying agents may be used to remove water from solvents and various adsorption techniques and chemical treatment may be required to free the solvent from specific impurities. Fractional distillation, often requiring several distillation steps, is the most important operation in solvent purification and recycle. It is used to separate solvents from impurities, water, and other solvents.

Physical Methods of Waste Treatment

This section addresses predominantly physical methods for waste treatment and the following section addresses methods that utilise chemical processes. It should be kept in mind that most waste treatment measures have both physical and chemical aspects. The appropriate treatment technology for hazardous wastes obviously depends upon the nature of the wastes. These may consist of volatile wastes (gases, volatile solutes in water, gases or volatile liquids held by solids, such as catalysts), liquid wastes (wastewater, organic solvents), dissolved or soluble wastes (water-soluble inorganic species, water-soluble organic species, compounds soluble in organic solvents), semi-solids (sludges, greases), and solids (dry solids, including granular solids with a significant water content, such as dewatered sludges, as well as solids suspended in liquids). The type of physical treatment to be applied to wastes depends strongly upon the physical properties of the material treated, including state of matter, solubility in water and organic solvents, density, volatility, boiling point, and melting point.

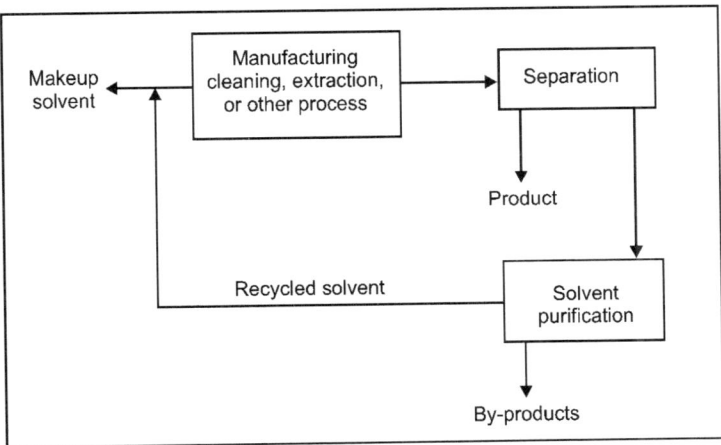

Fig. 3.7. Overall process for recycling solvents.

As shown in Fig. 3.8, waste treatment may occur at three major levels—primary, secondary, and polishing—somewhat analogous to the treatment of waste-water. Primary treatment is generally regarded as preparation for further treatment, although it can result in the removal of by-products and reduction of the quantity and hazard of the waste. Secondary treatment detoxifies, destroys, and removes hazardous constituents. Polishing usually refers to treatment of water that is removed from wastes so that it may be safely discharged. However, the term can be broadened to apply to the treatment of other products as well so that they may be safely discharged or recycled.

Methods of physical treatment

Knowledge of the physical behaviour of wastes has been used to develop various unit operations for waste treatment that are based upon physical properties. These operations include the following:

1. Phase separation	2. Phase transfer
Filtration	Extraction
3. Phase transition	Sorption
Distillation	4. Membrane separations
	Evaporation
Reverse osmosis	
Physical precipitation	Hyper-and ultrafiltration

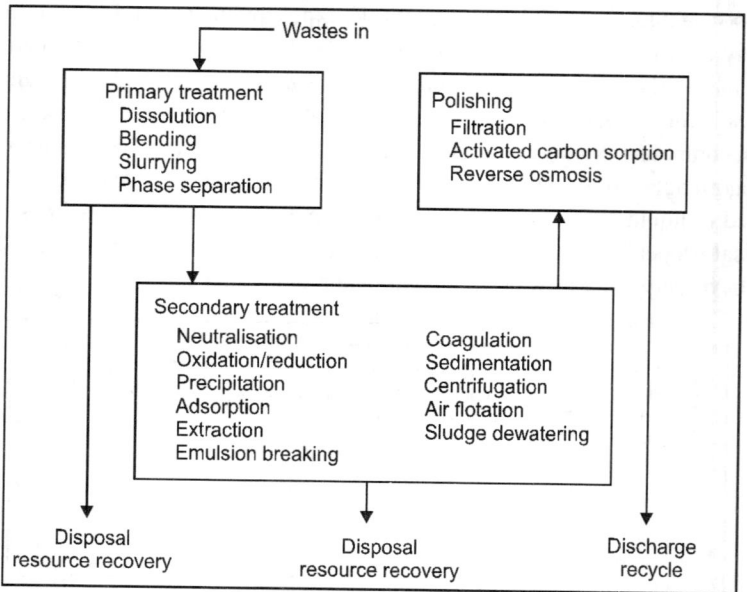

Fig. 3.8. Major phases of waste treatment.

Phase separations

The most straightforward means of physical treatment involves separation of components of a mixture that are already in two different phases. Sedimentation and decanting are easily accomplished with simple equipment. In many cases, the separation must be aided by mechanical means, particularly filtration or centrifugation. Flotation is used to bring suspended organic matter or finely divided particles to the surface of a suspension. In the process of dissolved air flotation (DAF), air is dissolved in the suspending medium under pressure and comes out of solution when the pressure is released as minute air bubbles attached to suspended particles, which causes the particles to float to the surface.

An important and often difficult waste treatment step is emulsion breaking in which colloidal-sized emulsions are caused to aggregate and settle from suspension. Agitation, heat, acid, and the addition of coagulants consisting of organic polyelectrolytes or inorganic substances, such as an aluminium salt, may be used for this purpose. The chemical additive acts as a flocculating agent to cause the particles to stick together and settle out.

Phase transition

A second major class of physical separation is that of phase transition in which a material changes from one physical phase to another. It is best exemplified by distillation, which is used in treating and recycling solvents, waste oil, aqueous phenolic wastes, xylene contaminated with paraffin from histological laboratories, and mixtures of ethylbenzene and styrene. Distillation produces distillation bottoms (still bottoms), which are often hazardous and polluting. These consist of unevaporated solids, semisolid tars, and sludges from distillation. Specific examples are distillation bottoms from the production of acetaldehyde from ethylene. Evaporation is usually employed to remove water from an aqueous waste to concentrate it. A special case of this technique is thin-film evaporation in which volatile constituents are removed by heating a thin layer of liquid or sludge waste spread on a heated surface.

Drying removal of solvent or water from a solid or semisolid (sludge) or the removal of solvent from a liquid or suspension is a very important operation because water is often the major constituent of waste products, such as sludges obtained from emulsion breaking. In freeze drying, the solvent, usually water, is sublimed from a frozen material. Hazardous waste solids and sludges are dried to reduce the quantity of waste, to remove solvent or water that might interfere with subsequent treatment processes, and to remove hazardous volatile constituents. Dewatering can often be improved with addition of a filter aid, such as diatomaceous earth, during the filtration step. Stripping is a means of separating volatile components from less volatile ones in a liquid mixture by the partitioning of the more volatile materials to a gas phase of air or steam (steam stripping). The gas phase is introduced into the aqueous solution or suspension containing the waste in a stripping tower that is equipped with trays or packed to provide maximum turbulence and contact between the liquid and gas phases. The two major products are condensed vapour and a stripped bottoms residue. Examples of two volatile components that can be removed from water by air stripping are benzene and dichloromethane. Air stripping can also be used to remove ammonia from water that has been treated with a base to convert ammonium ion to volatile ammonia.

Physical precipitation is used here as a term to describe processes in which a solid forms from a solute in solution as a result of a physical change in the solution, as compared to chemical precipitation in which a chemical reaction in solution produces an insoluble material. The major changes that can cause physical precipitation are cooling the solution, evaporation of solvent or alteration of solvent composition. The most common type of physical precipitation by alteration of solvent composition occurs when a water-miscible organic solvent is added to an aqueous solution, so that the solubility of a salt is lowered below its concentration in the solution.

Phase transfer consists of the transfer of a solute in a mixture from one phase to another. An important type of phase transfer process is solvent extraction, a process in which a substance is transferred from solution in one solvent (usually water) to another (usually an organic solvent) without any chemical change taking place. When solvents are used to leach substances from solids or sludges, the process is called leaching. Solvent extraction and the major terms applicable to it are summarised in Fig. 3.9. The same terms and general principles apply to leaching. The major application of solvent extraction to waste treatment has been in the removal of phenol from by-product water produced in coal coking, petroleum refining, and chemical syntheses that involve phenol.

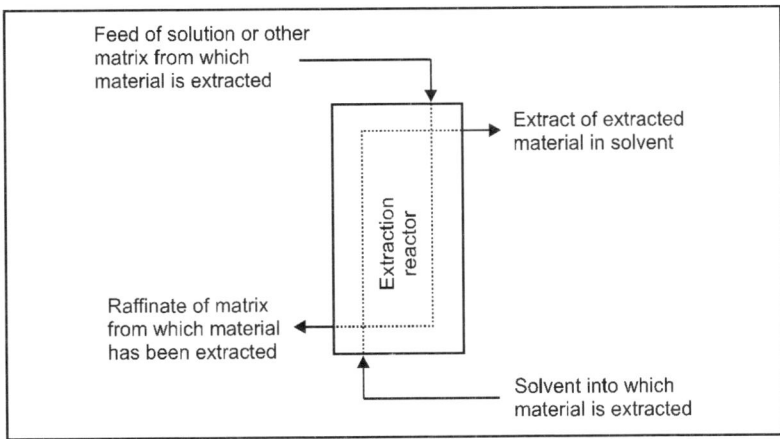

Fig. 3.9. Outline of solvent extraction/leaching process with important terms underlined.

One of the more promising approaches to solvent extraction and leaching of hazardous wastes is the use of supercritical fluids, most commonly CO_2, as extraction solvents. A supercritical fluid is one that has characteristics of both liquid and gas and consists of a substance above its critical temperature and pressure ($31.1°C$ and 73.8 atm, respectively for CO_2). After a substance has been extracted from a waste into a supercritical fluid at high pressure, the pressure can be released, resulting in separation of the substance extracted.

The fluid can then be compressed again and recirculated through the extraction system. Some possibilities for treatment of hazardous wastes by extraction with supercritical CO_2 include removal of organic contaminants from waste-water, extraction of organohalide pesticides from soil, extraction of oil from emulsions used in aluminium and steel processing, and regeneration of spent activated carbon. Waste oils contaminated with PCBs, metals and water can be purified using supercritical ethane.

Transfer of a substance from a solution to a solid phase is called sorption. The most important sorbent is activated carbon used for several purposes in waste treatment; in some cases it is adequate for complete treatment. It can also be applied to pretreatment of waste streams going into processes such as reverse osmosis to improve treatment efficiency and reduce fouling. Effluents from other treatment processes, such as biological treatment of degradable organic solutes in water can be polished with activated carbon. Activated carbon sorption is most effective for removing from water those hazardous waste materials that are poorly water-soluble and that have high molar masses, such as xylene, naphthalene, cyclohexane; chlorinated hydrocarbons, phenol, aniline, dyes, and surfactants. Activated carbon does not work well for organic compounds that are highly water-soluble or polar.

Solids other than activated carbon can be used for sorption of contaminants from liquid wastes. These include synthetic resins composed of organic polymers and mineral substance. Of the latter, clay is employed to remove impurities from waste lubricating oils in some oil recycling processes.

Molecular separation

A third major class of physical separation is molecular separation, often based upon membrane processes in which dissolved contaminants or solvent pass through a size-selective membrane under pressure. The products are a relatively pure solvent phase (usually water) and a concentrate enriched in the solute impurities. Hyper filtration allows passage of species with molecular masses of about 100 to 500, whereas ultrafiltration is used for the separation of organic solutes with molar masses of 500 to 1,000,000. With both of these techniques, water and lower molar mass solutes under pressure pass through the membrane as a stream of purified permeate, leaving behind a stream of concentrate containing impurities in solution or suspension. Ultrafiltration and hyperfiltration are especially useful for concentrating suspended oil, grease, and fine solids in water. They also serve to concentrate solutions of large organic molecules and heavy metal ion complexes.

Reverse osmosis is the most widely used of the membrane techniques. Although superficially similar to ultrafiltration and hyperfiltration, it operates on a different principle in that the membrane is selectively permeable to water and excludes ionic solutes. Reverse osmosis uses high pressures to force permeate through the membrane, producing a concentrate containing high levels of dissolved salts.

Electrodialysis, sometimes used to concentrate plating wastes, employs membranes alternately permeable to cations and to anions. The driving force for the separation is provided by electrolysis with a direct current between two electrodes. Alternate layers between the membranes contain concentrate (brine) and purified water.

Chemical Treatment

The applicability of chemical treatment to wastes depends upon the chemical properties of the waste constituents, particularly acid-base, oxidation-reduction, precipitation, and complexation behaviour; reactivity; flammability/combustibility; corrosivity; and compatibility with other wastes. The chemical behaviour of wastes translates to various unit operations for waste treatment that are based upon chemical properties and reaction. These include the following:

1. Acid/base neutralisation.
2. Chemical extraction and leaching.
3. Ion exchange.
4. Chemical precipitation.
5. Oxidation.
6. Reduction.

Some of the more sophisticated means available for treatment of wastes have been developed for pesticide disposal.

Acid/Base neutralisation

Waste acids and bases are treated by neutralisation

$$H^+ + OH^- \rightarrow H_2O \qquad \qquad ... (3.2)$$

Although simple in principle, neutralisation can present some problems in practice. These include evolution of volatile contaminants, mobilisation of soluble substances, excessive heat generated by the neutralisation reaction, and corrosion to apparatus. By adding too much or too little of the neutralising agent, it is possible to get a product that is too acidic or basic.

Lime, $Ca(OH)_2$, is widely used as a base for treating acidic wastes. Because of lime's limited solubility, solutions of excess lime do not reach extremely high pH values. Sulphuric acid, H_2SO_4, is a relatively inexpensive acid for treating alkaline waste. However, addition of too much sulphuric acid can produce highly acidic products; for some applications, acetic acid, CH_3COOH, is preferable. As noted above, acetic acid is a weak acid and an excess of it does little harm. It is also a natural product and biodegradable.

Neutralisation or pH adjustment, is often required prior to the application of other waste treatment processes. Processes that may require neutralisation include oxidation/reduction, activated carbon sorption, wet air oxidation, stripping, and ion exchange. Micro-organisms usually require a pH in the range of 6–9, so neutralisation may be required prior to biochemical treatment.

Chemical precipitation

Chemical precipitation is used in hazardous waste treatment primarily for the removal of heavy metal ions from water as shown below for the chemical precipitation of cadmium.

$$Cd^{2+}(aq) + HS^-(aq) \rightarrow CdS(s) + H^+(aq) \qquad \qquad ... (3.3)$$

Precipitation of metals

The most widely used means of precipitating metal ions is by the formation of hydroxides such as chromium (III) hydroxide.

$$Cr^{3+} + 3OH^- \rightarrow Cr(OH)_3 \qquad \qquad ... (3.4)$$

The source of hydroxide ion is a base (alkali), such as lime $Ca(OH)_2$, sodium hydroxide (NaOH) or sodium carbonate (Na_2CO_3). Most metal ions tend to produce basic salt precipitates, such as basic copper (II) sulphate, $CuSO_4.3Cu(OH)_2$, formed as a solid when hydroxide is added to a solution containing Cu^{2+} and SO_4^{2-} ions. The solubilities of many heavy metal hydroxides reach a minimum value, often at a pH in the range of 9–11, then increase with increasing pH values due to the formation of soluble hydroxo complexes, as illustrated by the following reaction:

$$Zn(OH)_2(s) + 2OH^-(aq) \rightarrow Zn(OH)_4^{2-}(aq) \qquad \text{... (3.5)}$$

The chemical precipitation method that is used most is precipitation of metals as hydroxides and basic salts with lime. Sodium carbonate can be used to precipitate hydroxides ($Fe(OH)_3.xH_2O$), carbonates ($CdCO_3$) or basic carbonate salts ($2PbCO_3.Pb(OH)_2$). The carbonate anion produces hydroxide by virtue of its hydrolysis reaction with water:

$$CO_3^{2-} + H_2O \rightarrow HCO_3^- + OH^- \qquad \text{... (3.6)}$$

Carbonate, alone, does not give as high a pH as do alkali metal hydroxides, which may have to be used to precipitate metals that form hydroxides only at relatively high pH values. The solubilities of some heavy metal sulphides are extremely low, so precipitation by H_2S or other sulphides can be a very effective means of treatment. Hydrogen sulphide is a toxic gas that is itself considered to be a hazardous waste. Iron(II) sulphide (ferrous sulphide) can be used as a safe source of sulphide ion to produce sulphide precipitates with other metals that are less soluble than FeS. However, toxic H_2S can be produced when metal sulphide wastes contact acid:

$$MS + 2H^+ \rightarrow M^{2+} + H_2S \qquad \text{... (3.7)}$$

Some metals can be precipitated from solution in the elemental metal form by the action of a reducing agent, such as sodium borohydride:

$$4Cu^{2+} + NaBH_4 + 2H_2O \rightarrow 4Cu + NaBO_2 + 8H^+ \qquad \text{... (3.8)}$$

or with more active metals in a process called cementation:

$$Cd^{2+} + Zn \rightarrow Cd + Zn^{2+} \qquad \text{... (3.9)}$$

Coprecipitation of metals

In some cases, advantage may be taken of the phenomenon of coprecipitation to remove metals from wastes. A good example of this application is the coprecipitation of lead from battery industry waste-water with iron hydroxide. Raising the pH of such a waste-water consisting of dilute sulphuric acid and contaminated with Pb^{2+} ion precipitates lead as several species, including $PbSO_4$, $Pb(OH)_2$, and $Pb(OH)_2.2PbCO_3$. In the presence of iron, gelatinous $Fe(OH)_3$ forms, which coprecipitates the lead, resulting in much lower values of lead concentration than would otherwise be possible. Effective removal of lead from battery industry waste-water to below 0.2 ppm has been achieved by first adding an optimum quantity of iron, adjustment of the pH to a range of 9 to 9.5, addition of a polyelectrolyte to aid coagulation, and filtration.

Oxidation/Reduction

Oxidation and reduction can be used for the treatment and removal of a variety of inorganic and organic wastes. Some waste oxidants can be used to treat oxidisable wastes in water and cyanides.

Ozone, (O_3), is a strong oxidant that can be generated onsite by an electrical discharge through dry air or oxygen. Ozone employed as an oxidant gas at levels of 1–2 wt % in air and 2–5 wt % in oxygen has been used to treat a large variety of oxidisable contaminants, effluents, and wastes including waste-water and sludges containing oxidisable constituents.

Electrolysis

As shown in Fig. 3.10, electrolysis is a process in which one species in solution (usually a metal ion) is reduced by electrons at the cathode and another gives up electrons to the anode and is oxidised there. In hazardous waste applications, electrolysis is most widely used in the recovery of cadmium, copper, gold, lead, silver, and zinc. Metal recovery by electrolysis is made more difficult by the presence of cyanide ion, which stabilises metals in solution as the cyanide complexes, such as $Ni(CN)_4^{2-}$.

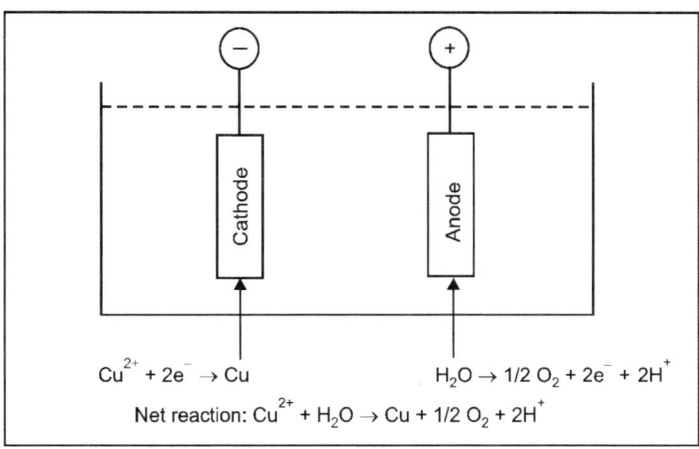

$$Cu^{2+} + 2e^- \rightarrow Cu \qquad H_2O \rightarrow 1/2\ O_2 + 2e^- + 2H^+$$
$$\text{Net reaction: } Cu^{2+} + H_2O \rightarrow Cu + 1/2\ O_2 + 2H^+$$

Fig. 3.10. Electrolysis of copper solution.

Electrolytic removal of contaminants from solution can be by direct electro-deposition, particularly of reduced metals, and as the result of secondary reactions of electrolytically generated precipitating agents. A specific example of both is the electrolytic removal of both cadmium and nickel from wastewater contaminated by nickel/cadmium battery manufacture using fibrous carbon electrodes. At the cathode, cadmium is removed directly by reduction to the metal:

$$Cd^{2+} + 2e^- \rightarrow Cd \qquad \qquad \text{... (3.10)}$$

At relatively high cathodic potentials, hydroxide is formed by the electrolytic reduction of water:

$$2H_2O + 2e^- \rightarrow 2OH^- + H_2 \qquad \qquad \text{... (3.11)}$$

or by the reduction of molecular oxygen, if it is present:

$$2H_2O + O_2 + 4e^- \rightarrow 4OH^- \qquad \qquad \text{... (3.12)}$$

If the localised pH at the cathode surface becomes sufficiently high, cadmium can be precipitated and removed as colloidal $Cd(OH)_2$. The direct electro-deposition of nickel is too slow to be significant, but it is precipitated as solid $Ni(OH)_2$ at pH values above 3.5.

Cyanide, which is often present as an ingredient of electroplating baths with metals such as cadmium and nickel, can be removed by oxidation with electrolytically generated elemental chlorine at the anode. Chlorine is generated by the anodic oxidation of added chloride ion:

$$2Cl^- \rightarrow Cl_2 + 2e^- \qquad \qquad \text{... (3.13)}$$

The electrolytically generated chlorine then breaks down cyanide by a series of reactions for which the overall reaction is the following:

$$2CN^- + 5Cl_2 + 8OH^- \rightarrow 10Cl^- + N_2 + 2CO_2 + 4H_2O \qquad \qquad \text{... (3.14)}$$

Hydrolysis

One of the ways to dispose of chemicals that are reactive with water is to allow them to react with water under controlled conditions, a process called hydrolysis. Inorganic chemicals that can be treated by hydrolysis include metals that react with water, metals carbides, hydrides, amides, alkoxidesm and halids; and non-metal oxyhalides and sulphides. Examples of the treatment of these classes of inorganic species are given in Table 3.2. Organic chemicals may also be treated by hydrolysis. For example, toxic acetic anhydride is hydrolysed to relatively safe acetic acid:

$$\underset{\substack{\text{Acetic anhydride}\\\text{(an acid anhydride)}}}{H-\overset{\overset{\displaystyle H}{|}}{\underset{\underset{\displaystyle H}{|}}{C}}-\overset{\overset{\displaystyle O}{\|}}{C}-O-\overset{\overset{\displaystyle O}{\|}}{C}-\overset{\overset{\displaystyle H}{|}}{\underset{\underset{\displaystyle H}{|}}{C}}-H} + H_2O \longrightarrow 2H-\overset{\overset{\displaystyle H}{|}}{\underset{\underset{\displaystyle H}{|}}{C}}-\overset{\overset{\displaystyle O}{\|}}{C}-OH \qquad \text{... (3.15)}$$

Chemical extraction and leaching

Chemical extraction or leaching in hazardous waste treatment is the removal of a hazardous constituent by chemical reaction with an extractant in solution. Poorly soluble heavy metal salts can be extracted by reaction of the salt anions with H^+ as illustrated by the following:

$$PbCO_3 + H^+ \rightarrow Pb^{2+} + HCO_3^- \qquad \text{... (3.16)}$$

Acids also dissolve basic organic compounds such as amines and aniline. Extraction with acids should be avoided if cyanides or sulphides are present to prevent formation of toxic hydrogen cyanide or hydrogen sulphide. Non-toxic weak acids are usually the safest to use. These include acetic acid, CH_3COOH, and the acid salt, NaH_2PO_4.

Table 3.2. Inorganic chemicals that may be treated by hydrolysis.

Class of chemical	Reaction with water
Active metals (calcium)	$Ca + 2H_2O \rightarrow H_2 + Ca(OH)_2$
Hydrides (sodium aluminium hydride)	$NaAlH_4 + 4H_2O \rightarrow 4H_2 + NaOH + Al(OH)_3$
Carbides (calcium carbide)	$CaC_2 + 2H_2O \rightarrow Ca(OH)_2 + C_2H_2$
Amides (sodium amide)	$NaNH_2 + H_2O \rightarrow NaOH + NH_3$
Halides (silicon tetrachloride)	$SiCl_4 + 2H_2O \rightarrow SiO_2 + 4HCl$
Alkoxides (sodium ethoxide)	$NaOC_5H_5 + H_2O \rightarrow NaOH + C_2H_5OH$

Chelating agents, such as dissolved ethylene diaminetetra acetate (EDTA, HY^{3-}), dissolve insoluble metal salts by forming soluble species with metal ions:

$$FeS + HY^{3-} \rightarrow FeY^{2-} + HS^- \qquad \text{... (3.17)}$$

Heavy metal ions in soil contaminated by hazardous wastes may be present in a coprecipitated form with insoluble iron(III) and manganese(IV) oxides, Fe_2O_3 and MnO_2, respectively. These oxides can be dissolved from soil by reducing agents, such as solutions of sodium dithuonate/citrate or hydroxylamine. This results in the production of soluble Fe^{2+} and Mn^{2+} and the release of heavy metal ions, such as Cd^{2+} or Ni^{2+}, which are removed with the water.

Ion exchange

Ion exchange is a means of removing cations or anions from solution onto a solid resin, which can be regenerated by treatment with acids, bases, or salts. The greatest use of ion exchange in hazardous waste treatment is for the removal of low levels of heavy metal ions from waste-water:

$$2H^+ \{CatExchr\} + Cd^{2+} \rightarrow Cd^{2+} \{CatExchr\}_2 + 2H^+ \qquad \dots (3.18)$$

Ion exchange is employed in the metal plating industry to purify rinse-water and spent plating bath solutions. Cation exchangers are used to remove cationic metal species, such as Cu^{2+}, from such solutions. Anion exchangers remove anionic cyanide metal complexed [for example $Ni(CN)_4^{2-}$] and chromium (VI) species, such as CrO_4^{2-}. Radionuclides may be removed from radioactive wastes and mixed waste by ion exchange resins.

A special ion exchange resin configuration has been described for the removal of chelatable heavy metals, such as copper, lead, nickel, and zinc, from otherwise innocuous sludge constituents.

Photolytic Reactions

Photolysis can be used to destroy a number of kinds of hazardous wastes. In such applications, it is most useful in breaking chemical bonds in refractory organic compounds. TCDD, one of the most troublesome and refractory of wastes, can be treated by ultraviolet light in the presence of hydrogen atom donors [H] resulting in reactions such as the following:

As photolysis proceeds, the H-C bonds are broken, the C-O bonds are broken and the final product is a harmless organic polymer. An initial photolysis reaction can result in the generation of reactive intermediates that participate in chain reactions that lead to the destruction of a compound. One of the most important reactive intermediates is free radical $HO^\bullet$. In some cases, sensitisers are added to the reaction mixture to absorb radiation and generate reactive species that destroy wastes.

Hazardous waste substances other than TCDD that have been destroyed by photolysis are herbicides (atrazine), 2,4,6-trinitrotoluene (TNT), and polychlorinated biphenyls (PCBs). The addition of a chemical oxidant, such as potassium peroxydisulphate, $K_2S_2O_8$, enhances destruction by oxidising active photolytic products.

Thermal Treatment Methods

Thermal treatment of hazardous wastes can be used to accomplish most of the common objectives of waste treatment — volume reduction; removal of volatile, combustible, mobile organic matter; and destruction of toxic and pathogenic materials. The most widely applied means of thermal treatment of hazardous wastes is incineration. Incineration utilises high temperatures, an oxidising atmosphere, and often turbulent combustion conditions to destroy wastes. Methods other than incineration that make use of high temperatures to destroy or neutralise hazardous wastes are discussed briefly at the end of this section.

Incineration

Hazardous waste incineration will be defined here as a process that involves exposure of the waste materials to oxidising conditions at a high temperature, usually in excess of 900°C. Normally the heat required for incineration comes from the oxidation of organically bound carbon and hydrogen contained in the waste material or in supplemental fuel

$$C \text{ (organic)} + O_2 \rightarrow CO_2 + heat \qquad \qquad \text{... (3.20)}$$

$$4H \text{ (organic)} + O_2 \rightarrow 2H_2O + heat \qquad \qquad \text{... (3.21)}$$

These reactions destroy organic matter and generate heat required for endothermic reactions, such as the breaking of C-Cl bonds in organochlorine compounds.

Incinerable wastes

Ideally, incinerable wastes are predominantly organic materials that will burn with a heating value of at least 5000 Btu/lb and preferably over 8000 Btu/lb. Such heating values are readily attained with wastes having high contents of the most commonly incinerated waste organic substances, including methanol, acetonitrile, toluene, ethanol, amyl acetate, acetone, xylene, methyl ethyl ketone, adipic acid, and ethyl acetate. In some cases, however, it is desirable to incinerate wastes that will not burn alone and which require supplemental fuel, such as methane and petroleum liquids. Examples of such wastes are non-flammable organochlorine wastes, some aqueous wastes or soil in which the elimination of a particularly troublesome contaminant is worth the expense and trouble of incinerating it. Inorganic matter, water, and organic hetero element contents of liquid wastes are important in determining their incinerability.

Hazardous waste fuel

Many industrial wastes, including hazardous wastes, are burned as hazardous waste fuel for energy recovery in industrial furnaces and boilers and in incinerators of non-hazardous wastes, such as sewage sludge incinerators. This process is called coincineration, and more combustible wastes are utilised by it than are burned solely for the purpose of waste destruction. In addition to heat recovery from combustible wastes, it is a major advantage to use an existing onsite facility for waste disposal rather than a separate hazardous waste incinerator.

Incineration systems

The four major components of hazardous waste incineration systems are shown in Fig. 3.11. Waste preparation for liquid wastes may require filtration, settling to remove solid material and water, blending to obtain the optimum incinerable mixture or heating to decrease viscosity. Solids may require shredding and screening. Atomisation is commonly used to feed liquid wastes. Several mechanical devices, such as rams and augers, are used to introduce solids into the incinerator.

The most common kinds of combustion chambers are liquid injection, fixed hearth, rotary kiln, and fluidised bed. These types are discussed in more detail later in this section.

Often the most complex part of a hazardous waste incineration system is the air pollution control system, which involves several operations. The most common operations in air pollution control from hazardous waste incinerators are combustion gas cooling, heat recovery, quenching, particulate matter removal, acid gas removal, and treatment and handling of by-product solids, sludges, and liquids.

Hot ash is often quenched in water. Prior to disposal, it may require dewatering and chemical stabilisation. A major consideration with hazardous waste incinerators and the types of wastes that are incinerated, is the disposal problem posed by the ash, especially in respect to potential leaching of heavy metals.

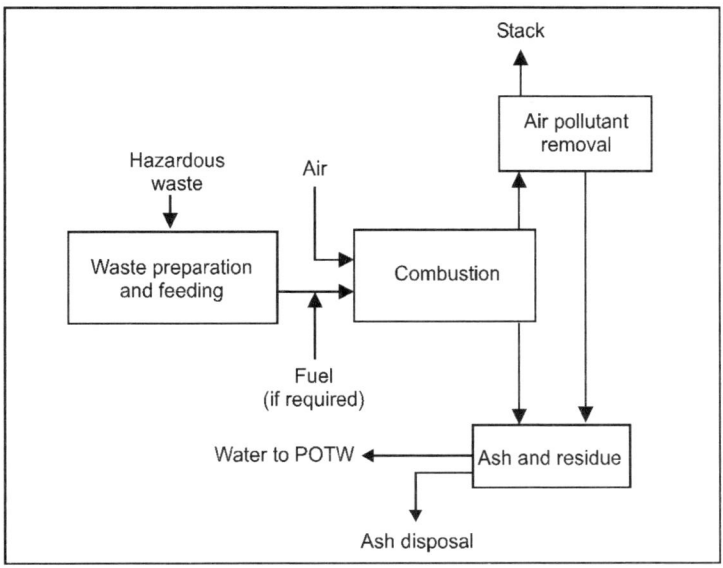

Fig. 3.11. Major components of a hazardous waste incinerator system.

Types of incinerators

Hazardous waste incinerators may be divided among the following, based upon type of combustion chamber:

1. Rotary kiln in which the primary combustion chamber is a rotating cylinder lined with refractory materials and an after burner downstream from the kiln to complete destruction of the wastes.
2. Liquid injection incinerators that burn pumpable liquid wastes dispersed as small droplets.
3. Fixed-hearth incinerators with single or multiple hearths upon which combustion of liquid or solid wastes occurs.
4. Fluidised-bed incinerators that have a bed of granular solid (such as limestone) maintained in a suspended state by injection of air to remove pollutant acid gas and ash products.
5. Advanced design incinerators including plasma incinerators that make use of an extremely hot plasma of ionised air injected through an electrical arc; electric reactors that use resistance-heated incinerator walls at around 2200°C to heat and pyrolyse wastes by radiative heat transfer; infrared systems, which generate intense infrared radiation by passing electricity through silicon carbide resistance heating elements; molten salt combustion that uses a bed of molten sodium carbonate at about 900°C to destroy the wastes and retain gaseous pollutants; and molten glass processes that use a pool of molten glass to transfer heat to the waste and to retain products in a poorly leachable glass form.

Combustion conditions

The key to effective incineration of hazardous wastes lies in the combustion conditions. These require: (i) sufficient free oxygen in the combustion zone, (ii) turbulence for thorough mixing of waste, oxidant, and (where used) supplemental fuel, (iii) high combustion temperatures above about 900°C to ensure that thermally resistant compounds do react, and (iv) sufficient residence time (at least 2 seconds) to allow reactions to occur.

Effectiveness of incineration

EPA standards for hazardous waste incineration are based upon the effectiveness of destruction of the principal organic hazardous constituents (POHC). Measurement of these compounds before and after incineration gives the destruction removal efficiency (*DRE*) according to the formula.

$$DRE = \frac{W_{in} - W_{out}}{W_{in}} \times 100 \qquad \qquad ... (3.22)$$

where, W_{in} and W_{out} are the mass flow rates of the POHC input and output (at the stack downstream from emission controls), respectively.

Wet air oxidation

Organic compounds and oxidisable inorganic species can be oxidised by oxygen in aqueous solution. The source of oxygen usually is air. Rather extreme conditions of temperature and pressure are required, with a temperature range of $175°$–$327°C$ and a pressure range of 300–3000 psig (2070–20,700 kPa). The high pressures allow a high concentration of oxygen to be dissolved in the water and the high temperatures enable the reaction to occur.

Wet air oxidation has been applied to the destruction of cyanides in electroplating waste-waters. The oxidation reaction for sodium cyanide is the following:

$$2Na^+ + 2CN^- + O_2 + 4H_2O \rightarrow 2Na^+ + 2HCO_3^- + 2NH_3 \qquad \qquad ... (3.23)$$

A method has been described in which cyanide is oxidised on an aerated carbon bed. The method employs added copper (II) ion and sulphite as catalysts. In addition to destroying CN^-, it also oxidises highly stable complexed cyanide in species such as $Fe(CN)_6^{4-}$.

Organic wastes can be oxidised in supercritical water, taking advantage of the ability of supercritical fluids to dissolve organic compounds. Wastes are contacted with water and the mixture raised to a temperature and pressure required for supercritical conditions for water. Oxygen is then pumped in, sufficient to oxidise the wastes. The process produces only small quantities of CO, and no SO_2 or NO_x. It has reportedly been used to degrade PCBs, dioxins, organochlorine insecticides, benzene, urea, and numerous other materials.

UV-enhanced wet oxidation

Hydrogen peroxide (H_2O_2) can be used as an oxidant in solution assisted by ultraviolet radiation (hv). For the oxidation of organic species represented in general as $\{CH_2O\}$, the overall reaction is:

$$2H_2O_2 + \{CH_2O\} + hv \rightarrow CO_2 + 3H_2O \qquad \qquad ... (3.24)$$

The ultraviolet radiation breaks chemical bonds and serves to form reactive oxidant species, such as $HO^•$.

Biodegradation of Wastes

Biodegradation of wastes is their conversion by biological processes to simple inorganic molecules (mineralisation) and to a certain extent, to biological materials. Usually the products of biodegradation are molecular forms that tend to occur in nature and that are in greater thermodynamic equilibrium with their surroundings than are the starting materials. Detoxification refers to the biological conversion of a toxic substance to a less toxic species. Microbial bacteria and fungi possessing enzyme systems required for biodegradation of wastes are usually best obtained from populations of indigenous micro-organisms at a hazardous waste site where they have developed the ability to degrade particular kinds of molecules. Biological treatment offers a number of significant advantages and has considerable potential for the degradation of hazardous wastes, even *in situ*.

It must be kept in mind, however, that there are many factors that can cause biodegradation to fail as a treatment process. Often physical conditions are such that mixing of wastes, nutrients, and electron acceptor species (such as oxygen) is too slow to permit biodegradation to occur at a useful rate. Low temperatures may make reactions too slow to be useful. Toxicants, such as heavy metals, may inhibit biological activity, and some metabolites produced by the micro-organisms may be toxic to them.

Under the label of bioremediation, the use of microbial processes to destroy hazardous wastes is experiencing a period of very rapid growth. Doubts still exist about claims for its effectiveness in a number of applications. It has been suggested that performance standards are badly needed for judging the effectiveness of bioremediation.

Biodegradability

The biodegradability of a compound is influenced by its physical characteristics, such as solubility in water and vapour pressure, and by its chemical properties, including molar mass, molecular structure, and presence of various kinds of functional groups, some of which provide a 'biochemical handle' for the initiation of biodegradation. With the appropriate organisms and under the right conditions, even substances such as phenol that are considered to be biocidal to most micro-organisms can undergo biodegradation. Recalcitrant or biorefractory substances are those that resist biodegradation and tend to persist and accumulate in the environment. Such materials are not necessarily toxic to oraganisms, but simply resist their metabolic attack. However, even some compounds regarded as biorefractory may be degraded by micro-organisms adapted to their biodegradation; for example, DDT is degraded by properly acclimated *Pseudomonas*. Chemical pretreatment, especially by partial oxidation, can make some kinds of recalcitrant wastes much more biodegradable.

Properties of hazardous wastes and their media can be changed to increase biodegradability. This can be accomplished by adjustment of conditions to optimum temperature, pH (usually in the range of 6–9), stirring, oxygen level, and material load. Biodegradation can be aided by removal of toxic organic and inorganic substances, such as heavy metal ions.

Aerobic treatment

Aerobic waste treatment processes utilise aerobic bacteria and fungi that require molecular oxygen, O_2. These processes are often favoured by micro-organisms, in part because of the high energy yield obtained when molecular oxygen reacts with organic matter. Aerobic waste treatment is well adapted to the use of an activated sludge process. It can be applied to hazardous wastes such as chemical process wastes and landfill leachates. Some systems used powdered activated carbon as an additive to absorb organic wastes that are not biodegraded by micro-organisms in the system.

Contaminated soils can be mixed with water and treated in a bioreactor to eliminate biodegradable contaminants in the soil. It is possible in principle to treat contaminated soils biologically in place by pumping oxygenated, nutrient-enriched water through the soil in a recirculating system.

Anaerobic treatment

Anaerobic waste treatment in which micro-organisms degrade wastes in the absence of oxygen can be practised on a variety of organic hazardous wastes. Compared to the aerated activated sludge process, anaerobic digestion requires less energy; yields less sludge by-product; generates hydrogen sulphide (H_2S), which precipitates toxic heavy metal ions; and produces methane gas, CH_4, which can be used as an energy source.

The overall process for anaerobic digestion is a fermentation process in which organic matter is both oxidised and reduced. The simplified reaction for the anaerobic fermentation of a hypothetical organic substance, '$\{CH_2O\}$', is the following:

$$2\{CH_2O\} \rightarrow CO_2 + CH_4 \qquad \qquad ...(3.25)$$

In practice, the microbial processes involved are quite complex. Most of the wastes for which anaerobic digestion is suitable consist of oxygenated compounds, such as acetaldehyde or methylethyl ketone.

Reductive dehalogenation

Reductive dehalogenation is a mechanism by which halogen atoms are removed from organohalide compounds by anaerobic bacteria. It is an important means of detoxifying alkyl halides (particularly solvents), aryl halides and organochlorine pesticides, all of which are important hazardous waste compounds. It is the only means by which some of the more highly halogenated waste compounds are biodegraded; such compounds include tetrachloroethene, hexachlorobenzene, pentachlorophenol, and the more highly chlorinated PCB congeners.

The two general processes by which reductive dehalogenation occurs are hydrogenolysis, as shown by the example in equation:

$$\underset{}{\text{C}_6\text{H}_5\text{Cl}} \; \xrightarrow[\text{HCl}]{2\text{H}} \; \underset{}{\text{C}_6\text{H}_6} \qquad \qquad ...(3.26)$$

and vicinal reduction:

$$\underset{\substack{|\;\;|\;\;| \\ \text{H H H}}}{\text{H}-\overset{\substack{\text{H Cl H} \\ |\;\;|\;\;|}}{\text{C}-\text{C}-\text{C}}-\text{Cl}} \; \xrightarrow[2\text{Cl}]{2\text{e}} \; \text{H}-\overset{\text{H}}{\underset{\text{H}}{\text{C}}}-\overset{\text{H}}{\underset{\text{H}}{\text{C}}}=\text{C}\overset{\text{H}}{\underset{\text{H}}{}} \qquad ...(3.27)$$

Vicinal reduction removes two adjacent halogen atoms, and works only on alkyl halides, not aryl halides. Both processes produce innocuous inorganic halide (Cl^-).

Land Treatment and Composting

Land treatment

Soil may be viewed as a natural filter for wastes. Soil has physical, chemical, and biological characteristics that can enable waste detoxification, biodegradation, chemical decomposition, and physical and chemical fixation. Therefore, treatment of wastes may be accomplished by mixing the wastes with soil under appropriate conditions.

Soil is a natural medium for a number of living organisms that may have an effect upon biodegradation of hazardous wastes. Of these, the most important are bacteria, including those from the genera *Agrobacterium, Arhrobacteri, Bacillus, Flavobacterium* and *Pseudomonas*. *Actinomycetes* and fungi are important organisms in decay of vegetable matter and may be involved in biodegradation of wastes.

Micro-organisms useful for land treatment are usually present in sufficient numbers to provide the inoculum required for their growth. The growth of these indigenous micro-organisms may be stimulated by adding nutrients and an electron acceptor to act as an oxidant (for aerobic degradation), accompanied by mixing. The most commonly added nutrients are nitrogen and phosphorus. Oxygen can be added by pumping air underground or by treatment with hydrogen peroxide, H_2O_2. In some cases, such as for treatment of hydrocarbons on or near the soil surface, simple tillage provides both oxygen and the mixing required for optimum microbial growth.

Wastes that are amenable to land treatment are biodegradable organic substances. However, in soil contaminated with hazardous wastes, bacterial cultures may develop that are effective in degrading normally recalcitrant compounds through acclimation over a long period of time. Land treatment is most used for petroleum refining wastes and is applicable to the treatment of fuels and wastes from leaking underground storage tanks. It can also be applied to biodegradable organic chemical wastes, including some organohalide compounds. Land treatment is not suitable for the treatment of wastes containing acids, bases, toxic inorganic compounds, salts, heavy metals, and organic compounds that are excessively (soluble, volatile or flammable).

Composting

Composting of hazardous wastes is the biodegradation of solid or solidified materials in a medium other than soil. Bulking material, such as plant residue, paper, municipal refuse, or sawdust may be added to retain water and enable air to penetrate to the waste material. Successful composing of hazardous waste depends upon a number of factors. The first of these is the selection of the appropriate micro-organism or inoculum. Once a successful composting operation is underway, a good inoculum is maintained by recirculating spent compost to each new batch. Other parameters that must be controlled include oxygen supply, moisture content (which should be maintained at a minimum of about 40 per cent), pH (usually around neutral), and temperature. The composting process generates heat, so if the mass of the compost pile is sufficiently high, it can be self-heating under most conditions. Some wastes are deficient in nutrients, such as nitrogen, which must be supplied from commercial sources or from other wastes. Composting is discussed in detail in chapter 20 of this book

Preparation of Wastes for Disposal

Immobilisation, stabilisation, fixation, and solidification are terms that describe techniques whereby hazardous wastes are placed in a form suitable for long-term disposal. These aspects of hazardous waste management are addressed below.

Immobilisation

Immobilisation includes physical and chemical processes that reduce surface areas of wastes to minimise leaching. It isolates the wastes from their environment, especially groundwater, so that they have the least possible tendency to migrate. This is accomplished by physically isolating the waste, reducing its solubility, and decreasing its surface area. Immobilisation usually improves the handling and physical characteristics of wastes.

Stabilisation

Stabilisation means the conversion of a waste from its original form to a physically and chemically more stable material. Stabilisation may include chemical reactions that generate products that are less volatile, soluble, and reactive. Solidification, which is discussed below, is one of the most common means of stabilisation. Stabilisation is required for land disposal of wastes. Fixation is a process that binds a hazardous waste in a less mobile and less toxic form; it means much the same thing as stabilisation.

Solidification

Solidification may involve chemical reaction of the waste with the solidification agent, mechanical isolation in a protective binding matrix or a combination of chemical and physical processes. It can be accomplished by evaporation of water from aqueous wastes or sludges, sorption onto solid material, reaction with cement, reaction with silicates, encapsulation or imbedding in polymers or thermoplastic materials.

In many solidification processes, such as reaction with portland cement, water is an important ingredient of the hydrated solid matrix. Therefore, the solid should not be heated excessively or exposed to extremely dry conditions, which could result in diminished structural integrity from loss of water. In some cases, however, heating a solidified waste is an essential part of the overall solidification procedure. For example, an iron hydroxide matrix can be converted to highly insoluble, refractory iron oxide by heating. Organic constituents of solidified wastes may be converted to inert carbon by heating. Heating is an integral part of the process of vitrification.

Hazardous waste liquids, emulsions, sludges, and free liquids in contact with sludges may be solidified and stabilised by fixing onto solid sorbents, including activated carbon (for organics), fly ash, kiln dust, clays, vermiculite, and various proprietary materials. Sorption may be done to convert liquids and semi-solids to dry solids, improve waste handling, and reduce solubility of waste constituents. Sorption can also be used to improve waste compatibility with substances such as portland cement used for solidification and setting. Specific sorbents may also be used to stabilise pH and pE (a measure of the tendency of a medium to be oxidising or reducing).

The action of sorbents can include simple mechanical retention of wastes, physical sorption, and chemical reactions. It is important to match the sorbent to the waste. A substance with a strong affinity for water should be employed for wastes containing excess waster, and one with a strong affinity for organic materials should be used for wastes with excess organic solvents.

Thermoplastics and organic polymers

Thermoplastics are solids or semi-solids that become liquefied at elevated temperatures. Hazardous waste materials may be mixed with hot thermoplastic liquids and solidified in the cooled thermoplastic matrix, which is rigid but deformable. The thermoplastic material mostly used for this purpose is asphalt bitumen. Other thermoplastics, such as paraffin and polyethylene, have also been used to immobilise hazardous wastes. Among the wastes that can be immobilised with thermoplastics are those containing heavy metals, such as electroplating wastes. Organic thermoplastics repel water and reduce the tendency toward leaching in contact with groundwater. Compared to cement, thermoplastics add relatively less material to the waste. A technique similar to that described above uses organic polymers produced in contact with solid wastes to imbed the wastes in a polymer matrix. Three kinds of polymers that have been used for this purpose include polybutadiene, urea-formaldehyde, and vinyl ester-styrene polymers. This procedure is more complicated than is the use of thermoplastics but, in favourable cases, yields a product in which the waste is held more strongly.

Vitrification

Vitrification or glassification consists of imbedding wastes in a glass material. In this application, glass may be regarded as a high-melting-temperature inorganic thermoplastic. Molten glass can be used or glass can be synthesised in contact with the waste by mixing and heating with glass constituents–silicon dioxide (SiO_2), sodium carbonate (Na_2CO_3), and calcium oxide (CaO). Other constituents may include boron oxide, B_2O_3, which yields a borosilicate glass that is especially resistant to changes in temperature and chemical attack. In some cases, glass is used in conjunction with thermal waste destruction processes, serving to immobilise hazardous waste ash constituents. Some wastes are detrimental to the quality of the glass. Aluminium oxide, for example, may prevent glass from fusing. Vitrification is relatively complicated and expensive, the latter because of the energy consumed in fusing glass. Despite these disadvantages, it is the best immobilisation technique for some special wastes and has been promoted for solidification of radionuclear wastes because glass is chemically inert and resistant to leaching. However, high levels of radioactivity can cause deterioration of glass and lower its resistance to leaching.

Solidification with cement

Portland cement is widely used for solidification of hazardous wastes. In this application, portland cement provides a solid matrix for isolation of the wastes, chemically binds water from sludge wastes, and may react chemically with wastes (for example, the calcium and base in portland cement react chemically with inorganic arsenic sulphide wastes to reduce their solubilities). However, most wastes are held physically in the rigid portland cement matrix and are subject to leaching.

As a solidification matrix, portland cement is most applicable to inorganic sludges containing heavy metal ions that form insoluble hydroxides and carbonates in the basic carbonate medium provided by the cement. The success of solidification with portland cement strongly depends upon whether or not the waste adversely affects the strength and stability of the concrete product. A number of substances— organic matter such as petroleum or coal; some silts and clays; sodium salts of arsenate, borate, phosphate, iodate, and sulphides; and salts of copper, lead, magnesium, tin, and zinc— are incompatible with portland cement because they interfere with its set and cure and cause deterioration of the cement matrix with time. However, a reasonably good disposal form can be obtained by absorbing organic wastes with a solid material, which in turn is set in portland cement. This approach has been used with hydrocarbon wastes sorbed by an activated coal char matrix.

Solidification with silicate materials

Water-insoluble silicates, (pozzolanic substances) containing oxyanionic silicon such as SiO_3^{2-} are used for waste solidification. These substances include fly ash, flue dust, clay, calcium silicates, and ground-up slag from blast furnaces. Soluble silicates, such as sodium silicate, may also be used. Silicate solidification usually requires a setting agent, which may be portland cement, gypsum (hydrated $CaSO_4$), lime or compounds of aluminium, magnesium or iron. The product may vary from a granular material to a concrete-like solid. In some cases, the product is improved by additives, such as emulsifiers, surfactants, activators, calcium chlorides, clays, carbon, zeolites, and various proprietary materials.

Success has been reported for the solidification of both inorganic wastes and organic wastes (including oily sludges) with silicates. The advantages and disadvantages of silicate solidification are similar to those of portland cement discussed above. One consideration that is especially applicable to fly ash is the presence in some silicate materials of leachable hazardous substances, which may include arsenic and selenium.

Encapsulation

As the name implies, encapsulation is used to coat wastes with an impervious material so that they do not contact their surroundings. For example, a water-soluble waste salt encapsulated in asphalt would not dissolve, as long as the asphalt layer remains intact. A common means of encapsulation uses heated, molten thermoplastics, asphalt, and waxes that solidify when cooled. A more sophisticated approach to encapsulation is to form polymeric resins from monomeric substances in the presence of the waste.

Chemical fixation

Chemical fixation is a process that binds a hazardous waste substance in a less mobile, less toxic form by a chemical reaction that alters the waste chemically. Physical and chemical fixation often occur together. Polymeric inorganic silicates containing some calcium and often some aluminium are the inorganic materials most widely used as a fixation matrix. Many kinds of heavy metals are chemically bound in such a matrix as well as being held physically by it. Similarly, some organic wastes are bound by reactions with matrix constituents. For example, humic acid wastes react with calcium ion in a solidification matrix to produce insoluble calcium humates.

Ultimate Disposal of Wastes

Regardless of the destruction, treatment, and immobilisation techniques used, there will always remain from hazardous wastes some material that has to be put somewhere. This section briefly addresses the ultimate disposal of ash, salts, liquids, solidified liquids, and other residues that must be placed where their potential to do harm is minimised.

Disposal above ground

In some important respects disposal aboveground, essentially in a pile designed to prevent erosion and water infiltration, is the best way to store solid wastes. Perhaps its most important advantage is that it avoids infiltration by groundwater that can result in leaching and groundwater contamination common to storage in pits and landfills. In a properly designed aboveground disposal facility any leachate that is produced drains quickly by gravity to the leachate collection system, where it can be detected and treated. Above ground disposal can be accomplished with a storage mound deposited on a layer of compacted clay covered with impermeable membrane liners laid somewhat above the original soil surface and shaped to allow leachate flow and collection. The slopes around the edges of the storage mound should be sufficiently great to allow good drainage of precipitation, but gentle enough to deter erosion.

Landfill

Landfill historically has been the most common way of disposing of solid hazardous wastes and some liquids, although it is being severely limited in many nations by new regulations and high land costs. Landfill involves disposal that is at least partially underground in excavated cells, quarries or natural depressions. Usually fill is continued above ground to utilise space most efficiently and provide a grade for drainage of precipitation.

The greatest environmental concern with landfill of hazardous wastes is the generation of leachate from infiltrating surface water and groundwater with resultant contamination of groundwater supplies. Modern hazardous waste landfills provide elaborate systems to contain, collect, and control such leachate. Landfill is discussed in detail in chapter 22 of this book.

Chapter 4

Abiotic Treatment

INTRODUCTION

Selection of a treatment process depends on the nature of the waste-water and the quality of the effluent desired. Hazardous components of the waste-water may be either separated or converted to nonhazardous forms in order to permit the disposal of the waste-water effluent. Conversion processes can be done in one step or several in series. Hazardous components which are separated from the waste-water must be disposed of and may require additional processing.

LIQUID–SOLID SEPARATION

Separation of suspended matter from waste-water can be accomplished by a number of different processes. Large, heavy solids are easier to remove than finely divided, light solids.

Screening devices are used to remove large pieces of solid matter that would interfere with subsequent processing operations or would cause damage to equipment such as pumps. Coarse screening devices may consist of parallel bars, rods or wires, perforated plates, gratings or wire mesh.

Gravity sedimentation involves the containment of waste-water for a sufficient period of time to allow some or all of the suspended materials to either settle out or float to the surface of the waste-water. In its simplest form as a batch process, a given volume of waste-water is transferred to a vessel and held there until nearly all the settleable and floatable matter separates. The floating matter can be skimmed off and the waste-water decanted for discharge or further treatment. Sludge may be allowed to collect until several batches of waste-water have been processed. Then, it is removed. The vessel may have a conical bottom so that the sludge can be removed via a valve. Large settling ponds may be constructed which are drained periodically to permit sludge removal. Solids contact or sludge-blanket clarifiers are useful for sludges that are flocculent and of low density. They are designed with large mixing and reaction zones that, coupled with the sludge blanket, account for greater efficiency in solids removal. Gravity sedimentation works like the clarifiers but more time is taken, for the settling.

Dissolved-air flotation is useful for suspended matter that does not sink or float in a reasonable period of time. Separation is brought about by the introduction of finely divided gas bubbles which become attached to the particulate matter, causing it to float to the surface where it is removed by skimming. Introduction of the gas bubbles is usually accomplished by reducing the pressure of the waste-water, thus causing dissolved gases to be released. This is commonly used to separate greasy or oily matter from industrial wastes.

Granular-media filters or deep-bed filtration is a polishing step that removes small amounts of suspended solids and produces a highly clarified water. Chemical coagulation and sedimentation typically precede this stage. Graded sand and pulverised coal are commonly used in the filter beds. Conventional operation is usually by downflow. The ability of the granular-media filter beds to produce a clear effluent results from the straining action and adhesion, which remove particles finer than the pore space.

Surface filters make use of a fine medium such as a cloth or close-mesh screen. In a rotary vacuum filter, the medium is in the form of a continuous belt that rotates over a perforated drum partially submerged in the slurry to be filtered. Water is pulled through the filter cake that forms on the belt to the inside of the drum, where it is transferred to the vacuum system.

Centrifugation is a useful alternative to filtration for sticky sludges that do not dewater rapidly on a filter. They operate by a rapid rotation of a liquid suspension, which induces a much greater force than gravity to hasten the separation of the suspended matter.

CHEMICAL TREATMENT

Chemical treatment is a widely used process for the destruction or separation of hazardous constituents in waste-water. This can be done by neutralisation of acidic or alkaline waste-water until a suitable pH is obtained. Precipitation/coagulation/flocculation is useful for the removal of heavy metals. Precipitation refers to the formation of a solid phase, coagulation is where the contaminant is trapped by the formation of a precipitate, and flocculation is the agglomeration of a coagulating chemical.

Oxidation-reduction, or the redox processes, are used for converting toxic pollutants to harmless or less toxic materials that are more easily removed. These processes involve the addition of chemical, reagents to waste-waters, causing changes in the oxidation states of substances both in the reagents and in the waste-waters. In order for one substance to be oxidised, another must be reduced.

Ozone is a powerful oxidising agent that is usually used to eliminate traces of organics. Wet oxidation utilises oxygen at temperatures and pressures up to 350 and 180 atmospheres to treat organic wastes.

Ion exchange involves a change in the chemical form of a compound, the exchange of ions in solution with other ions held by mixed anionic or cationic groups or charges. Typically, a waste solution is percolated through a granular bed of the ion exchanger, where certain ions in solution are replaced by ions contained in the ion exchanger. If the exchange involves cations, the exchanger is called a cation exchanger, correspondingly, an anion exchanger is one that involves anion exchange.

PHYSICAL METHODS

Several methods are available for separating pollutants from waste-water: activated carbon, steam stripping, evaporation, reverse osmosis, and solvent extraction. The chemical and physical characteristics of the pollutant are important in the selection of the physical removal method.

Steam stripping is effective for substances that have an appreciable vapour pressure at the boiling point of water, whereas evaporation is effective for those chemicals that will not volatilise. Soluble, small organic molecules are adsorbed by activated carbon, large ions are separated by reverse osmosis. In activated carbon adsorption the inorganic and organic chemicals are adsorbed onto activated carbon. Usually hydrophobic chemicals are more likely to be removed. The degree of adsorption is linked to the molecular weight. Methanol-water coefficient or solubility are also linked to the recalcitrance and/or toxicity. Activated carbon adsorption is applicable to the treatment of dilute aqueous wastes, which should also be treated to remove suspended solids, oil, and grease. Temperature and pH are important for various compounds to be treated. The carbon is either disposed of or regenerated. Carbon has also been added directly to biological treatment effluent in a contacting basin. The advantages of this are that the sludge toxicity is reduced by selectively removing the toxic organics from solution and that the carbon adsorption capacity is extended by bioregeneration of the 'biocompatible' species adsorbed on the surface. For aqueous solvent waste-containing contaminants in concentrations up to 10,000 mg/l, the activated sludge process has been proposed as a potential applicable treatment. However, these concentrations may be toxic to the sludge or they may be easily stripped to the atmosphere, thereby creating another hazard. The sludge may also contain recalcitrant waste, due to sorption of the contaminants, and may be difficult to dispose of.

Evaporation is the process that heats the liquid, vents the vapours to the atmosphere, and concentrates the pollutants into a slurry.

Osmosis is the process where a solvent (e.g. water) moves from an area of low concentration to high across a semipermeable membrane which does not allow the dissolved solids to pass. In reverse osmosis, a pressure greater than the osmotic pressure is applied so the flow is reversed. Pure water will then flow through the membrane from the concentrated solution.

Solvent extraction is a process whereby a dissolved or adsorbed substance is transferred from a liquid or solid phase to a solvent that preferentially dissolves that substance. For the process to be effective, the extracting solvent must be immiscible in the liquid and differ in density so that gravity separation is possible and there is minimal contamination of the raffinate with the solvent. The hydrophobic solutes are more likely to be extracted. Solvent extraction can be performed as a batch process or by the contact of the solvent with the feed in staged or continuous equipment.

Steam stripping is where water vapour at elevated temperatures is used to remove volatile components of a liquid. Countercurrent flow is generally used to promote gas–liquid contact, thus allowing soluble, gaseous organics from the liquid waste to be continuously exchanged with molecules within the stripping gas. Again, this is useful only for waste with low water solubilities.

INCINERATION

Incineration is a high-temperature oxidation process that converts the principal elements in most organic compounds (carbon, hydrogen, and oxygen) to carbon dioxide and water. Given the problems of land disposal, incineration may take a lead role in waste treatment. However, it is not without its problems. There is fear among the general public about the nature of the stack emissions, but it is an efficient method. The destruction of the molecular structure usually eliminates the toxicity of the chemical. But, the existence of other elements in a waste may result in the production of particulate pollutants that require removal in off-gas treatment systems. There are several types of incinerators available.

Liquid injection incinerators operate by spraying the combustible waste mix with air into a chamber where flame oxidation takes place. The purpose of spraying is to atomise the waste into small droplets which present a large surface area for rapid heat transfer thereby increasing the rate of vapourisation and mixing with air to promote combustion. Air is supplied to provide the necessary mixing and turbulence. These incinerators are widely used for the destruction of liquid organic wastes.

Rotary kiln incinerators are designed to process solids and tars that cannot be processed in the liquid incinerator. The rotary kiln is a cylindrical shell lined with refractory material that is horizontally mounted at a slight incline. It is rotated from 5 to 25 times an hour at high temperatures 1500° to 3000°F with excess air and the residence time is varied depending on the nature of the waste. The rotation causes a tumbling action that mixes the waste with air. The primary function is to convert, through partial burning, and volatilisation, solid wastes to gases and ash/residue. If the ash is free of dangerous levels of hazardous wastes, it is put in a landfill. Large quantities of organic vapour fumes are produced by many industries, including fat rendering, metal painting and varnishing, and various types of printing. These fumes are handled with fume incineration. The vapours are generally mixes of hydrocarbons, alcohols and acetates. The mixes may not be acutely toxic but they do cause odour problems.

The multiple-hearth incinerator is used for wastes that are difficult to burn or that contain valuable metals that can be recovered. It consists of a refractory-lined circular steel shell, with refractory hearths located one above the other. Solid waste or partially dewatered sludge is fed to the top of the unit, where a rotating plow rake plows it across the hearth to drop holes. The uncombusted material falls to the next hearth and the process is repeated until the combustion is complete.

Fluidised bed incinerators (FBI) are applicable to the destruction of halogenated organic waste streams. This type of incinerator consists of a vessel in which inert granular particles are fluidised by a low velocity air stream which is passed through a distributor plate below the bed. An FBI consists of a windbox, through which combustion air is introduced to the reactor, a reactor zone, containing a bed of sand, waste injection, and removal ports. Temperatures are in the range of 1300° to 2100°F; gas residence times are usually a few seconds. They have been used to treat municipal sewage sludge, low quality fuels, pulp and paper effluents; food processing wash, refinery waste, radioactive waste, and miscellaneous chemical waste. A molten salt incinerator uses a molten salt such as a sodium carbonate as a heat transfer and reaction medium. In the process, waste material along with air is added below the surface of the bed so that any gases formed during combustion are forced to pass through the melt. Reaction temperatures in the bed range from 1500° to 2000°F, and residence times are less than a second. Any acidic gases formed are neutralised by the alkalinity of the bed. This can change the fluidity of the bed and so it needs replacement frequently.

Plasma arc incineration is based on the concept of reducing or pyrolysing waste molecules to the atomic state using a thermal plasma field. The system uses very high energy at temperatures near 10,000°C to break bonds of hazardous waste chemical molecules down to the atomic state. An electrode assembly ionises air molecules which create a plasma field. Hazardous waste mixtures interact with the field, forming simple molecules such as carbon dioxide, hydrogen, hydrogen chloride, and other minor matrix compounds such as acetylene and ethane. Westinghouse Electric has a mobile plasma arc unit called pyroplasma that reportedly treats liquid wastes at the rate of 3 gal/min. The high temperatures decompose PCB and other wastes in an oxygen-deficient atmosphere. Hydrogen chloride is treated with sodium to form water and salt.

A cement kiln is basically a large rotary kiln in which raw materials are fed countercurrent to combustion gas flow. The wet process kilns use a 30 per cent water slurry feed and are the most suitable for hazardous waste destruction. The products formed are alkaline and so act as a scrubber, removing acidic gases formed during combustion. This system operates at 2800°F, resulting in very efficient removal of wastes.

WET AIR OXIDATION

Wet air oxidation involves the aqueous phase oxidation of organic materials at high temperature and pressure. A major advantage over other incineration methods is that the water in the waste stream is kept in the liquid state. Water is pumped into the reactor along with oxygen, which is heated by the hot effluent. Two types of reactors are used, a bubble tower reactor and a stirred tank cascade reactor. This process is good for wastes that are too dilute to incinerate but too toxic for biological methods. The products usually contain acetic acid and carbon dioxide.

SOLIDIFICATION TECHNIQUES

There are several innovative nonthermal processes that have been developed under the SITE program that immobilise wastes by vitrification or other types of solidification. The SITE program is the superfund innovative technology program developed to encourage the private development and demonstration of new technologies for cleaning up hazardous wastes. For example, the researchers at Battelle Pacific Northwest Laboratories have developed an *in situ* vitrification process (which was originally designed for the containment of nuclear wastes) in which electrodes are sunk into a contaminated area and attached to a diesel-powered generator. The currently produced temperatures of about 3600°F are much higher than the fusion temperature of soil. An exhaust hood is placed over the site to collect and treat any combustion products. The results is a massive glasslike product consisting of completely immobilised organics, inorganics, steel drums, and other components that are essentially locked up and inert. The time taken to complete the process depends on the electrode depth and frequency. Another solidification process uses a reagent called Urrichem that immobilises slurried hazardous components. The contaminated soil is excavated and mixed with the Urrichem off-site. After blending, the slurry is pumped out of the mixer and hardens into a concrete-like mass within 29 hours. Chemfix is a process developed by Chemfix Technologies. In this technique, a proprietary blend of soluble silicates and additives is used to convert high molecular weight organic and inorganic slurries into a cross-linked, clay-like matrix.

Alternative methods have also included landfills, deep-well injection disposal, and ocean dumping (which is no longer legally viable). Landfills were developed because it was believed that by placing waste in designated ground areas, there would be a natural decomposition over time. Unfortunately the water table rises in a landfill and this mounding effect means that a leaching of the water containing toxics occurs since the water flow is always from areas of high to low head, i.e. there is a gradient setup that favours water movement originating in the landfill and away from it thus contaminating groundwater supplies. This is a special problem for an area like Long Island, New York that relies on groundwater and not surface water for its domestic/industrial supplies. Deep injection wells have similar problems. When the aquifers in which the deep well sit are pumped, the contaminants are drawn into the well. Ocean dumping has upset the delicate ecosystem balance in some areas. Thus, there is a need for effective techniques that industry can safely use to try and combat the nation's hazardous waste problems. Biological treatment may be such a technique.

BIOLOGICAL TREATMENT OPTIONS

Biological treatment of wastes is a realistic option in the treatment strategies for hazardous wastes and waste-waters. An advantage is that the cost can be much less than other technologies, such as incineration. A disadvantage is the length of process time involved in biological treatment. There are different mechanisms that we can use to degrade compounds. Selection is based on the type of pollutant, type of degradation (i.e. aerobic or anaerobic), and various other parameters.

Suspended growth systems such as activated sludge, aerated lagoons, and anaerobic digesters are where micro-organisms are suspended. Fixed film processes which include filters and rotating biological contactors are where micro-organisms grow on a fixed surface. Since toxin degradation is a relatively slow process, the fixed film processes afford high microbial retention time. In aerobic degradation, the major portion of the cost is the aeration. Wastes in any form—solid, aqueous, and gaseous—can be treated biologically. The treatment can be done by several methods such as applying the waste to the soil, by composting, using a hybrid liquid/solid treatment technique, treating the wastes *in situ*, using soil filter gases, and with waste-water treatment. The biological treatment must be thoroughly assessed since non degradable pesticides and other organics may be affected by the treatment or toxic intermediates may be formed and therefore a further finishing process such as activated carbon may be needed.

Land Treatment

Land treatment is the biological option most widely used to treat hazardous wastes. Many of the industrial waste treatment sites using biological treatment are at petroleum refineries. Steps in the process typically are: spread the waste on land, let it dry, till the soil to mix the waste in, control moisture, and if necessary, add nutrients so that waste-destroying bacteria will grow.

In a land treatment system, bottom liners under the treatment zone may generally not be needed. Physical, chemical, and biological processes in the soil degrade, immobilise, or transform the wastes to environmentally acceptable forms. Most pollutants are captured and transformed in the top 6–12 inches in the soil, though the treatment zone may extend down five feet.

Land treatment is suitable with wastes such as petroleum refinery sludges, creosote sludges and waste-waters, and processing sludges from wood, paper, and textile manufacturing. It is also widely used for municipal sludges and waste-waters, and food processing sludges and waste-waters. Oil, metals, and other constituents of environmental concern are successfully controlled by land treatment. Most land treatment sites are at operating facilities.

Costs for unlined land treatment systems are a little cheaper but needs arise to protect the groundwater systems; extensive site testing is necessary to obtain federal RCRA permits. As a result of the Hazardous and Solid Waste Amendments of 1984 and the current land disposal restrictions rule, USEPA has attempted to develop achievable treatment techniques to be used as an alternative to the land disposal of hazardous wastes.

Composting

Composting uses less space than land treatment and controls gases and leachate. In composting, piles of the waste 3–6 feet high are treated. Aeration is provided either by turning the piles mechanically or through a forced aeration system. Bulking agents such as wood chips are sometimes added to facilitate mixing and oxygen transfer. These bulking agents are typically screened out at the completion of composting, and are mixed in with the next batch.

Composting has been used in treating municipal sewage sludge, but it has also been used to treat several industrial solid wastes, such as industrial waste-water treatment sludge, food processing wastes, and some industrial wastes containing low levels of pesticides.

Liquids/Solids Treatment Systems

These methods are hybrids, intermediate between land treatment and conventional water-suspended biological systems. The waste is not in a solid form as in land treatment, nor in water as is conventional municipal waste-water treatment, but halfway between. Wastes are in a suspended solid, slurry, or sludge form.

Such systems reduce the level of contaminants in a waste primarily by dissolving the organics, biodegrading the dissolved organics into less toxic and less environmentally significant forms, and releasing the products as gases to the atmosphere. The key step, dissolving the wastes from solid to liquid, can result from microbial action and/or physical-chemical action.

There are two basic process types: single-reactor, and two-stage. In either case it is vital to have sufficient mixing of the liquid/solid mixture to achieve effective mass transfer of both organics in the liquid phase, adequate oxygen transfer to the micro-organisms, and to keep the suspension of solids in solution. The single-batch reactor is the simpler process and the most widely used. Here both the dissolving and biodegradation steps occur together in a single tank, pit or lagoon. Air is introduced to provide mixing for the extraction process and oxygen for biological reasons. Aeration and mixing may be provided by submerged aerators, floating aerators or compressor/sparger (lance) systems.

At completion of treatment, the aerators and mixing are stopped and the solids are allowed to settle. The treated liquid is decanted off the top, while the treated solids are treated further, by conventional land treatment or by on-site disposal. If the process is done in a lagoon or surface impoundment, it may be possible to decant the liquid and leave the solids in place.

Soil Biofilters

Hazardous wastes in gaseous form can be successfully biologically treated. On-site biofilters can reduce nuisance odours, volatile organic carbon compounds, and particulates in waste gases. The soil biofilter involves two mechanisms: mechanical filtration and treatment by bacteria on the soil particles.

A biofilter consists of a distributing pipe network underneath soil or gravel. The gas to be treated is pumped through the media and escapes through holes in the pipe wall. It rises through the media bed, which is usually 3–10 feet deep. Pollutants captured on the biofilter can be removed by filtration, physical and chemical absorption, chemical oxidation, and structural treatment.

The technique is not a new one, filters have been used for many years to remove odours from agricultural, food and waste-water processing operations, and emissions from paper and chemical manufacturing plants. The soil filters have successfully removed low concentrations of H_2S, SO_2, NH_3, and NO_2, as well as particulates in these gases. Other potential applications include the treatment of exhaust gases from air stripping units for waste-water and solvents, from mechanical technologies such as belt presses for oily sludges, from aerated waste treatment systems, and from *in situ* soil stripping of volatile organic compounds (VOCs) using vacuum and forced aeration methods.

WASTE-WATER TREATMENT

Biological treatment has been very successful in the removal of organic pollutants and colloidal organics from waste-water. Activated sludge, biological filters, aerated lagoons, oxidation ponds, and aerobic fermentation are some of the methods available for waste-water biodegradation. In the removal of toxic waste, more care is needed since the bacteria can be destroyed from shock loading or increases of toxic material fed in without allowing time for the population to grow large enough to deal with it. Biodegradation occurs because bacteria are able to metabolise the organic matter through enzyme systems to yield carbon dioxide, water, and energy. The energy is used for synthesis, motility, and respiration. Simple dissolved matter is taken into the cell and oxidised, but with more complex inorganics, enzymes are secreted extracellularly to hydrolyse the proteins and fats into a soluble form which can then be taken into the cell and oxidised. Hence, more complex matter takes longer to process. Some organic compounds are 'refractory', they cannot be oxidised while others are toxic to the bacteria at high concentrations.

The purpose of biodegradation is to convert the waste into end-products and material that will settle and can be removed as sediment. Again, biodegradation may not be 100 per cent. Also, toxic by-products may be formed. Further treatment by chemical methods or dilution may be needed to get the contaminant to a concentration prescribed as safe.

Nitrogen and phosphorous are essential in the oxidation process for the synthesis of new cells. Trace amounts of potassium and calcium are also required. The former are sometimes deficient, so nitrogen is added in the form of ammoniacal nitrogen (nitrite and nitrate are not readily used by bacteria). Biochemical oxygen demand (BOD), measures the strength of the organics present and is defined as the amount of oxygen needed by the bacteria for oxidation. The more concentrated the organic material, the higher the BOD. A BOD:N:P ratio of 100:5:1 is thought to be the optimum ratio of nutrients needed by bacteria.

Activated Sludge Process

This process involves the generation of a suspended mass of bacteria in a reactor to degrade soluble and finely suspended organic compounds. In this method, the waste-water with its organic compounds is fed into the aeration tank. This is supplied with air and is vigorously mixed to allow maximum contact of bacteria and waste.

The contents, referred to as mixed liquor suspended solids (MLSS), are then fed to a sedimentation tank where the treated solids settle to the bottom and the top liquid layer is treated and discharged. Part of the biological solids are recycled back to the aeration tank to maintain the correct mix; the remainder is waste. This method is flexible and can be used on many types of biological wastes.

Trickling Filter Process

Here, waste-water is distributed by a flow distributor over a fixed bed of medium on which the bacteria grow, forming a slime layer to which oxygen is supplied. The waste-water flows down over the slime layer which absorbs organic materials and nutrients, releasing the oxidised end-products to the drainage system underneath. Eventually some of the layer will detach with the waste-water, and then some additional separation will be necessary.

Stabilisation Ponds

In this process, waste-water is stabilised by the actions of bacteria in shallow ponds. There are basically two types of ponds: one, where there is a natural supply of oxygen from algal photosynthesis (oxidation ponds), and other mechanically supplied-oxygen ponds (aerated lagoons, as shown in Fig. 4.1). The bacteria metabolise the wastes and the solids settle at the bottom as sludge. Also, there is anaerobic decomposition where the bacteria at the bottom will degrade the waste without the presence of oxygen. Or, there can be a lagoon that has both aerobic and anaerobic decomposition, with an interchange of products between the two layers of bacteria in a symbiotic relationship.

BIOLOGICAL DEGRADABILITY

Terms that are important in understanding the fate of wastes in biological systems include: biodegradable, persistent, recalcitrant, and mineralisation. A compound that is biodegradable can be changed by the action of micro-organisms to another compound. This does not necessarily mean that the product is less toxic than the parent compound or that intermediate toxic compounds are formed that may inhibit the degraders, or that the product will not be toxic to the next degraders or to humans. In land disposal, the pH of the soil also plays a role since at a low pH, fungi dominate the ecosystem and are more likely to form epoxides (associated with mutagenicity) than bacteria, which are dominant at higher pHs. So, there are some possible types of reactions that are not acceptable.

Fig. 4.1. Aeration basin for biological treatment.

Information is required so that we can select favourable degradation reactions so as to reduce some types of toxic waste. Recalcitrant means that the compound cannot be biodegraded under any circumstances, that is, it's the compound itself that is inherently resistant and not the treatment system that has failed to account for some vital fact. Persistence is a 'conditional' property of the biodegradable compound in that it may be biodegraded if the correct circumstances favouring biodegradation are present. Mineralisation is the complete conversion of an organic compound to the end-products of carbon dioxide and water. Primary biodegradation is the single transformation of a compound and partial biodegradation is somewhere between primary degradation and mineralisation.

BASIS FOR BIODEGRADATION

Some compounds are degraded by micro-organisms because the organisms gain energy needed for growth, repair, reproduction, and other biological functions needed for survival. Food sources contain oxygen in hydroxyl (OH) or carboxylic acid (COOH) groups. Oxidation reactions take place where electrons are transferred along an electron chain with compounds accepting and passing on electrons to a terminal electron acceptor. The coenzymes may be NADH and NAD, which get reduced and then coupled with the electron transport chain to produce high-energy bonds in ATP. Many compounds are biochemically inert such as alkanes, saturated ring structure, and unsubstituted benzene. They are devoid of oxygen and not subject to dehydrogenation reactions. The ability of bacteria to utilise these compounds lies in the fact, that they can catalyse oxidation using oxygen. Other bacteria have enzymes which work without oxygen and coenzymes are needed.

Bacteria can only do things for which they have a genetic capability. They must be able to produce the right enzymes to do the job, and the right environment must exist for them to be able to produce the right enzymes. If a chemical is present in concentrations either, too low or too high, then it may not be biodegraded: too low and the enzymes will not be induced, too high and the compound may be toxic. One reason why bacteria are robust at biodegradation is that they may still be able to degrade different compounds to the ones that they normally utilise if the active site of the molecule has not been altered. A xenobiotic still needs to be able to induce production enzymes or the reaction will stop. The bacteria will not be able to metabolise it and will die.

This use of a different substrate is fortunate and has been termed 'gratuitous' metabolism. Complicating biodegradation further is 'cometabolism', where two substrates are needed, i.e. one compound cannot fulfil all the bacterial needs and acts as the sole carbon and energy source. A second compound is needed as a growth substrate.

Biological degradation treatment systems have to be designed to take care of another problem. Bacteria need a, continuous carbon source for growth and yet if too much substrate is added, initially the bacteria are unable to metabolise it. They need a period of 'acclimation', where they are growing, strengthening, and perhaps even undergoing genetic changes.

If the intermediate is toxic, the bacteria may be killed or the formation of the next strain that metabolises it may be stopped so that it accumulates and destroys the system. Bacteria often work with a succession of strains so that complete mineralisation needs more then one organism which does not alone have, the required genetic capability.

A less desirable aspect of the ability of bacteria to degrade compounds is noted for example in some pesticides that are degraded quickly by micro-organisms before they have eliminated the pests. This phenomenon is called accelerated pesticide degradation. Evolution appears to be the factor here; individual bacteria may acquire the ability to utilise a pesticide by the evolution of specific enzymes. Several organisms may have to coexist in a community to metabolise a pesticide.

Assays for specific enzymatic activities of micro-organisms shown to metabolise certain pesticide substrates indicate that cross-adaptation of the micro-organisms for degradation of chemically similar pesticides in the soil environment may exist. Typical microbial metabolic reactions of importance to biological treatment include:

1. Anaerobic non-photosynthetic reactions:
   (a) Nitrate reduction (denitrification):
   $$5CH_3COOH + 8NO_3^- \rightarrow 10CO_2 + 4N_2 + 6H_2O + 8OH^-$$
   $$5S + 6NO_3^- + H_2O \rightarrow 5SO_4^= + 3N_2 + 4H^+$$
   (b) Sulphate reduction:
   $$2CH_3CHOHCOOH + SO_4^= \rightarrow 2CH_3COOH + H_2S + 2OH^-$$
   $$4H_2 + SO_2^= \rightarrow 2H_2O + H_2S + 2OH^-$$
   (c) Organic carbon reduction (fermentation):
   $$CH_3COOH \rightarrow CH_4Y + CO_2$$
   $$4CH_3OH \rightarrow 3CH_4 + CO_2 2H_2O$$
   $$C_6H_{12}O_6 \; bacteria > 3CH_3COOH$$
   $$C_6H_{12}O_6 \; yeast > 2CH_3CH_2OH + 2CO_2$$
   (d) Carbon dioxide reduction:
   $$2CH_3CH_2OH + CO_2 \rightarrow 2CH_3COOH + CH_4$$
   $$4H_2 + 2CO_2 \rightarrow CH_3COOH + 2H_2O$$

2. Anaerobic nonphotosynthetic bacterial reactions:
   (a) Oxygen-limited reactions:
   $$CH_3CH_2OH + O_2 \rightarrow CH_3COOH + H_2O$$
   $$2CH_3CHO + O_2 \rightarrow 2CH_3COOH$$
   $$2CH_3CHOHCH_3O_2 \rightarrow 2CH_3COOH + 2H_2O$$
   (b) Complete oxidation:
   $$CH_3COOH + 2O_2 \rightarrow 2CO_2 + 2H_2O$$
   $$2H_2 + O_2 \rightarrow 2H_2O$$
   (c) Nitrification:
   $$2NH_3 + 3O_2 \rightarrow 2NO_2^- + 2H^+ + 2H_2O$$
   $$2NO_2^- + O_2 \rightarrow 2NO_3^-$$
   (d) Sulphur oxidation:
   $$2H_2S + O_2 \rightarrow 2S + 2H_2O$$
   $$2S + 2H_2O + 3O_2 \rightarrow 2SO_4^= + 4H^+$$
   $$S_2O_3^= + H_2O + 2O_2 \rightarrow SO_4^= + 2H^+$$
   (e) Nitrogen fixation:
   $$N_2 \rightarrow \text{Nitrogenous organics.}$$

Chapter 5

Treatment of Sewage Sludge

INTRODUCTION

The amount of sludge generated by waste-water treatment plants has been increasing at a rapid pace in recent years due to stricter Federal and State regulations regarding the amount of pollutants allowed to be discharged into receiving waterways and the environment. Regulations requiring secondary treatment significantly increased the total amount of sludge generated. As regulations continue to require more total solids to be removed from the effluent, the end result is the generation of more sludge that requires disposal. The problem of how to effectively handle the sludge generated within the waste-water treatment plant in an environmentally acceptable manner, a cost effective manner, and also in a manner that meets the approval of the community has become a very challenging problem to the waste-water treatment industry. Finding an economical solution to the management of sludge is important since the total cost of the sludge processing and handling can account for up to 50 per cent of a treatment facility's total operating costs.

There are approximately 15,300 publicly owned sewage treatment plants (POTWs) in the US that generate approximately eight million dry tons of sludge a year. The processing and disposal of sludge is often considered the most complex problem facing the engineer in the waste-water treatment field. Some of the reasons that make sludge management difficult include:

1. Solids only make-up a small portion of sludge.
2. Sludge is composed largely of the substances responsible for the offensive odours of untreated waste-water.
3. The portion of sludge produced by biological treatment is composed of the organic matter contained in the waste-water, but it is now in another form however, this material can still decompose and produce offensive odours.
4. The presence of heavy metals concentrated in sludge, and public acceptance of the sludge disposal option chosen or proposed.

Figure 5.1 shows potential pathogens to humans from sludge. The issue of sludge disposal is one that is, extremely sensitive to the public. Community acceptance has become one of the top priorities in determining sludge management alternatives. It is important to keep the community well-informed and to let them take an active role in the sludge disposal selection process. The best practice is to keep a very open line of communication between the public and the engineers/municipal waste-water management personnel.

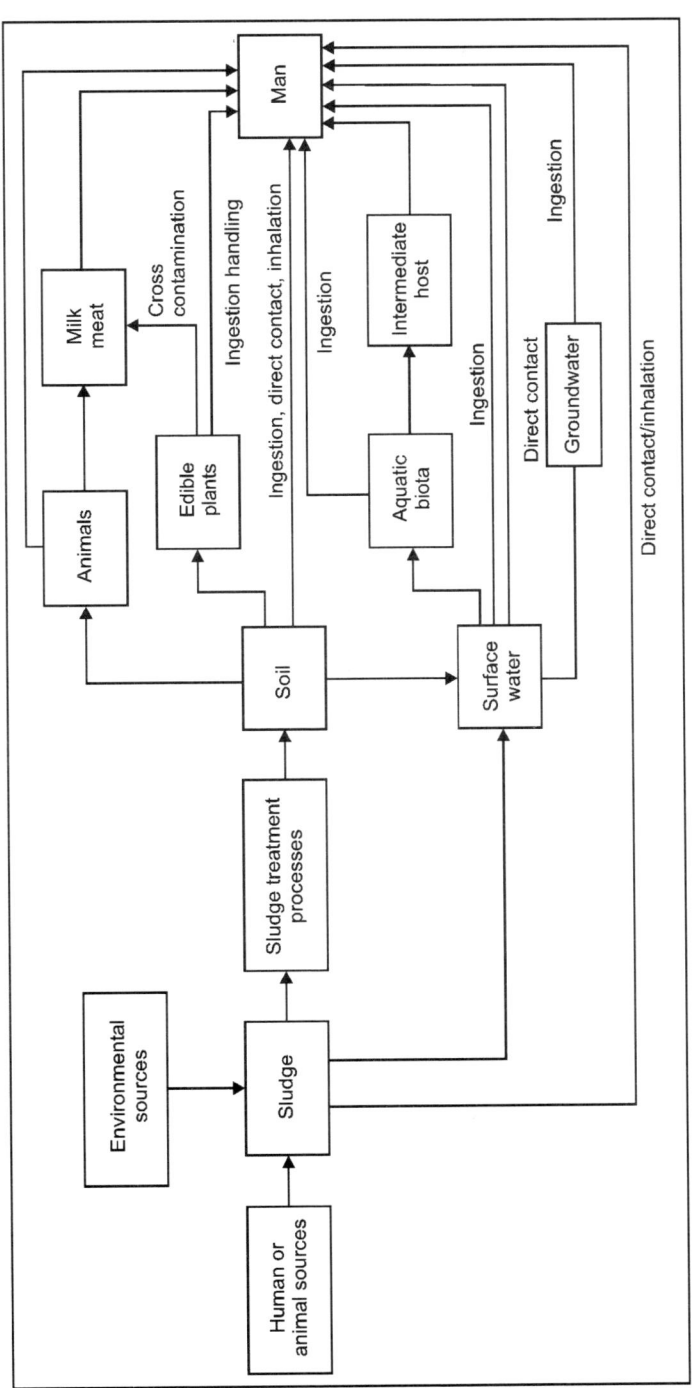

Fig. 5.1. Potential pathogen pathways to human from sludge.

IDENTIFYING FACTORS IN SELECTING THE BEST SYSTEM

Sludge management must meet the prime function of reliability as well as satisfy the following conditions:

1. Must be legally acceptable.
2. Processing and disposal options must be readily available.
3. Environmental impacts and health, risks must, satisfy public as well as regulatory requirements.
4. Must be cost competitive.
5. Necessary equipment and reliability must be available.
6. A system must be operational almost immediately after implementation.
7. Must be straightforward with assured financing.

It is important for a management system to be able to accept all of the sludge all of the time, under all circumstances. Factors involved in selecting the best sludge management system must be identified in any consideration of determining which is the right system to use. Potential criteria to be considered are included in the following checklist:

1. System flexibility—ability to respond to new technology; changes in regulations; changes in capacity/loads.
2. Reliability—probable fail rate; backup requirements; required operator attention; vulnerability to labour.
3. Production of significant related environment problems.
4. Performance data.
5. Technology available.
6. Financing.
7. Computability—with land use; solid waste/air pollution; existing treatment facilities.
8. Energy requirements—operation/construction/energy recovery; credits for product use.
9. Costs and benefits—capital costs; operating and maintenance; revenues; cost/benefits.
10. Public health—pathogenic organisms; toxic organics; heavy metals; soil effects; water quality impacts such as groundwater; surface water and marine environment.
11. Effects on air quality—odours; aerosols.
12. Effects on natural resources and the environment—depletion of resources; effects on the ecosystem; depletion of environmental resources.
13. Safety—effects of sludge transportation and operation of a sludge management operation on the public.
14. Administrative burdens—level of effort; marketing; jurisdictional disputes; public relations.

Candidate systems can be viewed from the array indicated in Fig. 5.2.

SLUDGE TREATMENT AND DISPOSAL PROCESSES

The solids-handling stream within the overall waste-water treatment process is quite involved and usually consists of several individual processes prior to final disposal of the sludge. The different processes used, the specific equipment/techniques used, and the combination of processes used varies greatly from treatment plant to treatment plant. There are many factors to consider when a municipality is determining the best way to deal with the sludge generated by the POTW.

Before going into a detailed analysis of some of the sludge treatment and disposal methods commonly used and some that are in the experimental stages, an overview of the sludge treatment process is given in the sections that follow. In general, the goal of the waste-water treatment plant operator is to reduce the water content of the sludge (hence the volume) and stabilise the sludge for final disposal. Figure 5.2 shows sludge treatment processes and their functions. Figure 5.3 shows a flowsheet for a typical wastewater treatment plant and how sludge is generated.

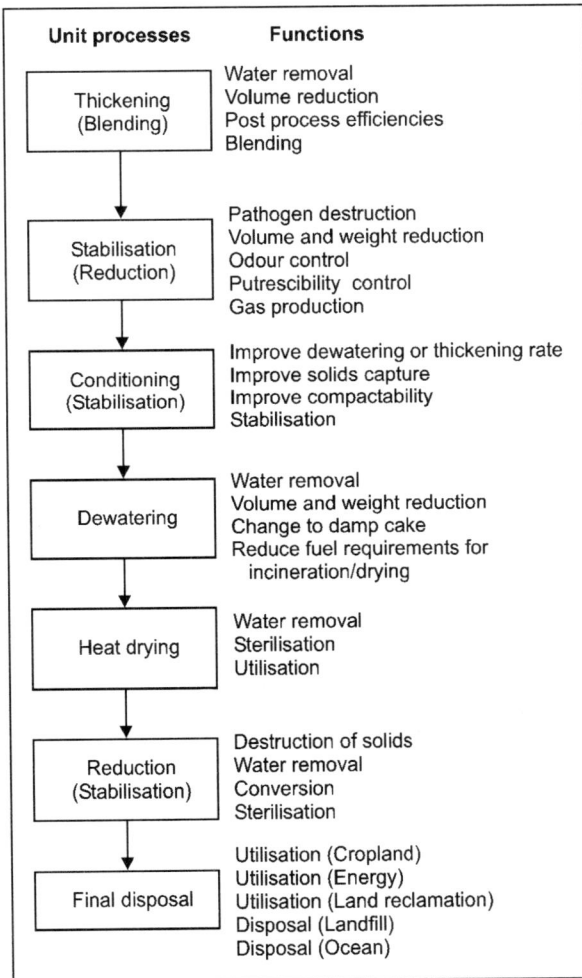

Fig. 5.2. Enumeration of sludge treatment processes and their functions.

Sludge treatment processes can be divided into the following eight basic functional groups:
1. Thickening.
2. Stabilisation.
3. Conditioning.
4. Disinfection.
5. Dewatering.
6. Thermal processes.
7. Composting.
8. Ultimate disposal.

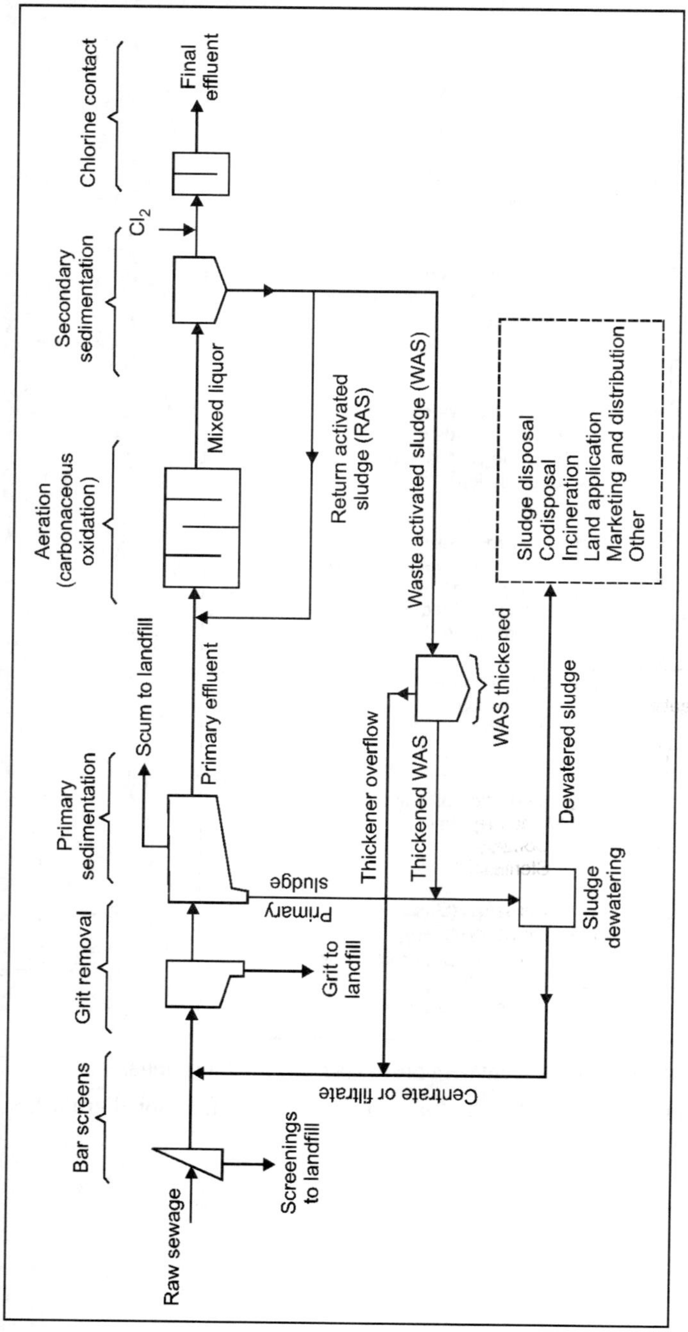

Fig. 5.3. Typical waste-water treatment plant flowsheet.

Using specific processes within the functional group, one develops a specific sludge treatment train. All the above groups will not necessarily be utilised in any one specific treatment train and very often one process may accomplish the goals of two or more of the functional groups listed above.

Thickening

As mentioned previously, one of the major goals of sludge treatment is to reduce the amount of water in the sludge. The first major process within the sludge treatment train is typically thickening of sludge. Primary sludge typically has a solids concentration around 4 per cent and secondary waste-activated sludge usually has a solids concentration around 1 per cent.

The main purpose of the thickening process is to reduce the volume of sludge. By reducing the water content in the sludge, the capital and operating costs of the remaining sludge treatment processes may be significantly reduced.

Thickening is achieved by the following processes:
1. Gravity thickening.
2. Dissolved air flotation (DAF) centrifugal thickening.
3. Gravity belt thickening.
4. Rotary drum thickening.

Stabilisation

The purpose of sludge stabilisation is to reduce the amount of odours emitted by the sludge and to reduce the number of pathogenic organisms within the sludge. When the method of ultimate disposal chosen for the sludge is land disposal, it is extremely important that the sludge be relatively free of odours and pathogens. Some of the stabilisation processes commonly used are:
1. Aerobic digestion.
2. Anaerobic digestion.
3. Dual digestion.
4. Vertical tube reactors.
5. Chemical stabilisation (lime or chlorine).

Conditioning

Conditioning of sludge is performed prior to the dewatering process. The main purpose of sludge conditioning is to improve the dewaterability of the sludge prior to sludge dewatering. As mentioned earlier, the production of a dry sludge cake is extremely important in the overall treatment and disposal process of sludge. Some conditioning processes also accomplish some of the goals of the other functional groups. For example, some conditioning processes deodourise the sludge (stabilisation process) or disinfect the sludge. Sludge conditioning processes may also physically alter the sludge, improve solids recovery, or reduce the solids content. Some of the basic processes typically used to condition sludge are:
1. Chemical addition.
2. Thermal conditioning.
3. Freezing.

Disinfection

Sludges generated from the treatment of waste-water contain a large number of pathogenic organisms including viruses, bacteria, parasites, and fungi. When the method of ultimate sludge disposal is to utilise the sludge as a soil conditioner or as a fertiliser, the elimination or significant reduction of these pathogens becomes extremely important.

Disinfection is the process by which pathogens are destroyed or deactivated so that they can no longer reproduce. The sludge stabilisation processes also serve as a disinfection process and significantly reduce the number of pathogenic organisms in the sludge. Other processes that are used for are disinfection of sludge include:

1. Heat treatment.
2. Complete composting.
3. Air drying or heat drying.
4. Incineration.
5. Pasteurisation.
6. Long-term storage high-energy, radiation.

Dewatering

The purpose of dewatering sludge is to remove a substantial part of the water present in the sludge. Sludge solids concentrations of 20 to 40 per cent are typically obtained by using the dewatering processes. The importance of removing as much water as possible out of the sludge cannot be stressed enough. The reduction of the water content in the sludge is usually required prior to incineration and composting, is required when the sludge is going to be landfilled so as to reduce the amount of leachate at the landfill, and is required to reduce sludge trucking costs since the volume to be transported will be significantly reduced by dewatering the sludge.

There are many different dewatering processes available and choosing the right one for the particular, treatment plant is very important. There are many factors to consider including: available space, chemical demand and cost, volumetric flow of influent sludge, type of previous upstream processes, suspended solids, and dissolved solids concentrations.

Some of the most common methods used to dewater sludge include:

1. Drying beds.
2. Lagoons.
3. Vacuum filters.
4. Pressure filters.
5. Belt filter presses.
6. Centrifuges.
7. Solar inclined beds.
8. Perched beds.
9. Solar drying.

Thermal Processes

Thermal processes are those that use heat to reduce the moisture content of the sludge or those that reduce the volume of the sludge by the evaporation of water as well as by destruction of organic matter. Thermal processes also significantly reduce the number of pathogenic organisms present in the sludge. One of the advantages of the thermal processes is the relatively small amount of space required. One of the problems with these processes is the high capital cost of the equipment that must also include extensive pollution control equipment.

The major thermal processes used in treatment of sludge are:
1. Heat drying.
2. Multiple-hearth incineration.
3. Fluidised bed incineration.
3. Co-incineration with solid wastes.
4. Pyrolysis (starved air combustion).

Composting

Composting is a specific type of stabilisation process (reduces odours and pathogens), as well as the last step in producing a sludge that can now be reused in a variety of ways. The composting process decomposes the organic matter within the sludge under aerobic thermophilic conditions. The end-product is a relatively stable humus-like material that is musty in odour and is relatively pathogen-free.

The sludge compost has many nutrients which contain especially high concentrations of nitrogen and phosphorus making it ideal for use as a fertiliser or soil conditioner. The three basic methods of composting are:
1. Windrow composting.
2. Aerated static pile composting.
3. Mechanical composting.

Ultimate Disposal

Landfilling of sludge in municipal solid waste landfills was historically the most common method of sludge disposal. This option is still in use; however, due to the passage of the Federal Clean Water Act and Resource Conservation and Recovery Act, a strong emphasis has been put on the utilisation of the sludge in a beneficial manner instead of disposal such as in a landfill or the past practice of ocean disposal of sludge (which has since been banned). Many beneficial uses of sludge have been discovered/rediscovered in recent years as new ideas and modifications of past sludge disposal methods are put into use. Some of the beneficial uses include using sludge as a source of energy, industrial raw material, a material of construction, soil amendment, and landfill topping. The biggest problems with finding uses for sludge are the pathogens, heavy metals, and other contaminants present in the sludge that represent a possible health threat to humans and the environment.

SLUDGE DISPOSAL REGULATIONS

Sludge disposal is regulated by either the Clean Water Act (CWA), Section 405(d) or jointly under the CWA and the Resource Conservation and Recovery Act (RCRA) depending on the sludge disposal options chosen. The CWA and the RCRA both require that every reasonable effort be made by wastewater treatment plants to implement treatment/disposal methods that utilise resource recovery for sludge in some manner. Establishment of programs that effectively utilise the resource values of sludge on a consistent basis have raised a number of concerns that include the following:
1. Reliability of process.
2. Cost effectiveness of process.
3. Community acceptance of treatment process and end-product.
4. Consistent production of an environmentally safe end-product that is free of toxic elements, heavy metals, organic compounds or disease-producing organisms.
5. Finding a reliable market for the end-product.

There are many beneficial uses of sludge that have been successfully implemented, but each waste-water treatment plant has its own unique set of circumstances and variables that must be carefully analysed before choosing a sludge disposal management plan. The presence of significant quantities of heavy metals, which is very common in the industrial sections of the US, especially in the Northeast, creates additional concerns and problems that limit some of the potential sludge disposal options.

The New Jersey Department of Environmental Protection and Energy (NJDEPE) classifies sludge into three land use categories: Class A, B, and C sludges with Class A sludge being of the highest quality.

The general definitions of the different classes of sludge are as follows:

1. Class A sludge: Acceptable for land application at agricultural rates for 40 years before cumulative metals-loading limits are reached.
2. Class B sludge: Acceptable for land application for a period of 20 years before cumulative metals limits are reached.
3. Class C sludge: Lowest quality and is generally not suitable for land application.

Successfully finding, and implementing beneficial uses for sludge is the implementation of strict industrial pretreatment standards to reduce the amount of heavy metals and toxic material that enter the POTWs, which eventually end-up in the sludge. Improving the overall quality of sludge generated at municipal treatment plants would greatly aid in implementation of beneficial uses of sludge without endangering the public or environment.

SLUDGE TREATMENT PROCESSES

Chemical Stabilisation Processes

The presence of heavy metals in municipal sludge limits the number of sludge disposal options available to be used in which a beneficial use/resource from the sludge can be utilised. Several sludge disposal methods that provide a beneficial use of a sludge that contains heavy metals include: chemical fixation of sludge for use as landfill cover, incineration, and co-incineration.

Incineration

The incineration process reduces the volume of sludge by combusting the volatile matter contained in the sludge and by evaporating the water contained in the sludge. Incineration of sludge provides the maximum reduction of sludge volume out of all sludge disposal alternatives. Sludge combustion destroys or reduces most toxic materials and also offers energy recovery potential. The multiple-hearth furnace (MHF) and fluidised bed furnace (FBF) are the most widely used types of incinerators used for sludge combustion. The use of incinerators for sludge disposal can represent a beneficial reuse alternative.

To obtain the best performance from both FBF- and MHF-type sludge combustion incinerators, it is important that a consistent, near homogeneous, and uniform feed sludge with desirable combustion characteristics be fed into the incinerator furnaces. In designing a sludge incinerator it is also imperative to avoid oversizing the furnace, since an oversized incinerator will result in higher unit operation and maintenance costs as well as frequent operational problems since the unit will not be operating at design conditions.

Incinerator design must meet strict regulatory requirements. The critical operating parameters that must be maintained to achieve complete combustion are burn time and temperature. Incinerators must be equipped with pollution control equipment in order to comply with strict Federal and State air quality discharge limits. The air pollution control equipment typically installed for incinerators includes either a wet or dry scrubber and either bag houses or electrostatic precipitators.

Fluid Bed Furnace System

The recent popularity of fluidised bed furnaces (FBF) for the incineration of waste-water sludges has been key to the resurgence of the incineration of sludge. Multiple-hearth furnaces (MHFs) in the past dominated the sludge incineration market; however, the vast majority of new sludge incinerators utilise the FBF technology.

Advantages that FBFs have over MHFs, are:

1. Lower capital costs.
2. Lower operating cost.
3. Ease and flexibility of operation.
4. Compliance with emission requirements without an afterburn (MHFs require an afterburn step to meet air emission standards).

The dewatered waste-water sludge enters the incinerator within a bed of fluidised silica sand where the sludge is combusted. Air from the windbox located below the fluidised sand bed constantly blows, combustion air through the fluidised sand bed to keep the bed in a fluidised state. A turbulent zone is established within the fluidised sand bed where the sludge violently mixes with the combustion air and fuel. Almost immediately upon entering the turbulent zone, the sludge particles disintegrate.

The remaining combustion of the sludge is undertaken in the freeboard zone of the incinerator located immediately above the fluidised sand bed. Within the freeboard zone, combustion of the remaining organics, including odorous substances, is achieved prior to the exhaust gases exiting the reactor. The residence time within the freeboard area is typically 4–5 seconds and the area performs basically as a built-in afterburner.

FBFs should be operated near design loads since under these conditions the incineration of sludge is most efficiently achieved. The most efficient level of operation of FBFs is achieved near design loads because the fluidised bed is sized so that the air needed for the combustion process is also sufficient to maintain fluidisation of the bed. The plant operator should therefore adjust the hours of operation of the incinerator based on sludge processing requirements. Fortunately, the sand bed retains the heat generated previously for a long period of time avoiding the necessity of reheating the bed prior to start-up after a shutdown of several hours, such as overnight. The heat, retaining properties of the fluidised sand bed allows the plant operator to efficiently schedule the incineration of sludge based on current sludge processing demands.

Almost all new FBF systems, utilise a high-pressure venturi scrubber in series with an impingement tray scrubber/cooler compartment to reduce emissions released to the atmosphere. The EPA considers this combination to be the 'best available control technology' that will meet current EPA emission standards. The EPA is developing new risk-based emission standards for sludge incinerators. The inclusion of wet electrostatic precipitators to provide additional removal of particulates may become typical pollution control equipment for FBFs in the near future.

In FBFs, all the ash produced in the combustion process is carried out with the exhaust gases. The exhaust scrubbing system produces a slurry solution that contains the ash which requires proper disposal. In most waste-water treatment plants, the ash slurry is discharged to an ash lagoon for long-term storage and gradual thickening.

The ash lagoon may be periodically emptied out as required. The ash removed from the lagoon is typically landfilled. At treatment plants that lack space for an ash lagoon or opt not to choose this route, the ash slurry is dewatered prior to off-site disposal. The ash slurry is typically dewatered using a combination of gravity thickening with a vacuum filter.

A few plants have recently started using small belt filter presses to dewater the ash slurry prior to trucking the ash end-product off-site for final disposal. A solids content of 50 to 60 per cent of the ash slurry is typically obtained.

The ash product produced by FBFs is suitable for disposal since it is inert and usually has less than 1.0 per cent volatile solids content.

One of the primary goals of FBF designers and operators is to minimise the amount of auxiliary fuel required. The best sludge fed into an FBF in order to achieve autogenous operation (operation without the use of auxiliary fuel) is a sludge that is high in volatile solids and has a low moisture content (high per cent solids content). Typically, a hot windbox (air supplied to furnace preheated to 1000°–1200°F) FBF system requires about 2700 BTU for each pound of water applied to the system in order for combustion to take place. For cold windbox (air supplied to furnace at ambient temperature) FBF systems, this requirement is increased to 4000 BTU per pound of water. The majority of this demand is usually supplied by the heating value contained within the sludge. If the natural heating value of the sludge equals or exceeds the total BTU value mentioned above required for combustion, the system will operate autogenously.

However, if the natural heating value of the sludge does not equal the values mentioned above required for combustion, an auxiliary fuel must be added to the fluidised sand bed in order for combustion of the sludge to occur.

Autogenous operation for hot windbox FBF systems is generally achieved when the feed sludge has a dry solids content of 25 to 30 per cent. Modern belt filter presses typically obtain dry solids content in this range.

Multiple-Hearth Furnace System

There are approximately 350 multiple-hearth furnace (MHF) systems throughout the US used for sludge combustion. A schematic of a typical MHF system is shown in Fig. 5.4. The furnace has a cylindrical steel shell which houses a series of horizontal fire brick hearths positioned in vertical layers. The dewatered sludge is fed into the uppermost hearth and the sludge moves downward. Ash is discharged from the bottom of the incinerator and exhaust gases exit from the top of the furnace. Operating temperatures in the combustion zone vary from about 1400° to 1800°F.

The complete MHF system is composed of several distinct vertically stacked MHF zones which may consist of one or more fire brick hearths where specific functions are carried out. The functions carried out in the specific MHF zones are typically:

1. Drying of wet sludge.
2. Combusting volatiles.
3. Complete fixed carbon burning.
4. Cooling ash.

The number of hearths in each MHF zone varies depending on the feed sludge characteristics, furnace design, and burner system.

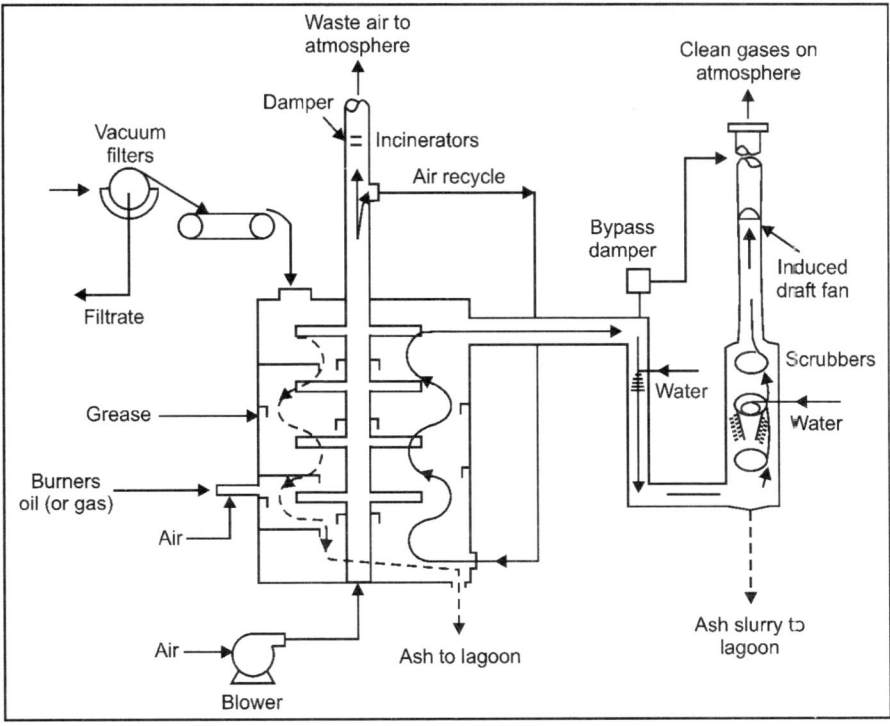

Fig. 5.4. Multiple-hearth system for sludge disposal.

Co-incineration

An alternative to incineration is co-incineration. Co-incineration is the process where sludge is incinerated in a solid waste incinerator. Typically one part sludge is combined with ten parts solid waste. The advantage of this sludge disposal option is the cost savings realised. There are no large capital expense expenditures to be concerned with, but instead, a relatively constant amount of money must be paid to the owner/municipality of the solid waste incinerator for the service of incinerating the sludge.

In many cases, the co-incineration sludge disposal option is not available to sludge managers. The first requirement is that there must be a solid waste incinerator in close proximity to the waste-water treatment plant and it must be one that will provide total incineration of the sludge. The sludge will have to meet the incineration plant specifications for moisture content, heavy metals, and other constituent standards.

Composting

Composting gained interest as a sludge disposal alternative in the early 1970s. The three main methods of composting are: the extended aerated static pile process, the conventional windrow process, and the aerated windrow process.

All composting methods basically consist of the same steps. Initially the sludge must be dewatered. A bulking agent such as sawdust, wood chips or dried compost is mixed in with the dewatered sludge to bring the solids content up to approximately 40 to 50 per cent. The temperature within the pile of compost composed of the sludge and bulking agent is brought up to a temperature of from 130° to 150°F at which time the decomposition of the organic matter begins. At this point, pathogenic organisms are destroyed and the moisture content is reduced. After a lengthy period of time, the sludge becomes adequately stabilised and reaches a suitable moisture content so that the compost can be ultimately disposed of as desired. In some cases, the bulking agent will be recovered and used again. Some of the major drawbacks to composting are the large amount of open space required to carry out such an operation as well as the environmental concerns with the heavy metals that become concentrated in sludge. Odours produced during the composting process also represent a situation that sludge managers and engineers must deal with.

In the aerated static pile method dewatered sludge is mixed with a bulking agent and then this mixture is placed in static piles and aerated for at least 21 days. Induced aeration is provided during active composting and sometimes during the curing and drying stages.

Conventional and Aerated Windrow Processes

In both the conventional and aerated windrow composting processes, dewatered sludge is mixed with a bulking agent such as finished compost which is sometimes supplemented with an external amendment. This mixture is then placed into long windrows. The windrows are turned periodically every two or three days for a period of at least 30 days during the active windrow composting period. This is followed by a compost curing period of at least 30 days. The only difference between the aerated and conventional windrow composting processes, is that in the aerated process, the windrows are aerated to accelerate and enhance the composting process and to accelerate the drying process of the composted material.

Chapter 6

Managing Risks and Hazards

INTRODUCTION

India, with its large population, is vulnerable to undesired toxic chemical releases and accidents. The situation here deems a crucial need for effective risk and hazard management. It also makes good business sense to identify risks and hazards of an operation, and take appropriate steps.

The benefits of implementing risk and hazards management plans and the requirements for preparing such plans are dealt with in this chapter. With the help of tables and figures, the potential risk vs. cost of prevention is examined to show that developing, implementing, monitoring and continually upgrading a practical risk and hazard management plan is a necessary part of running an industrial operation.

The twentieth century has been a period of phenomenal industrial growth worldwide. This has resulted in a surge in the number of manufacturing units, which are more and more co-located with commercial and residential areas. Managing risks and hazards of industrial operations is, therefore, fast becoming an integral part of doing business, not only as a response to growing regulatory and public pressures but also as a means of reducing financial and legal risks.

A clear understanding of what we mean or imply by a hazard and a risk is important. A hazard can be referred to as a condition or activity which has the potential to cause adverse consequences on life, property and the environment. Risk is the probability of the occurrence of an adverse consequence of a specified nature and magnitude resulting from a potential hazard.

The post-mortem of hundreds of accidents and analyses of lessons learned have shown that a majority of accidents are preventable and that risks and hazards of industrial operations can be minimised, if not eliminated. Commitment and diligence in executing an effective risk and hazard management plan are essential to achieving risk and hazard minimisation.

It is important to note at the outset that the type and level of risk and hazard management will differ significantly with the diversity and specific nature of the operation. The main objective of this chapter is to demonstrate that it makes good business sense to identify risks and hazards of an operation, and take appropriate steps to manage them effectively and meet business objectives of the operation. Developing and implementing an effective risk and hazard management plan also contributes to other requirements and standards that are practiced by industry, e.g. ISO 14000, country specific industrial safety and emergency response regulations, and healthy, community relations.

DEVELOPMENT OF RISK AND HAZARD MANAGEMENT SYSTEM

The challenge has been in accepting and integrating the discipline of risk and hazard management of industrial operation as a part of good business practice. Items such as operational risks, hazards, safety, environment, etc. have been traditionally considered as purely technical and scientific support disciplines to an operation and a cost burden. The prime objective of making profit focused on the cost and financial management of the basic operation of converting raw materials to products and selling the products. Risk management focused on financial and commercial aspects. Risks and hazards of the operations resulting in accidents and other consequences were treated as items covered under normal casualty insurance and a cost of doing business.

Recent experiences and consequences of accidents and undesired releases, evolving regulations and public pressures, are changing the way businesses are managing risks and hazards. Industry is beginning to recognise and accept that the management of business as well as operational risks need to be considered to remain competitive. Besides the main benefits of minimising injuries, deaths and environmental impacts, industry leaders report that effective risk and hazard management plans yield the following side benefits: (i) improved productivity, (ii) improved product quality, (iii) improved employee teamwork and morale, and (iv) better company image.

Key Elements

Most of the present day risk and hazard management programs are derived from industry guidelines and practices. In some instances, regulations requiring a risk management plan have adopted industry practice, logic and experience in formulating guidelines and requirements of items to be addressed. An effective program typically addresses the following elements:

1. Objective and purpose.
2. Employee participation.
3. Process and materials safety information.
4. Plant/process hazards and risk analysis.
5. Operating procedures and practice.
6. Employee training.
7. Contractors.
8. Pre-startup safety review.
9. Plant mechanical integrity.
10. Permits for work on operational/energised systems.
11. Management of change.
12. Incident investigations.
13. Emergency preparedness and response.
14. Self audits of risk management plans.

A useful and successful risk management plan must be:
1. With clear objectives.
2. Credible.
3. Simple and easy to understand and implement.
4. Thorough and accurate.
5. Well-publicised and accessible.

Preparation of such a plan requires

1. A team of senior, experienced and qualified individuals, who believe in the usefulness of the plan under the leadership of a strong champion committed to its success.
2. Clear and achievable objectives, scope and budget.

3. A definitive and committed schedule.
4. Commitment of resources to achieve the scope, budget and schedule.

Figure 6.1 depicts the key activity path for preparing a risk management plan at an existing plant. If the plan is for a new facility, all elements will be developed in parallel with the detailed design, engineering, procurement and construction phases. If the technology is being transferred from another similar plant, useful information on the existing plant can be retrieved and tailored to the needs of the new plant.

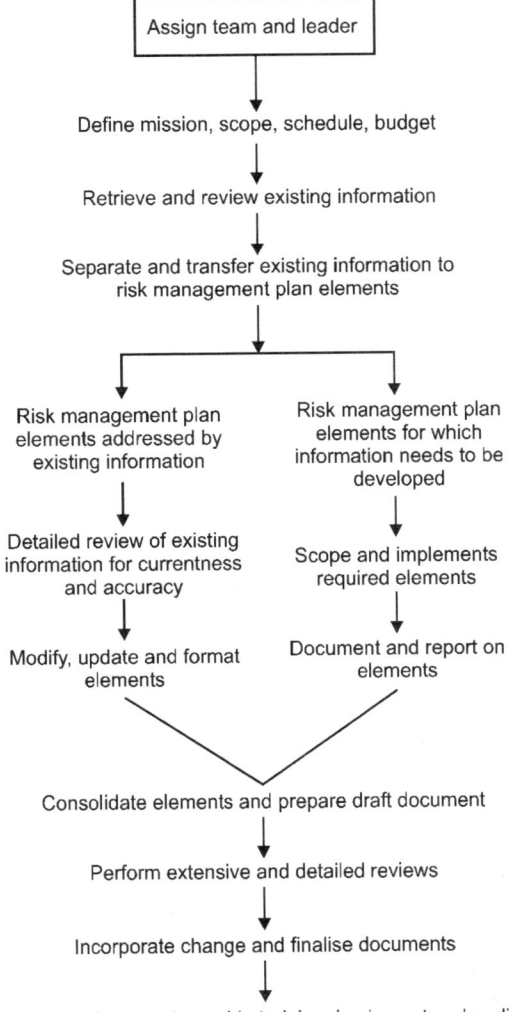

Fig. 6.1. Key activity path for preparing a risk management plan for an existing operation.

Experience in developing risk management plans has shown that the most important activity at the heart of the whole process is a thorough and objective hazard and risk identification and analysis. This element has a relationship with every element of the overall risk management plan, as shown in Fig. 6.2.

Fig. 6.2. Relationships of hazard and risk analysis in a risk management plan.

HAZARD AND RISK ANALYSIS METHODS

There are seven methods that have evolved and have been developed for performing hazard and risk analysis in the chemical industry:

1. Process/system checklists (PSCL).
2. Safety review (SR).
3. Relative ranking: Dow and Mond Indices (RR).
4. Preliminary hazard analysis (PHA).
5. What if analysis.
6. Hazard and operability studies (HAZOPs).
7. Failure modes, effects and criticality analysis (FMCAs).

Methods 1 through 5 tend to be more experienced based, qualitative and limited in prediction. Methods 6 and 7 are more advanced and more quantitative. The method chosen depends on the nature of the industrial operation, the local plant culture and the desired level of quantification and prediction of risk.

HAZOPs are popular in petroleum, chemical and allied process plants, and have found increasing application in the last decade.

Experience shows that the usefulness of a hazard analysis is independent of any of the above methods used. Of greater importance are:

1. Experience and in-depth knowledge of the hazard analysis team, with regard to the facility and industrial operation.

2. Throughness in consideration and evaluation of every possible deviation leading to a hazard and its cause, consequence, risk and response/mitigation.

3. Using a method or combinations of methods that are appropriate to the type, size and complexity of the operation under study.

It must be emphasised that the quality of the hazard and risk evaluation is not dictated by the method but by the performing teams' depth of experience in the nature of the process, materials, engineering, construction, operations and maintenance of the plant. Risk and hazard analysis requires experience; a forward-thinking open mind; the ability to consider all potential hazard and risk situations; and prioritisation of the most risky situations that require mitigation and/or response training and preparedness.

Finally, the ability to usefully apply the findings of the risks in their daily and overall business management should be the main focus. Information from the following elements is necessary for performing a hazard and risk analysis:

1. Process and materials safety.
2. Operating procedures and practice.
3. Plant mechanical integrity and maintenance.
4. Incident history and investigation findings.
5. Pre-start-up and safety review.
6. Existing level of employee training.
7. Existing emergency preparedness and response.

The findings of the hazard and risk analysis is then useful in enhancing all of the above and developing the following elements:

1. Employee participation/training.
2. Contractor qualification and training.
3. Management of change.
4. Emergency response.

The integration of the upgraded and developed elements provide a complete risk and hazard management plan. The next challenge is to roll out and implement the plan. This responsibility lies with the leader and the members of the risk and hazard management plan team. A risk management plan enables the analysis of potential hazards. The next step is the prioritisation of the high risk situations that can impact the operation of the plant as well as the overall company business. Integrating the financial risks of operations into the overall business risks seems logical and essential to today's management. At the operational level, the management can review the benefits of corrective actions against the potential financial and business risks, and act to prevent and minimise risks—a form of added insurance. In almost every case, performing a thorough hazard analysis identifies potential hazards and risks as well as their penalties and adverse impacts. From that knowledge, the management can evaluate the integrated business risk of that operation and implement preventive steps of added insurance of safety. It also enables to prepare and train for response to unanticipated incidents and accidents. Developing, implementing, monitoring and continually upgrading a practical risk and hazard management plan is fast becoming a necessary part of running an industrial operation. It is never too late to start.

HAZARDOUS WASTE INCINERATION—TRENDS AND PROSPECTS

Incineration has been widely adopted the world over as a long-term disposal option for hazardous waste over the last three decades. The process continues to undergo development to cover new applications, meet more stringent regulations and overcome problem areas.

Incineration: Concepts and Systems

The term 'incineration' needs no introduction to many of us who are in some way concerned with the disposal of organic hazardous wastes. In simple terms, it involves combustion under controlled conditions to convert waste into harmless end-products. The technology includes apart from proper combustion, other aspects such as processing and preparation of the waste, suitable feeding devices, flue gas cooling/heat recovery, flue gas cleaning and proper disposal of residues, if any. A generalised scheme is given in Fig. 6.3. While this technique offers a great flexibility to handle changes in physical and chemical nature of waste, it is possible to achieve desired efficiency and get maximum availability only if proper attention is given to following areas while planning the facility:

1. Waste characterisation.
2. Performance criteria.
3. Feeding system selection.
4. Incinerator selection.
5. Operation conditions and controls.
6. Safeties and interlocks.
7. Heat recovery feasibility.
8. Flue gas cleaning system.
9. Material selection.

Some of these are discussed here:

Incineration selection

Main criteria which decide the type of incinerator offered are, physical state of waste such as solid, semi-solid, liquid, gas, etc. firing rate or thermal capacity required; properties of waste on heating such as low or high melting, polymerising, easily pumpable, atomisable, etc. and of fusing properties of the ash including presence of low melting salts.

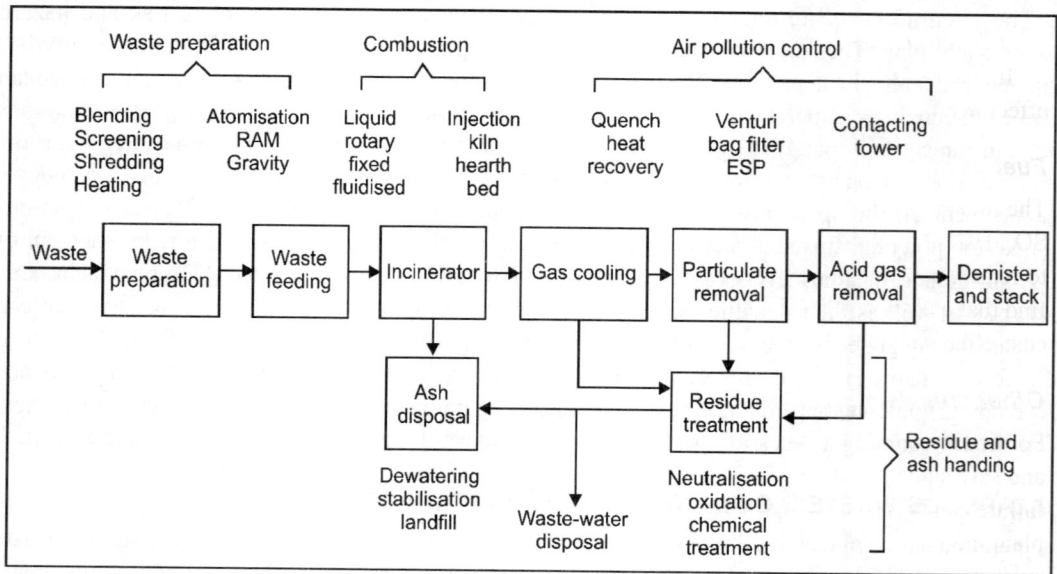

Fig. 6.3. General orientation of an incineration system.

It is possible that more than one design may be suitable for a particular application, in which case, flexibility and total system cost will dictate the final choice. As is widely known, all incinerators have to be designed with due regard to 3Ts, namely time, temperature and turbulence, to achieve required DRE. It is important to know if these aspects are taken care of adequately. Dead zones in the incinerator can create inadequate turbulence, giving rise to undesirable by-products. The same holds good for other parameters.

Rotary kiln incinerators, popular in the world due to their versatility, find a limited market in India, since they are not cost effective for the smaller quantity of waste generated in our industries. Maintenance cost of bricks in the kiln is high since they are subject to harsher conditions compared to other types of incinerators. Hence, other designs such as fixed hearth or fluidised bed have more applicability in our situation.

Heat recovery feasibility

Waste heat from flue gases can be recovered to generate steam, pre-heat air to minimise fuel or in other forms such as thermic fluid heating. Following aspects are important while taking the decision:

1. Waste heat boiler is best run on continuous basis for steady operation and preventing dew point corrosion.
2. Smoke tube boiler, commonly employed for clean flue gas, is feasible when steam generation exceeds 1 ton/hour. For high ash content, water tube boiler is preferred. It is not economical below 2 tons/hour capacity.
3. Smoke tube boilers require shut-down at regular intervals for cleaning ash deposits.
4. Pre-heating of air or water cannot be considered when flue gas contains acidic gases like SO_x, HCl with water vapour as the various acids corrode by condensing.
5. Air pre-heat is important when wastes contain high moisture as in bio-sludges or when they have low heating value as in the case of exhaust air carrying solvent vapours. It achieves substantial reduction in the fuel requirement.

In each case, one needs to check if fouling or corrosion of waste heat recovery systems does not effect availability of the total system.

Fuel gas cleaning system

The fuel gas needs to be cleaned if it contains pollutants in the form of particulates and/or gases such as SO_x, HCl above stack emission requirements. Typically, the system would involve quenching by water to reduce the temperature followed by one or more stages of wet scrubbing. In Indian conditions, we find that quenching in spray column followed by a venturi scrubber and/or absorber column provides a cost-effective gas cleaning system.

Other critical areas

For trouble-free operation and performance at constant efficiency, auto control of important parameters and safe shutdown through safety interlocks has to be considered. Forced shutdowns due to corrosion failure can be avoided by selection of suitable materials at all stages of the system. As in any chemical plant, adherence to operating procedure and regular preventive maintenance improves availability of system. Above areas are equally important as selection of equipment and our experience indicates that any compromise can result in frequent shutdowns and increase in overall cost of the plant.

Application Areas

Waste organics, where other methods have limitations, are disposed of by incineration. In India, the average size of industry being small, waste quantities are 500–2000 kg/day. Most commonly used incinerators are liquid injection, fixed hearth or fluid bed type due to economy of scale. Apart from this, following constraints may have to be considered:
1. Waste is containerised and it is preferable to handle it in that form in batches.
2. Semi-manual operation can be considered.
3. Operation has to be limited to maximum two shifts.
4. Instrumentation is to be kept to minimum essential to keep the investment low.

Some trolley loaded incinerators have been developed for this application. Apart from the risk of explosion, they cannot guarantee efficient burnout. In such cases, a common facility for a group of industries can provide a viable solution. An outcome of Hazardous Waste Rules, 1989, is that many industries are looking at incineration as a possible solution for disposal of waste included in some categories of waste.

For some waste, such as tank bottom sludge, ETP sludge, effluents containing toxic/high COD as well as recoverable salts, incineration can prove a better long-term option with possibility of heat and material recovery as an added benefit. When industries plan their environmental management program for complying with ISO 14000 requirements, incineration along with other techniques will play a key role in management of waste.

Applications

Types of waste
1. Exhaust air containing VOC's.
2. By-product gases/vapours.
3. Organic liquid streams.
4. Aqueous waste containing dissolved organics with or without salts.
5. Slurries and high moisture containing sludges.
6. Organic sludges and semi-solids.
7. Granular solids, filter cakes.
8. Distillation bottom tars (low and high melting) (along with non-hazardous general factory waste in some cases).

Generating processes
1. Pesticides.
2. Pharmaceuticals.
3. Bulk drugs.
4. Oil refineries.
5. Petrochemicals.
6. Organic chemicals manufacturing.
7. Automobiles (paint shop).
8. Coating, printing, laminating operations.
9. Pulping mills.
10. Dyes and dye intermediates.

Liquid Injection Incinerator

Advantages

1. No secondary combustion chamber.
2. Accepts wide range of hazardous liquid and gaseous waste.
3. Capable of high turndown.
4. Simple construction, no moving parts, hence low maintenance.

Limitations

1. Can burn only atomisable liquid wastes.
2. Cannot burn slurries with large size solids.

Fixed Hearth Incinerator

Advantages

1. High fuel efficiency.
2. Low particulate emissions.
3. Well suited for bulky solids.
4. Simple construction and operation.
5. Lowest cost design for applicable range.

Limitations

1. Suitable only for low capacity.
2. Not suited for wastes requiring high temperatures, high in ash content.
3. Auto deashing difficult.

Rotary Kiln Incinerator

Advantages

1. Any type of hazardous wastes can be fed and incinerated.
2. Continuous ash removal possible.
3. Adjustable residence time for solids by changing kiln speed.
4. High temperature possible.

Limitations

1. High installation, maintenance cost.
2. Perfect sealing at kiln ends difficult.
3. Poor thermal and fuel efficiency.
4. Low refractory life.
5. Economical only at high capacities.

Fluidised Bed Incinerator

Advantages

1. High heat transfer efficiencies.
2. Uniform and lower bed temperature for same efficiencies.

3. Highly stable operation for water containing wastes.
4. In bed neutralisation using lime possible.
5. In bed heat removal possible.

Limitations

1. Agglomeration possible when salts are present.
2. Solid wastes need shredding or crushing before feeding.
3. Large particulate carryover, hence dust collecting is a must.
4. Replenishment of sand required from time to time.

Table 6.1 highlights the application of incineration to waste types.

Table 6.1. Applicability of incineration to waste types.

Waste types	Liquid injection	Rotary kiln	Fixed hearth	Fluid bed
Solids				
Granular homogenous		X	X	X
Irregular bulky		X	X	
Low melting type	X	X	X	X
with fusible ash		X		
Large, bulky material		X		
Gases				
Organic vapour laden	X	X	X	X
Liquids				
Aqueous wastes	X	X	X	X
Organic liquids	X	X	X	X
Solids/liquids				
Halogenated compound	X	X		X
Sludges		X		X

HAZARD MANAGEMENT IN CHEMICAL PROCESS INDUSTRY

This section provides an overview of the various essential components of risk management technology applicable to the chemical process industries (CPI), such as the techniques used for early detection of hazards and effectiveness of process safety management systems. Examples drawn from hazard analysis and risk assessment studies are used to illustrate the wide spectrum of process engineering, design, operation and maintenance practices which are subjected to a close scrutiny.

The increase in industrial chemical activity throughout the world provides us with better quality of products creating a better standard of living but, in turn, increases the potential for major disasters of the kind witnessed during the eighties. Managers of chemical technology are, therefore, expected to ensure that enough care is exercised in assessing the consequence of an unexpected event and abstain from employing a technology until it is quantitatively evaluated in all respects. This means that the responsibility of quantifying, controlling and minimising the risks lies with the management and, as per the prevalent legislative framework, they have to demonstrate that their plans are properly designed, constructed and safely operated. Simultaneously a legal framework has been evolved which satisfactorily fulfils the expectations with respect to siting of new industrial units, pollution control, environmental impact assessment, on-site and off-site emergency plans, and disaster mitigation measures.

Chemical process industry (CPI) has over the years developed several methodologies aimed at early detection of hazards and assessment of consequences of an untoward accidental release of hazardous materials in order to effectively attenuate residual risks.

The comprehensive criteria developed for acceptability of risks provides a satisfactory means of evaluating the technology at various stages of implementation. Also, CPI recognises that adoption of a safe technology alone is not adequate and that management controls are needed to ensure that safety systems in place work. This approach, which brings into focus the three major aspects — the technology, the personnel who operate and manage it and the design and maintenance of facilities — has enabled CPI to fulfil its societal and legal obligations.

Early Detection of Hazards

Process technology passes through discrete phases of development and implementation, which offer unique opportunities for detection of potential hazards. CPI utilises these for selection of process routes, reduction in inventories of hazardous materials, elimination of contaminants, incorporation of suitable safety devices/systems, etc. At the design stage, choice of proper process equipment and instruments along with an appropriate processes control strategy results in inherently safe technology. For example, it is possible, at the design phase, to answer specific queries, such as 'Is the cost of more reliable equipment which needs less maintenance, justified?' Process safety reviews are conducted at predetermined stages throughout the project implementation to provide the most economical way of avoiding potential operational hazards.

New processes

Process involving mature technologies and experienced contractors benefit immensely from a policy of stagewise safety reviews. In the case of new technologies, opportunities for devising an inherently safe process are availed right from the bench scale/laboratory scale studies, i.e. at the process development stage. For example, process routes involving less hazardous materials or hazardous material in less hazardous state are selected. Similarly, thermal runaway risks are examined carefully with reference to undesirable chemical activity and to keep the actual plant parameters from reaching critical condition. In this context, the presence of undesirable but likely corrosion products and other potential contaminants in process streams is examined. Accelerated rate calorimetry (ARC) and differential scanning calorimetry (DSC) have been extensively used for quantifying such thermal effects. These techniques permit simulation of thermochemical activity of not only the desired reactions but also of those which are likely to be encountered during routine maintenance of process equipment, such as chemical cleaning, or inadvertant deviations of process parameters from the desired/optimum values, etc.

Commercial exploitation of a process technology involves development of a conceptual design, based on which process design and engineering of the industrial plant are undertaken. This phase of project implementation is utilised by the CPI for detection and mitigation of potential hazards since the cost of incorporating safety measures during the installation of plant and machinery is the lowest. Process plant layouts, instrumentation and process control, piping designs and the type of equipment are subjected to a detailed scrutiny in order to make the commercial operation as safe as reasonably possible. Attention is focused at this stage on inadvertant deviations in temperatures, pressures, levels and flow rates, either to augment the control systems or to select the right types of sensors, or safety devices to meet the requirements.

Regular safety reviews

Many CPI operators follow a standard schedule of safety reviews at three well-defined stages of project. The first review is undertaken at the end of basic and process design, when most of the parameters such as plant capacity, location, feed stock, general layout, etc. are known. At this stage, the safety aspects of the project in relation to other units of the site or nearby are considered. During the second review, preferably halfway through the detailed engineering phase, the piping and instrumentation diagrams are examined by standard techniques such as HAZOP fault and even tree analyses, to ensure that no other options which might result in greater safety of the plant have been ignored. The third review is normally at an advanced stage in the construction phase. It is dedicated to evaluate/quantify the residual risks for comparison with other similar installations and to take any corrective steps.

In addition to the above three reviews, a more comprehensive pre-start-up safety review is normally organised based on checklists specifically compiled to account for safety aspects which have been identified during the design and construction phases. The review team examines the operating and safety manuals that give details of the major potential hazards, including toxic, hazardous and inflammable materials. Alarms, interlocks, sizing of safety relief valves, reaction exotherms and fire exposure of critical units are particularly studied. The team conducts interviews with operating, maintenance, medical, fire fighting and safety personnel to satisfy themselves. Inspection of process equipment and facilities forms part of this review. Following the interviews and the inspection, the team summarises its findings and priority-ranked recommendations are made section-wise. Safety management reviews, with safety audit as its essential component, are a regular feature of major operating complexes and plants. Conducted objectively and at regular intervals, the audit reports help CPI in keeping the facilities well-maintained as well as updated with respect to the design and operation of safety systems, and to ensure adequate protection to the plant and personnel.

A general checklist approach, applicable to the above safety reviews, has been frequently used for early detection of hazards. This was initially proposed by DuPont and later modified to reflect specific project requirements. It covers several categories of plant and machinery and materials. Under each category, several 'items' are listed. In each case, the aspects that need to be investigated are identified. A sample of the checklist is shown in Table 6.2 covering materials, material handling, storage, reactions and equipment.

Hazard Management

Practical implementation of design and operational measures with respect to safety and a framework to monitor their efficacy are the essential elements of hazards and risk management (HRM). Some of the typical scenarios that prompt formal organisation of hazard and risk management are: occurrence of a major accident in a similar operating plant, which results in considerable damage, excessive non-routine maintenance episodic material or energy releases and frequent 'near-misses'. Once established in any CPI, its major objectives encompass identification, evaluation and analysis of risks, determination of measures to eliminate/minimise unacceptable risks, establishment of a baseline and continuous improvement in safety standards and performance. The most important measure of a good HRM system is the evolution of guidelines for organisational set-up that results in recognition of safety as a 'line' responsibility which cannot be delegated.

Table 6.2. General checklist.

Category/Item	Subjects to be investigated
Materials	
Raw materials	Toxicity, flammability
Interim materials	Reactions, decompositions
End-products	Corrosivity
By-products	Long-term storage behaviour
Waste	Total amount, possible reductions
Material handling	
Transportation container	Overfilling protection
Pumping	Spill collection
Road/rail transportation	Leak detection
Ship transport	Cleaning/inspection procedure
Crane handling	Dropped load and potential targets
Conveyor belts	Stop devices, guards
Reactions	
Hazardous reactions	Wrong materials/contaminants
Combustible mixtures	Wrong proportions
Runaway reactions	Deviation of process parameter
	Unknown kinetics
	Pump/agitator failure
	Flow blockage isolation to stop reaction
	Depressurisation/draining to stop reaction
Equipment	
Vessels	Design, size
Columns	Material selection (corrosion)
Heat exchangers	Overpressure protection
Piping	Level, temperature protection
Ducts	Reverse flow protection
Valves	Emergency isolation (remotely)
Machinery	Emergency depressurisation (remotely)
	Isolation for maintenance
	Potential leaks: Glass components
	Small-bore connections
	Inspection and maintenance
	Compliance with codes certificates
Piping	Thermal stress, movement, support freeze protection, possibility for flushing
Valves	Maintenance: Accessibility, by-pass and isolation,
	Fail safe in case of power failure
	Function testing
	Interlock against unintentional opening/closing.

In order to be most effective, an HRM system is headed by a senior manager reporting directly to the top management/managing director. HRM works with the management to:

1. Establish a safe philosophy.
2. Establish and implement the risk management.
3. Coordinate the safety efforts in the CPI system.
4. Monitor the safety performance.

Several elements of HRM have been identified which cover a wide range of activities related to process safety. Some of these are listed and elaborated in Table 6.3.

Table 6.3. Elements and components of hazard and risk management.

Process knowledge and documentation

    Process Definition and Design Criteria

    Process and Equipment Design

    Company Memory (management information)

    Documentation of Risk Management Decisions

    Normal and Upset Conditions

    Chemical and Occupational Health Hazards

Process and equipment integrity

    Reliability of Engineering

    Materials of Construction

    Fabrication and Inspection Procedures

    Installation Procedures

    Preventive Maintenance

    Process, Hardware, and Systems Inspections and Testing (pre-start-up safety review)

    Maintenance Procedures

    Alarms and Instrument Management

    Demolition Procedures

Training and performance

    Definition of Skills and Knowledge

    Training Programs (e.g. new employees, contractors, technical employees)

    Design of Operating and Maintenance Procedures

    Initial Qualification Assessment

    Ongoing Performance and Refresher Training

    Instructor Programs

    Records Management

Incident investigation

    Major Incidents

    Near-miss Reporting

    Follow-up and Resolution

    Communication

    Incident Recording

    Third-party Participation as Needed

(Contd ...)

Standards, codes and laws

    Internal Standards, Guidelines Practices

    (past history, flexible performance standards, amendments, and upgrades)

    External Standards, Guidelines, and Practices

Audits and Corrective Actions

    Process Safety Audits and Compliance Reviews

    Resolutions and Close-out Procedures

Enhancement of Process Safety Knowledge

    Internal and External Research

    Improved Predictive System

    Process Safety Reference Library

Risk-Reducing Measures

Risk-reducing measures are the outcome of any of the following activities that are undertaken in a process plant:

1. Periodical safety audit.
2. Project safety reviews.
3. Quantitative risk assessment.

The measures recommended as a result of any of the above reviews are commensurate with type and magnitude of potential hazards responsible for the unaccepted levels of risk. Potential hazards are either high probability moderate consequence hazards. Rupture of pipelines carrying hazardous materials or leakage through flanged joint, etc. constitute the first category while fire and subsequent rupture/BLEVE of LPG spheres or release of toxic material like MIC in large quantities are classified in the second category. Some of the areas specifically covered for achieving reduction in risk levels are:

1. Disposal of effluents containing toxic/hazardous materials through vents and drains.
2. Protective measures for plant and personnel.
3. Safety measures during major emergencies.

Following examples of risk reducing measures illustrate the efficacy of hazard analysis and safety reviews conducted for the CPI.

Secondary containment for toxic gases

The example relates to the modifications incorporated in a prototype plant which handles large quantities of aqueous sodium cyanide as starting material and gaseous cyanogen chloride as an intermediate. A safety review was carried out at the final stages of the detailed engineering phase This consisted of identification of vulnerable process operations, preliminary HAZOP and consequence analysis. The results indicated that in the event of an unexpected failure of the reactor, substantial quantity of toxic gases will be released into the atmosphere. An estimate of the damage distance was made taking into consideration the probable weather conditions, threshold concentration, workforce deployed for plant operation and other activities on the site. The consequence were deemed to be unacceptable. It was decided to isolate the reactor by providing a well-ventilated enclosure around it. A blower with sufficient capacity was installed to evacuate the enclosure at a rate of 6–10 times per hour. Ducts carrying the evacuated gases are regularly monitored for presence of toxic contaminants before they are taken to a destruction system. Thus, the enclosure around the reactor has provided protection to operators and others from an undesirable and unexpected exposure to highly toxic gases.

Loading and unloading of hydrocarbon products

This operation constitutes a probable source of leakage and as such it is carefully scrutinised to avoid generation of an unexpected vapour cloud. It was recommended that the beginning and completion of each of these operations should be clearly indicated to alert truck drivers to restrict the movement of the vehicles during the operation. Announcement of loading and unloading operations through appropriate signals/communications also helps the fire safety system to provide the necessary back-up support with regard to activation of remotely controlled sprinkler systems, etc. Since each installation is characterised by a certain type of loading and unloading manifolds and fittings, it was recommended that only fully-skilled operators should be employed. Prior training of the operators through simulated exercises, pipelines drawings and models were also recommended in some cases.

Liquid level control

Charging and discharging operations in the process plant are monitored through level indicators, controllers, switches and alarms. The reasons for installation of alarms are specific. However, once installed, these devices must be optimally utilised in order to prevent process upsets upstream or downstream of the equipment. Recommendations for reducing risks should, therefore, not only ensure maintenance of levels in a particular unit but evaluate the effects of level change in a wider sense.

High and low-level alarms are generally installed to alert the operator about any impending upset in process conditions. The time gap between the alarm and the actual process upset needs to be reviewed periodically to make the level control system effective. In some cases, it was necessary to recommend initiation of plant shutdown procedure in order to avert a major disaster if the operators response to critical level alarms is found to be inadequate.

Cascade effects

Damage distance calculations under the most unfavourable conditions indicate if a cascade effect analysis is required. For example, if it is found that the radiation effect in an hypothetical incident would be such that autoignition temperature of hydrocarbons stored nearby will be exceeded, then the cascade effect as indicated. Appropriate precautions are recommended in this case to prevent the initiation of a cascade effect. These include relocation of the hydrocarbon storage vessel, construction of barrier to control the radiation effects, installation of an automatic sprinkler system which gets activated by the surface temperature of the storage vessel, etc.

Thus, the CPI has developed and adopted various procedures for effective detection of hazard potential of process plant and associated facilities. Periodic safety reviews are organised during design and implementation stages to minimise risk levels. Such reviews are also conducted on operating plants. A system of hazard and risk management is devised which effectively monitors and augments safety standards.

ELEMENTS OF A BALANCED APPROACH TOWARDS HAZARDOUS WASTE MANAGEMENT

Since the start of the industrial revolution, industrial waste management practices have primarily involved the disposal and use of either on-site or off-site landfills. Industrial wastes were often co-disposed using typical so-called sanitary landfill practices. Many of these land disposal sites were not designed or operated to handle these industrial residues. As a result, leachate resulting from these wastes was not properly contained or controlled and has migrated out of the landfill into underlying soils and contaminated groundwater or nearby surface water and drinking water supplies.

Several hazardous waste management alternatives are available other than land disposal and these include:

1. Recovery and reuse.
2. Waste minimisation and reduction.
3. Low-waste generation technology.
4. Destruction via incineration.
5. Biological degradation and treatment.
6. Chemical detoxification.
7. Thermal treatment and destruction.
8. Stabilisation/solidification.
9. Above-ground, long-term storage.

Most of these technologies will generate a process residue (i.e. ash, sludge) which may ultimately require land disposal. However, this residue is typically non-hazardous, less leachable or less toxic, and is in a smaller quantity than the original waste material. A properly designed, secure landfill may be an acceptable option for handling these types of treatment residues. To complement the advancement in landfill designs, emphasis must be placed on pretreatment of waste, including stabilisation through the use of cements, pozzolanic or bituminous materials, thermal treatment, dewatering and moisture removal, and detoxification. Emphasis must be placed on minimising the amount of liquids contained in the waste material, which is deposited in the landfill, along with reducing overall toxicity of the disposed of materials. Many new and effective technologies are being developed for addressing the environmental problems attributable to past landfilling practices. In general, they fall into two broad categories of source control and migration control. This is the key to avoiding many of the environmental problems that have plagued many industrialised countries. Some of the key elements of such an approach include the following:

1. Promulgation of realistic and achievable environmental regulations that are based on cost-effective, environmentally sound expectations.
2. Focusing on means of avoidance. This includes selection of low-waste generating processes, in-plant recycling, re-use, closed-loop materials handling, etc.
3. Integrating waste management, treatment and disposal with the economic product life cycle to avoid a negative drain on productivity and the economy.
4. Application of less complex and rudimentary technology that requires less capital and is easier to operate with average skilled labour.
5. Remediation of highly polluting facilities at earlier stages to avoid environmental degradation and costly remediation in the future.

Special emphasis must be placed on the application of clean technologies that produce no waste or hazardous by-products and on waste minimisation and control at the source of its generation. Hazardous waste management practices for major industries have undergone significant changes in the last five years. These changes have been driven by public demand for environmental regulations, increasing cost at off-site disposal facilities, closing of commercial facilities and increasing awareness of environmental problems caused by current and past disposal practices. It is expected that this trend of major changes in the waste management field will continue for the next 10 years or more. Traditional methods for managing industrial waste are being phased out or are undergoing significant upgrading/modification in response to the more stringent regulations. Industrial firms are also working to minimise their waste and anticipate potential future liabilities due to environmental and public health concerns related to hazardous waste management practices. Examples of several major industrial sectors confronted with significant changes in the management of their hazardous wastes include petroleum refining, petrochemicals, pharmaceuticals and electronics. In the petroleum-refining industry, wastes typically fall into both the inorganic and organic categories. Inorganic residues include spent catalysts from process reactor units. Past disposal practices for these spent catalysts typically utilised uncontrolled landfill or recovery/regeneration. Other inorganic residues include spent caustic, lime sludge, boiler ash, and filtering materials.

Organic residues have typically been treated by discharging to the process sewer, which historically has treated the waste-waters through oil separation and/or an effluent settling pond system. Major upgrading of the waste-water treatment systems have been undertaken by the refining industry in response to clean water laws and now include biological treatment, physical/chemical treatment and enhanced oil separation and recovery techniques. Organic sludges such as API separator sludge are generated from the waste-water system and these have typically been land-disposed or land-applied. Other organic sludges include tank bottoms from tank cleaning operations. Current trends in managing these organic residues are by either thermal destruction or land application for biological degradation.

The petrochemical industry is also undergoing major changes with respect to hazardous waste management. The types of waste generated by the petrochemical manufacturer vary widely from plant to plant and are highly dependent upon the type of process and plant production. A portion of the wastes has been handled via the process sewer and a waste-water system consisting of oil/water separation, and settling of solids in a pond/lagoon system. Major upgrading of waste-water treatment facilities is underway to incorporate biological treatment in addition to physical/chemical treatment processes, such as filtration and carbon adsorption. Residues are generated from the waste-water treatment system along with the plant process units that are typically produced in the form of residues and sludges. Historically, these have been handled via land disposal; however, the current trend is toward thermal destruction, biological degradation and stabilisation. Thermal destruction processes involve incinerators capable of handling both liquid and solid materials, and typically incorporate heat recovery units. Chlorinated and non-chlorinated solvents are major waste streams in the petrochemical industry, and the current trends for managing these materials involve thermal destruction or solvent recovery.

The major waste stream from the pharmaceuticals industry is typically waste-water from production operations. This has historically been handled either through discharge to a municipal sewer or through an on-site facility. Current practices and trends involve installation of pre-treatment facilities prior to sewer discharge and upgrading of the existing treatment facilities. This upgrading involves the construction of physical/chemical treatment modules or biological treatment. Other pharmaceutical waste materials are generated by laboratory chemicals from research, organic sludges from waste-water treatment or fermentation steps, spent solvents, off-spec products and other biological materials from research. Current management trends for these materials include thermal destruction, solvent recovery, destruction via incineration and waste reduction.

The proper application of technology and institutional measures, including environmental regulations, is the key to protecting our environmental resources for future generations.

HAZARDOUS WASTE MANAGEMENT IN INDIA: POLICY ISSUES AND PROBLEMS

The Bhopal tragedy will be remembered for many years. Often, in cases of industrial accidents, such as that occurred at Bhopal, the significance of warning signs are not fully appreciated until after the disaster has struck. In the past few years, such localised disasters have taken place in India but the basic reasons for their occurrence has not been scientifically analysed nor any corrective measures outlined for action by public, government agencies, social and political organisations. Chlorine leakage in the suburbs of

Mumbai, sulphuric acid fumes from a Delhi industry, killed fishes in the Tungabhandra and Gomti rivers which are some of the examples of mini disasters and in the absence of emergent action by all concerned would have led to a larger scale disasters. One can learn from these occurrences; and the results, if disseminated widely, be of immense importance to those planning and co-ordinating disaster management activities elsewhere as well as to those who become the first victims of these disasters.

Some policy issues together with the problems encountered in hazardous waste generation and disposal/management are addressed here.

Background

India has primarily an agriculture base with more than 70 per cent of the population living in small villages. There is still illiteracy in many parts of the country, but people are becoming more knowledgeable and conscious of their rights with the help of information acquired through radio and television media (computers/internet). Small-scale industries do not give much thought to pollution they may generate and more often than not, pollution protection equipment is more costly than that of the production unit itself.

Even sewage collection and treatment facilities are inadequate and together with industrial pollution place considerable stress on the natural environment. Sludges and solid wastes from industrial units and households are eyesores and a cause for concern. Uncontrolled sources of pollution like household grates, cattle sheds, slum areas and automotive exhausts are also matters requiring immediate attention. As most of the larger towns in India are situated along the river banks, their run-offs and also those from agricultural operations find their way to the rivers, making them unfit for drinking purposes. Drinking water quality is of prime importance and every effort is being made to maintain it for all freshwater resources, such as rivers, lakes, underground water. Poverty among people is the biggest source of pollution in India. Where existence itself is a struggle and illiteracy widespread, environmental protection is not a priority. This is where the United Nations with its poverty alleviation programs has substantially contributed to environmental conservation.

Perspectives

India has missed the two historical industrial revolutions, it cannot afford to miss the third. In fact, some making up is in order. Technologies have to be acquired, developed, absorbed and digested. Hazardous materials will have to be handled, wastes produced and treated. There are already innunmrable small- and medium-scale industrial units consuming and producing hazardous materials. The methods of their safe handling and disposal have developed by experience and innovation. The one sad experience has been with a turnkey project in which the problems were kept secret along with the operations. With a full flow of scientific and technological information and with sincere efforts, many of these mishaps can be avoided.

Problems

For hazardous waste management in India and other developing countries in the region, the following problems exist:

1. Lack of information on the hazardous waste.
2. Lack of co-ordination of different agencies.
3. Lack of adequate resources.
4. Lack of specific legislation and standards for hazardous wastes.
5. Lack of incentives and the will to enforce.

Information

In India, data are being collected on the type, nature and sources of hazardous wastes from different process industries, the quantities produced inventories, methods and sites of disposal. These data will be invaluable for the formulation of rational policies and pragmatic disposal/management procedures. In addition, data on growth in the industrial sectors envisaged in the planned documents will serve as indicators on the specificity of hazardous wastes and overall implications as regards air and water pollution.

Co-ordination

The data outlined above are being collected by a number of organisations and establishments, including government and semi-government bodies. Co-ordination among these organisations is of utmost necessity in order to avoid duplicity of effort and expense. Hazardous waste management operates with the same controlling procedures as those hitherto applied by municipal and local bodies to sewage treatment and solid disposal methods. There is scope for improved co-ordination among different organs of the government, such as the Ministries of Health, Chemicals, Industry, Energy, Environment and so on. The Ministry of Environment has initiated the overall policy guidance to improve co-ordination among different ministries and to improve follow-up measures.

Legislation

The legislation for the containment of hazardous waste is still embodied in a variety of other laws relating to, for example, health protection and factory safety. There are no specific laws to deal with hazardous wastes. The criteria used to define hazardous wastes are still under discussion. An exclusive list (and/or inclusive) for classifying hazardous wastes has been visualised and work started thereon. One specific area which required exclusive attention in the recent past relates to the transportation of hazardous waste/products through rail, road and waterways. The responsibilities are being clearly delegated and accountability ensured.

Penalties/Incentives

More often, the penalties imposed for non-compliance with measures under the existing laws are too small to warrant any additional expenditure on treatment of wastes. It is more cost-effective and expedient to continue to pay fines than incur the cost of corrective measures. Incentives such as tax rebates, low interest rates and selective subsidies have been more effective in bringing about a change in attitudes and providing at the same time necessary finance.

Waste Management

Reduction at source

Experience gained on the pollution abatement in process industries has shown that preventive policies turn out to be most cost-effective in the long run, even though initially there is reluctance for changing over from conventional processes and practices. In existing installations, however, the change over to improved processes has not been possible in view of the high costs and potential risks involved.

Treatment and disposal

Reducing dependence on land disposal, though a technical possibility, has always presented problems. Wastes containing organic compounds have, in some cases, been completely destroyed for hazardous constituents; the inorganic wastes, however, can at best be solidified to produce environmentally acceptable treatment residues. Wastes containing mixtures of organic and inorganic constituents often present technical difficulties. Many treatment options commercially available are shown in Table 6.4.

Table 6.4. Some commercially available treatment options.

Treatment	Typical applications
Physical	
Centrifugation	Separates liquids and solids
Filter presses	Removes moisture from solids, sludges
Distillation	Solvent purification
Carbon absorption	Removes organics
Reverse osmosis	Removes metals and organics
Chemical	
Precipitation	Removes metals
Oxidation	Destroys organics
Reduction-dechlorination	Reduces chlorine content of hydrocarbons
Photolysis	Destroys dioxin and cyanide
Biological	
Aerobic/anaerobic	Removes metals and organics
Land treatment	Degradation of organic sludges
Thermal	
Liquid injection	Destroys organics in liquid wastes
Rotary kiln	Destroys organics in sludges and solids
Stabilisation/solidifcation	
Sorption	Uses variety of material to solidify inorganic liquids (e.g. fly ash, lime, clays and carbon)
Pozzolanic reactions	Uses lime-fly ash or portland cement to solidify inorganic wastes

Chemical treatment options have been available for a long time and attention has mainly focused on new applications and operations, and the efficient use of the reactions taking place. Some commercial applications include oxidation — reduction reactions and degradation of trace organic compounds, photolysis to destroy cyanides and dioxins, and precipitation of metals for recovery and/or disposal.

Biological processes are sensitive to the presence of toxic elements and non-biodegradables but some degree of success for hazardous wastes, essentially for refinery sludges, has been achieved by this method. Traditional incineration methods are useful primarily for organics with fairly high calorific value. They are efficient but expensive as compared with physical, chemical and biological methods.

Solidification/stabilisation methods are best used for wastes containing inorganic constituents. There is no change in toxicity of the waste but the potential mobility of the constituents is substantially reduced.

New technologies

New technologies are being developed for handling hazardous wastes. A recent review by the Hazardous Waste Consultant identifies these innovative technologies, some beyond the laboratory stage of development and considered ready for pilot and commercial development. They are outlined in Table 6.5.

Table 6.5. New technological developments.

Process	Process description
Enzyme destruction	Biological destruction of organics; does not involve living organisms; can be maintained in immobilised systems or applied directly to wastes or contaminated material
uv photolysis	Used to detoxify liquids containing dioxin, being developed for application on contaminated solids; dioxin mobilised by surfactants and subjected to uv photolysis; can reduce concentrations by 90 to 99%
Pyroplasma processes	Breakdown of waste fluids to element constituents; being developed as a mobile unit; tested for destruction of chlorinated organics; low power consumption and rapid start-stop mode.
Plasma dust process	Recovery of metals from iron and steel mill baghouse dust; reduces metal oxide to elemental forms; iron removed with molten slag; zinc and lead removed as gas; tests resulted in yields of 96% for iron, zinc and lead
Plasma arc	Destruction of PCBs and PCB-contaminated equipment; destruction and removal efficiency of 99.9999%; possibility for metal recovery from molten slag
Circulating bed	High heat-transfer and turbulence allow operation at incineration temperatures lower than incinerator traditional incinerators; accommodates solid and liquid wastes; complete destruction of organics at relatively low temperatures; no need for scrubber system to remove acid gases; particularly cost-efficient for homogeneous wastes from oil and petrochemical processes
High-temperature fluid	Most suitable for contaminated soil; liquid wastes wall reactor require a carrier; pyrolyse organics wall reactor to carbon, carbon monoxide and hydrogen, equipment not attacked by inorganic components; mobile units possible; reaches destruction of efficiencies of 99.9999%
Penberthy Pyro-converter	Glass-melting furnace technology adapted for converter destruction of organics; suitable for liquids, vapours, solids and sludges; solid residues (inorganics) incorporated into glass matrix; current use for production of HCl and destruction of chlorinated organics
Pyrolysing rotary	Operates in oxygen-free environment and at lower temperatures than conventional kiln; produces gas suitable for energy recovery or further treated to recover condensed hydrocarbons; recovery of metals possible without volatisation; reduced need for air pollution control; need to verify destruction efficiencies of hazardous constituents
Rollins rotary reactor	Suitable for viscous and high-solids content wastes; no need for supplemental fuel; reduced gas scrubbing requirements; high-transfer efficiencies may increase destruction efficiencies at lower temperatures
Supercritical water	Oxidises organics to carbon dioxide and water; high oxidation pressure steam or electricity oxidation produced, inorganic salts precipitated; especially efficient with highly concentrated organic wastes; for water containing 10% organics, destruction efficiency greater than 99.99%; suitable for chlorinated solvents and PCBs
Wet oxidation	Suitable for dilute aqueous waste that cannot be incinerated or biologically treated; destruction efficiencies expected in range of 99% or 99.99%; oxidises organics and inorganics; not appropriate for halogenated aromatics
Vertical-tube reactor	Adaptation of wet oxidation into 1-mile deep well system; operates at lower pressure than conventional process; currently applied to municipal waste-water

Thus, it may appear that the current level of industrialisation in India does not warrant the handling of large quantities of hazardous waste. The problems lie with the small- and medium-scale industries, which generate small but significant quantities of waste that cannot be subjected to uniform treatment. In the future, dependence on land disposal must be reduced and, therefore, it is essential that innovative approaches and new technologies are developed to deal with the increasing quantities of waste being generated by industry.

SECTION III

Recycling of Wastes

Chapter 7

Recycling of Solid Wastes: A Review

INTRODUCTION

Recycling is processing used materials (waste) into new products to prevent waste of potentially useful materials, reduce the consumption of fresh raw materials, reduce energy usage, reduce air pollution (from incineration) and water pollution (from landfilling) by reducing the need for 'conventional' waste disposal, and lower greenhouse gas emissions as compared to virgin production. Recycling is a key component of modern waste reduction and is the third component of the 'Reduce, Reuse, Recycle' waste hierarchy. Recyclable materials include many kinds of glass, paper, metal, plastic, textiles, and electronics. Although similar in effect, the composting or other reuse of biodegradable waste — such as food or garden waste — is not typically considered recycling. Materials to be recycled are either brought to a collection center or picked up from the curbside, then sorted, cleaned, and reprocessed into new materials bound for manufacturing.

In the strictest sense, recycling of a material would produce a fresh supply of the same material — for example, used office paper would be converted into new office paper, or used foamed polystyrene into new polystyrene. However, this is often difficult or too expensive (compared with producing the same product from raw materials or other sources), so 'recycling' of many products or materials involves their reuse in producing different materials (e.g. paperboard) instead. Another form of recycling is the salvage of certain materials from complex products, either due to their intrinsic value (e.g. lead from car batteries or other reusable products from computer components) or due to their hazardous nature (e.g. removal and reuse of mercury from various items). Critics dispute the net economic and environmental benefits of recycling over its costs, and suggest that proponents of recycling often make matters worse and suffer from confirmation bias. Specifically, critics argue that the costs and energy used in collection and transportation detract from (and outweigh) the costs and energy saved in the production process; also that the jobs produced by the recycling industry can be a poor trade for the jobs lost in logging, mining, and other industries associated with virgin production; and that materials such as paper pulp can only be recycled a few times before material degradation prevents further recycling. Proponents of recycling dispute each of these claims, and the validity of arguments from both sides has led to enduring controversy.

LEGISLATION

Supply

For a recycling program to work, having a large, stable supply of recyclable material is crucial. Three legislative options have been used to create such a supply: mandatory recycling collection, container deposit legislation, and refuse bans. Mandatory collection laws set recycling targets for cities to aim for, usually in the form that a certain percentage of a material must be diverted from the city's waste stream by a target date. The city is then responsible for working to meet this target.

Container deposit legislation involves offering a refund for the return of certain containers, typically glass, plastic, and metal. When a product in such a container is purchased, a small surcharge is added to the price. This surcharge can be reclaimed by the consumer if the container is returned to a collection point. These programs have been very successful, often resulting in an 80 per cent recycling rate. Despite such good results, the shift in collection costs from local government to industry and consumers has created strong opposition to the creation of such programs in some areas.

A third method of increase supply of recyclates is to ban the disposal of certain materials as waste, often including used oil, old batteries, tyres and garden waste. One aim of this method is to create a viable economy for proper disposal of banned products. Care must be taken that enough of these recycling services exist, or such bans simply lead to increased illegal dumping.

Government-Mandated Demand

Legislation has also been used to increase and maintain a demand for recycled materials. Four methods of such legislation exist: minimum recycled content mandates, utilisation rates, procurement policies, recycled product labelling. Both minimum recycled content mandates and utilisation rates increase demand directly by forcing manufacturers to include recycling in their operations. Content mandates specify that a certain percentage of a new product must consist of recycled material. Utilisation rates are a more flexible option: industries are permitted to meet the recycling targets at any point of their operation or even contract recycling out in exchange for tradeable credits. Opponents to both of these methods point to the large increase in reporting requirements they impose, and claim that they rob industry of necessary flexibility.

Governments have used their own purchasing power to increase recycling demand through what are called 'procurement policies'. These policies are either 'set-asides', which earmark a certain amount of spending solely towards recycled products, or 'price preference' programs which provide a larger budget when recycled items are purchased. Additional regulations can target specific cases: in the United States, for example, the environmental protection agency mandates the purchase of oil, paper, tyres and building insulation from recycled or re-refined sources whenever possible.

The final government regulation towards increased demand is recycled product labelling. When producers are required to label their packaging with amount of recycled material in the product (including the packaging), consumers are better able to make educated choices. Consumers with sufficient buying power can then choose more environmentally conscious options, prompt producers to increase the amount of recycled material in their products, and indirectly increase demand. Standardised recycling labelling can also have a positive effect on supply of recyclates if the labelling includes information on how and where the product can be recycled.

PROCESS

Collection

A number of different systems have been implemented to collect recyclates from the general waste stream. These systems lie along the spectrum of trade-off between public convenience and government ease and expense. The three main categories of collection are 'drop-off centres', 'buy-back centres' and 'curbside collection'.

Drop-off Centres

Drop-off centres require the waste producer to carry the recyclates to a central location, either an installed or mobile collection station or the reprocessing plant itself. They are the easiest type of collection to establish, but suffer from low and unpredictable throughput.

Buy-Back Centres

Buy-back centres differ in that the cleaned recyclates are purchased, thus providing a clear incentive for use and creating a stable supply. The post-processed material can then be sold on, hopefully creating a profit. Unfortunately government subsidies are necessary to make buy-back centres a viable enterprise, as according to the United States Nation Solid Wastes Management Association it costs on average US$50 to process a ton of material, which can only be resold for US$30.

Curbside Collection

Curbside collection encompasses many subtly different systems, which differ mostly on where in the process the recyclates are sorted and cleaned. The main categories are mixed waste collection, commingled recyclables and source separation. A waste collection vehicle generally picks up the waste.

At one end of the spectrum is mixed waste collection, in which all recyclates are collected mixed in with the rest of the waste, and the desired material is then sorted out and cleaned at a central sorting facility. This results in a large amount of recyclable waste, paper especially, being too soiled to reprocess, but has advantages as well: the city need not pay for a separate collection of recyclates and no public education is needed. Any changes to which materials are recyclable is easy to accommodate as all sorting happens in a central location. In a Commingled or single-stream system, all recyclables for collection are mixed but kept separate from other waste. This greatly reduces the need for post-collection cleaning but does require public education on what materials are recyclable.

Source separation is the other extreme, where each material is cleaned and sorted prior to collection. This method requires the least post-collection sorting and produces the purest recyclates, but incurs additional operating costs for collection of each separate material. An extensive public education program is also required, which must be successful if recyclate contamination is to be avoided.

Source separation used to be the preferred method due to the high sorting costs incurred by commingled collection. Advances in sorting technology, however, have lowered this overhead substantially—many areas which had developed source separation programs have since switched to commingled collection.

Sorting

Once commingled recyclates are collected and delivered to a central collection facility, the different types of materials must be sorted. This is done in a series of stages, many of which involve automated processes such that a truck-load of material can be fully sorted in less than an hour. Some plants can now sort the materials automatically, known as single-stream recycling. A 30 per cent increase in recycling rates has been seen in the areas where these plants exist (Fig. 7.1).

Initially, the commingled recyclates are removed from the collection vehicle and placed on a conveyor belt spread out in a single layer. Large pieces of corrugated fibreboard and plastic bags are removed by hand at this stage, as they can cause later machinery to jam.

Next, automated machinery separates the recyclates by weight, splitting lighter paper and plastic from heavier glass and metal. Cardboard is removed from the mixed paper, and the most common types of plastic, PET and HDPE, are collected. This separation is usually done by hand, but has become automated in some sorting centers: a spectroscopic scanner is used to differentiate between different types of paper and plastic based on the absorbed wavelengths, and subsequently divert each material into the proper collection channel.

Fig. 7.1. Early sorting of recyclable materials: glass and plastic bottles in Poland.

Strong magnets are used to separate out ferrous metals, such as iron, steel, and tin-plated steel cans ('tin cans'). Non-ferrous metals are ejected by magnetic eddy currents in which a rotating magnetic field induces an electric current around the aluminium cans, which in turn creates a magnetic eddy current inside the cans. This magnetic eddy current is repulsed by a large magnetic field, and the cans are ejected from the rest of the recyclate stream. Finally, glass must be sorted by hand based on its colour: brown, amber, green or clear.

COST-BENEFIT ANALYSIS

There is some debate over whether recycling is economically efficient. Municipalities often see fiscal benefits from implementing recycling programs, largely due to the reduced landfill costs. A study conducted by the Technical University of Denmark found that in 83 per cent of cases, recycling is the most efficient method to dispose of household waste. However, a 2004 assessment by the Danish Environmental Assessment Institute concluded that incineration was the most effective method for disposing of drink containers, even aluminium ones (Table 7.1).

Table 7.1. Environmental effects of recycling.

Material	Energy savings	Air pollution savings
Aluminium	95%	95%
Cardboard	24%	–
Glass	5–30%	20%
Paper	40%	73%
Plastics	70%	–
Steel	60%	–

Fiscal efficiency is separate from economic efficiency. Economic analysis of recycling includes what economists call externalities, which are unpriced costs and benefits that accrue to individuals outside of private transactions. Examples include: decreased air pollution and greenhouse gases from incineration, reduced hazardous waste leaching from landfills, reduced energy consumption, and reduced waste and resource consumption, which leads to a reduction in environmentally damaging mining and timber activity. About 4000 minerals have been identified, of these around 100 can be called common, another several hundred are relatively uncommon, and the rest are rare. At current rates, current known reserves of phosphorus will be depleted in the next 50 to 100 years. Without mechanisms such as taxes or subsidies to internalise externalities, businesses will ignore them despite the costs imposed on society. To make such non-fiscal benefits economically relevant, advocates have pushed for legislative action to increase the demand for recycled materials. The United States Environmental Protection Agency (EPA) has concluded in favour of recycling, saying that recycling efforts reduced the country's carbon emissions by a net 49 million metric tons in 2012. In the United Kingdom, the Waste and Resources Action Program stated that Great Britain's recycling efforts reduce CO_2 emissions by 10–15 million tons a year. Recycling is more efficient in densely populated areas, as there are economies of scale involved.

Why is Recycling Not Mandatory in All Countries of World

Mandatory recycling is a hard sell in the United States, where the economy runs largely along free market lines and landfilling waste remains inexpensive and efficient. When the research firm Franklin Associates examined the issue a decade ago, it found that the value of the materials recovered from curbside recycling was far less than the extra costs of collection, transportation, sorting and processing incurred by municipalities.

Recycling often costs more than sending waste to landfills

Plain and simple, recycling still costs more than landfilling in most locales. This fact, coupled with the revelation that the so-called 'landfill crisis' of the mid-1990s may have been overblown — most of our landfills still have considerable capacity and do not pose health hazards to surrounding communities — means that recycling has not caught on the way some environmentalists were hoping it would.

Education, logistics and marketing strategies can lower recycling costs

However, many cities have found ways to recycle economically. They have cut costs by scaling back the frequency of curbside pickups and automating sorting and processing. They've also found larger, more lucrative markets for the recyclables, such as developing countries eager to reuse our cast-off items. Increased efforts by green groups to educate the public about the benefits of recycling have also helped. Today, dozens of US cities are diverting upwards of 30 per cent of their solid waste streams to recycling.

Recycling is mandatory in some US cities

While recycling remains an option for most Americans, a few cities, such as Pittsburgh, San Diego and Seattle, have made recycling mandatory. Seattle passed its mandatory recycling law in 2006 as a way to counter declining recycling rates there. Recyclables are now prohibited from both residential and business garbage. Businesses must sort for recycling all paper, cardboard and yard waste. Households must recycle all basic recyclables, such as paper, cardboard, aluminum, glass and plastic.

Mandatory recycling customers fined or denied service for non-compliance

Businesses with garbage containers 'contaminated' with more than 10 recyclables are issued warnings and eventually fines if they don't comply. Household garbage cans with recyclables in them are simply not collected until the recyclables are removed to the recycling bin. Meanwhile, a handful of other cities, including Gainesville, Florida and Honolulu, Hawaii, require businesses to recycle, but not yet residences.

New York City: A Case Study for Recycling

In perhaps the most famous case of a city putting recycling to the economic test, New York, a national leader on recycling, decided to stop its least cost-effective recycling programs (plastic and glass) in 2002. But rising landfill costs ate up the $39 million savings expected.

As a result, the city reinstated plastic and glass recycling and committed to a 20-year contract with the country's largest private recycling firm, Hugo Neu Corporation, which built a state-of-the art facility along South Brooklyn's waterfront. There, automation has streamlined the sorting process, and its easy access to rail and barges has cut both the environmental and transportation costs previously incurred by using trucks. The new deal and new facility have made recycling much more efficient for the city and its residents, proving once and for all that responsibly run recycling programs can actually save money, landfill space and the environment.

Chapter 8

Life Cycle Analysis of Material Recycling

INTRODUCTION

The concept of conducting a detailed examination of the life cycle of a product or a process is a relatively recent one which emerged in response to increased environmental awareness on the part of the general public, industry and governments. The immediate precursors of life cycle analysis and assessment (LCAs) were the global modelling studies and energy audits of the late 1960s and early 1970s. These attempted to assess the resource cost and environmental implications of different patterns of human behaviour.

LCAs were an obvious extension, and became vital to support the development of eco-labelling schemes which are operating or planned in a number of countries around the world. In order for eco-labels to be granted to chosen products, the awarding authority needs to be able to evaluate the manufacturing processes involved, the energy consumption in manufacture and use, and the amount and type of waste generated. To accurately assess the burdens placed on the environment by the manufacture of an item, the following of a procedure or the use of a certain process, two main stages are involved. The first stage is the collection of data, and the second is the interpretation of that data.

A number of different terms have been coined to describe the processes. One of the first terms used was life cycle analysis, but more recently two terms have come to largely replace that one: Life cycle inventory (LCI) and life cycle assessment (LCA). These better reflect the different stages of the process. Other terms such as cradle to grave analysis, eco-balancing, and material flow analysis are also used.

Whichever name is used to describe it, LCA is a potentially powerful tool which can assist regulators to formulate environmental legislation, help manufacturers analyse their processes and improve their products, and perhaps enable consumers to make more informed choices. Like most tools, it must be correctly used, however, a tendency for LCAs to be used to 'prove' the superiority of one product over another has brought the concept into disrepute in some areas.

WHAT IS A LIFE CYCLE ANALYSIS?

Taking as an example the case of a manufactured product, an LCA involves making detailed measurements during the manufacture of the product, from the mining of the raw materials used in its production and distribution, through to its use, possible re-use or recycling, and its eventual disposal. LCAs enable a manufacturer to quantify how much energy and raw materials are used, and how much solid, liquid and gaseous waste is generated, at each stage of the product's life. Such a study would normally ignore second generation impacts, such as the energy required to fire the bricks used to build the kilns used to manufacture the raw material.

However, deciding which is the 'cradle' and which the 'grave' for such studies has been one of the points of contention in the relatively new science of LCAs, and in order for LCAs to have value there must be standardisation of methodologies, and consensus as to where to set the limits. Much of the focus worldwide to date has been on agreeing the methods and boundaries to be used when making such analyses, and it seems that agreement may have now been reached.

While carrying out an LCA is a lengthy and very detailed exercise, the data collection stage is — in theory at least — relatively uncomplicated, provided the boundary of the study has been clearly defined, the methodology is rigorously applied, and reliable, high-quality data is available. Those of course are fairly large provisos.

INTERPRETATION

While such a record is helpful and informative, on its own it is not sufficient. Having first compiled the detailed inventory, the next stage should be to evaluate the findings. This second stage — life cycle assessment — is more difficult, since it requires interpretation of the data, and value judgements to be made. A life cycle inventory will reveal — for example, how many kilos of pulp, how much electricity, and how many gallons of water, are involved in producing a quantity of paper. Only by then assessing those statistics can a conclusion be reached about the product's environmental impact overall. This includes the necessity to make judgements based on the assembled figures, in order to assess the likely significance of the various impacts.

PROBLEMS

It is here that many of the problems begin. Decisions, without scientific basis, such as whether three tons of emitted sulphur is more or less harmful than the emission of just a few pounds of a more toxic pollutant, are necessarily subjective.

1. How can one compare heavy energy demand with heavy water use: which imposes greater environmental burden?
2. How should the use of non-renewable mineral resources like oil or gas (the ingredients of plastics) be compared with the production of softwoods for paper?
3. How should the combined impacts of the landfilling of wastes (air and groundwater pollution, transport impacts, etc.) be compared with those produced by the burning of wastes for energy production (predominantly emissions to air)?

Some studies attempt to aggregate the various impacts into clearly defined categories, for example, the possible impact on the ozone layer, or the contribution to acid rain. Others go still further and try to add the aggregated figures to arrive at a single 'score' for the product or process being evaluated. It is doubtful whether such simplification will be of general benefit. Reliable methods for aggregating figures generated by LCA, and using them to compare the life cycle impacts of different products, do not yet exist. However, a great deal of work is currently being conducted on this aspect of LCAs to arrive at a standardised method of interpreting the collected data.

CONTRADICTIONS

Many LCAs have reached different and sometimes contradictory conclusions about similar products. Comparisons are rarely easy because of the different assumptions that are used, for example in the case of food packaging, about the size and form of container, the production and distribution system used, and the forms and type of energy assumed.

To compare two items which are identically sized, identically distributed, and recycled at the same rate is relatively simple, but even that requires assumptions to be made. For example, whether deliveries were made in a 9 tons truck or a larger one, whether it used diesel or petrol, and ran on congested city centre roads where fuel efficiencies are lower or on country roads or motorways where fuel efficiencies might be better. Comparisons of products which are dissimilar in most respects can only be made by making even more judgements and assumptions. Preserving the confidentiality of commercially-sensitive raw data without reducing the credibility of LCAs is also a major problem. Another is the understandable reluctance of companies to publish information which may indicate that their own product is somehow inferior to that of a competitor. It is not surprising that many of the studies which are published, and not simply used internally, endorse the views of their sponsors.

RECYCLING

Recycling introduces a further real difficulty into the calculations. In the case of materials like steel and aluminium which can technically be recycled an indefinite number of times (with some melt losses), there is no longer a 'grave'. And in the case of paper, which can theoretically be reprocessed four or five times before fibres are too short to have viable strength, should calculations assume that it will be recycled four times, or not? What return rates, for example, should be assumed for factory-refillable containers?

For both refillable containers and materials sent for recycling, the transport distance in each specific case is a major influence in the environmental impacts associated with the process. An LCA which concludes that recycling of low-value renewable materials in one city is environmentally preferable may not hold good for a different, more remote city where reprocessing facilities incur large transport impacts.

LCA IN WASTE MANAGEMENT

LCA has begun to be used to evaluate a city or region's future waste management options. The LCA or environmental assessment, covers the environmental and resource impacts of alternative disposal processes, as well as those other processes which are affected by disposal strategies such as different types of collection schemes for recyclables, changed transport patterns and so on.

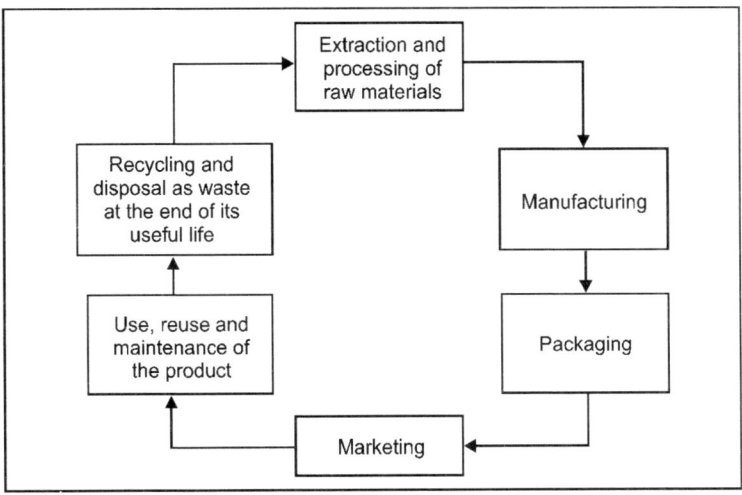

Fig. 8.1. The complexity of the task, and the number of assumptions.

The complexity of the task, and the number of assumptions which must be made, is shown by the simplified Fig. 8.1 showing some of the different routes which waste might take, and some of the environmental impacts incurred along the way. Those shown are far from exhaustive.

WHY PERFORM LCAs?

LCAs might be conducted by an industry sector to enable it to identify areas where improvements can be made, in environmental terms. Alternatively the LCA may be intended to provide environmental data for the public or for government. In recent years, a number of major companies have cited LCAs in their marketing and advertising, to support claims that their products are 'environmentally friendly' or even 'environmentally superior' to those of their rivals. Many of these claims have been successfully challenged by environmental groups.

All products have some impact on the environment. Since some products use more resources, cause more pollution or generate more waste than others, the aim is to identify those which are most harmful. Even for those products whose environmental burdens are relatively low, the LCA should help to identify those stages in production processes and in use which cause or have the potential to cause pollution, and those which have a heavy material or energy demand. Breaking down the manufacturing process into such fine detail can also be an aid to identifying the use of scarce resources, showing where a more sustainable product could be substituted.

INCONCLUSIVE

In most situations it is impossible to prove conclusively using LCAs that any one product or any one process is better in general terms than any other, since many parameters cannot be simplified to the degree necessary to reach such a conclusion. It seems likely that, in the case of manufactured goods, the most important time for LCA information to be taken into consideration is at the design stage of new products. Where LCA is used to evaluate procedures rather than products, the information can help ensure appropriate choices are made.

TOOL

Life cycle analysis must be used cautiously, and in the interpretation of the inventory, care must be taken with subjective judgements. When first conceived, it was predicted that LCA would enable definitive judgements to be made. That misplaced belief has now been discredited. In combination with the trend towards more open disclosure of environmental information by companies, and the desire by consumers to be guided towards the least harmful purchases, the LCA is a vital tool.

Chapter 9

Paper Recycling

INTRODUCTION

Paper recycling is the process of recovering waste paper and remaking it into new paper products. There are three categories of paper that can be used as feedstocks for making recycled paper: mill broke, pre-consumer waste, and post-consumer waste. Mill broke is paper trimmings and other paper scrap from the manufacture of paper, and is recycled internally in a paper mill. Pre-consumer waste is material which left the paper mill but was discarded before it was ready for consumer use. Post-consumer waste is material discarded after consumer use, such as old corrugated containers (OCC), old magazines, old newspapers (ONP), office paper, old telephone directories, and residential mixed paper (RMP). Paper suitable for recycling is called 'scrap paper'. The industrial process of removing printing ink from paperfibres of recycled paper to make deinked pulp is called deinking.

RATIONALE FOR RECYCLING

Industrialised paper making has an effect on the environment both upstream (where raw materials are acquired and processed) and downstream (waste-disposal impacts). Recycling paper reduces this impact.

Today, 90 per cent of paper pulp is made of wood. Paper production accounts for about 35 per cent of felled trees, and represents 1.2 per cent of the world's total economic output. Recycling one ton of newsprint saves about 1 ton of wood while recycling 1 ton of printing or copier paper saves slightly more than 2 tons of wood. This is because kraft pulping requires twice as much wood since it removes lignin to produce higher quality fibres than mechanical pulping processes. Relating tons of paper recycled to the number of trees not cut is meaningless, since tree size varies tremendously and is the major factor in how much paper can be made from how many trees. Trees raised specifically for pulp production account for 16 per cent of world pulp production, old growth forests 9 per cent and second — and third — and more generation forests account for the balance. Most pulp mill operators practice reforestation to ensure a continuing supply of trees. The Program for the Endorsement of Forest Certification (PEFC) and the Forest Stewardship Council (FSC) certify paper made from trees harvested according to guidelines meant to ensure good forestry practices. It has been estimated that recycling half the world's paper would avoid the harvesting of 20 million acres (81,000 km^2) of forestland. A flow diagram (Fig. 9.1) is showing the various recycling operations of paper.

Energy

Energy consumption is reduced by recycling, although there is debate concerning the actual energy savings realised. The energy information administration claims a 40 per cent reduction in energy when paper is recycled versus paper made with unrecycled pulp, while the Bureau of International Recycling (BIR) claims a 64 per cent reduction. Some calculations show that recycling one ton of newspaper saves about 4000 kW · h (14 GJ) of electricity, although this may be too high. This is enough electricity to power a 3-bedroom European house for an entire year, or enough energy to heat and air-condition the average North American home for almost six months. Recycling paper to make pulp may actually consume more fossil fuels than making new pulp via the kraft process, however, since these mills generate all of their energy from burning waste wood (bark, roots) and by-product lignin. Pulp mills producing new mechanical pulp use large amounts of energy; a very rough estimate of the electrical energy needed is 10 gigajoules per ton of pulp (2500 kW · h per short ton), usually from hydroelectric generating plants. Recycling mills purchase most of their energy from local power companies, and since recycling mills tend to be in urban areas, it is likely that the electricity is generated by burning fossil fuels.

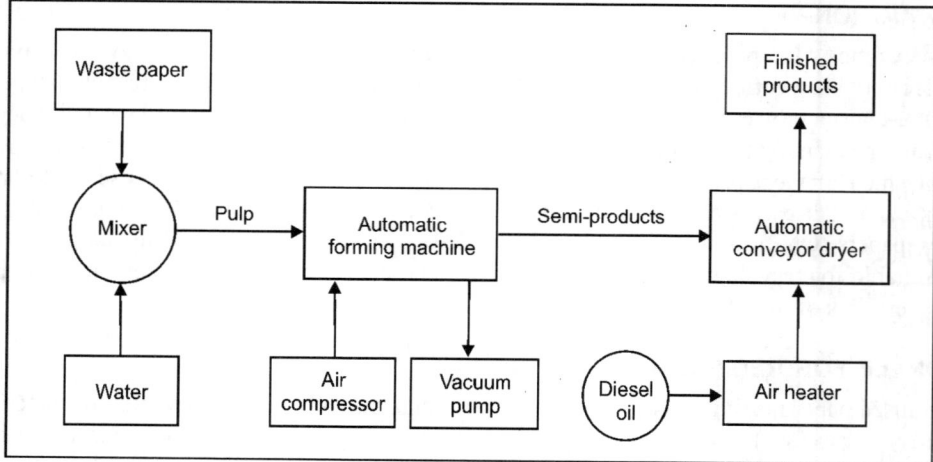

Fig. 9.1. Flow diagram for recycling of paper.

Landfill Use

About 35 per cent of municipal solid waste (before recycling) by weight is paper and paper products.

Water and Air Pollution

The United States Environmental Protection Agency (EPA) has found that recycling causes 35 per cent less water pollution and 74 per cent less air pollution than making virgin paper. Pulp mills can be sources of both air and water pollution, especially if they are producing bleached pulp. Modern mills produce considerably less pollution than those of a few decades ago. Recycling paper decreases the demand for virgin pulp and thus reduces the overall amount of air and water pollution associated with paper manufacture. Recycled pulp can be bleached with the same chemicals used to bleach virgin pulp, but hydrogen peroxide and sodium hydrosulphite are the most common bleaching agents. Recycled pulp or paper made from it, is known as PCF (process chlorine free) if no chlorine-containing compounds were used in the recycling process. However, recycling mills may have polluting by-products, such as sludge. De-inking at Cross Pointe's Miami, Ohio mill results in sludge weighing 22 per cent of the weight of wastepaper recycled.

Criticism

Some of the claimed benefits of paper recycling have fallen under criticism, such as the claim that recycling saves trees, reduces energy consumption, reduces pollution, creates desirable jobs, and saves money.

RECYCLING FACTS AND FIGURES

In the mid-19th century, there was an increased demand for books and writing material. Up to that time, paper manufacturers had used discarded linen rags for paper, but supply could not keep up with the increased demand. Books were bought at auctions for the purpose of recycling fibre content into new paper, at least in the United Kingdom, by the beginning of the 19th century.

Internationally, about half of all recovered paper comes from converting losses (pre-consumer recycling), such as shavings and unsold periodicals; approximately one-third comes from household or post-consumer waste.

Some statistics on paper consumption:

1. The average per capita paper use worldwide was 110 pounds (50 kg).
2. It is estimated that 95 per cent of business information is still stored on paper.
3. Recycling 1 short ton (0.91 T) of paper saves 17 mature trees, 7 thousand US gallons (26 m^3) of water, 3 cubic yards (2.3 m^3) of landfill space, 2 barrels of oil (84 US gal or 320 l), and 4100 kilowatt-hours (15 GJ) of electricity—enough energy to power the average American home for six months.
4. Although paper is traditionally identified with reading and writing, communications has now been replaced by packaging as the single largest category of paper use at 41 per cent of all paper used.
5. 115 billion sheets of paper are used annually for personal computers. The average web user prints 28 pages daily.
6. Most corrugated fibreboard boxes have over 25 per cent recycled fibres. Some are 100 per cent recycled fibre.

Paper Recycling by Region

European union

Paper recovery in Europe has a long history and has grown into a mature organisation. The European papermakers and converters work together to meet the requirements of the European Commission and national governments. Their aim is the reduction of the environmental impact of waste during manufacturing, converting/printing, collecting, sorting and recycling processes to ensure the optimal and environmentally sound recycling of used paper and board products. In 2007 the paper recycling rate in Europe was 54.6 per cent or 45.5 million short tons (41.3 Mt). The recycling rate in Europe reached 64.5 per cent in 2009, which confirms that the industry is on the path to meeting its voluntary target of 66 per cent by 2010.

Japan

Municipal collections of paper for recycling are in place. However, according to the Yomiuri Shimbun, in 2008, eight paper manufacturers in Japan have admitted to intentionally mislabelling recycled paper products, exaggerating the amount of recycled paper used.

United States of America

Recycling has long been practiced in the United States. The history of paper recycling has several dates of importance:

1. 1690: The first paper mill to use recycled linen was established by the Rittenhouse family.
2. 1896: The first major recycling center was started by the Benedetto family in New York City, where they collected rags, newspaper, and trash with a pushcart.
3. 1993: The first year when more paper was recycled than was buried in landfills.

Today, over half of the material used to make paper is recovered waste. Paper products are the largest component of municipal solid waste, making up more than 40 per cent of the composition of landfills. In 2006, a record 53.4 per cent of the paper used in the US (or 53.5 million tons) was recovered for recycling. This is up from a 1990 recovery rate of 33.5 per cent. The US paper industry has set a goal to recover 55 per cent of all the paper used in the US by 2012. Paper packaging recovery, specific to paper products used by the packaging industry, was responsible for about 77 per cent of packaging materials recycled with more than 24 million pounds recovered in 2005.

By 1998, some 9000 curbside programs and 12,000 recyclable drop-off centers had sprouted up across the US for recycles collection. As of 1999, 480 materials recovery facilities had been established to process the collected materials. In 2008, the global financial crisis resulted in the price of old newspapers to drop in the US from \$130 to \$40 per short ton (\$140/T to \$45/T) in October.

Mexico

In Mexico, recycled paper, rather than wood pulp, is the principal feedstock in papermills accounting for about 75 per cent of raw materials.

Recycle Carbon Paper

Most of the places in world carbon paper are not recycled. Because of its complexity it is difficult to recycle carbon paper. However with new technologies in some part of the world there is a tedious process to remove wax carbon coating from carbon paper before it is recycled.

The process starts with removing wax carbon coating from carbon paper by giving it aqueous bath under high temperature. This process separates wax from carbon paper. The floating waxes is collected and filtered out and remaining paper is moved to next level through a squeezer. This process is repeated couple of time to separate the wax from carbon paper. Next wet paper is given rinse bath and again squeezed and separated in tank. Once carbon paper has reached to the stage where no wax or less pigments are left on paper, it is moved to next level for recycling carbon paper.

Recycling Corrugated Cardboard

Recycling old corrugated cardboard (OCC) can be economically beneficial for business and industry, and it can be easily targeted in any recycling program. North Carolina companies are at various stages in establishing cardboard recycling programs: while many have well-established programs, others are just beginning to initiate a program or are evaluating whether to bale cardboard for better cost control.

In general, recycling markets for OCC are well established, and restrictions such as landfill ordinances increase the urgency to recycle. Below are some options for managing waste cardboard and general guidelines for baling cardboard on-site.

OCC problem

Old corrugated cardboard makes up a significant percentage of landfilled waste. Studies in Wake County show that OCC comprised 26 per cent of the commercial/industrial/institutional waste stream and 18 per cent of all municipal solid waste entering the county's landfills in 1992. In efforts to reduce landfilled waste, several counties and cities have enacted ordinances to ban or place surcharges on waste loads containing OCC. Other municipalities are planning similar restrictive ordinances. The Office of Waste Reduction can provide current restriction ordinances or resolutions in place for the local government units.

Managing OCC

Even if there is no local ordinance concerning OCC, a company avoids disposal costs and keeps recyclables out of the landfill by recycling. When it comes to managing cardboard, consider the following:
1. Start a waste reduction program for OCC; look for ways to reduce and reuse corrugated cardboard.
2. Examine the available markets for OCC.
3. Consider the pros and cons of baling OCC on-site.
4. Even if you already have an OCC recycling program in place, look for ways to improve/expand it.

Establishing a cardboard recycling program

There are many ways to establish an OCC recycling program. The cardboard may be collected loose or baled. It can be picked up or delivered to a local recycler. An OCC recycler will work with you to set up logistics that best meet your needs.

To locate an OCC recycler: (i) contact the local municipality or county recycling coordinator, (ii) look under 'recycling centers' in the telephone book, and (iii) contact the office of waste reduction.

Determining the amount of OCC generated

Many companies are surprised to find out how much cardboard they generate. The chart below lists estimated weights of loose OCC in containers of different sizes at 100 per cent of capacity (Table 9.1).

Table 9.1. Estimated weights of Loose, flattened cardboard boxes.

Container (yards)	Size 100% full (pounds)
40	6000
30	4500
20	3000
8	1200
6	900

Bale or not to bale

While many businesses recycle loose OCC, the economics of baling the cardboard should be evaluated. The first step in making such a determination is to get acceptable bale sizes and purchase prices from the local OCC recycler. If the cardboard is currently selling at $0.50 per 100 pounds ($10 per ton) in a loose form, this same cardboard in baled form may bring $20 per ton. Savings may also result from avoided hauling costs, by baling several materials for recycling or by sharing a baler with another company.

The following are some general guidelines:

1. Determining baler size: Baler size is application-specific and is based on storage space constraints, OCC collection and handling methods, and buyer specifications. Assuming that it takes 40 minutes to load and strap a bale from a vertical baler (300 to 1000 pound bales) and that all the OCC is at the baler location, one employee will be needed to load the baler and one or two to strap the bale. Unless the facility is generating very high volumes of OCC (greater than 25 tons a month), a vertical baler should have sufficient capacity.

2. Baler and bale volumes. Table 9.2 gives information concerning bale volumes.

Table 9.2. Information related to baler and bale volumes.

Bale weight (pounds)	Vertical balers Feed opening (in.)	Bale volume (ft^3)
300	36 × 15	15
800	48 × 28	47.5
1000	60 × 28	50
1200	72 × 28	60
1500	72 × 32	84
	Horizontal balers	
1200	28 × 50	52
1500	46 × 50	57
2000	45 × 60	64

Note that other recyclable materials such as plastic film wrap, textile scraps, and other plastics can also be baled. The equipment investment may well pay for itself with the added value you will receive from baled materials.

Chapter 10

Glass Recycling

INTRODUCTION

Glass recycling is the process of turning waste glass into usable products. Glass waste should be separated by chemical composition, and then, depending on the end use and local processing capabilities, might also have to be separated into different colours. Many recyclers collect different colours of glass separately since glass retains its colour after recycling. The most common types used for consumer containers are colourless glass, green glass, and brown/amber glass.

Glass makes up a large component of household and industrial waste due to its weight and density. The glass component in municipal waste is usually made up of bottles, broken glassware, light bulbs and other items. Adding to this waste is the fact that many manual methods of creating glass objects have a defect rate of around forty per cent. Glass recycling uses less energy than manufacturing glass from sand, lime and soda. Every metric ton of waste glass recycled into new items saves 315 additional kilograms of carbon dioxide from being released into the atmosphere during the creation of new glass. Glass that is crushed and ready to be remelted is called cullet.

GLASS REUSE

Reuse of glass containers is preferable to recycling according to the waste hierarchy. Refillable bottles are used extensively in many European countries, Canada and until relatively recently, in the United States. In Denmark 98 per cent of bottles are refillable and 98 per cent of those are returned by consumers. A similarly high number is reported for beer bottles in Canada. These systems are typically supported by container deposit laws and other regulations. In some developing nations like India and Brazil, the cost of new bottles often forces manufacturers to collect and refill old glass bottles for selling carbonated and other drinks.

Glass Collection

Glass collection points, known as bottle banks are very common near shopping centres, at civic amenity sites and in local neighbourhoods in the United Kingdom. The first Bottle Bank was introduced by Stanley Race CBE, then president of the Glass Manufacturers' Federation and Ron England in Barnsley on 6 June 1977.

Bottle banks commonly stand beside collection points for other recyclable waste like paper, metals and plastics. Local, municipal waste collectors usually have one central point for all types of waste in which large glass containers are located. There are now over 50,000 bottle banks in the United Kingdom.

Most collection points have separate bins for clear, green and amber/brown glass. Glass reprocessors require separation by colour as the different colours of glass are usually chemically incompatible. Heat-resistant glass like Pyrex or borosilicate glass should not be disposed of in the glass container as even a single piece of such material will alter the viscosity of the fluid in the furnace at remelt.

GLASS RECYCLING BY COUNTRY

Germany

In 2004, Germany recycled 2,116,000 tons of glass. Reusable glass or plastic (PET) bottles are available for many drinks, especially beer and carbonated water as well as softdrinks (Mehrwegflaschen). The deposit per bottle (Pfand) is €0.08–€0.15, compared to €0.25 for recyclable but not reusable plastic bottles. There is no deposit for glass bottles which do not get refilled.

United Kingdom

Seven lakh fifty two thousand (7,52,000) tons of glass are now recycled annually in the United Kingdom. Glass is an ideal material for recycling and where it is used for new glass container manufacture it is virtually infinitely recyclable. The use of recycled glass in new containers helps save energy. It helps in brick and ceramic manufacture, and it conserves raw materials, reduces energy consumption, and reduces the volume of waste sent to landfill. In the United Kingdom, the waste recycling industry cannot consume all of the recycled container glass that will become available over the coming years, mainly due to the colour imbalance between that which is manufactured and that which is consumed. The UK imports much more green glass in the form of wine bottles than it uses, leading to a surplus amount for recycling.

The resulting surplus of green glass from imported bottles may be exported to producing countries, or used locally in the growing diversity of secondary end uses for recycled glass. Cory environmental are presently shipping glass cullet from the UK to Portugal.

The use of the recycled glass as aggregate in concrete has become popular in modern times, with large scale research being carried out at Columbia University in New York. This greatly enhances the aesthetic appeal of the concrete. Recent research findings have shown that concrete made with recycled glass aggregates have shown better long-term strength and better thermal insulation due to its better thermal properties of the glass aggregates. Secondary markets for glass recycling may include:

1. Glass in ceramic sanitary ware production.
2. Glass as a flux agent in brick manufacture.
3. Glass in astroturf and related applications (e.g. top dressing, root zone) material or golf bunker sand.
4. Glass in recycled glass countertops.
5. Glass as water filtration media.
6. Glass as an abrasive.

Mixed glass waste streams can also be recycled and converted into an aggregate. Mixed waste streams may be collected from materials recovery facilities or mechanical biological treatment systems. Some facilities can sort out mixed waste streams into different colours using electro-optical sorting units.

United States

Rates of recycling and methods of waste collection vary substantially across the United States because laws are written on the state or local level and large municipalities often have their own unique systems.

Many cities do curbside recycling, meaning they collect household recyclable waste on a weekly or bi-weekly basis that residents set out in special containers in front of their homes.

Apartment dwellers usually use shared containers that may be collected by the city or by private recycling companies which can have their own recycling rules. In some cases, glass is specifically separated into its own container because broken glass is a hazard to the people who later manually sort the co-mingled recyclables. Sorted recyclables are later sold to companies.

In 1971 the state of Oregon passed a law requiring buyers of carbonated beverages (such as beer and soda) to pay five cents per container as a deposit which would be refunded to anyone who returned the container for recycling. This law has since been copied in nine other states including New York and California. The abbreviations of states with deposit laws are printed on all qualifying bottles and cans. In states with these container deposit laws, most supermarkets automate the deposit refund process by providing machines which will count containers as they are inserted and then print credit vouchers that can be redeemed at the store for the number of containers returned. Small glass bottles (mostly beer) are broken, one-by-one, inside these deposit refund machines as the bottles are inserted. A large, wheeled hopper (very roughly 1.5m by 1.5m by 0.5m) inside the machine collects the broken glass until it can be emptied by an employee.

BENEFITS OF GLASS RECYCLING: WHY RECYCLE GLASS

Glass recycling is both simple and beneficial. Let's start with the benefits of glass recycling:

Glass Recycling is Good for the Environment

A glass bottle that is sent to a landfill can take up to a million years to break down. By contrast, it takes as little as 30 days for a recycled glass bottle to leave your kitchen recycling bin and appear on a store shelf as a new glass container.

Glass Recycling is Sustainable

Glass containers are 100 per cent recyclable, which means they can be recycled repeatedly, again and again, with no loss of purity or quality in the glass.

Glass Recycling is Efficient

Recovered glass from glass recycling is the primary ingredient in all new glass containers. A typical glass container is made of as much as 70 per cent recycled glass. According to industry estimates, 80 per cent of all recycled glass eventually ends up as new glass containers.

Glass Recycling Conserves Natural Resources

Every ton of glass that is recycled saves more than a ton of the raw materials needed to create new glass, including: 1300 pounds of sand; 410 pounds of soda ash; and 380 pounds of limestone.

Glass Recycling Saves Energy

Making new glass means heating sand and other substances to a temperature of 2600° Fahrenheit, which requires a lot of energy and creates a lot of industrial pollution. One of the first steps in glass recycling is to crush the glass and create a product called 'cullet'. Making recycled glass products from cullet consumes 40 per cent less energy than making new glass from raw materials, because cullet melts at a much lower temperature.

Recycled glass is useful

Because glass is made from natural materials such as sand and limestone, glass containers have a low rate of chemical interaction with their contents. As a result, glass can be safely reused. Besides serving as the primary ingredient in new glass containers, recycled glass also has many other commercial uses — from creating decorative tiles and landscaping material to rebuilding eroded beaches.

Glass recycling is also simple

Glass recycling is also simple because glass is one of the easiest materials to recycle. For one thing, glass is accepted by almost all curbside recycling programs and municipal recycling centers. About all most people have to do to recycle glass bottles and jars is to carry their recycling bin to the curb, or maybe drop off their empty glass containers at a nearby collection point. If you need an extra incentive to recycle glass, how about this: Several US states offer cash refunds for most glass bottles, so in some areas glass recycling can actually put a little extra money in your pocket.

How to Make Lamps and Vases from Glass Bottles?

All you need to convert an old bottle into a lamp is a bottle adapter lamp kit which you can find at your local hardware/lighting store, and a lamp shade. The easiest method is to leave the lamp cord running from the socket instead of being concealed in the base.

Your kit should include drilled corks in different sizes, a spindle, harp, sock with cover, lamp cord with plug, and nuts and washers to put them all together. If the bottle is tall and slender you will want to fill it with marbles, small stones, shells, sand, etc. to give it some ballast. Another option is to detach the harp from the socket assembly and use a lamp shade that will snap directly over the light bulb. That will be the lamp's center of gravity.

Select the drilled cork that fits your bottle neck. Insert the spindle and secure washers and knurled nuts on both ends. Then twist the cork into the bottle. Place the harp cradle over the spindle. Screw the socket base over the spindle and tighten the setscrew. Insert the cord through the socket base and strip the ends. Attach the wire ends to the socket connections, wrapping clockwise before tightening the terminal screws. Install the socket cover and snap the assembly into the socket base. Unscrew the finial on the harp, place a shade over the spindle, and replace the finial.

To hide the cord inside the bottle: With a hacksaw, cut a 4″ length from the threaded part of the spindle. Screw it into the hole in the metal arm attached to the socket, leaving enough room for the wires that will pass through. Push the bottom of the spindle through a 1″ washer, the drilled cork, and another washer. Secure with a nut. Drill a hole in the base of the bottle. Push the sripped end of the electric cord through the hole and up through the neck of the bottle, then through bottom of the spindle and out the top. Connect the wires to the screws on the socket. Pull the wire taut and replace the socket cover.

How to cut glass without a diamond cutter?

Method 1: Dip a piece of common string in alcohol or kerosene and squeeze dry or as dry as it will get without dripping. This string should then be placed on the already marked glass and tied tight. Light the string and let it burn off. Immediately, while the glass is still hot, plunge it into cold water. Be sure the container of water is large enough to let the glass go completely under as well as your arm up to the elbow, so as to deaden the vibration when you strike the glass. Strike the glass with your other hand outside the line of cutting using a stick of wood and hitting a sharp stroke. This quick, sharp stroke will break the glass where it has been weakened by the burning string into a clean cut as if done by a regular glass cutter. This method may be used to cut bottles in any shape and to make vases and to perform many such cuttings on glass.

Method 2: Here is a method that rarely fails to break the glass clean in the place you want it broke. First, scratch the glass with the corner of a file or sharp graver. Have a piece of wire bent to the desired shape you want to cut the glass. Heat the wire red hot and lay it upon the scratch. Sink the glass into cold water just deep enough to come on a level with the wire, not quite covering it. The glass will break clean.

How to cut glass with scissors?

To do this you must place the glass under water completely, then with a pair of ordinary scissors, proceed to cut the glass as you would paper or cloth. This method is, of course, not as smooth as job as the methods described above. The edges will not be as smooth, but for getting a piece of glass down to size and where the edges are not needed to be smooth, this method is satisfactory.

How to drill glass?

Get a piece of steel wire and file to shape of drill. This must be tempered as follows: Heat the end of the drill on a flame until it is dull red, then place it in metallic mercury. This drill, tempered in this manner will bore through glass as easily as through soft metal. When using in glass, always use oil of turpentine with a little camphor added to lubricate the drill. As you drill, be careful not to drill clear through from one side as you will break the glass this way. Drill partly, or almost through, then start from the opposite side and finish the hole. Or if you cannot do this, as when you are drilling bottles, etc. fill this bottle with water or place the glass in water.

Chapter 11

Recycling of Metals

INTRODUCTION

Metals play an important part in modern societies and have historically been linked with industrial development and improved living standards. Society can draw on metal resources from earth's crust as well as from metal discarded after use in the economy. Inefficient recovery of metals from the economy increases reliance on primary resources and can impact nature by increasing the dispersion of metals in ecosystems. Though the practice of recovering metals for their value dates back to ancient civilisations, today, the protection of earth's resource endowments and ecosystems adds to the incentive for recovering metals after use.

Industrial society values metals for their many useful properties. Their strength makes them the preferred material to provide structure, as girders for buildings, rails for trains, chassis for automobiles, and containers for liquids. Metals are also uniquely suited to conduct heat (heat exchangers) and electricity (wires), functions that are indispensable to industrial economies. Finally, metals and their compounds are used for their chemical properties as catalysts for chemical reactions, additives to glass, electrodes in batteries, and many other applications. The basic and unique properties of metals, including the ability to work them into complex shapes (i.e. ductility), insure that long-term demand for metals will certainly grow. Opinions on long-term metals demand range from predictions that growth in demand will pace the global economy to the position that the ascent of knowledge-based industries as economic drivers, competition from other materials, greater consumption of lighter more sophisticated metal products, and more efficient use in the economy will slow the rate of future growth.

Metals can be recycled nearly indefinitely. Unlike polymer plastics, the properties of metals can be restored fully, though not always easily, regardless of their chemical or physical form. Nevertheless, the ability to recover metals economically after use is largely a function of how they are used initially in the economy and their chemical reactivity. The success of secondary metals markets depends on the cost of retrieving and processing metals embedded in abandoned structures, discarded products, and other waste streams and its relation to primary metal prices.

Demand for scrap metals depends on industry structure and the availability of production technologies that accommodate scrap feeds to yield value added products. The New England study concluded that 95 per cent or more of the recycled metal remains within the scrap system. Despite this apparent high efficiency, the 5 per cent losses compounded over decades introduce a significant amount of lost metal either bound in scrap heaps or dispersed into the environment. The model of industrial ecology emphasises the containment and reuse of wastes generated by society as an overarching guideline for improving environmental quality. To realise this model, industry and society should work together to recover metals by recirculating metal from all secondary sources and losing a minimum amount of material from the industrial/social system.

FERROUS METALS

Ferrous metals are able to be recycled with steel being one of the most recycled materials in the world, Ferrous metals contain an appreciable percentage of iron and the addition of carbon and other substances creates steel. The most commonly recycled items are containers, cans, automobiles, appliances, construction materials, structural steel and automobiles. A typical appliance is about 75 per cent steel by weight and automobiles are about 65 per cent steel and iron.

The steel industry has been actively recycling for more than 150 years, in large part because it is economically advantageous to do so. It is cheaper to recycle steel than to mine iron ore and manipulate it through the production process to form new steel. Steel does not lose any of its inherent physical properties during the recycling process, and has drastically reduced energy and material requirements compared with refinement from iron ore. The energy saved by recycling reduces the annual energy consumption of the industry by about 75 per cent, which is enough to power eighteen million homes for one year.

Basic oxygen steelmaking (BOS) uses between 25 and 35 per cent recycled steel to make new steel. BOS steel usually has less residual elements in it, such as copper, nickel and molybdenum and is therefore more malleable than electric arc furnace (EAF) steel so it is often used to make automotive fenders, soup cans, industrial drums or any product with a large degree of cold working. EAF steelmaking uses almost 100 per cent recycled steel. This steel contains more residual elements that cannot be removed through the application of oxygen and lime so it is used to make structural beams, plates, reinforcing bar and other products that require little cold working. Downcycling of steel by hard to separate impurities such as copper or tin can only be prevented by well-aimed scrap selection or dilution by pure steel. Recycling 1000 kilograms of steel saves 1100 kilograms of iron ore, 630 kilograms of coal, and 55 kilograms of limestone.

IRON AND STEEL RECYCLING

Consumption of iron and steel scrap and the health of the scrap industry depend directly on the health of the steelmaking industry. The United States, as well as most of the world, is expected to consume increasing amounts of scrap as a steadily increasing population demands more steel products. World resources of scrap should be sufficient for the foreseeable future.

Iron, which includes its refined product steel, is the most widely used of all the metals. Iron and steel products are used in many construction and industrial applications, such as appliances, bridges, buildings, containers, highways, machinery, tools, and vehicles. The recycling of iron and steel scrap (ferrous scrap) is an important activity worldwide. Obsolete iron and steel products and the ferrous scrap generated in steel mills and steel-product manufacturing plants are collected because it is economically advantageous to recycle iron and steel products by melting and recasting them into semifinished forms for use in the manufacture of new steel products. The steel scrap market is mature and highly efficient.

Iron and steel scrap is more than just economically beneficial to steelmakers; ferrous scrap recycling is part of wise management of iron resources. Recovery of 1 metric ton of steel from scrap conserves an estimated 1030 kilograms (kg) of iron ore, 580 kg of coal, and 50 kg of limestone. Each year, steel recycling saves the energy equivalent required to electrically power about one-fifth of the households in the United States (about 18 million homes) for 1 year (Steel Recycling Institute). In the production of steel, 99.9 per cent of scrap melted is consumed in the new steel while producing negligible environmentally undesirable waste. This materials flow study, as summarised in Fig. 11.1, describes the materials cycle of pig iron, direct reduced iron (DRI), and scrap used in the manufacture of iron and steel products; the recycling of scrap; and the losses of iron and steel during the steelmaking and product fabrication processes. The flow diagram shows the quantities of iron present at stages of steel product manufacture, shipping, and recycling. In a free-market economy, scrap prices react quickly to changes in supply and, especially, demand. When demand for steel mill and foundry products is low, demand for scrap is low, and prices fall. Dealers cannot influence sales of scrap if mills and foundries do not need it to charge their furnaces. Although prices of scrap depend upon the market conditions for new products, the scrap industry uses inventory to absorb price differentials; that is, inventories increase as scrap prices decrease. Prices are also influenced by technological changes in mills, processing of scrap, the use of scrap substitutes, environmental controls and other Government regulations, and export demand.

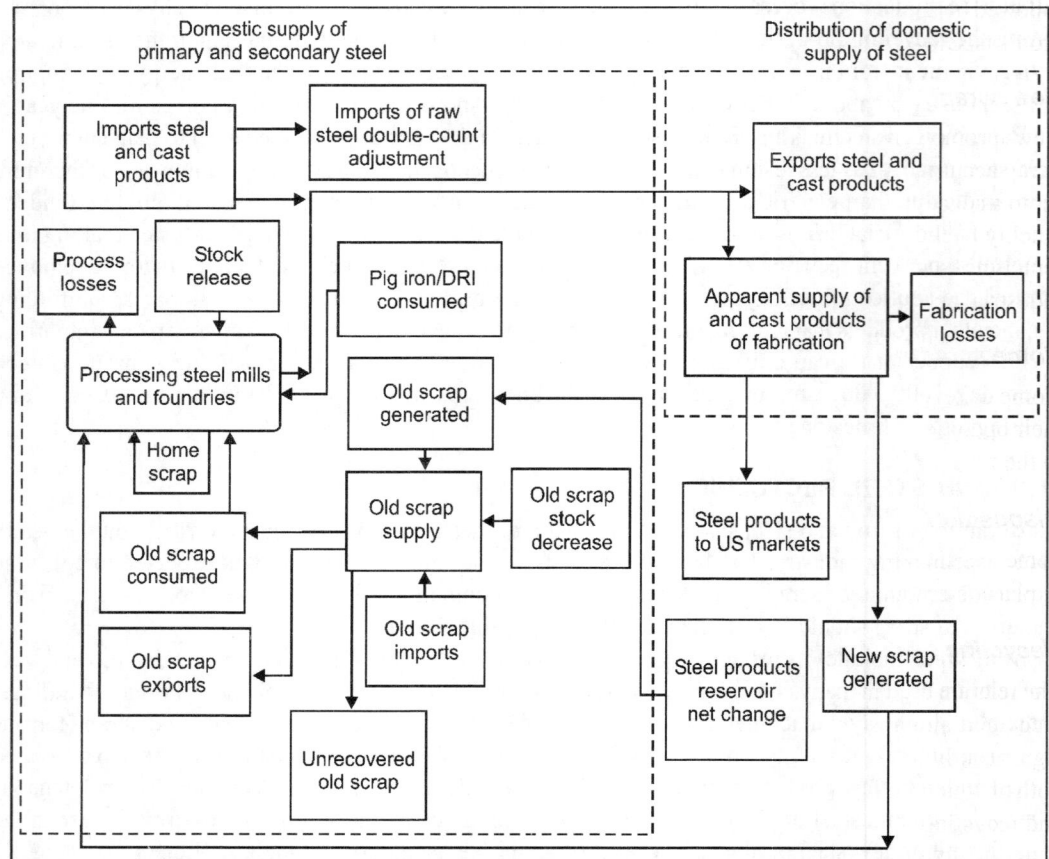

Fig. 11.1. Steel scrap material flow (Numbers are in million metric tons; DRI, direct reduced iron).

Sources of Iron and Steel Scrap

Sources of different types of scrap are key features of the flow diagram (Fig. 11.1). Ferrous scrap available for recycling comprises home, new, and old scrap. Home or mill, scrap is generated within the steel mill during production of iron and steel. Trimmings of mill products and defective products are collected and quickly recycled back into the steel furnace because their chemical compositions are known. The availability of home scrap has been declining as new and more-efficient methods of casting have been adopted by the industry. Old scrap includes metal articles that have been discarded after serving a useful purpose. Because of the wide variety of chemical and physical characteristics, old scrap often requires significant preparation, such as sorting, de-tinning, and de-zincing, prior to consumption in mills. Statistical data for domestic consumption of home and old scrap were collected by the US Geological Survey (USGS). The amount of new scrap is an estimated 15 per cent of apparent consumption of steel mill products.

Old scrap generated

Old, obsolete, or postconsumer scrap are available for recycling. The largest source is junked automobiles followed by appliances, machinery, worn out railroad cars and tracks, demolished steel structures, and other products. Old scrap estimated as available for recycling were products in circulation that became obsolete.

New scrap

New, prompt, or industrial scrap is generated from manufacturing plants that make steel products. Scrap accumulates when steel is cut, drawn, extruded, or machined. The casting process also produces scrap as excess metal. Its chemical and physical characteristics are known, and it is usually transported quickly back to steel plants through scrap processors and dealers or directly back to the steel plant for remelting to avoid storage space and inventory control costs. The supply of new scrap is a function of industrial activity. When activity is high, more industrial scrap is generated.

Home scrap

Home or revert, scrap consists of scrap that is produced in steel mills and foundries as a by-product of their operations, as well as old plant scrap. This scrap has a known composition and is always recycled to the furnace for remelting.

Disposition of iron and steel scrap

Some scrap is lost to the environment and is unrecoverable. Scrap that is discarded to scrap yards or abandoned in place is considered to be temporarily unrecoverable and may be recycled at some future date.

Recycling efficiency for old scrap

The relation between the amount of scrap that is theoretically available for recycling and what is actually recovered and reused is called recycling efficiency. Recycling efficiency is not expected to increase significantly because ferrous scrap competes with direct reduced iron and pig iron as a raw material, both of which are readily available and tend to hold down scrap prices, thereby limiting scrap availability and recycling efficiency.

Processing of iron and steel scrap

Using a variety of equipment, scrap dealers collect and process scrap into a physical form and chemical composition that steel mill furnaces can consume. The type and size of equipment they use depends on the types and volume of scrap available in the area and the requirements of their customers. The largest and most expensive piece of equipment is the shredder. The shredder can fragment vehicles and other discarded steel objects into fist-size pieces of various metals, glass, rubber, and plastic. These materials are segregated before shipment by using fans, magnets, air ducts, hand pickers, and flotation equipment. Hydraulic shears, which have cutting knives of chromium-nickel-molybdenum alloy steel for hardness, slice heavy pieces of ship plate, railroad car sides, and structural steel into chargeable pieces. Baling presses are used to compact scrap into manageable bundles thereby reducing scrap volume and shipping costs. Scrap dealers must carefully sort the scrap they sell, and steelmakers must be careful to purchase scrap that does not contain unacceptable levels of undesirable elements.

Steel mills melt scrap in basic oxygen furnaces (BOF), electric arc furnaces (EAF), and, to a minor extent, blast furnaces. The proportion of scrap in the charge in a BOF is limited to less than 30 per cent, whereas that in an EAF can be as much as 100 per cent. Steel and iron foundries use scrap in EAF's and cupola furnaces.

Fabrication of new steel products produces new steel scrap that is relatively clean, chemically and physically, and of known chemical composition. For this reason, most scrap consumers prefer new scrap to old scrap. Preparation of new scrap is usually limited to cutting, cleaning, and baling prior to rapid transport back to the steelmaker for recycling.

Summary and Outlook for Ferrous Scrap Recycling Flow

Consumption of ferrous scrap and the growth or decline of the scrap industry depends directly on the health of the steelmaking industry. Most regions of the world will see a marked increase in steel consumption during the next 5 years, according to the International Iron and Steel Institute. In the United States, a steadily increasing population and a growing economy in the long-term should assure that demand for steel products, and the scrap used to make them, will also increase. Steel and scrap consumption will continue to be in strong demand in the automotive and consumer appliance sectors. New highway and bridge projects supported by increased Federal funding will require structural and reinforcement bar products. The use of steel framing is increasing in the construction of multifamily developments, retirement homes, and single-family residences. Steel demand in can production, which includes aerosol cans, food, and paint, should remain strong in the long-term. A thriving steel industry is dependant on plentiful inexpensive energy. As energy costs increase, the demand for steel pipe and tubular goods used in oil and gas industry will increase for new drilling projects. Foundries are an important market for tin-bearing scrap from recycled cans.

The EAF contribution to the total production of steel has risen dramatically, and the proportion of EAF steel produced should continue to increase, perhaps at a rate of 4 per cent per year during the next 10 years. The EAF may be the primary steel production method in the world by 2010. The EAF has evolved in minimills from the small unit limited in use for speciality steel production to the large-capacity unit used to produce a wide range of steels, which includes flat product sheet and plate, long product bars, structural shapes, tubulars, and wire.

The availability of scrap and operating and capital cost advantages have made EAF growth possible. Locations of new minimills in areas of increasing population growth and manufacturing activity in the Southern and Western States, and away from the traditional 'rust belt' States have, to a large extent, satisfied demand for construction steel products and products used by the oil and gas industry. The EAF process is flexible in its raw material requirements and sources and can operate with considerable flexibility in making products depending on market requirements. Steelmaking by the EAF will continue to grow because of the capital and operating cost advantages relative to those of the BOF, an increasingly wide range of steel products that it makes, and its environmental cleanliness. The use of the EAF is the most effective way of reducing carbon dioxide emissions because of the lower energy needed to melt scrap than to smelt ore. Use of EAF's will increase as new minimills are built and EAF's may replace operating BOF's. Ferrous scrap will remain the most important raw material used, but reduced iron in the form of DRI and hot briquetted iron (HBI) will become a larger component in the raw materials mix for EAF steel production. Increasing availability from domestic producers of DRI and HBI will be a factor in this trend as will the increasing need for low-residual feedstock for the production of high-quality flat steel and special-bar quality steels required to compete in the higher end markets.

E-commerce has been making progress at different rates in the steel industry, as well as many sectors of the economy. Selling scrap through e-commerce has potential but adoption of e-commerce marketing by the ferrous scrap industry will probably be a slow evolutionary process. For e-commerce to be adopted, scrap processors, dealers, and brokers must realise the value of this new Internet technology and see potential gains in profits and efficiency.

Recycling a large amount of scrap steel reduces the total energy needed to produce steel. Nevertheless, the steel industry uses about 3 per cent of the energy consumed in the United States and more than 10 per cent of that used by the industrial sector. Energy purchases represent nearly 20 per cent of the total manufacturing cost of steel. Of particular concern is the recent trend toward declining availability of and higher prices for oil, gas, and electricity. To control these significant production costs, the industry will use and promote, as much as possible, new technologies to conserve energy, such as scrap preheating. Scrap preheating may increase threefold in new furnaces owing to energy conservation, shorter cycle times, and reduced operating costs. Another technology advance is the development of strip casting that reduces energy usage by as much as 50 per cent by casting steel into its final thickness and shape with minimal further hot or cold rolling. Of immense concern to the scrap and steelmaking industries is the threat of accidental melting of radioactive scrap. Steel mills that receive ferrous scrap have been exposed to radioactive materials without warning. Such accidents can be extraordinarily expensive to steelmakers.

Recycling Steel and Iron Used in Automobiles

Steel's importance in automobiles

We rely on automobiles to transport us from place to place. We also rely on automobiles to keep us safe. Fortunately auto manufacturers depend on steel to protect their customers. In addition to its strength, durability and dependability, steel is also recyclable and contains recycled steel.

Recycling efforts

Automobiles are the most recycled consumer product. Each year, the steel industry recycles more than 14 million tons of steel from end-of-life vehicles. This is equivalent to nearly 13.5 million automobiles. When comparing the amount of steel recycled from automobiles each year to the amount of steel used to produce new automobiles that same year, automobiles maintain a recycling rate of nearly 100 per cent.

Recycled content of automobiles

By weight, the typical passenger car consists of about 65 per cent steel and iron. The steel used in car bodies is made with about 25 per cent recycled steel. Many internal steel and iron parts are made using even higher percentages of recycled steel. All steel products contain recycled steel because steel scrap is a necessary ingredient in the production of new steel. Steel scrap is derived not only from automobiles but also from steel cans, appliances and construction material.

Basics of recycling automobiles

Old cars are typically hauled to an automobile dismantler, where reusable parts are removed. After removing the reusable parts and other items like batteries, tyres and fluids, the hulks are usually shipped to ferrous scrap processors where they are weighed for payment and unloaded. At a scrap yard, the automobiles enter the shredder. The shredding process, which typically handles one car every 45 seconds, generates three streams: iron and steel; nonferrous metal; and fluff (fabric, rubber, glass, etc.). The iron and steel are magnetically separated from the other materials and recycled. The iron and steel is then shipped to end markets or steel mills where it is recycled to produce new steel.

Environmental benefits

Recycling steel saves energy and natural resources. The US steel industry alone annually saves the equivalent energy to power about 18 million households for a year. Recycling one ton of steel conserves 1134 kilograms of iron ore, 635 kilograms of coal and 54 kilograms of limestone.

NON-FERROUS METAL RECYCLING

Aluminium Recycling

Aluminium recycling is the process by which scrap aluminium can be reused in products after its initial production. The process involves simply re-melting the metal, which is far less expensive and energy intensive than creating new aluminium through the electrolysis of aluminium oxide (Al_2O_3), which must first be mined from bauxite ore and then refined using the Bayer process. Recycling scrap aluminium requires only 5 per cent of the energy used to make new aluminium. For this reason, approximately 31 per cent of all aluminium produced in the United States comes from recycled scrap.

A common practice since the early 1900s and extensively capitalised during World War II, aluminium recycling is not new. It was, however, a low-profile activity until the late 1960s when the exploding popularity of aluminium beverage cans finally placed recycling into the public consciousness. Sources for recycled aluminium include aircraft, automobiles, bicycles, boats, computers, cookware, gutters, siding, wire, and many other products that require a strong light weight material or a material with high thermal conductivity. As recycling does not damage the metal's structure, aluminium can be recycled indefinitely and still be used to produce any product for which new aluminium could have been used.

Advantages

The recycling of aluminium generally produces significant cost savings over the production of new aluminium even when the cost of collection, separation and recycling are taken into account. Over the long-term, even larger national savings are made when the reduction in the capital costs associated with landfills, mines and international shipping of raw aluminium are considered.

Energy savings

Recycling aluminium uses about 5 per cent of the energy required to create aluminium from bauxite, because the latter requires a lot of electrical energy to electrolyse aluminium oxide into aluminium. Just how much is vividly shown when aluminium oxidises, in thermite and ammonium perchlorate composite propellant.

Environmental savings

If energy directly equated to carbon dioxide, then recycled aluminium could be said to create 5 per cent of the carbon dioxide produced in the creation from raw materials. In practice, this cannot be assumed.

Electrolysis can be done by electricity from non-fossil-fuel sources, such as nuclear, geothermal, hydroelectric or solar. Aluminium production is attracted to sources of cheap electricity. Canada, Brazil, Norway, and Venezuela have 61 to 99 per cent hydroelectric power, and are major aluminium producers.

The vast amount of aluminium used means that even small percentage losses are large expenses, so the flow of material is well monitored and accounted for financial reasons. Efficient production and recycling benefits the environment as well.

Process of Recycling Aluminium Beverage Cans

Aluminium beverage cans are usually recycled in the following basic way:

1. Cans are first divided from municipal waste, usually through an eddy current separator.
2. Cans are cut into little, equal pieces to lessen the volume and make it easier for the machines which separate them.
3. Pieces are cleaned chemically/mechanically.
4. Pieces are blocked to minimise oxidation losses when melted (The surface of aluminium readily oxidises back into aluminium oxide when exposed to oxygen).
5. Blocks are loaded into the furnace and heated to $750°C \pm 100°C$ to produce molten aluminium.
6. Dross is removed and the dissolved hydrogen is degassed. (Molten aluminium readily disassociates hydrogen from water vapour and hydrocarbon contaminants.) This is typically done with chlorine and nitrogen gas. Hexachloroethane tablets are normally used as the source for chlorine. Ammonium perchlorate can also be used, as it decomposes mainly into chlorine, nitrogen, and oxygen when heated.
7. Samples are taken for spectroscopic analysis. Depending on the final product desired, high purity aluminium, copper, zinc, manganese, silicon, and/or magnesium is added to alter the molten composition to the proper alloy specification. The top 5 aluminium alloys produced are apparently 6061, 7075, 1100, 6063, and 2024.
8. The furnace is tapped, the molten aluminium poured out, and the process is repeated again for the next batch. Depending on the end product it may be cast into ingots, billets, or rods, formed into large slabs for rolling, atomised into powder, sent to an extruder or transported in its molten state to manufacturing facilities for further processing.

Ingot Production Using Reverberatory Furnaces

The scrap aluminium is separated into a range of categories, i.e. irony aluminium (engine blocks, etc.), alloy wheels, 'clean aluminium' depending on the specification of the required ingot casting will depend on the type of scrap used in the start melt. Generally the scrap is charged to a reverberatory furnace (other methods appear to be either less economical and/or dangerous) and melted down to form a 'bath', the molten metal is tested using spectroscopy on a sample taken from the melt to determine what refinements are needed to produce the final casts. After the refinements have been added the melt may be tested several times to be able to fine tune the batch to the specific standard. Once the correct 'recipe' of metal is available the furnace is tapped and poured into ingot moulds, usually via a casting machine. The melt is then left to cool, stacked and sold on as cast silicon aluminium ingot to various industries for reuse.

Secondary Aluminium Recycling

White dross from primary aluminium production and from secondary recycling operations still contains useful quantities of aluminium which can be extracted industrially. The process produces aluminium billets, together with a highly complex waste material. This waste is difficult to manage. It reacts with water, releasing a mixture of gases (including, among others, hydrogen, acetylene, and ammonia) which spontaneously ignites on contact with air; contact with damp air results in the release of copious quantities of ammonia gas. Despite these difficulties, however, the waste has found use as a filler in asphalt and concrete.

Recycling Aluminum Aerospace Alloys

For decades, thousands of obsolete aircraft have been sitting in 'graveyards', while the demand for recycled aluminum continues to increase. The discarded aircraft provide a large source of valuable metal. However cost-effective recycling of aircraft alloys is complex because aircraft alloys are: (i) typically relatively high in alloying elements, and (ii) contain very low levels of impurities to optimise toughness and other performance characteristics.

Thus recycling of aluminum aerospace alloys represents a major challenge to both the aluminum and aerospace industries. While the recycling of high percentages of aluminum from packaging and automotive applications has been commercialised and become economically attractive, the unique compositions and performance requirements of aerospace alloys have resulted in delaying directly addressing techniques for cost-effectively recycling those alloys.

The purpose of the study described herein was to identify the most attractive means of cost-effectively recycling aluminum alloys used in the production of private, civil, and military aircraft. We will endeavour to define and illustrate the practicality of any new approaches needed, as well as identify opportunities for related alloy development and product evaluation that might make aerospace structure recycling even more attractive.

Aluminum remains the most economically attractive material from which to make aircraft and space vehicles, and new construction proceeds at a prodigious rate. However, the development of newer aircraft structures has proceeded at such a pace that thousands of obsolete civil and military aircraft stand idle in 'graveyards' around the USA. Yet it has been impractical to reuse the metal in these planes because of the combination of the differences in compositions of older obsolete aircraft and those of new aircraft, often having special performance requirements requiring specialised alloy compositions.

Driving forces for this study

The driving forces for this study to enable large-scale recycling of aluminum aircraft alloys are very clear and very strong:

1. The production of aluminum as 'secondary metal' (i.e. producing it by recycling) requires only about 2.8 kWh/kg of metal produced while primary aluminum production requires about 45 kWh/kg of metal produced. The 95 per cent energy saving are a powerful economic incentive.
2. The ecological driving force is great too, as recycling results in the emission of only about 4 per cent as much CO_2 as does primary production. Clearly, it is in the industry's and in the nation's best advantage to maximise the amount of recycled metal that can be regained from obsolete aircraft.

Ideal aircraft recycling process

Today, in order to meet the performance requirements of aerospace alloy and product specifications, all alloys are produced utilising primary metal. Typically the specifications for these alloys require that such strict controls on impurities are maintained that recycled metal cannot be used without additional processing. It is timely to consider a new paradigm in which obsolete aircraft, like obsolete beverage cans and obsolete automotive vehicles, are recognised as valuable sources of aluminum, and an appropriate new commercial scenario developed.

The most desirable recycling scenario would include the following:

1. An aircraft-recycling center would be established, and as aircraft become obsolete they would be flown or delivered to this facility.
2. To the degree feasible economically, the major components of the aircraft would be disassembles and major non-aluminum components would be removed.
3. To extent readily practical, the aluminum aircraft components would be pre-sorted by alloy type, most importantly by 2xxx and 7xxx series alloys.
4. The remaining structure would be automatically shredded, sorted, and remelted to provide metal in the most valuable form for reuse.
5. The recycled metal would be cast into ingot or billet of one of a useful set of high-strength aluminum alloy compositions available for a wide variety of nonfracture critical aerospace components, and subsequently fabricated into new end products that meet established performance requirements.

Challenges in achieving cost-effective aircraft recycling

The principal challenges that must be dealt with in creating this ideal aircraft-recycling scenario include the following:

1. Identifying decision options for dismantling aircraft to simplify recycling.
2. Identifying and optimising technologies for automated shredding, sorting, and remelting of those 2xxx and 7xxx alloys with relatively high levels of alloying elements (sometimes in excess of 10 per cent).
3. Identifying the range of representative compositions likely to be obtained from recycling aircraft components.
4. Identifying the combination of performance requirements and compositions that would make useful aircraft components from recycled metal, even though they may not achieve the highest achievable levels of toughness.
5. Identifying useful by-products to handle elemental residual unable to be used in recycled metal, e.g. iron.

Some useful progress has already being made in addressing items 2 and 5 on this list of challenges. Examples include laser-induced breakdown spectroscopy (LIBS), developed and applied by Huron Valley Steel Corp. (HVSC), which is already being applied to the shredding and sorting of some aluminum alloys, and the use of high iron containing aluminum for deoxidising steel (de-ox).

In general, however, little or nothing has yet been done to apply recycling technology to shredding, sorting, and reuse of recycled metal from obsolete aircraft and space vehicle components. Focusing upon that area was the purpose of this study.

Some options for consideration

In the following sections, we will consider:

1. Dismantling and pre-sorting strategies.
2. Automated shredding, sorting, and remelting.
3. Identifying the resulting compositions of recycling aircraft components.
4. Options for reuse of the metal from recycled aircraft components in new aircraft.
5. Options for reuse of the metal from recycled aircraft components in non-aircraft applications.
6. Options for reuse of the metal from recycled aircraft components in aluminum castings.

Dismantling and pre-sorting strategies

One of the first things to be considered is the degree to which dismantling prior to shredding is helpful and cost-effective. And assuming that it may be, what are the most useful strategies that might be employed. To a large extent, aircraft alloys fall into two series, the Al-Cu or 2xxx series and the Al-Zn-Mg or 7xxx series. While automated sorting techniques applied after shredding will unquestionably work, anything that can be readily done to pre-sort those alloys would be helpful.

One technique that seems practical would be to dismantle aircraft into certain logical component groups, as these typically are made of similar alloys of the same series. As example, landing gears, engine nacelles, tail sections, and flaps could be presorted, and wings separated from fuselages. Such separations may be desirable anyway to permit removal of non-aluminum components before shredding.

Guidance in such dismantling and pre-sorting should be available from the aircraft manufacturers who can identify the alloys used in various components of specific aircraft produced over the years. The availability of such manufacturer information will be very useful in establishing procedures for dismantling aircraft components. An additional consideration, especially if/when manufacturing information is not available, is that it may be possible to take the approach of identifying the various metallic constituents and their chemistry prior to dismantling the aircraft with devices like handheld mobile spectrometers. Such devices are available that may be used for *in situ* identification of alloy types of specific components. Manufacturers include Vericheck Technical Services, Thermo Electron and Spectro. Non-aluminum components may also be readily identified using this technique.

Automated shredding and sorting

The remainder of this study assumes that the laser induced breakdown spectroscopy (LIBS) technology developed and applied by Huron Valley Steel Corp. (HVSC), scaled up for handling large aircraft components, will provide useful sorting of the aircraft alloys regardless of the degree to which they are presorted. The challenge is determining more-precisely the level of sorting possible: for example, is the process sufficiently discriminating to separate 2014 from 2024? Or separate 7055, 7075, and 7085, for example?

An even more challenging feature that will influence reuse of the recycled metal is to what degree will the LIBS process sort alloys of different impurity levels, defined for example as the level of Fe, e.g. 2024 from 2124 and 2324 or 7075 from 7175 and 7475. The higher toughness alloys 2124, 7175 and 7475 typically have Fe levels in the 0.05–0.20 per cent range while 2024 and 7075 have Fe in the range of 0.35–0.50 per cent. If such differences can be sorted automatically at high speed for an element like Fe, that would greatly add to the flexibility and cost-effectiveness of the reuse of recycled metal.

Identifying the compositions of recycled aircraft alloys

Dependent upon the levels of discrimination achievable in shredding and sorting, a variety of different compositions may result from these operations. A more detailed study is needed to define the representative levels, including some in which sorting is minimal and others where it is the best achievable. This will lay the groundwork for extended studies of reuse of the metal.

Of several basic things that we can be certain of: the metal from recycled 2xxx alloys will be high in Cu, Mg, Mn and Si and the metal from 7xxx alloys will be high in Zn, Cu, and Mg. The compositions of some alloys used for many years in aircraft structures are shown in Table 11.1; in older aircraft likely to be found in many graveyards, 2024 has been the most widely used 2xxx alloy, 7075 the most widely used of the 7xxx series. Newer aircraft will have more high-purity alloys like 2124, 2324, 7050, 7175, and 7475.

Table 11.1. Nominal Compositions of Some 2xxx and 7xxx Alloys.

Alloy	Al	Cu	Fe	Mg	Mn	Si	Zn
2014	~93	4.4	0.7 max	0.50	0.8	0.8	0.15 max
2214	~93	4.4	0.3 max	0.50	0.8	0.8	0.15 max
2024	~93	4.4	0.5 max	1.5	0.6	0.5 max	0.25 max
2324	~94	4.1	0.12 max	1.5	0.6	0.1 max	0.15 max
7050	~89	2.3	0.15 max	2.2	0.1 max	0.12 max	6.2
7075	~90	1.6	0.5 max	2.5	0.3 max	0.4 max	5.6
7475	~90	1.6	0.12 max	2.2	0.06 max	0.1 max	5.7
7178	~89	2.0	0.5 max	2.8	0.3 max	0.4 max	6.8

While a high level of discrimination and thus of sorting may possibly be achieved based upon Fe levels in recycled aircraft components, it is more logical at this point to assume that sorting is limited to identifying only 2xxx and 7xxx series alloys, in which case the compositions of recycled metal are likely to represent something like the following in Table 11.2.

Table 11.2. Potential compositions of some recycled aircraft alloys assuming pre-sorting.

Alloy	Al	Cu	Fe	Mg	Mn	Si	Zn	Others
R2xxx	~93	4.4	0.5	1.0	0.7	0.5	0.1	0.2
R7xxx	~90	2.0	0.4	2.5	0.2	0.2	6.0	0.2

If this is correct and if 2xxx and 7xxx alloys can be sorted successfully leading to compositions as in Table 11.2, there would appear to be some opportunities to reuse the recycled metal in a 2024 like alloy from the former and a 7075 type alloy from the latter. The properties of these alloys are likely to resemble those of 2024 and 7075, and subject to more thorough performance evaluation there is every reason to conclude that such metal might be utilised in non-fracture-critical aerospace components.

If, on the other hand, 2xxx and 7xxx alloys cannot be separated before melting, and assuming an approximately equal amount of 2xxx and 7xx alloys, the recycled metal composition is more likely to look like that in Table 11.3.

Table 11.3. Potential composition of some recycled aircraft alloys assuming no pre-sorting.

Alloy	Al	Cu	Fe	Mg	Mn	Si	Zn
R2 + 7xx	~92	3.0	0.4	1.8	0.4	0.4	3.0

The characteristics of this composition are difficult to estimate as they do not match any existing registered alloy.

Two other factors should be noted at this stage:

1. First, it should be recognised that at least small quantities of several other wrought aluminum alloys like 2219 and 6061 and cast alloys like 201.0, A356.0, and A357.0 may go into the recycle mix.
2. Second, aircraft alloys typically have grain-refining elements such as Cr, Zr, and V present in small quantities (~ 0.1 per cent or less), and the potential buildup of such elements in addition to Fe, Mg, and Si needs to be the subject of further study. This second factor will become increasingly important as newer aircraft employing later alloys such as 2124, 2048, 7050, 7055, and 7085 enter the obsolete mix.

A further in-depth analysis and mass balance is needed to more precisely determine the compositions of recycled aircraft metal.

Options for reuse of the recycled aircraft components in aircraft

Assuming that the estimated composition in Table 11.2 based upon pre-sorting of 2xxx and 7xxx alloys are reasonably correct, as noted above it would appear that the resultant alloys could be used for a number of non-critical aircraft components, such a stiffeners, flaps, and other relatively low-to-moderately stressed components made of sheet, plate or extrusions. These might be used in private, civil, and many military aircraft. Typically these would be components that are not designed based upon fracture mechanics concepts employing fatigue crack growth rates and fracture toughness parameters. The alloys utilised in fracture critical areas may still have to be fabricated using primary metal.

Therefore another key question to be addressed with further study is whether the percentage of aircraft components that do not require fracture-critical design is broad enough to justify the reuse of compositions likely to result from recycling.

Options for reuse of the recycled aircraft components in non-aircraft applications

If the use of these compositions for non-fracture critical components in new aircraft is too tightly limited, i.e. if the number of non-fracture critical components in civil and military aircraft is not large enough to justify reuse of the recycle compositions, it is useful to look at what other opportunities may exist for use of the compositions in Tables 11.2 and 11.3.

To aid in addressing that question, the compositions of several wrought 2xxx and 7xxx alloys used in other applications (including 2014, also an aircraft alloy) are presented in Table 11.4.

Table 11.4. Nominal compositions of some 2xxx and 7xxx alloys used in non-aerospace applications.

Alloy	Application	Cu	Fe	Mg	Mn	Si	Zn
2014	RR; truck bodies	4.4	0.7 max	0.50	0.8	0.8	0.15 max
2017	Rivets	4.0	0.7 max	0.60	0.7	0.5	0.25 max
7129	Auto bumpers	0.7	0.3 max	1.6	0.1 max	0.15 max	4.7

Comparing compositions in Tables 11.2 and 11.4 illustrates that it may not be too much of a stretch to reuse recycled aircraft metal in certain other products. Further study of this option is justified.

Options for reuse of the recycled aircraft components in castings

Other opportunities for the use of recycled metal from aircraft may include aluminum alloy castings, especially those of the 2xx.0 and 7xx.0 series, Al-Cu and Al-Zn respectively. Examples of some such alloys include the following in Table 11.5.

Table 11.5. Some aluminum casting alloy compositions.

Alloy	Al	Cu	Fe	Mg	Mn	Si	Zn	Others
201.0	~95	4.6	0.15 max	0.35	0.35	0.10	0.25	0.05 max
242.0	~94	4.1	0.6 max	1.4	0.10 max	0.6 max	0.10	0.05 max
295.0	~94	4.5	1.0 max	0.03	0.35 max	1.1	0.25	0.05 max
710.0	~93	0.5	0.5 max	0.7	0.05 max	0.15 max	6.5	0.05 max
713.0	~91	0.7	1.1 max	0.35	0.6 max	0.25	7.5	0.05 max

Even these relatively tolerant limits pose some challenge for direct use of recycled metal reuse. Nevertheless, some opportunity for study of some new alloy options remains, and the properties of the alloys in Table 11.2 when produced as casting should be studied.

Alternative products for use of excess alloying content

It is appropriate to address the fact that as aluminum alloys are recycled there is a trend for the Fe content to increase gradually, primarily through pickup from scrap handling equipment. While aircraft alloys may not be recycled very frequently, as are beverage cans, for example, it is a factor to consider.

With only a few exceptions, Fe is an impurity in all wrought alloys today, and is an ideal candidate for an alternative product if/when it exceeds desirable levels. An excellent example of such a product is the use of high Fe-bearing aluminum as a deoxidising agent for steel production. Maximisation of this capability will benefit both the aluminum and steel industries and add to the life cycle benefits to aluminum operations. Another possible approach to the increased Fe content is to make use of the affinity of Zr for Fe, resulting in a heavy particle that sinks to the bottom of crucibles during processing. Combining this Fe-Zr product with Mg and perhaps other undesirable impurities like Ni and V may also improve their impact on resulting recycle content.

Alloys designed with aircraft recycling in mind

As noted above, an ideal component of maximisation of resources in aircraft recycling would be the availability of several new aluminum alloys that would take advantage of the unique characteristics of recycled aircraft metal. Such an approach may call for some 'tailored' alloys, enabling broader specification limits on alloying elements likely to be found in recycled aircraft metal, notably the high Cu in 2xxx alloys and Zn in 7xxx alloys. Adopting the approach of alloy optimisation for aircraft recycling requires several steps, potentially phases in a development program:

1. Define the range of expected current and future recycled metal alloy content, utilising collaborative studies with organisations such as HVSC that are already capitalising on the economics of recycling. Perform a mass balance to the extent practical indicating the relative volumes of various scrap compositions to be expected.
2. Identify 3–4 basic candidate compositions of alloys that would accept recycled aircraft metal directly, including those listed in Table 11.2.
3. Determine the performance characteristics of such candidate alloys for a wide variety of applications, including non-critical aircraft components, structural components for bridges and buildings, high-temperature applications, and architectural usage including the following determinations:
   (a) Atmospheric corrosion resistance.
   (b) Stress-corrosion crack growth.
   (c) Toughness, with tear tests and/or fracture toughness tests (for thick sections).
   (d) Formability tests, with bulge, minimum bend and hemming tests.

Conclusions and Looking Ahead

There are strong economic and environmental driving forces for aggressively pursuing the recycling of obsolete aircraft, thousands of which exist throughout the world. Based upon the preliminary evaluations from the preliminary study conducted to date, it is recommended that the following steps be taken to establish the cost-effectiveness of recycling obsolete aircraft:

1. Conduct an in-depth study to determine the quantity and character of the obsolete aircraft readily available for recycling, including the alloys utilised in producing those aircraft.

2. Determine the practicality of disassembling and pre-sorting aircraft components by alloy and product form.
3. Conduct trials of the application of technologies such as LIBS for automatically sorting aircraft alloys collaboratively with an organisation like HVSC.
4. Carry out a mass balance to estimate from the results of Items 1 and 2 what likely and/or potential compositions may result from various approaches to shredding and sorting components from these aircraft.
5. Define a selection of 2–4 alloys based upon the foregoing mass balance that should be produced, either by a sample remelting of aircraft parts or in the laboratory, and their characteristics thoroughly evaluated, including:
    (a) Design strengths.
    (b) Atmospheric corrosion resistance.
    (c) Stress-corrosion crack growth.
    (d) Toughness, with tear tests and/or fracture toughness tests (for thick sections).
    (e) Formability tests, with bulge, minimum bend, and hemming tests.
6. Generate a representative set of applications for which the properties and performance generated in Item 5 would be adequate, including:
    (a) Non-fracture critical aircraft components.
    (b) Railroad and highway vehicle construction.
    (c) Highway structures.

RECYCLING OF COPPER

For thousands of years, copper and copper alloys have been recycled. This has been a normal economic practice. In the Middle Ages it was common that after a war the bronze cannons were melted down to make more useful items. In times of war even church bells were used to produce cannon.

The entire economy of the copper and copper alloy industry is dependent on the economic recycling of any surplus products. There is a wide range of copper based materials made for a large variety of applications. To use the most suitable and cheapest feedstock for making components gives the most economic cost price for the material.

Scrap Value—Copper

The usual commercial supplies of pure copper are used for the most critical of electrical applications such as the production of fine and superfine enamelled wires. It is essential that purity is reproducibly maintained in order to ensure high conductivity, consistent annealability and freedom from breaks during rod production and subsequent wire drawing. Since the applied enamel layers are thin but have to withstand voltage, they must have no surface flaws; consequently the basis copper wire must have an excellent surface quality. Primary copper of the best grade is used for producing the rod for this work. Uncontaminated recycled process scrap and other scrap that has been electrolytically refined back to grade 'A' quality may also be used.

The copper used for power cables is also drawn from high conductivity rod but to a thicker size than fine wires. The quality requirements are therefore slightly less stringent. The presence of any undesirable impurities can cause problems such as hot shortness which gives expensive failures during casting and hot rolling. For the same reason, scrap containing such impurities can only be used for this purpose if well diluted with good quality copper.

For non-electrical purposes, copper is also used to make large quantities of plumbing tube, roofing sheet and heat exchangers. High electrical conductivity is not mandatory and other quality requirements are not so onerous. Secondary copper can be used for the manufacture of these materials, though still within stipulated quality limits for impurities.

Where scrap copper is associated with other materials, for example after having been tinned or soldered, it will frequently be more economic to take advantage of such contamination than try to remove it by refining. Many specifications for gunmetals and bronzes require the presence of both tin and lead so this type of scrap is ideal feedstock. Normally it is remelted and cast to ingot of certified analysis before use in a foundry. Scrap of this type commands a lower price than uncontaminated copper.

Scrap Value—Brasses

The recycling of brass scrap is a basic essential of the economics of the industry. Brass for extrusion and hot stamping is normally made from a basic melt of scrap of similar composition adjusted by the addition of virgin copper or zinc as required to meet the specification before pouring. The use of brass scrap bought at a significantly lower price than the metal mixture price means that the cost of the fabricated brass is considerably less than it might otherwise be.

The presence in brass of some other elements such as lead is often required to improve machinability so such scrap is frequently acceptable. Besides the common free-machining brasses, there are many others made for special purposes with properties modified to give extra strength, hardness, corrosion resistance or other attributes, so strict segregation of scrap is essential.

Brass scrap arising from machining operations can be economically remelted but should be substantially free from excess lubricant, especially those including organic compounds that cause unacceptable fume during remelting. When brass is remelted, there is usually some evolution of the more volatile zinc. This is made up in the melt to bring it back within specification The zinc is evolved as oxide that is drawn off and trapped in a baghouse and recycled for the manufacture of other products.

Brass to be made into sheet, strip or wire form must be significantly free of harmful impurities in order to retain ductility when cold. It can then be rolled, drawn, deep drawn, swaged, riveted, spun or otherwise cold formed. It is normal therefore to make it substantially from virgin copper and zinc, together with process scrap arising from processing that has been kept clean, carefully segregated and identified.

Scrap Value—Other Copper Alloys

Copper alloys such as phosphor bronzes, gunmetals, leaded bronzes and aluminium bronzes are normally made to closely controlled specifications in order to ensure fitness for demanding service. They are normally made from ingots of guaranteed composition together with process scrap of the same composition that has been kept carefully segregated. Where scrap has become mixed or is of unknown composition, it is first remelted by an ingot maker and analysed so that the composition can be suitably adjusted to bring it within grade for an alloy.

Good quality high conductivity copper can be recycled by simple melting and check analysis before casting, either to finished shape or for subsequent fabrication. However, this normally only applies to process scrap arising within a copper works. Where copper has been contaminated and it is required to re-refine it, it is normally remelted and cast to anode shape so that it can be electrolytically refined. If, however, the level of impurities in the cast anode is significant, it is unlikely that the cathode produced will then meet the very high standards required of grade 'A' copper used for the production of fine wires.

Where copper and copper alloy scraps are very contaminated and unsuitable for simple remelting, they can be recycled by other means to recover the copper either as the metal or to give some of the many copper compounds essential for use in industry and agriculture. This is the usual practice for recovery of usable copper in slag, dross or mill scale arising from production processes or from life-expired assemblies of components containing useful quantities of copper.

Environmental considerations

Copper is an essential trace element needed for the healthy development of most plants, animals and human beings. In general, moderate excess quantities of copper are not known to cause problems. Every care is taken to avoid wasting copper and it is recycled where possible. Excess copper is not allowed to escape into the atmosphere as fume, nor into discharged process cooling water, all of which is generally treated to keep within agreed limits.

Other metals associated with copper alloys are generally not in a form that is dangerous. However, when fume is generated, for example by melting or welding, it may be necessary to use fume extraction equipment. Beryllium is sometimes used as an alloying element in copper to make some of the strongest copper alloys known, being invaluable for the production of heavy duty springs. When alloyed with copper and in the solid state this presents no health hazard. However, if present in the atmosphere, beryllium can cause a health hazard and should be controlled.

Product Value

If the scrap is pure copper and has not been contaminated by anything undesirable, a high quality product can be made from it. Similarly, if scrap consists only of one alloy composition it is easier to remelt to a good quality product, although there may have to be some adjustment of composition on remelting. If scrap is mixed, contaminated or includes other materials such as solder then, when remelted, it will be more difficult to adjust the composition within the limits of a chosen specification. Where lead or tin have been included, but no harmful impurities, it is usually possible to adjust composition by the addition of more lead or tin to make leaded bronzes. For some scrap contaminated with undesirable impurities it is sometimes possible to dilute it when melting so that the impurity level comes within an acceptable specification. All these techniques retain much of the value of the scrap.

Where scrap has been contaminated beyond acceptable limits it is necessary to re-refine it back to pure copper using conventional secondary metal refining techniques that provide a useful supplement to supplies of primary copper.

Battery Recycling

Most batteries contain heavy metals which is the main cause for environmental concern. Disposed of incorrectly, the heavy metals may leak into the ground when the battery erodes. This contributes to soil and water pollution and endangers wildlife. Some components in batteries can be toxic to fish and make them unfit for human consumption.

Each year Australians discard about 8000 tons of used batteries. The Australian Bureau of Statistics stated in a report published in November 2007 (4602.0 environmental issues people's views and practices). Batteries are the most common form of hazardous waste disposed of by Australian Households, with 97 per cent of those disposing of them via their usual rubbish collection.

To the average Australian consumer, a battery is a battery. They are not always aware of its chemistry, its ability to be recycled or the effects that battery can have on our environment if disposed of incorrectly.

In Australia, except for lead acid type chemistries, all other battery disposal collection for recycling overseas is carried out by MRI Australia. At present MRI export batteries to recycling facilities in France and Asia.

However, scientists, industry, environmentalists, government and recyclers are presently investigating the feasibility of building Australia's first plant for recycling consumer product batteries, e.g. Alkaline household batteries.

Mobile phone battery recycling

The mobile phone industry, because of its desire to maintain high environmental standards, has voluntarily developed the Mobile Phone Industry Recycling Program (MPIRP). The program aims to ensure that potentially toxic components in mobile phones and batteries do not end up in landfill, but rather are recycled.

The MPIRP is a voluntary scheme where the participating members provide the necessary funding by paying a levy on each handset sold into the Australian market.

The collected phone handsets, batteries and accessories are recycled under contract by MRI Australia. The recycling process prevents the reformation of environmentally damaging compounds such as dioxins and furans in the exhaust gas stream. It provides a complete breakdown of chemical compounds and is suitable for all phones and batteries including the newer Lithium Ion and Lithium Polymer types.

However in a 2007 published report, the handset manufacturing industry's peak representative body, the Australian Mobile Telecommunications Association (AMTA), revealed that the collection scheme has only caught a fraction of the estimated 5.5 million mobiles retired since it started in 1999.

Current battery recycling procedures

Normal household batteries-alkaline

Since the early 1990's nearly all alkaline batteries have been manufactured with 'no mercury added'. These batteries are considered non-hazardous waste and are safe for disposal in the normal municipal waste stream. Recycling of alkaline batteries is still considered too expensive to be a commercial reality, however an Australian company is working on building the first alkaline battery recycling plant which will allow the recovery of up to 30 per cent of the battery. Currently these batteries are sent to landfill by usual means or if in large volumes may be encapsulated in concrete.

Carbon-zinc

These batteries are considered non-hazardous waste and are safe for disposal in the normal municipal waste stream. Recycling of carbon zinc batteries is still considered too expensive to be a commercial reality. These batteries are sent to landfill by usual means or if in large volumes may be encapsulated in concrete.

Industrial batteries—lithium

Lithium (metal) batteries contain no toxic metals but there is a possibility of fire if the metallic lithium is exposed to moisture upon cell corrosion, so it is recommended that these are returned to Battery World stores for recycling.

Laptop batteries—lithium-ion (Li-Ion)

Li-ion batteries do not contain metallic lithium and therefore are not an environmental risk. These batteries do however contain recyclable materials and are accepted for recycling by MRI.

Car batteries—lead acid and sealed lead battery (SLA)

Lead acid batteries are recoverable to 96 per cent and the materials extracted are used in remanufacturing of batteries, plastic moulding applications and the acid is neutralised and discharged.

Rechargeable batteries

Nickel cadmium (NiCd)

The toxic cadmium content renders these types of batteries hazardous to the environment. Returning them for recycling to the manufacturer or battery retailer is considered non-careless disposal. The Melbourne based company MRI are specialists in NiCd battery disposal.

Nickel-metal hydride (Ni-MH)

Although Ni-MH batteries are considered environmentally friendly, this type of battery chemistry can be recycled. The Nickel component is semi-toxic and electrolyte in large amounts can be hazardous to the environment.

Chapter 12

Plastic Recycling

INTRODUCTION

Plastic recycling is the process of recovering scrap or waste plastics and reprocessing the material into useful products, sometimes completely different in form from their original state. For instance, this could mean melting down soft drink bottles and then casting them as plastic chairs and tables. Typically a plastic is not recycled into the same type of plastic, and products made from recycled plastics are often not recyclable.

CHALLENGES

When compared to other materials like glass and metal materials, plastic polymers require greater processing to be recycled. Plastics have a low entropy of mixing, which is due to the high molecular weight of their large polymer chains. A macromolecule interacts with its environment along its entire length, so its enthalpy of mixing is large compared to that of an organic molecule with a similar structure. Heating alone is not enough to dissolve such a large molecule; because of this, plastics must often be of nearly identical composition in order to mix efficiently.

When different types of plastics are melted together they tend to phase-separate, like oil and water, and set in these layers. The phase boundaries cause structural weakness in the resulting material, meaning that polymer blends are only useful in limited applications. Another barrier to recycling is the widespread use of dyes, fillers and other additives in plastics. The polymer is generally too viscous to economically remove fillers, and would be damaged by many of the processes that could cheaply remove the added dyes. Additives are less widely used in beverage containers and plastic bags, allowing them to be recycled more frequently. The use of biodegradable plastics is increasing. If some of these get mixed in the other plastics for recycling, the reclaimed plastic is not recyclable because the variance in properties and melt temperatures.

Processes

Before recycling, plastics are sorted according to their resin identification code, a method of categorisation of polymer types, which was developed by the Society of the Plastics Industry in 1988. Polyethylene terephthalate, commonly referred to as PET, for instance, has a resin code of 1. They are also often separated by colour. The plastic recyclables are then shredded. These shredded fragments then undergo processes to eliminate impurities like paper labels. This material is melted and often extruded into the form of pellets which are then used to manufacture other products.

Monomer recycling

Many recycling challenges can be resolved by using a more elaborate monomer recycling process, in which a condensation polymer essentially undergoes the inverse of the polymerisation reaction used to manufacture it. This yields the same mix of chemicals that formed the original polymer, which can be purified and used to synthesise new polymer chains of the same type. Du Pont opened a pilot plant of this type in Cape Fear, North Carolina, USA, to recycle PET by a process of methanolysis, but it closed the plant due to economic pressures.

Thermal depolymerisation

Another process involves the conversion of assorted polymers into petroleum by a much less precise thermal depolymerisation process. Such a process would be able to accept almost any polymer or mix of polymers, including thermoset materials such as vulcanised rubber tyres and the biopolymers in feathers and other agricultural waste. Like natural petroleum, the chemicals produced can be made into fuels as well as polymers. A pilot plant of this type exists in Carthage, Missouri, USA, using turkey waste as input material. Gasification is a similar process, but is not technically recycling, since polymers are not likely to become the result.

Heat compression

Yet another process that is gaining ground with startup companies (especially in Australia, United States and Japan) is heat compression. The heat compression process takes all unsorted, cleaned plastic in all forms, from soft plastic bags to hard industrial waste, and mixes the load in tumblers (large rotating drums resembling giant clothes dryers). The most obvious benefit to this method is the fact that all plastic is recyclable, not just matching forms. However, criticism rises from the energy costs of rotating the drums, and heating the post-melt pipes.

Difference Between a Polymer and a Plastic

The term 'plastics' is used to describe a wide variety of resins or polymers with different characteristics and uses. Polymers are long chains of molecules, a group of many units, taking its name from the Greek 'poly' (meaning 'many') and 'meros' (meaning 'parts' or 'units').

The term 'polymer' is often used as a synonym for plastic, but many other types of molecules— biological and inorganic—are also polymeric. While all plastics are polymers, not all polymers are plastic. Polymers are rarely useful in themselves and are most often modified or compounded with additives (including colours) to form useful materials. The compounded product is generally termed a plastic. Most people have little contact with 'polymers' because most articles that they come across are actually modified and coloured and therefore are actually plastics. Polymers can be classified in many ways, based on how they are developed and perform. For this discussion of recycling, an understanding of two basic types of polymers is helpful:
1. Thermoplastic polymers can be heated and formed, then heated and formed again and again. The shape of the polymer molecules are generally linear or slightly branched. This means that the molecules can flow under pressure when heated above their melting point.
2. Thermoset polymers undergo a chemical change when they are heated, creating a three-dimensional network. After they are heated and formed, these molecules cannot be re-heated and re-formed.

Comparing these types, thermoplastics are much easier to adapt to recycling.

Plastic identification; recycling code

When working with plastics there is often a need to identify which particular plastic material has been used for a given product. Most consumers recognise the types of plastics by the numerical coding system created by the Society of the Plastics Industry in the late 1980s. There are six different types of plastic resins that are commonly used to package household products. The identification codes listed below can be found on the bottom of most plastic packaging.

1. PETE polyethylene terephthalate (PET): Soda and water containers, some water proof packaging. Recycling PET is similar to the polyethylenes (PE). Bottles may be colour sorted and are ground up and washed. Unlike polyethylene, PET sinks in the wash water while the plastic caps and labels are floated off. The clean flake is dried and often repelletised. Recycled PET has many uses and well established market for this useful resin. By far, the largest usage is in textiles. Carpet companies can often use 100 per cent recycled resin to manufacture polyester carpets in a variety of colours and textures. PET is also spun like cotton candy to make fibre filling for pillows, quilts and jackets. PET can also be rolled into clear sheets or ribbon for VCR and audio cassettes. In addition a substantial quantity goes back into the bottle market.

2. HDPE high-density polyethylene: Milk, detergent and oil bottles, Toys and plastic bags. HDPE is called natural since that is its natural colour, and it is the most valuable because it can be made into any colour when it is recycled. Other products are often packed in brightly coloured bottles which are mixed together at recycling plants into mixed colour or rainbow bales. Most of this material is later dyed black after it is processed. Recycling HDPE is a pretty simple process. The bales are broken apart and ground into small flakes. These flakes are then washed and floated to removed and heavy (Sinkable) contaminants. This cleaned flake is then dried in a stream of hot air and may be boxed and sold in that form. More sophisticated plastic plants may reheat these flakes, add pigment to change the colour and run the material through a pelletiser. This equipment forms little beads of plastic that can then be reused in injection moulding presses to create new products. Some end uses for recycled HDPE are plastic pipes, lumber, flower pots, trash cans or formed back into nonfood application bottles.

3. Polyvinyl chloride (PVC): Food wrap, vegetable oil bottles, blister packages.

4. LDPE low-density polyethylene: Many plastic bags. Shrink wrap, garment bags. It is chemically similar to HDPE but it is less dense and more flexible. Most polyethylene film is made from LDPE which you often see as plastic bags and grocery sacks. This scrap may be clear or pigmented and it is hand sorted and baled at recycling processing plants. Recycling LDPE is verry similar to HDPE except special grinders are used to handle the thin films. The films are often washed and repelletised or used directly to make new products. Some end uses for recycled LDPE are plastic trash bags and grocery sacks, plastic tubing, agricultural film, and plastic lumber.

5. PP polypropylene: Refrigerated containers, some bags, most bottle tops, some carpets, some food wrap.

6. PS polystyrene: Throwaway utensils, meat packing, protective packing.

7. Other usually layered or mixed plastic: No recycling potential-must be landfilled.

These symbols are meant to indicate the type of plastic, not its recyclability. Types 1 and 2 are commonly recycled. Type 4 is less commonly recycled. The other types are generally not recycled, except perhaps in small test programs. Common plastics polycarbonate (PC) and acrylonitrile-butadiene-styrene (ABS) do not have recycling numbers. Chemical engineers will say that there are many more types and uses for polymers. But most debate in recycling focuses on these seven categories.

Uncoded plastics

Plastic consumer goods not identified by code numbers are not usually collected. Plastic tarps, pipes, toys, computer keyboards, and a multitude of other products simply do not fit into the numbering system that identifies plastics used in consumer containers. There are actually thousands of different varieties of plastic resins or mixtures of resins. These are developed to suit the needs of particular products. There is limited recycling of some of these specific plastic products in truckload quantities from industrial sources. No one has entered the business of collecting a variety of these plastics in small quantities.

Problem with plastics recycling

When glass, paper and cans are recycled, they become similar products which can be used and recycled over and over again. With plastics recycling, however, there is usually only a single reuse. Most bottles and jugs don not become food and beverage containers again. For example, pop bottles might become carpet or stuffing for sleeping bags. Milk jugs are often made into plastic lumber, recycling bins, and toys.

A recent development has been the bottles-to-bottles recycling of 'regenerated' pop bottles. Though it is technologically possible to make a 100 per cent recycled bottle, there are serious economic questions. Also, some critics claim that the environmental impact of the regeneration process is quite high in terms of energy use and hazardous by-products.

Currently only about 3.5 per cent of all plastics generated is recycled compared to 34 per cent of paper, 22 per cent of glass and 30 per cent of metals. At this time, plastics recycling only minimally reduces the amount of virgin resources used to make plastics. Recycling papers, glass and metal, materials that are easily recycled more than once, saves far more energy and resources than are saved with plastics recycling (Fig. 12.1).

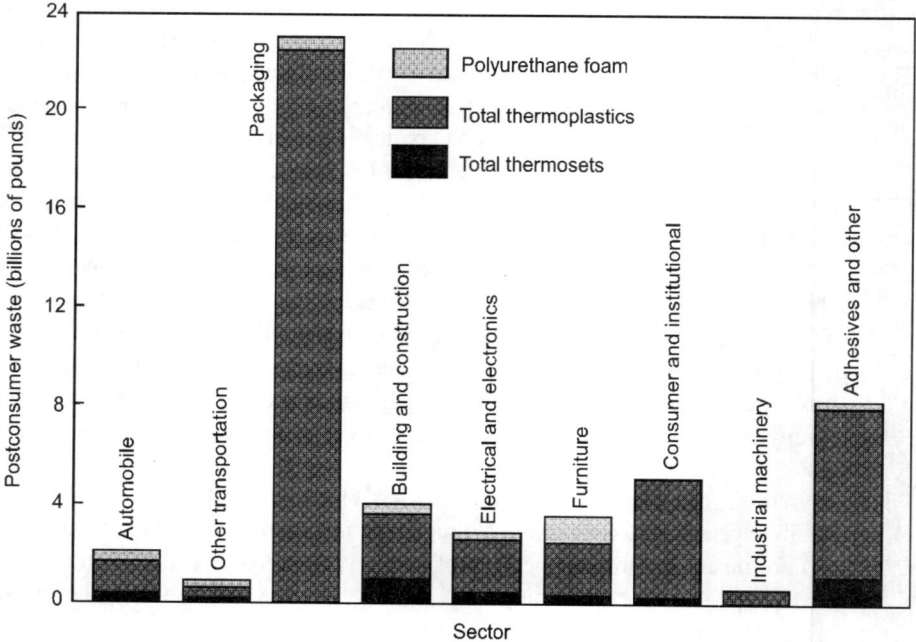

Fig. 12.1. A projection of post-consumer plastic waste is shown for different sectors.

Consider this example: polyvinyl chloride (PVC) bottles are hard to tell apart from PET bottles, but one stray PVC bottle in a melt of 10,000 PET bottles can ruin the entire batch. It is understandable why purchasers of recycled plastics want to make sure that the plastic is sorted properly. Equipment to sort plastics is being developed, but currently most recyclers are still sorting plastics by hand. That's expensive and time consuming. Plastics also are bulky and cumbersome to collect. In short, they take up a lot of space in recycling trucks.

Problem with PVC

PVC is used for packaging and other short-life consumer products, furnishings and long-life goods, mostly construction material such as window frames and pipes. Short-life products, disposed of within a few years, have caused serious PVC waste problems, especially when incinerated. The average life span of the long-life products is around 34 years. Long-life PVC goods produced and sold since the 1960s are now just starting to enter the waste stream. We are now only seeing the first stages of an impending PVC waste mountain.

There are currently over 150 million tons of long-life PVC materials in existence globally, used mostly in the construction sector, which will constitute this waste mountain in coming decades. Taking into account the ongoing growth in production, by the year 2012 this amount will double and the world will have to deal with approximately 300 million tons of PVC starting to enter the waste stream. The amount of PVC waste arising in industrialised countries is already expected to grow faster than PVC production. Of even more concern is the fact that the PVC industry is rapidly expanding in Latin America and Asia, so that eventually a growing waste mountain will be generated in these parts of the world.

In the late 1980s, PVC recycling was promoted by the vinyl industry in order to make PVC more acceptable to the public and to prevent government action to limit PVC production and use. As a result, the general public and decision-makers are now accepting recycling as a technical solution to the environmental problems associated with PVC. This is especially the case in countries with advanced recycling policies, like Denmark, Germany, the Netherlands and the USA.

Independent research shows that by the year 2012, it will only be possible to mechanically recycle 15–30 per cent of PVC consumed, and at a very high cost. It is virtually impossible to separate, collect and recycle the remaining 70–85 per cent. Thus for 70–85 per cent of PVC waste, recycling is not even an option for the mid- to long-term. A major problem in the recycling of PVC is its high chlorine content of raw PVC—56 per cent of the polymer's weight and the high levels of hazardous additives added to the polymer to achieve the desired material quality. Additives may comprise up to 60 per cent of a PVC product's weight. Of all plastics, PVC uses the highest proportion of additives (Fig. 12.2).

As a result, PVC requires separation from other plastics and sorting before mechanical recycling. PVC recycling is particularly problematic because of high separation and collection costs, loss of material quality after recycling, the low market price of PVC recyclate compared to virgin PVC and, therefore, the limited potential of recyclate in the existing PVC market. Feedstock recycling of PVC is hardly feasible at present, from an economic or an environmental perspective, and it is doubtful whether it will ever play a significant role in PVC waste management. The PVC industry seems to acknowledge that PVC recycling is no solution for PVC waste and it therefore is not surprising that industry is now lobbying for PVC incineration as a recovery option (for energy, hydrochloric acid and/or salt) in Western Europe and Japan and for landfilling in the USA and Australia. This forces local authorities to shoulder the burden of pollution and costs from PVC consumption.

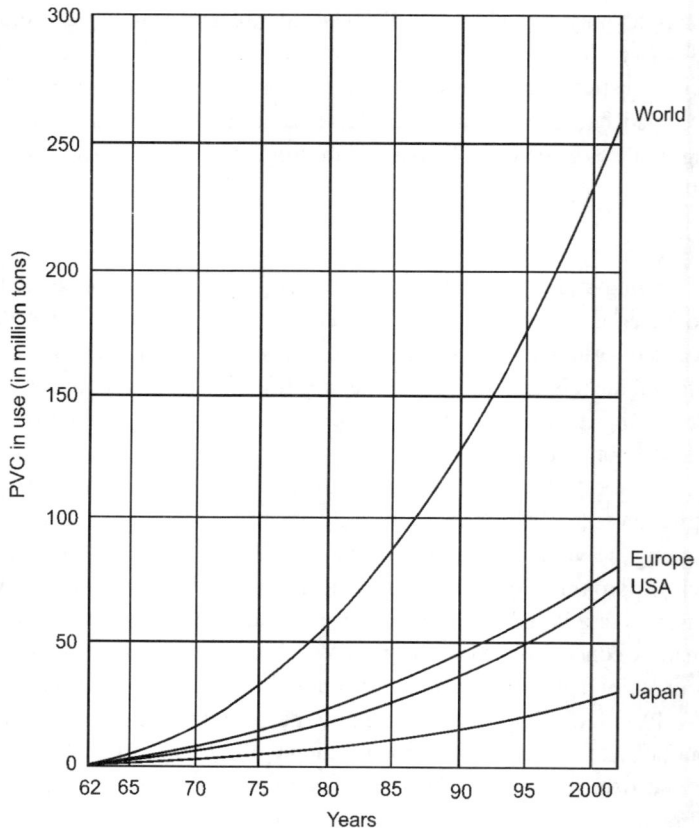

Fig. 12.2. Estimated PVC accumulation in long-life goods 1962–2005 (in million tons). These PVC products will enter society in the next three to four decades as waste.

Incineration is not a sustainable option for dealing with waste. Less energy is generated from burning the plastic than was used to make it, and incineration also means that the carbon contained within it is emitted as CO_2 — a greenhouse gas. Toxic substances are also emitted, and large amounts of solid wastes are produced as slag, ash, filter residues and neutralisation salt residues. Part of this needs to be disposed of as hazardous waste.

Despite these concerns, PVC production is still increasing, especially in developing economies where PVC consumption is being encouraged. PVC waste is exported from the USA, Europe and Australia to developing countries, often for recycling into lower quality products such as shoes and low quality pipes, or 'downcycling'. According to the Indonesian Environment Minister, up to 40 per cent of the plastic waste imported into Indonesia is not recycled but directly disposed of, partly as hazardous waste. Downcycled products will eventually be dumped or burned since downcycling simply delays the inevitable need to dispose of PVC plastic waste. In light of the large volume of long-life PVC products due to become waste in the coming decades, and the projected increase in PVC production, it becomes apparent that an international PVC phase-out is urgently required. Only this will put a halt to a growing, dangerous and intractable waste problem. Political frameworks for PVC phase-outs already exist. The North Sea Ministers Conference agreed in 1995 to stop environmental emissions of hazardous substances within one generation. According to the Swedish Chemical Committee, PVC has no place in a sustainable society and should be phased out for all uses by the year 2012. Denmark has proposed restrictions on the use of softeners, lead and other additives used in PVC plastic and is questioning the recycling potential claimed by the PVC industry. The Czech Republic agreed to phase-out production, imports and use of PVC packaging from 2001 onwards and Switzerland has banned PVC drinking bottles in 1991.

POLYMER RECYCLING: OPPORTUNITIES AND LIMITATIONS

The disposal of polymer solid waste by means other than landfilling is necessary. The various approaches source reduction, incineration, degradation, composting, and recycling—all have their roles and must be employed in an integrated manner. Where appropriate, recycling has ecological advantages, but its application is dependent upon the feasibility of collection, sorting, and/or compatibilisation of resulting mixtures to produce economically viable products.

The practice should be encouraged by societal or legislative pressure which recognises that the cost of disposal should be a factor in determining the cost of a product. The disposal of trash is a problem confronting our society. Our available landfills are becoming exhausted. While polymers compose only about 8 per cent by weight (20 per cent by volume) of landfills, there is much focus on polymer accumulation because of their high visibility. This has stimulated the formulation of considerable restrictive legislation regulating polymer use. A response has been the consideration of alternatives for polymer disposal. The principal means are (i) source reduction, (ii) incineration, (iii) bio- or photodegradation, (iv) composting, and (v) recycling.

Source Reduction

Means for source reduction are apparent. These involve such measures as the elimination of unnecessary packaging and the packaging of products as concentrates. One approach is the replacement of polymers by alternative materials. This should be done with care, since the replacement is not always ecologically desirable and sometimes functionally inadequate.

One estimate, for example, suggests that the abandonment of plastics in packaging would result in a 404 per cent increase in the weight of waste, a 201 per cent increase in energy consumption in making the alternatives, and a 212 per cent increase in cost. Alternative materials are sometimes heavier, more permeable, more water absorbant, and less strong than their polymer counterparts and thus may not function as well. A case in point is the replacement of Styrofoam 'hamburger shells' by MacDonald's with a paper-based wrapping. We are faced with the decision 'paper or plastic' at the supermarket check-out. Several studies have contested the environmental superiority of paper as compared with plastic and the choice is not as simple as is commonly portrayed.

Another illustration involves the replacement of synthetic fibres such as nylon and Dacron with 'natural degradable' fibres such as cotton and wool. A proper analysis of the ecological effect would require consideration of the agricultural implications of growing the required amount of cotton and raising the sheep, the fertiliser needed, the fuel for the tractors, differences in energy requirements for processing, and differences in the care required for maintenance and laundering. The choice is indeed not a simple one.

Incineration

Incineration is widely used in Asia, necessitated by the limited space for landfills. It has not been popular in the United States, primarily because of concerns about toxic fumes and ash. These problems could probably be avoided through use of current technology in incinerator design and by employing some degree of separation of feedstock so as to eliminate 'bad actors'. The acceptance of incineration depends upon the success of these measures.

A polymer of concern for incineration is poly(vinyl chloride) (PVC), where hydrochloric acid can be produced with improper incineration. This has prompted possible legislation to restrict its use. However, a principal use for PVC is for pipe which does not represent a significant waste problem.

Of course, a principal product of incineration is carbon dioxide, which may contribute to the global warming problem. However, since a relatively small fraction of the petroleum supply is used to produce polymers and only a fraction of that would be incinerated, this source should not be a major factor in comparison with carbon dioxide production arising from the burning of fossil fuels. Incineration consumes about 15 per cent of today's solid waste, and a goal of 25–30 per cent by the year 2012 has been suggested.

Biodegradation and Photodegradation

There is skepticism as to whether the employment of degradable polymers will be effective in reducing the buildup of landfills. Under usual conditions, degradation rates in landfills are too slow. Modification of polymers so as to increase degradation rates often leads to the problem of their degrading under normal conditions of use. Furthermore, degradation is in opposition to possibilities for future recycling of the polymer. A usual product of degradation is carbon dioxide, so it parallels incineration in this respect (without the advantage of possible energy recovery). The contribution of toxic residues to the environment, which is of concern in the consideration of incineration, also must be considered here, since when a polymer containing such residues degrades, these are also released.

There is a definite role for degradable polymers. While ideally, articles such as 'six-pack rings' and old fish line and nets should not be carelessly discarded, some such practices will always occur, so rendering such articles to be degradable will reduce the possibility of their being harmful to marine and wildlife. The state of Maine has recently banned the use of six-pack rings. Furthermore, in addition to the environmental harm arising from littering, there are cosmetic advantages to its reduction through use of degradable polymers. While the amount of polymer rubbish is not large, it is very visible and often leads to unsightly appearance of beaches and public areas. Thus, improvement may result from making commonly discarded articles degradable. There is a price to pay, however, in that degradation may also occur in normal use, so that means for monitoring this are necessary so as to avoid their failure under these conditions. Educational efforts to reduce littering through instilling good habits are essential and may even be more effective than rendering the litter degradable.

One should distinguish between intrinsically degradable polymers and those to which a degradable material is added. An example of the latter type is polyethylene to which starch has been added. The polymers do not disappear on degradation of the additive; they just fall apart into small bits. This may have cosmetic value or may serve to release the contents of a plastic bag to exposure for composting, but it should not be considered as a means for disposal of polymers. Intrinsically degradable polymers include poly (lactic acid) and bacterially synthesised polyalkonates. These convert completely to nonpolymeric products on degradation. So far, their physical properties are not as good as those of conventional polymers, but it seems likely that these may be improved. The principal drawback is cost. Unless this can be significantly lowered, the use of this class of materials will be limited to very specialised applications. A promising development is the formation of intrinsically biodegradable polymers by the thermoforming of starch. The starting material is sufficiently cheap that there is hope of producing an economically competitive material. The amount of degradable synthetic polymer in use today is negligible, probably under 1 per cent. This could conceivably be increased to a few per cent, which would have a minor effect on landfill growth but which could be important if applied to critical situations.

Composting

While degradation rates are low in landfills, the carrying out of degradation in compost piles appears to be a reasonable prospect. Yard and agricultural waste constitutes a much larger part of the solid waste burden than polymers and composting seems the right approach for these. It would appear feasible to add biodegradable polymers to the composting mixture. Also cellulosic wastes such as disposable diapers could be accommodated. An infrastructure is needed to collect wastes of these types in a manner so that they are separated from nondegradable material. To have a major effect such composting would have to be carried out under controlled conditions in centralised facilities. Several studies are in progress concerning how this might be done.

Recycling

Recycling is the main theme of this chapter. This implies reuse of the polymer as polymer. A classification is (i) recycling of industrial scrap (sometimes called 'prompt scrap'), and (ii) post-consumer recycling.

The former primarily involves recovering and reusing waste polymer resulting from processing operations. This has long been practiced by many industrial groups. It is simpler than the latter in that polymers are usually of a particular kind and they are collected in a central location by experienced people. The practice is motivated by economics.

Post-consumer recycling of commingled plastics

Postconsumer recycling is more difficult. Here, plastics of a variety of types are in the hands of consumers after being used for a variety of purposes. The consumer may be motivated to recycle these articles by good citizenship, monetary rewards, or legislation. He/she can usually be able to separate polymers from nonpolymeric components. If this is not done, a certain degree of separation may be accomplished after rubbish collection, making use of the fact that polymers normally have lower densities than other components and are usually nonmagnetic and nonconducting. Such separation usually results in obtaining commingled plastic as opposed to separated. Such commingled plastics can be fabricated to make articles such as 'plastic lumber', fence posts, and traffic barriers. However, they compete with fairly cheap materials, but their cost may be two or three times as much. For some applications, such as picnic tables, costs may be more competitive. Thus, the potential market is probably limited and may become saturated as collection of plastic for recycling increases.

A problem with applying commingled plastics to more demanding applications where their cost would be justified is that the interfaces between different polymers are often weak as a consequence of their usual thermodynamic immiscibility leading to little intermolecular penetration. However, there are many less demanding applications where high mechanical performance is not required.

The value of commingled plastics may be enhanced through strengthening such interfaces by the following: (i) copolymerisation or grafting, since it is known that copolymers are more miscible than homopolymers. Grafting is often done with in normal polymer practice, as in the preparation of ABS (acrylonitrile/butadiene/styrene) or HIPS (highimpact polystyrene), by grafting a rubbery component to polystyrene so as to introduce microphase separation of the rubbery component which results in energy-dissipating mechanisms which contribute to their impact strength. One may 'impact modify' by this means by carrying out chemical reactions on the polymer(s) or else by adding another component leading to microphase separation. This benefit can sometimes arise, to some extent, in mixing different polymers, provided the disadvantage of weak interfaces do not dominate, (ii) Functionalisation of components of commingled plastics by introducing chemical groups onto components which attract each other through hydrogen bonding, donor-acceptor interactions, and (iii) The addition of compatibilisers such as block copolymers which reside at interfaces and act like emulsifying agents to strengthen the interaction between components.

These approaches are often costly, so it is doubtful whether they can provide an economically viable major impact on the plastic solid waste problem. However, they do have a limited role, and such efforts should be encouraged.

Separated plastics

The market for clean, separated plastics is much greater. While currently it usually costs more to use these than virgin polymer, increasing costs for alternative disposal and legislation will undoubtedly render their use more favourable. The problem of separation is resolved for readily identifiable objects such as soda bottles and milk jugs. For soda bottles, the poly(ethylene terephthalate) (PET) polymer is a relatively expensive one, and the return of the bottles may be encouraged by 'bottle bills' requiring deposits and refunds and by legislation prohibiting their disposal along with other rubbish. Redemption machines which read bar codes and make refunds and then grind up the bottle are beginning to appear in supermarkets.

There are problems even with PET bottles. They are sometimes coloured, and caps, labels, and bases are often made of material other than PET. While some degree of separation is possible using techniques such as floatation, the presence of small amounts of contaminants can detract from the value of recycle feedstock. It would be desirable, by either cooperation or legislation, to design such bottles so as to minimise the number of different components and choose them so as to facilitate separation. FDA restricts the use of recycled PET to non-food applications. However, there is currently an ample market for applications such as fabrication of carpets and fibrefill.

Milk and water jugs have proved a good source of recyclable high-density polyethylene (HDPE). There has been public reluctance toward imposing deposits on these, so their return rate is not as good as for soda bottles. As with PET, such recycled HDPE cannot be used for food containers, but they have been extensively used, for example, for containers for motor oil. (The restriction on food use may be reasonable, since it has been reported that there is enough oil absorption from the milk by HDPE to act as a plasticiser and affect future recycling and may be a consideration in its further use in contact with food.) For other polyethylene sources such as shopping bags, the economics of recycling is less favourable, since the amount of polymer per article is small, so the cost of collection represents a greater fraction of the value of the recycled polymer. Where possible, reuse of such articles should be encouraged. Multiple use of a shopping bag, for example, is probably more effective use of the polymer than would be possible by collection and recycling.

For certain applications, recycling of polyethylene may be feasible. For example, it is often used as a 'mulch' in agriculture or as a temporary covering in building construction. Collection and recycling may be effective for such large volume use by knowledgeable consumers. Photodegradation is also a viable disposal approach for polyethylene mulch. A similar situation may exist for other recycled polymers. Polypropylene is extensively used industrially and in agriculture as strapping or binding. It is also used as outdoor carpet. Nylon is used as indoor carpet. A problem may occur because of the employment of different kinds of nylon and the presence of dyes and other additives.

Polystyrene is another polymer for which recycling has met with some success. Styrofoam is commonly used for coffee cups, hamburger shells, insulation, cafeteria trays, etc. Recycling of this is feasible if sources of ready collection can be identified. School and other cafeterias have often proved quite cooperative in cleaning and stacking such trays for centralised collection. The contaminants are usually food residues and paper, which can be readily separated by washing and floatation.

In addition to the ecological advantages of recycling Styrofoam, the educational value of enlisting the efforts of school children in cafeteria recycling procedures cannot be ignored. A mission should be to train a generation of children with positive environmental habits.

It is evident that even if polymers are segregated by chemical species, variations will occur when such polymers from different sources are mixed. For example, they may be of differing molecular weight, branching or tacticity. This will affect the rheology of the molten mixture, which is important for its processing. Furthermore, if the polymer crystallises, the crystallisation kinetics and morphology will be affected. Crystallisation can be greatly affected by the presence of nucleating agents which may have been added to some of the sources or may be inadvertently present. Thus, the processing behaviour and product properties of polymers produced from a variable feedstock may be inconsistent and difficult to control. This suggests the need for analytical methods to monitor properties and to consider possibilities for modifying these by adding virgin polymer to the recycle feedstock.

Other than in a relatively few such cases, the separation of polymers into constituents may prove difficult. The need for sophisticated separation techniques can be minimised if recycling is a consideration in product design. For example, the ability for ready disassembling of a polymer-containing product into basic constituents should be a factor in their design. The contribution of polymeric parts of junked automobiles to the solid waste stream is an increasing burden. Dealing with this quantity of material requires means for recovering the plastic from a junked car in reasonable time. It has been demonstrated that a Volkswagen Passet 83 can be disassembled within 20 min to yield 14.5 kg of polypropylene, 8.7 kg of polyethylene, and 5.4 kg of ABS. Of course, an infrastructure is necessary, probably a computer database, to provide information to the disassembler about procedures and constituents for each of the many models of cars and instructions for dealing with the recovered polymers.

While efforts are being and should be made to develop better separation techniques, it seems evident that there will be combinations of polymers which may contain nonpolymeric additives, fillers, and reinforcing fibres such that separation is not possible or economically feasible. This source will grow with the increasing use of polymer-based composites in automobiles, housing, etc. Thus alternatives need be considered for dealing with such materials.

Reduction to low molecular weight species

Polymers may be degraded to low molecular weight species by processes such as pyrolysis, hydrolysis, or methanolysis. This could be done for mixed polymers leading to mixtures of low molecular weight materials. These may be of sufficiently low viscosity so that they may be separated from insoluble nonpolymeric additives by means such as filtration. The low molecular weight mixture could be regarded as an organic feedstock which could be separated into components by procedures such as fractional distillation.

Polymer dissociation is an endothermic process and one converting a low entropy polymer molecule into higher entropy dissociation products. Since energy conservation should deal with conservation of free energy, which implies minimising entropy production, this aspect is ecologically undesirable. However, it is a price to pay for permitting separation.

Today, about 69 per cent of the plastic solid waste in the United States ends up in landfills. This cannot continue. There is no one alternative which will solve the problem, so multiple approaches must be taken. Source reduction, incineration, employing degradable polymers, and recycling all have their place, and increases in all of these measures are required. Source reduction and recycling are preferable in that they conserve resources and minimise pollution. Rapid progress in recycling post-consumer plastics is being made with a growth of 45 per cent in 10 plastics markets between 1989 and 1990. Composting may be preferred to incineration in that it does not lead to air pollution (but could lead to some sanitation concern). Landfilling is the disposal means of 'last resort' to be used when none of the other more preferable methods are applicable. Through these means, it appears that the amount of landfilled polymer could be decreased by a factor of 2 by year 2013. It appears as though meeting these goals will require a cooperative effort of the industrial and environmental communities with motivation and regulation provided by government. Measures must be consistent with technological and economic limitations and a reasoned analysis of the factors involved is essential for making the proper choices.

Processes

A process has also been developed in which many kinds of plastic can be used as a carbon source in the recycling of scrap steel.

Applications

PET

Post-consumer polyethylenes are sorted into different colour fractions, cleaned, and prepared for processing. This sorted post-consumer PET waste is crushed, chopped into flakes, pressed into bales, and offered for sale.

One use for this recycled PET that has recently started to become popular is to create fabrics to be used in the clothing industry. The fabrics are created by spinning the PET flakes into thread and yarn. This is done just as easily as creating polyester from brand new PET. The recycled PET thread or yarn can be used either alone or together with other fibres to create a very wide variety of fabrics. Traditionally these fabrics were used to create strong, durable, rough, products, such as jackets, coat, shoes, bags, hats, and accessories. However, these fabrics are usually too rough on the skin and could cause irritation. Therefore, they usually are not used on any clothing that may irritate the skin, or where comfort is required. But in today's new eco-friendly world there has been more of a demand for 'green' products. As a result, many clothing companies have started looking for ways to take advantage of this new market and new innovations in the use of recycled PET fabric are beginning to develop. These innovations included different ways to process the fabric, to use the fabric or blend the fabric with other materials. Some of the fabrics that are leading the industry in these innovations include Billabong's Eco-Supreme Suede, Livity's Rip-Tide III, Wellman Inc's Eco-fi (formerly known as EcoSpun), and Reware's Rewoven. Some additional companies that take pride in using recycled PET in their products are Crazy Shirts and Playback.

PVC

PVC- or Vinyl Recycling has historically been difficult to perfect on the industrial scale. But within the last decade several viable methods for recycling or upcycling PVC plastic have been developed.

HDPE

The most-often recycled plastic, HDPE, is downcycled into plastic lumber, tables, roadside curbs, benches, truck cargo liners, trash receptacles, stationery (e.g. rulers) and other durable plastic products and is usually in demand.

Other plastics

The white plastic foam peanuts used as packing material are often accepted by shipping stores for reuse. Successful trials in Israel have shown that plastic films recovered from mixed municipal waste streams can be recycled into useful household products such as buckets.

Similarly, agricultural plastics such as mulch film, drip tape and silage bags are being diverted from the waste stream and successfully recycled into much larger products for industrial applications such as plastic composite railroad ties. Historically, these agricultural plastics have primarily been either landfilled or burned on-site in the fields of individual farms.

Consumer education

Low national plastic recycling rates have been due to the complexity of sorting and processing, unfavourable economics, and consumer confusion about which plastics can actually be recycled. Part of the confusion has been due to the recycling symbol that is usually on all plastic items. This symbol is called a resin identification code. It is stamped or printed on the bottom of containers and surrounded by a triangle of arrows. The intent of these arrows was to make it easier to identify plastics for recycling. The recycling symbol doesn't necessarily mean that the item will be accepted by residential recycling programs.

Chapter 13

Mechanical Recycling of PVC Waste

INTRODUCTION

PVC has been subject to a controversial debate amongst environmental groups (e.g. Greenpeace), governments, the public and industry for many years now. A number of environmental issues associated with the production, use and disposal of PVC have been addressed. In Europe, the debate has focused on a number of countries.

A major reason of concern has been the disposal of PVC wastes. A number of environmental issues have been discussed. Additionally, PVC waste quantities are projected to increase significantly in the next years: A major part of PVC is used for long-life products in the construction sector (e.g. pipes, window frames, floor coverings) which are still in use.

This chapter assess the environmental, technical and economic aspects of the mechanical recycling of PVC and the evaluation of measures for improvements. In detail the objective includes the following aspects:

1. Quantitative and qualitative assessment of existing PVC waste recycling systems.
2. Identification of environmental, technical and economic problems involved in the recycling of PVC wastes.
3. Analysis of the impact of the presence of PVC on the recycling of other plastics.
4. Identification of community and national measures to improve the recycling of PVC wastes.

Mechanical recycling refers to recycling processes where the material is treated mechanically (e.g. grinding, seeving, screening). There exist other recovery and recycling processes, so-called 'feedstock recycling' processes like, e.g. the controlled incineration with recovery of HCl which can be reused for the production of chlorine (feedstock for PVC) or the so-called 'Vinyloop' process. All these processes involve a chemical treatment of the PVC wastes. The incineration process includes a thermal decomposition and the 'Vinyloop' process includes the dissolution of PVC wastes in a solvent with the subsequent recovery of pure PVC. As this chapter deals with mechanical processes only these processes are not considered. Nevertheless, they may provide additional potentials for the recovery of PVC wastes since they allow for the processing of PVC wastes with a comparatively high level of contaminations.

Various steps in PVC recycling are given below:

1. Preparations.
2. Assessment of existing PVC waste recycling systems.
3. Identification of environmental, technical and economic problems.
4. Assessment of the impact of PVC on the recycling of other plastics.

5. Outline of future scenarios of PVC recycling.
6. Assessment of measures to improve the recycling of PVC wastes.

DEVELOPMENT OF PVC WASTES—GENERAL CONSIDERATIONS

In order to assess the mechanical recycling of PVC, it is necessary to distinguish between the different PVC products and waste types respectively. The opportunities and limits of recycling are different depending on the product group. In order to develop a realistic future scenario of PVC recycling, it is also necessary to have a general knowledge of the major factors influencing the recycling quantities.

Classification of PVC Wastes

Like for other plastics, the recycling potentials of PVC are to a large extent determined by the degree of contamination which must be accepted for the collected wastes. The production of high-quality recyclates is the easier, the purer the collected PVC material is. 'Degree of contamination' refers to two criteria:
1. The degree to which PVC is mixed with other materials when collected.
2. Differences in the composition of the collected PVC material itself.

As for the second aspect, it has to be taken into account that the PVC used in products does not consist of pure PVC but of PVC compounds which contain different quantities of additives, such as softeners, filling agents, stabilisers and others. One major difference in the material composition exist between rigid PVC applications with lower additive contents and soft PVC applications which may contain more than 50 per cent of additives. Even in the same application (e.g. window profiles, pipes, films) the composition of the PVC material differs between different PVC converters having their own specific PVC compounds and between different production years, due to technological advances. For example, in cable insulations the content of additives (plasticisers, fillers, stabilisers) ranges from 50– 60 per cent with different mixtures and compounds being used.

The production of high-quality recyclates with defined technical specifications (e.g. strength, elasticity, colour) requires input materials with a defined quality, i.e. pure PVC in terms of the contents of other materials and composition of the PVC compounds.

The degree of contamination which can be achieved for collected PVC wastes depends to a large extent on:
1. The type of waste in which the PVC products end up.
2. The PVC application (product group).

Therefore, in this chapter PVC wastes will be classified depending on these criteria (Table 13.1).

Table 13.1. Classification of PVC wastes.

PVC applications[a]
Construction products
Cables (F)
Flexible films (F)
Flooring calandered (F)
Flooring paste (F)
Roofing membranes (F)
Profiles and hoses (F)

(Contd. .)

PVC applications[a]

    PVC wall papers (F)

    Air inflated structures, container, marquee (F)

    Varnishes—coil coating (F)

    Pipes (R)

    Window profiles (R)

    Profiles—cable trays (R)

    Other profiles (R)

    Pipe insulation films (R)

    Sheets (R)

Packaging products

    Flexible films (F)

    Cans (F)

    Rigid films (R)

    Bottles (R)

Furniture components

    Flexible films (F)

    Flexible profiles (F)

    Rigid films, kitchens (R)

    Rigid films, drawers (R)

    Other rigid films (R)

Other consumer and commercial products

    Bags, luggage, cushions (F)

    Office supply, books, photo articles (F)

    Camping, leisure, toys (F)

    Misc. plasticised films (F)

    Garden hoses (F)

    Drinking hoses (F)

    Other industrial hoses (F)

    Other flexible profiles (F)

    Artificial leather (F)

    Conveyor belts (F)

    Miscellaneous coatings (F)

    Rotational mouldings (F)

    Slush mouldings (F)

    Misc. organo-/plastisols (F)

    Shoes, soles (F)

    Miscellaneous (F)

    Office supply (R)

    Printing films (R)

(Contd...)

PVC applications[a]
Credit cards (R)
Computer disks (R)
Other technical applications (R)
Sheets, chemical equipments (R)
Miscellaneous sheet products (R)
Miscellaneous rigid profiles (R)
Vinyl records (R)
Other rigid products (R)
Electric/electronics
Cables (F)
Adhesive tapes (F)
Flex. profiles, hoses (F)
Injection moulding parts (F)
Rigid profiles
Automotive
Cars cables (F)
Instrument panels and other films (F)
Cabletapes and cablebinders (F)
Hoses, flexible profiles (F)
Foamed films/artificial leather (F)
Tarpaulins for lorries (F)
Underfloor protection (F)
Others, injection moulding (F)
Rigid profiles (R)
Battery separators (R)
Other products
Agricultural films (F)
Medical products (F)

[a]F = Flexible PVC applications; R = Rigid PVC applications.

With regard to the PVC waste types two major groups must be distinguished: Preconsumer wastes are generated in the production of PVC final and intermediate products (production wastes) and installation wastes from the handling or installation of PVC products: The processing of PVC to final products takes one to more than three production steps, each of them may be carried out by a different company. For example, the production of packagings starts with the production of films from PVC compounds in calanders followed by the thermoforming of the films to packagings in a second step. In each step production wastes are generated (e.g. cut-offs in the calandering of films). Some of the final products have to be handled or installed to reach their final purpose, resulting in additional installation wastes. Cut-offs from the laying of cables or floorings are examples. A part of the preconsumer wastes is recycled at the PVC processors in-house (production wastes like the cut-offs from the production of films can be used directly as raw material in the same process), the other part is collected by recyclers.

The collection of installation wastes especially is carried out by recycling companies which return the material to the PVC processors after mechanical treatment. PVC preconsumer wastes as a group are comparatively easy to recycle, since they can be collected separately in defined qualities. This is why recycling of PVC preconsumer wastes is applied to a large extent in practice.

The recycling of post-consumer wastes is generally more difficult to realise since they occur in form of products (end-of life products such as pipes, windows, packagings) and hence in more or less mixed waste fractions or as a part of composite materials. Depending on the specific products, PVC in wastes can occur as a more or less pure material fraction (in 'mono fractions') which can be extracted from the waste stream by sorting (e.g. bottles, pipes, some films, some profiles). Alternatively, PVC can form a part of composite products or materials which must be subjected to disassembling or mechanical treatment processes in order to extract PVC (e.g. windows, car components, floorings, cables). Both PVC 'mono fractions' and composite products/materials can be collected separately (i.e. in product specific collection systems, e.g. bottle, window or cable collection systems) or in mixed fractions together with other materials (e.g. packaging wastes, municipal solid wastes).

For the post-consumer wastes the different PVC product groups determine to some extent in which specific waste flow the PVC occurs. It is also the waste flow (not the material as such) which determines how easy or difficult PVC can be separated out as a pure fraction. And it is only the waste flow which can be influenced by waste management measures and policies. We distinguish five different product groups:

1. Construction products (pipes, windows, flooring, etc.) which end up in construction and demolition wastes—many products arrive at mixed waste streams today but a separate collection is feasible, a part of it even as 'mono fractions' (pipes and some profiles).
2. Consumer and technical products (packagings, rigid film applications, etc.) arrive at (mixed) municipal solid wastes (from households, industry and commerce) or (mixed) packaging wastes; a separate collection is feasible for few products only.
3. Vehicle components (e.g. dashboard elements, cables, coatings) which unless disssambled before shredding end up in the shredder residues.
4. Electric/electronic products forming the so-called electro/electronics waste whose major share arrives at municipal solid wastes, but a separate collection is feasible.
5. Other products ending up in special waste flows (e.g. hospital and agricultural wastes).

Factors Influencing PVC Recycling

In order to analyse, to forecast and to improve the mechanical recycling of PVC it is necessary to have a general knowledge of the factors which determine the recycled quantities.

The absolute quantity of recycled PVC per year can be thought as a result of (i) the total annual quantity of PVC in wastes, and (ii) the recycled fraction of it ('recycling rate').

The total quantity of PVC in wastes is a function of PVC consumption: The higher the PVC consumption the higher will be the quantity of PVC in wastes. In contrast to most other commodity plastics, (especially polyethylene and polypropylene), the major part of PVC production is converted into long-life products in the construction sector (pipes, windows, etc.) with an expected lifetime of up to 50 years and more. This is why there is a considerable 'time lag' between PVC consumption and PVC in wastes. The PVC production consumption took off to reach significant market shares in the 1970s. The production quantities of many large volume products such as window profiles reached an order of magnitude near today's production levels not before the beginning of the 1980s. So, with an average lifetime of around 30 years for PVC products as a rule of thumb, the quantity of PVC in wastes is still very small compared to PVC consumption. The 'big push' of PVC waste quantities can be expected to start around 2012 only.

The total waste arising of PVC has an impact on PVC recycling not only because it determines the absolute amounts of recyclable PVC but also due to the fact that the feasibility of a recycling system requires a minimum quantity of wastes. This is due to the fact that recycling plants must reach a minimum capacity to allow for a technical and economic feasible operation. Also the geographical area supplying one recycling plant must not exceed a certain size in order to keep transport distances and costs in a reasonable range. Additionally, the PVC content in mixed wastes must be high enough to make the operation of separate collection system or specific separation and sorting processes feasible.

The part of total PVC wastes which is going to recycling ('recycling rate') depends on four major factors (Fig. 13.1):

1. Technical factors, mainly the achievable quality of the recyclates in relation to the required quality in the possible applications; this is in turn determined by the degree of contamination of the collected PVC wastes or the relevant waste streams respectively.

2. Legal and organisational factors, including recycling regulations (e.g. minimum recycling quota), statutory requirements limiting or discouraging the use of the 'nonrecycling' waste disposal routes (especially landfilling and incineration), voluntary agreements or commitments of industry to establish (and finance) collection and recycling systems and finally technical standards and regulations limiting the application of the recyclates (e.g. certification systems, food contact laws).

3. Economic factors, especially the overall (net-)cost of recycling (collection + logistic + sorting + treatment – credits for produced recyclates), which is inter alia, influenced by the price of virgin PVC and the technical factors (degree of contamination).

4. Ecological factors, especially the achievable savings of resources and emissions to the environment due to the substitution of virgin PVC and other materials in relation to emissions and resource consumption of the recycling processes (collection, transport, treatment/processing, etc.); the achievable savings depend on the products/materials which can be substituted by the recyclates, which in turn depends on the achievable quality of the recyclates (i.e. high-quality recyclates can substitute virgin PVC, lowquality recyclates or mixed plastics recyclates can substitute concrete, wood or other non-plastics only).

It must be pointed out that there is a close connection between the different factors. Especially, the economic and environmental performance of PVC recycling is closely linked to the technical factors (degree of contamination, separate collection, etc.). Therefore, for the assessment of PVC recycling the whole picture must be taken into account.

General analysis of the impact of the factors influencing PVC recycling

At this stage a general analysis of the impact of the different factors described above will be given. A more specific analysis has been elaborated in the description of the existing PVC recycling systems taking also into account country-specific circumstances.

Technical factors

The technical potentials of the mechanical PVC recycling are determined by the achievable quality of the PVC recyclates. To be used for the production of new products, recyclates must comply with a set of technical specifications which at last refer to the contamination and the composition of the recyclates.

These specifications take account of the specific characteristic of PVC that the composition of the material differs depending on the specific application:

1. Much more than other commodity plastics-such as polyethylene and polypropylene—PVC is a compound material, i.e. it does not consist of polymer PVC alone but includes also a variety of additives such as stabilisers (to avoid degradation of the PVC), plasticisers (in flexible PVC), fillers, impact modifiers, pigments and processing agents.
2. Each PVC application has its specific material composition (Table 13.2).
3. Also for a specific PVC application, the composition of the PVC compounds can differ depending on the producer or processor. Furthermore, the composition of the PVC compounds for a specific application has changed in time due to technological changes, e.g. today window profiles are produced from different PVC compounds than window profiles 20 years ago.

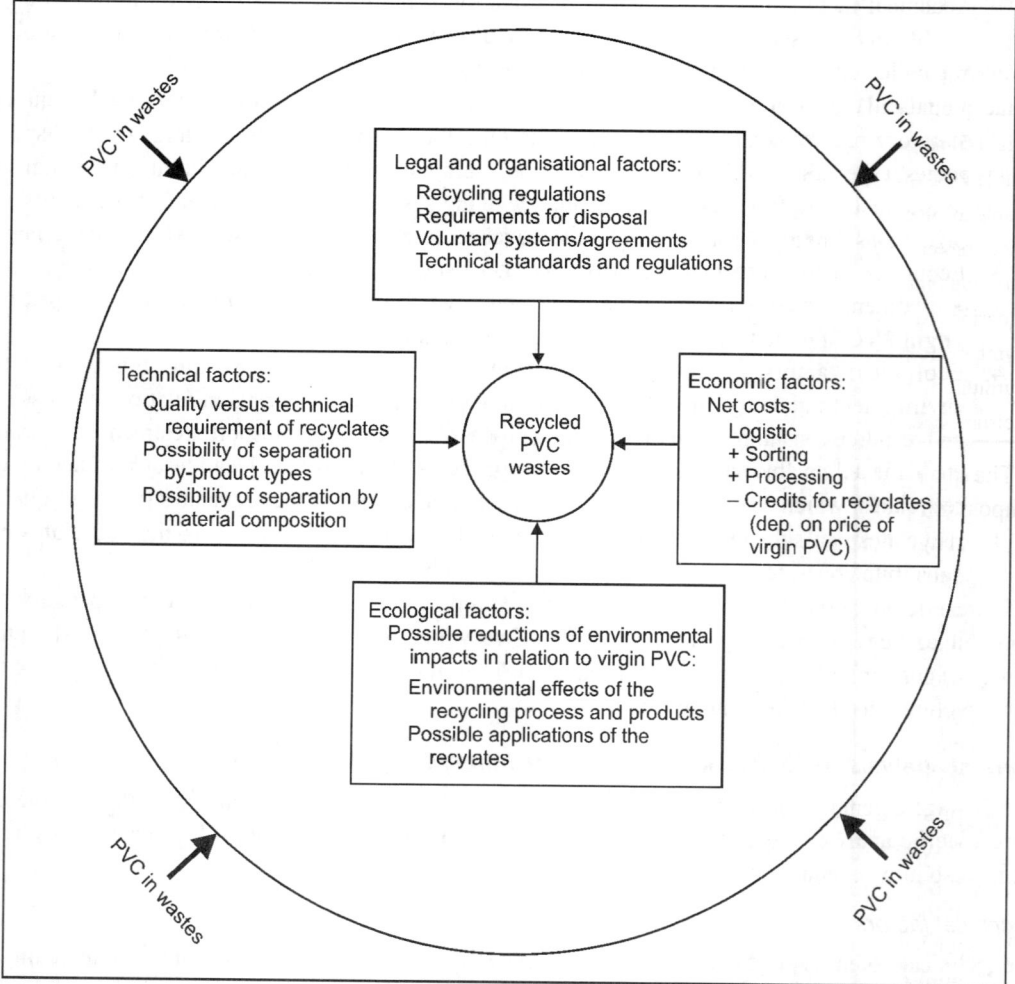

Fig. 13.1. Factors influencing recycled PVC waste quantities.

Therefore, even by separate collection of PVC wastes by type of product it is hardly possible to gain PVC material of an exactly uniform composition. For preconsumer wastes it may be possible to recover material of a defined composition (if for example a cable layer returns cut-offs to his specific supplier). This is however not the case for post-consumer wastes. As a consequence, at least for post-consumer wastes, a 1:1 substitution of virgin PVC by recycled PVC is not feasible.

Nevertheless, in some applications like window frames PVC wastes of different compositions can be mixed in practice and recycled as separate material layers.

Table 13.2. Typical composition of PVC compounds.

Application	Share of the components (weight-%)				
	PVC polymer	Plasticiser	Stabiliser	Filler	Others
Rigid PVC applications					
Pipes	98	–	1–2	–	–
Window profiles (lead stabilised)	85	–	3	4	8
Other profiles	90	–	3	6	1
Rigid films	95	–	–	–	5
Flexible PVC applications					
Cable insulation	42	23	2	33	–
Flooring (calander)	42	15	2	41	0
Flooring (paste, upper layer)	65	32	1	–	2
Flooring (paste, inside material)	35	25	1	40	–
Synthetic leather	53	40	1	5	1
Furniture films	75	10	2	5	8
Leisure articles	60	30	2	5	3

The quality of the recyclates is determined by the degree of contamination and the variation of the composition of the collected material. We distinguish between two major groups of recyclates:

1. 'High-quality recyclates' from a specific PVC application can be reused in the same application due to their low degree of contamination and similar composition. Due to the differences in the composition of the PVC compounds, the recycling material can be used as a separate layer in the new products (e.g. core of window profiles, medium layer in pipes) in most cases. One problem is that the recycled products are of different colours, so the recycling process must provide for a separation by colour or the collection must be separated by colour which in many cases is not feasible in practice. As a minimum requirement for high-quality-recyclates, soft PVC recyclates cannot be used in rigid PVC applications. Also recyclates from rigid PVC products are generally not applied for soft PVC applications since the material has to be reformulated, i.e. plasticisers and other additives have to be added. An exchange of material inside each group, soft and rigid PVC applications, is feasible to a limited extent.

2. If these requirements cannot be met by the recycling system, 'low-quality recyclates' are produced which due to a higher degree of contamination and a mixture of PVC material from different applications cannot be used but as a substitute for 'non-PVC-materials' only (e.g. general plastics, concrete or wood products). This type of recycling is generally referred to as 'down-cycling'.

The assessment of the existing PVC recycling activities will show which quality is achieved for the recyclates in practice. It should be mentioned that the quality issue of the recyclates is only partially specific to PVC. It applies also for the recycling of other plastics, where the collection and separation of pure fractions is the major bottleneck.

The achievable quality of the recyclates depends greatly on the achievable degree of contamination of the collected PVC wastes. In order to produce high-quality recyclates it is necessary to have the PVC wastes collected by type of application (pipes, windows, floorings, etc.). With this in mind, the recycling potentials of PVC wastes can be roughly classified as follows (Fig. 13.2):

1. The highest-quality PVC recyclates can be achieved from PVC production wastes: The wastes occur at PVC converters where PVC wastes of defined compositions (i.e. additive contents) are produced which can be used nearly as an equivalent to virgin PVC.

2. The (technical) recycling potential of cut-off wastes from the handling or installation of the different PVC products is also high. However, depending on the product, logistic conditions and the collection of PVC charges with specified compositions are more difficult than for production wastes, due to a disperse distribution of the 'waste producers' (e.g. large number of small workshops or enterprises producing windows or laying floorings).

3. The technical recycling potential of post-consumer wastes is generally lower than the recycling potential of preconsumer wastes since the collection of fractions with defined material compositions is not feasible in most PVC applications. Thus lower-quality recyclates are produced or expensive sorting or separation processes have to be applied. The highest recycling potentials of PVC post-consumer wastes can be expected for 'mono fractions' which can be collected separately. This applies for pipes, (rigid) profiles, bottles, a smaller part of rigid film applications, some car components (which can be disassembled) agricultural films and some medical products.

	Purity of PVC recyclates		
	PVC fraction of a homogenous composition	Mix of different PVC compounds	
1. Pre-consumer wastes			Bigger
(a) Production wastes	√	√	
(b) Cut-offs	(√)	√	
2. Post-consumer wastes			Recycling potentials
(a) PVC 'mono fraction'			
(i) Separate collection	–	√	
(ii) Mixed collection	–	(√)	
(b) Composite production			
(i) Separate collection	–	(√)	
(ii) Mixed collection	–	–	Smaller
√ = Possible			
(√) = Limited	Smaller	Recycling potentials	Bigger
– = Not possible			

Fig. 13.2. General technical recycling potentials of PVC wastes.

4. Moderate recycling potentials can be attributed to PVC-'mono fractions' in mixed wastes (e.g. profiles or pipes in mixed construction wastes, packaging films in mixed packaging wastes)

and composite materials which can be collected separately (e.g. windows and cables). In order to gain higher-quality recycling materials the first group of PVC wastes must undergo a sorting process to extract PVC, whilst the second group of PVC wastes must be treated in a mechanical separation process to separate PVC from the other materials in the related products.

5. PVC in composite products which cannot be collected separately have the lowest recycling potentials. In many cases a mechanical recycling is not feasible at all, in some cases a recycling in mixed plastics fractions may be possible yielding low-quality materials with a limited application spectrum ('down-cycling').

It must be taken into account that with recovery processes other than mechanical recycling the potentials to recover composite products and materials from mixed collections may be increased significantly. Such processes include, e.g. the 'Vinyloop' process which is based upon the dissolution of PVC wastes in a solvent, allowing for the processing of commingled PVC wastes to obtain comparatively pure PVC recyclates. All these processes are based upon chemical operations and are thus not included in this chapter.

The economic profitability of the recycling of PVC preconsumer wastes and PVC cable wastes is due to the following reasons:

1. PVC preconsumer wastes can be collected at low cost (using the distribution channels of the products, e.g. by combining delivery and take-back logistics) and in defined material qualities (separated by PVC compounds). Therefore, high-quality recyclates allowing for higher recyclate prices can be produced.

2. The economics of the recycling of PVC from cables is determined by the fact that it is a 'secondary waste', i.e. a waste from the mechanical treatment of cable wastes. Cable recycling is carried out to recover the precious copper mainly. Therefore, for economic considerations the recycling of the PVC waste fraction starts at the gate of the cable recycler, not including the collection and treatment of the cables. Thus, the recycling is profitable as soon as transportation costs plus/minus costs or credits for the processing (extrusion) of the material to new products are lower than incineration or landfill costs (including transportation). The costs of cable collection and treatment are not included in the PVC recycling costs but have to be covered by the proceeds from the marketing of the main product, i.e. copper and other metals.

For all the other waste types of post-consumer PVC, mechanical recycling is not profitable under present conditions. The existing recycling schemes for these wastes are either voluntary initiatives of industry or the result of the statutory requirements for the packaging sector, but have not been established for economic reasons. The high recycling costs are mainly due to the high cost of separate collection and sorting.

Environmental factors

Also when not being profitable in economic terms a promotion of recycling is justified when it provides environmental advantages. Mechanical recycling has been regarded as the ecologically most favourable waste management option. However, recent studies to assess the environmental performance of the mechanical recycling of plastics have shown that this does not apply principally but, depending on the recycling processes and the applications of the recyclates, the ecological advantages differ. Thus, the environmental advantages must be proven and significant in order to justify a promotion of recycling. There are two criteria which can be used to 'measure' the environmental advantages of mechanical recycling:

1. Life cycle assessments: The overall environmental impacts of the mechanical recycling must be smaller than the overall environmental effects of other waste management routes, landfilling

and incineration especially. To account for the indirect savings of resources, energy and emissions which are achieved by the substitution of 'virgin' materials by recyclates from mechanical recycling life cycle analysis is an appropriate method for this assessment.

2. Ecological and health risks: If mechanical recycling is favourable in terms of life cycle assessments, the possible exposure of humans and the environment by single toxic or eco-toxic substances must be controlled in the recycling processes.

Life cycle assessments on the recycling of PVC and plastics are available for a limited number of example cases of products and recycling routes only. Nevertheless, from the available results of selected recent studies (Table 13.3) it seems to be possible to come to the following general evaluation:

1. For production wastes, cut-offs and post-consumer wastes from which PVC can be separated easily mechanical recycling provides an environmental advantage.

2. Mechanical recycling of mixed plastics fractions provides environmental advantages only if it is feasible to sort out plastics materials which can be used in applications typical for plastics. The environmental performance of the recycling of mixed plastics for the production of products which substitute concrete, wood or other non-plastic applications is generally lower than the performance of other waste management routes such as energy recovery or feedstock recycling.

Table 13.3. Results of life cycle studies on the environmental advantages of mechanical recycling of plastics

Product/waste group	Results of related studies	
	Recycling system	Ecological advantages of mechanical recycling
Plastic production wastes	Recycling of collected production wastes compared to landfilling (incineration)	Savings: Energy: 66–90% CO_2: 89–97%
Separately collected post-consumer wastes		
PVC-windows	Window profiles with 70% recyclates compared to window profiles made of virgin PVC	Savings: Energy: 40-53% Air emissions (index): 56–69% Water emissions (index): 47–64%
PVC-windows	Window frames with PVC profiles with 70% recyclates compared to the window frames with profiles made of virgin PVC	Savings: Energy: 48% GWP^a: 42% Water emissions (index): 52%
PVC-pipes	Sewage system with multilayer pipes with 50% PVC recyclates compared to the sewage system with pipes from virgin PVC	Savings: Energy: 10% CO_2: 9% NO_x: 16% COD^b: 37%
Mixed post-consumer wastes		
Packaging wastes	Recycling of separate collected plastics according to the present situation in Austria from household packaging wastes compared to landfilling (incineration)	Savings: Energy: 37% (27%) CO_2: 12% (16%)

(Contd...)

Product/waste group	Results of related studies	
	Recycling system	*Ecological advantages of mechanical recycling*
Packaging wastes	Recycling of sorted plastics fractions from household packaging wastes compared to energy recovery	Use of the recyclates in cable pipes: mechanical recycling advantageous
		Use of the recyclates in waste bags: mechanical recycling not advantageous
Packaging wastes	Recycling of mixed plastics from household packaging wastes for use in products which substitute wood or concrete compared to energy recovery of the mixed plastics fraction	No advantage for mechanical recycling

[a]GWP = Contribution to the greenhouse warming potential.
[b]COD = Chemical oxygen demand (water).

With regard to possible ecological and health risks associated with the mechanical recycling of PVC the general situation can be summarised as follows:

1. Collection, sorting and treatment of plastics wastes is not associated with specific 'new' risks related with the exposure of workers and environment to hazardous substances. General risks like accidents in transportation processes or accidental fires in material stores do exist. However they are not specific for mechanical recycling but represent general risks existing in other waste management routes as well.

2. Possible specific risks of mechanical PVC recycling are related with toxic substances in the recycling material. There are two major issues:

    (a) Heavy metals and other additives: Some PVC products like window frames, pipes and cables contain heavy metal stabilisers which (as single substances) are toxic (cadmium and lead compounds especially). A special matter of concern has been the cadmium stabilisers in window frames. In recent years the use of cadmium has been reduced significantly. However, it is still applied. Notwithstanding this development the old windows to be disposed of contain cadmium in significant amounts. When they are recycled mechanically, the cadmium stabilisers will be brought into new products. The evaluation of the associated risks has been a matter of controversial discussions. Since the heavy metal compounds are fixed in the PVC matrix a release of the toxic substances to the environment is not possible but in the production of the stabilisers, the compounding of PVC, waste disposal (incineration, landfill) and accidental fires. In general the quantities which can be released in this way are low compared to other sources of heavy metal emissions. Therefore the environmental and health risks of the stabilisers are regarded as not relevant by some experts. Others argue that for precautionary reasons toxic and persistent substances like heavy metals should be extracted from the technosphere principally and disposed of safely to avoid risks to health and the environment.

    Generally the risks must be regarded as less critical in 'product-to-product' recycling systems (i.e. recyclates from window profiles are exclusively used in new window profiles) than in 'open' systems where the recycling material is used in a variety of other products, thus having no control over the substance flow. However also in the latter case the respective potential releases of heavy metals could be considered as low compared to other sources of heavy metals emissions.

(b) PCB in the PVC fraction from cable recycling: In the past, polychlorinated biphenyls (PCB) were added to PVC cable compounds for some high voltage cables to increase the insulation performance and for low voltage cables as flame retardant and plasticisers. A fraction of the cables contained in electric/electronic devices will be recycled in recycling systems for electronics wastes. Other sources of PCB and other toxic substances in electric and electronic wastes are transformer oils or condensators. As a consequence PVC recyclates from cable recycling and electric/electronic wastes recycling can be contaminated with PCB, which is brought into the products produced with the related recyclates. In contrast to heavy metals which are fixed in the plastics matrix PCB can be released from the plastics, thus constituting a chronic risk potential for health. PCB in products is subject to statutory regulations. Recyclers and users of the recyclates are controlling the PCB content of the materials and are able to comply with the legal concentration limits. However, in Germany especially there are discussions and proposals to reduce the existing concentration limits down to a level where compliance of the PVC recyclates maybe not feasible. This would effect PVC recycling immediately. However, since PCB is not used any more, the restriction would be effective temporarily only, until the PCB-free materials will become wastes.

3. There are some other issues related with PVC waste disposal and recycling which will not be discussed in this chapter since they are not connected with mechanical recycling (e.g. potential dioxin formation in thermal waste incineration or recovery).

As a conclusion of this issue, the environmental impacts of mechanical recycling can be given for the different PVC applications (Table 13.4), taking into account the product-specific composition of the PVC compounds and the product-specific general potentials for a separate recovery.

Table 13.4. General assessment of the environmental benefits of mechanical PVC recycling by PVC applications.

PVC applications[b]	Environmental benefits of mechanical PVC recycling[a]	
	Life cycle improvements	Control of toxic dispersion
Construction materials		
Pipes (R)	+	–
Window profiles (R)	+	–
Other profiles (R)	+	–
Floorings (S)	(+)[c]	–
Roofing membranes (S)	–?	–
Cables (S)	–?	–
Consumer and other short-life products		
Bottles (R)	+	+
Packagings (R)	–	+

(Contd...)

PVC applications[b]	Environmental benefits of mechanical PVC recycling[a]	
	Life cycle improvements	Control of toxic dispersion
Other applications (separate PVC recovery)	+	+
Other applications (no separate PVC recovery)	–?	+
Cars		
Components (R + S)	+	+
Others (cable, coating)	–	–
Electric/electronics	–	–
Other products (health, agriculture)	–?	+

[a] + = benefits through mechanical recycling; – = no benefits through mechanical recycling; ? = information lacking, no benefits expected through mechanical recycling.
[b] S = Soft PVC applications; R = Rigid PVC applications.
[c] Contaminations (like sand) can cause problems for the recycling.

DESCRIPTION OF EXISTING PVC RECYCLING SYSTEMS

The term 'recycling system' includes the whole material chain starting with the PVC waste at the place where it arises and ending with the recyclates which are used for the production of new products (Fig. 13.3).

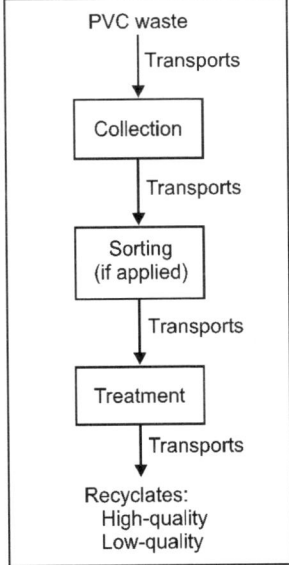

Fig. 13.3. Recycling system for PVC wastes (schematically).

The material flow can include the following steps:
1. Collection of the wastes, e.g. in pick-up systems or bring systems, in pure or mixed fractions.
2. If the PVC wastes are collected in mixed fractions (e.g. mixed packaging wastes) some fractions may be separated in sorting plants to obtain pure materials for further treatment.
3. Mechanical treatment processes aiming at the (more or less) automatic separation of pure fractions of PVC and other materials and the production of recyclates with a defined particle size; generally

the mechanical treatment process consists of shredding units for size reduction, separation units to extract specific sizes or materials from the main material flow (e.g. magnetic drums to separate ferrous metals) and mills and extruders to convert the separated plastics fractions into re-granulates.

Between each of these steps transports may be necessary depending on the organisation and the location of the related plants.

With regard to the recyclates produced we distinguish between:

1. 'High-quality recyclates' which due to their low degree of contamination can be reused in the production of the same products.
2. 'Low-quality recyclates' which can be used as a substitute for 'non-PVC-materials' only, generally referred to as 'downcycling' (e.g. general plastic, concrete or wood products).

The 'recycling system' is not only characterised by the material flow and the related technical collection, sorting and treatment methods but also by its organisation. This includes the way the material flows are managed (e.g. specific recycling organisations or free market) and the type of financing (e.g. waste fees or fees on the price of the related products) especially.

In order to describe the different recycling systems the following criteria have been used:

1. Methods and technologies:
   (a) Collection.
   (b) Sorting.
   (c) Treatment.
2. Capacities and quantities: capacities of the recycling plants and annual quantities of recycled PVC wastes.
3. Input: origin and composition of the recycled wastes (post-consumer/preconsumer; mixed/pure).
4. Output: quality (composition, degree of contamination) and use of the recyclates.
5. Geographical area covered, transport intensity.
6. Organisation: free market system or specific recycling organisation; duties and competences of the recycling organisation.
7. Financing: who pays the recycling costs?
8. Costs: overall cost of the recycling chain (gross cost and net cost including credits for recyclates).

Some of the systems which have been analysed will be described subsequently.

Description of Selected Recycling Systems

Mechanical recycling or preconsumer PVC wastes

For the mechanical recycling of preconsumer PVC wastes a variety of 'recycling systems' exist. Cut-offs and production wastes are either given back directly to the relevant PVC converters or they are treated mechanically to produce recyclates of a defined size and quality. The recycling is not organised by special recycling organisations but there exists a market for PVC preconsumer wastes which depends on price considerations.

The free market recycling structure consists of different, mainly small and medium enterprises which are occupied with one or several of the major operations in the recycling cycle: collection, grinding, compounding and processing (i.e. users of the recycling materials). There is no specific structure for PVC wastes, but the whole spectrum of plastics recycling is covered. For example in Italy about 20 grinders and 60 compounders recycle PVC. The number of PVC processors is 600–700.

There is a close linkage between the recycling market and the market for virgin PVC: The recycling activities fluctuate with the price for virgin PVC to some extent. If virgin PVC prices are down the economic feasibility of some of the recycling activities ceases and they are stopped until the virgin PVC prices exceed a certain level again. In Italy, the production of PVC recyclates fell from about 1,20,000 tons to 1,10,000 tons between 1997 and 1998, as a result of the drop of PVC prices. It follows a description of the mechanical recycling of preconsumer PVC wastes by using the above mentioned criteria as far as sensible.

Methods and technologies

1. The PVC wastes are collected by the PVC processors (production wastes) and the users of PVC intermediate products (e.g. the packaging industry using PVC films) and the handicraft enterprises installing PVC floorings, roofing membranes and other products.
2. Depending on the specific application the PVC wastes must be treated in a mechanical process (grinding) to produce regranulates of a defined size and composition. Some of the plants for the mechanical treatment of preconsumer wastes simultaneously process PVC post-consumer wastes.
3. The recycling material can be bought by compounders which often blend it with virgin PVC to produce compounds of a defined quality.

Input

The economic feasibility to recycle a specific type of preconsumer wastes depends on the degree of contamination (e.g. sand, dirt), the defined formulation of the material and the deliverable quantities. Most production wastes can be collected in a defined specification, in some cases (e.g. recycling of cut-offs of a packaging producer using the products of a specific PVC film producer) even in a defined formulation. The recycling of cut-off wastes from the laying of floorings, cables, pipes, etc. is often more difficult, for most are mixtures of PVC compounds from different PVC processors and they may be contaminated or they are collected together with other plastics (e.g. plastic pipes) making additional washing and/or sorting necessary before grinding.

Output

In general the recycling of preconsumer wastes yields high-quality recyclates: A part of the recyclates can be used by the PVC compounders or processors as an equivalent for virgin PVC. The other part can be used for example as separate layers in coextruded PVC-products (e.g. profiles with a core of recyclates) or as backing material of floorings, i.e. it replaces virgin PVC, but not '1:1'.

Organisation

There is no specific recycling organisation but the recycling activities are driven by the 'free market'.

Financing

As a result of the 'free market' system the recycling is ultimately 'financed' by the waste producers. Ideally, every actor in the material chain (recycler, collection service, transport service) charges cost prices for his services. The recycling is finally profitable if the net recycling costs (collection plus transport plus treatment minus proceeds for recyclates per ton of waste collected) can compete with waste disposal prices (esp. landfilling or incineration) — or from the other point-of-view — if the gross recycling costs per ton of recyclate output (at a given fee on the waste arising) are lower than the achievable prices for the recyclates.

Costs

The recycling costs for preconsumer wastes vary depending on the specific case and type of wastes. Thus no general valid cost figure can be given but only an order-of-magnitude of the recycling costs.

Mechanical recycling of PVC cable insulations

As for the mechanical recycling of preconsumer PVC wastes, the recycling of cable insulations is not carried out by specific recycling organisations but there exists a 'free market' for these wastes. PVC cable insulations are recycled for pure economic reasons, i.e. the recycling is a competitive or near-competitive waste management option. Therefore, like in the case of preconsumer wastes, the recycling activities fluctuate with the price for virgin PVC and the prices for alternative waste disposal options for the PVC wastes. Post-consumer PVC cable insulations arise as a waste fraction in the mechanical recycling of cables. It is a mixed plastics waste which contains also other plastics and contaminations. As mentioned, being a secondary waste, it constitutes a special case amongst the post-consumer PVC wastes. The material is already available in its final form for extrusion into new products. The primary objective of cable recycling is the recovery of the copper content of the cables. Therefore, for the cable recycler the costs of cable collection and treatment are attributed to the primary outputs of cable recycling, i.e. copper and other metals (e.g. aluminium). The PVC waste fraction is a cost factor only. Thus its recycling is profitable as soon as transportation costs plus/minus costs or credits for the processing (extrusion) of the material to new products are lower than the cost for incineration, landfilling or possibly energy recovery (including transportation). It follows a general description of the existing cable recycling activities in the EU (Austria, Belgium, Denmark, France, Germany, the Netherlands, UK, Italy and Spain).

Methods and technologies

1. Collection: The cable insulation material arises as an output of cable recycling. Hence there is no collection as such, but only the transportation from the cable recycler to the user of the recyclates (plastics processors).
2. Treatment: Generally the material is used by plastics processors, e.g. for the extrusion or injection moulding of plastics products, without extensive prior mechanical operations.

Material quality and use

The PVC cable insulation material is a mixed plastics fraction containing about 80 per cent PVC compounds (approx. 50 per cent of pure PVC and 50 per cent plasticisers, fillers and other additives, e.g. lead stabilisers). The remaining 20 per cent are other plastics such as polyethylene including about 2 per cent of contaminations (e.g. residual metal content). The material is used for applications similar to the products of mixed plastics recycling, e.g. poles for roads, industrial floorings and other products substituting concrete and wood products. The presence of metals residues in the plastics fraction prohibits the recycling as cable insulations. The plastics material can contain PCB due to the recycling of older cables where PCB has been used as an additive to PVC compounds. Therefore regular measurements are being carried out. According to the interviewed recyclers the limit value of 50 mg PCB per kg can be achieved generally. However there are discussions to reduce the standard down to 5 mg/kg, in Germany especially. According to the recyclers, compliance with this value will hardly be possible. Another environmental issue is the content of toxic lead additives.

Organisation

There is no specific recycling organisation but the recycling activities are driven by the 'free market'. The actors in the market are cable recyclers, trading companies and plastics processors.

Financing

As a result of the 'free market' system the recycling is ultimately 'financed' by the waste producers, i.e. the cable recyclers on the basis of economic considerations (recycling fee versus fee for other waste management options).

Costs

Exact cost calculations have not been made available by the recyclers.

Mechanical recycling of PVC window frames in Germany

In Germany two recycling systems for PVC windows have been established:

1. The 'VEKA system', an investment of the window profile producer VEKA AG who has established a recycling plant for preconsumer and post-consumer window profile wastes in Thuringia.
2. The 'FREI system', an organisation founded by 14 profile producers which has established a system of collection points at local recycling companies and cooperates with a recycling plant in North Rhine-Westfalia.

The following description refers to the VEKA system mainly, whose recycling plant has been visited. Major differences to the FREI system are pointed out.

Methods and technologies

1. There exist two major pick-up systems: Collection by the transport services of VEKA at window producers/assemblers and collection by independent container services at construction sites or window producers.

   (a) In the first case the collection of old windows has been integrated into the take-back system for cut-off wastes (preconsumer wastes). Based upon co-operation contracts, VEKA takes back the cut-offs together with the old windows if a minimum threshold quantity is reached. To date the old windows are no more than a marginal position in the system, since the major quantities are cut-off wastes. For each old window a charge of 25 DM is raised, i.e. approx. 500 DM or 255 Euro per ton. This system is used for old windows from renovation projects mainly.

   (b) In the second case, the old windows are picked-up by independent container services, which co-operate with VEKA but work on their own account. This system is used mainly for the pick-up at construction sites, e.g. in case of demolition projects or for small window workshops who do not produce sufficient quantities to fill a complete lorry. In this case the container services are charged between 70 and 250 DM/t, depending on the degree of contamination and whether the cartload contains window panes or not. The owner of the old windows pays directly to the container services who has to cover the VEKA charge, the transportation costs and his margin.

   The FREI recycling system is organised as a bring-system: A network of more or less centralised collection points has been established (about 110 collection points distributed over Germany), there the old windows can be delivered. The transport to the collection point is organised by the owner of the old window on his own account. The charge for the old windows f.o.b. collection points is 15 DM per window (approx. 300 DM or 150 Euro per ton). The charge for delivery at the recycling plant is 200–300 DM per ton.

2. Sorting and mechanical treatment: The VEKA recycling plant has been constructed on the principle of maximum automatisation. The mechanical treatment process consists of the following major unit operations (Fig. 13.4):

   (a) Shredder unit.
   (b) Magnetic separation of Fe-metals.
   (c) Screening unit where the coarser fractions are fed back to the shredder (directly or after passing a non-ferrous metal and rubber separator).

(d) The fine fraction passes separation units for glass and residual metals.

(e) Wet grinding unit with subsequent rubber separation.

(f) Colour separation (in a coloured and a white fraction).

(g) Extrusion to regranulates with microfiltration to remove residual contaminations.

One speciality of the process is the colour separation. However, this unit is still in the process of technical optimisation. At present an improved system is being installed.

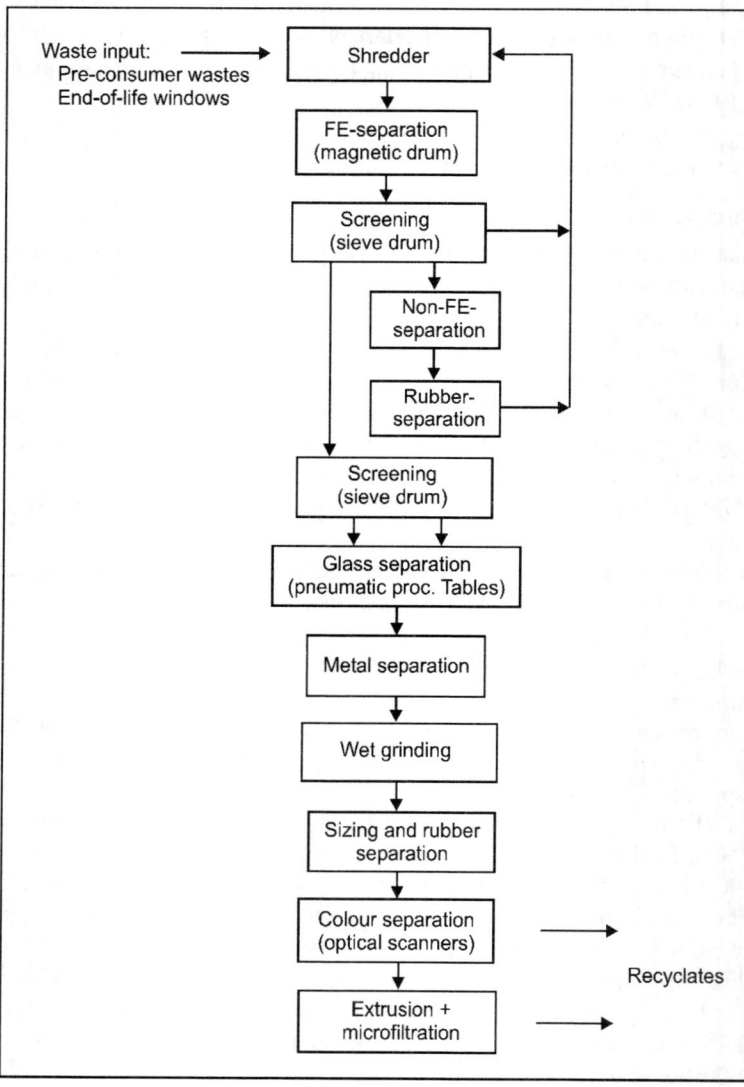

Fig. 13.4. Recycling plant for PVC window profiles (VEKA).

Compared to this process the recycling plant of the FREI system is a 'low-tech solution'. A major part of the separation and sorting (disassembling, separation of glass and rubber, separation by colours) is done manually. So the mechanical unit operations are restricted to shredder, grinding and metal separation before extrusion with microfiltration.

Capacities and quantities: The utilisation of the recycling plants is dominated by preconsumer wastes from the production and installation of window profiles:

1. The VEKA plant has processed about 14,000 tons in 1998, of which only 4000 tons where end-of-life windows (corresponding to about 2000 tons of PVC compound). The capacity of the plant is 20,000 tons per year (after removal of bottlenecks).

Mechanical recycling of PVC pipes in Netherlands

In the Netherlands the FKS association has been founded by the six national producers of plastics pipes (Draka Polva, Dyka, Martens, Omniplast, Viplex, Wavin) to organise the recycling of plastics pipes in the Netherlands.

Methods and technologies

1. Collection: About 50 collection points have been set up all over the country where used pipes can be delivered free of charge. In parallel, rental containers have been installed at specific customers. Container rent and transportation costs are charged to the customer (about 100 Euro per ton on an average).
2. Sorting and mechanical treatment: There exists one recycling plant at the company Wavin (Fig. 13.5). The mechanical treatment process consists of the following major unit operations:
   (a) Manual sorting line, there PP- and PE-pipes are separated from PVC pipes; this is possible due to the different colours (there is no labelling for different plastic types).
   (b) Shredder unit.
   (c) Separation units for rubber, Fe-metals and non-ferrous metals.
   (d) Sieve where sand and a coarse PVC fraction which is returned to the shredder are separated.
   (e) Extrusion to regranulates with microfiltration to remove residual contaminations.
3. Quantities: According to ECVM about 3000 tons of PVC pipes have been recycled by the FKS system in 1996/1997. Due to the co-treatment with used PE and PP pipes as well as preconsumer wastes, the capacity of the recycling plant is higher.
4. Input material: The input material to the recycling plant is used plastics pipes (PVC, PE and PP) together with preconsumer wastes (e.g. cut-offs). The contamination of this material is about 4 per cent.
5. Output material/recyclates: The recycling plant produces high-quality recyclates. In the Netherlands the recyclates can be used for the production of new multilayer pipes in a co-extrusion process where the recyclates constitute the middle layer in the pipe wall and the inner and outer layers are made of virgin PVC. In other countries like Germany the existing technical EN standards did not allow for the use of recyclates in new pipes by now. Here, the recyclates are used in applications with low material standards, e.g. cable channels. However, the relevant standards are in the process of revision. In future, recyclates from used pipes and other products can be used for pipe production provided that the specification of the used materials is known. This is certainly the case for used pipes, but not necessarily for other plastics wastes.

A possible technical limitation for the use of the recyclates may be residual contaminations whose complete removal requires extensive technological measures (information of the German recycling organisation for pipes).

Impact of PVC on the Recycling and Recovery of Plastics and Other Materials

In the recycling of mixed plastics wastes (containing also PVC) only metals and stony material is separated before the process to avoid abrasions and damages of the shredder and the grinder. After shredding and grinding, the agglomerated or just grinded mixed plastic is plastified in an extruder. When assessing the impact of PVC in mixed plastics recycling processes two cases must be distinguished:

1. The recycling of plastics wastes with a low PVC content.
2. The recycling of 'PVC-rich' plastics wastes.

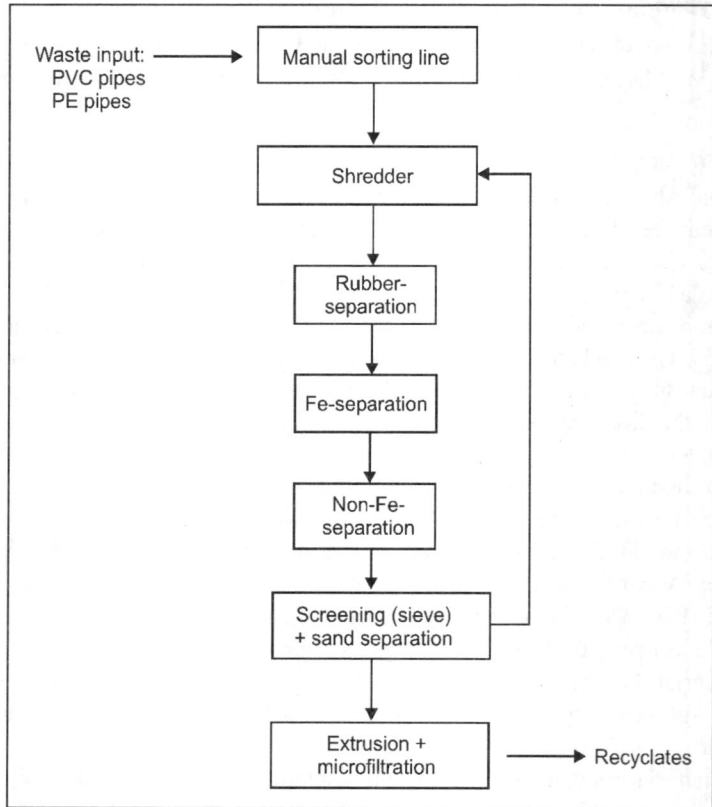

Fig. 13.5. Recycling plant for plastic pipes (Wavin).

In practice the most important materials in mixed plastics fractions are polyolefins (PE and PP, e.g. in plastics packaging wastes). The achievable quality of the products of the mixed plastics recycling depends on the melting temperature, on possible chemical reactions and on the rheological properties (flowing properties) in the extruders. PVC has a comparatively low processing temperature with a narrow 'temperature window'. If the processing temperature of PVC is exceeded, hydrochloric acid (HCl) is released, material structures can be destroyed, and the risk for corrosion of the facilities (especially in the extruder) increases.

As a consequence, in the recycling of mixed plastics fractions with low PVC contents the polyolefin content must be at least 70 per cent, in order to avoid that the PVC disturbs the processing and degrades the material quality. To avoid these negative impacts the HCl can be absorbed with lime or limestone. However, if the PVC content is small (e.g. less than 5 per cent like in packaging wastes) in most cases a dehydrogenation unit is installed prior to the extrusion where the chlorine is removed in a thermal and chemical process. In the processing of plastics fractions with a high PVC content and limited amounts of other plastics the processing temperature of the mixed fraction must be in the range of pure PVC processing, i.e. 180°–210°C. Since they provide more flexible 'temperature windows' than PVC polyolefins do not disturb the processing of these PVC-rich fractions in the same way as PVC disturbs the recycling of polyolefin-rich fractions as described above.

However, the presence of PET and rubber disturbs the extrusion process of PVC-rich plastics fractions resulting in a poor product quality.

As a conclusion it can be maintained that the processing of PVC together with other plastics is generally possible. However, normally this co-processing will lead to a lower quality. In mixed plastics fractions with low PVC contents PVC disturbs the co-processing, requiring additional measures where PVC or chlorine is removed from the process. Due to this fact it seems to be questionable to speak of 'PVC recycling' for this case. However, the related PVC quantities (e.g. PVC in packagings) are usually included in the recycling balances for PVC.

In addition to mixed plastics wastes PVC can disturb the treatment and recycling of other wastes too. PVC can contribute to the formation of dioxines in thermal waste treatment processes, especially in the presence of metals. An example is the recycling of metal scraps, e.g. from car shredder plants which can result in dioxine formation if the metal is coated with PVC (e.g. underground protection of cars).

ASSESSMENT OF PVC RECYCLING: LIMITS AND POTENTIALS

Based upon the experiences with existing PVC recycling systems and the general considerations concerning the future development of PVC waste quantities and PVC recycling, the general potentials and limits of PVC recycling can be determined. The analysis is carried out in two steps:

1. In the first step the technical recycling potentials are assessed by product group, based upon practical experiences with recycling schemes in general. The recycling potential is the maximum PVC quantity which can be recycled under practical conditions, i.e. taking into account the available collection systems as well as sorting and separation technologies.

2. In the second step it will be discussed to which extent these technical potentials can be realised in practice, taking into account the economic and environmental limits experienced in existing PVC and plastics recycling systems.

The major factors limiting PVC recycling are at the same time the starting-points for measures to improve PVC recycling.

Technical Potentials for Mechanical PVC Recycling

As mentioned above, the technical potential of mechanical PVC recycling refers to the maximum achievable PVC quantity under practical conditions. This represents not strictly a technical potential, but economic aspects are taken into consideration too: technical solutions which involve excessive costs, for example the separation of PVC from mixed waste streams with very small PVC contents (smaller than 5–10 per cent) in sorting plants, are not included. One example is PVC in mixed packaging wastes, where PVC is not sorted out as a separate fraction but recovered together with other plastics in a mixed fraction.

The assessment of these technical potentials is based upon the general considerations concerning the development of PVC waste quantities and the experiences with recycling schemes in general and the existing PVC recycling systems in particular. When determining the technical recycling potential one important point must be stressed that recycling potential depends on the quality of the recyclates to be achieved. So, in accordance with the general considerations two types of mechanical PVC recycling can be distinguished.

'High-quality recycling' yielding recyclates which can be reused in the same PVC applications. 'Low-quality recycling' or 'downcycling' yielding mixed materials of different types of PVC or different types of plastics. For both types of PVC recycling the recycling potential depends on the potentials to separate a pure PVC, mixed PVC or mixed plastics fraction respectively by separate collection, subsequent sorting or subsequent mechanical treatment.

In the following sections the recycling potentials for both 'high-quality' and 'low-quality' mechanical PVC recycling will be assessed product group by product group applying the following general procedure:

1. First the product or waste groups which are suitable for mechanical recycling will be determined based upon the potentials to separate out PVC fractions of an appropriate purity from the related waste streams.

2. Secondly, the potential recycling rates (percentage of PVC waste arising which can be recycled mechanically) will be determined by estimating the achievable collection rates and (in case of 'high-quality' recycling) the achievable absorption rates for recyclates in the different products.

3. Thirdly, the recycling quantity must exceed a certain minimum to be able to operate recycling plants of feasible sizes.

Technical potentials for 'high-quality' PVC recycling

High-quality recycling is applicable for those product groups only there a pure PVC fraction of a sufficient quantity can be separated; this may be by separated collection, subsequent sorting or subsequent mechanical treatment.

When checking this for the different PVC waste groups a first important distinction has to be made between preconsumer and post-consumer wastes:

1. Preconsumer wastes (production wastes or installation wastes like cut-offs) are suitable for mechanical recycling for more or less all PVC product groups: At least a part of the preconsumer waste arising in most product groups can be returned to the processors and reused directly without major sorting or pre-treatment. The main exceptions are PVC applications in the areas of coatings, organosols and plastisols where processing wastes are no more available for mechanical recycling.

2. As for post-consumer wastes, only a part of them can be separated. The following product or waste groups especially cannot be regarded as candidates for high-quality recycling:

   (a) Composite materials, applications of PVC pastes especially: For products like paste based floorings, wall coverings, car underfloor protection, artificial leather or coatings, the separation of the plastics content by mechanical treatment is not feasible. The same holds for composite products like furniture films or multilayer packaging films.

   (b) Mixed post-consumer plastics fractions from which PVC in an appropriate quality cannot be separated with the existing technologies: this applies to PVC from the recycling of cables, electric/electronic appliances and cars. In the mechanical treatment processes of these products (shredding with subsequent separation of different material fractions) PVC is recovered mixed with other plastics or materials. In packaging and other product groups like health care products, the separation of pure PVC is not feasible, due to the low PVC content (far below 10 per cent) in the related waste streams.

PVC post-consumer wastes with a potential for mechanical recycling

Based upon these criteria, the major part of PVC preconsumer wastes has a potential for mechanical recycling (Table 13.5).

Table 13.5. Recycling and absorption potentials of the major PVC product/waste groups suitable for high-quality mechanical recycling.

PVC product/waste group	Potential recycling rate (%)		Potential absorption rate[a] (%)
	Preconsumer wastes	Post-consumer wastes[b]	
Construction products			
Flooring calandered (F)	40	20–30 (25)	30
Profiles and hoses (F)	50	15–25 (20)	20
Pipes (R)	50	60–70 (65)	35
Window profiles (R)	75	50–60 (55)	45
Profiles—cable trays (R)	75	30–50 (40)	40
Other profiles (R)	55	30–50 (40)	30
Packaging products			
Bottles (R)	90	35–45 (40)	–
Furniture components	10–35[c]		0–5
Other consumer and commercial products			
Shoes, soles (F)	90	15–25 (20)	35
Miscellaneous (F)	90	5–15 (10)	–
Printing films (R)	80	30–40 (35)	–
Sheets, chemical equipments (R)	80	30–40 (35)	30
Miscellaneous sheet products (R)	70	20–40 (30)	30
Miscellaneous rigid profiles (R)	75	10–20 (15)	10
Other rigid products (R)	90	10–20 (15)	10
Electric/Electronics	10–90[c]	–	0–10
Automotive	15–90[c]	–	0–10
Other products	65–75[c]	–	–

F = Flexible PVC applications; R = Rigid PVC applications.

[a]Related to post-consumer wastes.

[b]In brackets = chosen rates.

[c]Absolute quantities are low; range of recycling rate depending on the product group.

The set of potential post-consumer PVC is much more limited. It comprises the following groups: The high-volume construction products are specifically suitable for recycling since they allow for comparatively favourable logistical conditions, i.e. they can be collected separately or separated in sorting plants for construction wastes. For floorings, pipes and window profiles, recycling systems with separate collection already exist in some EU Member States. Additional high-volume candidates are other rigid profiles (cable trays and shutters especially) as well as flexible profiles and hoses. The other high-volume construction products are not suitable for high-quality recycling; floorings produced from PVC pastes and wall papers are composite products and cable insulation wastes are mixed plastics wastes. By now the reuse of recyclates from pipe recycling for the production of new pipes is limited by technical standards. These standards are however in the process of revision in order to lift these restrictions. Thus we expect that the high-quality recycling of pipes will be possible before 2014.

In the group of consumer and commercial products (packaging products, furniture components, other consumer and commercial products) there is a large number of composite products (e.g. coating applications like artificial leather, organosol and plastisol applications or furniture components) and low-volume PVC applications which are not suitable for high-quality mechanically recycling. However, there is still a number of potential candidates remaining. Most of these products arrive at municipal solid wastes or packaging wastes collections by now. If they are to be recycled mechanically, it will not be feasible to have specific collection systems for PVC or the related PVC products respectively, but the collection will be organised for the related product-groups as a whole, together with other materials. The existing collection systems for packaging (separate collection of plastics or 'light' packaging wastes) are examples for such systems.

To achieve high-quality recyclates PVC must be sorted out in sorting plants or separated in mechanical treatment processes subsequently. The feasibility of these operations depends amongst others on the PVC content of the collected waste fractions. If it is too low, feasibility cannot be achieved. Therefore, at least for some of the product groups the recycling potentials will be limited due to the low PVC contents in the collected wastes. This is especially true for plastics packaging collections where polyolefins (PE and PP) are the dominant polymers and PVC packaging films are of minor importance (although constituting a significant PVC waste stream). The only packaging products suitable for high-quality recycling are PVC bottles in some Member States. They are collected together with PET bottles with subsequent separation of a pure PVC material fraction. However, PVC bottles are increasingly substituted by PET, so that a gradual phase-out of the PVC bottle recycling activities is expected. For the recycling of the non-packaging PVC candidates, separate plastics collection systems (with subsequent separation of the polymers) are a potential solution.

Waste streams from commercial and industrial sources are concerned here (printing films and sheets especially). Another potential waste stream is used footwear. A potential solution for their recycling is through collection systems for used textiles and clothes where PVC shoes can be sorted out as a separate fraction.

The present practice of the recycling of electric/electronics and cars is a mechanical treatment: After limited dismantling operations to remove hazardous substances or reusable components, the end-of-life products or vehicles respectively are shredded and in subsequent separation operations different material fractions are separated. For economic reasons the separation aims at the recovery of the metals mainly, whilst plastics end up in mixed fractions which are landfilled or incinerated to a large extent (shredder residues in the case of car shredders and mixed plastics fractions with contaminations in the case of electronic scrap recycling). So, the high-quality mechanical recycling of PVC in these products is limited to the preconsumer wastes arising in pure fractions.

For the remaining other products (medicine and agriculture) waste arisings are too low to allow for a feasible operation of a PVC recycling system.

Potential recycling rates

Table 13.5 includes also the estimated potential recycling rates for high-quality mechanical recycling.

The potential recycling rates for preconsumer wastes recycling have been oriented at the present recycling practice.

The potential recycling rates are the results of the combination of:

1. The potential collection rates, i.e. the percentage of the total waste arising per product group which can be separated from the related mixed waste streams (e.g. municipal solid wastes, mixed construction wastes) by separate collection systems; this can be either product-specific collection systems where PVC is collected together with other materials (like in the case of packaging waste) or PVC-specific collection systems (like in the case of windows).
2. The potential percentage of the PVC quantity contained in the separately collected wastes which can be separated as a pure PVC fraction (suitable for high-quality recycling); PVC fractions which cannot be separated in sorting or mechanical treatment processes as well as material losses in the recycling processes (sorting, mechanical treatment) are subtracted here.

As a matter of course, for preconsumer wastes higher recycling rates can be achieved than for post-consumer wastes. High recycling rates of more than 70 per cent can be achieved in those product groups where production wastes constitute the major part of the preconsumer wastes, e.g. in the cases of bottles, shoes/soles or injection moulding components for cars or electronic devices.

Lower recycling rates of less than 70 per cent can be achieved in those product groups where cut-offs from the installation of the products form the major part of the preconsumer wastes; here PVC is contaminated, a component of a composite product or mixed with other wastes. This applies e.g. for the major part of the building products.

The potential recycling rates for post-consumer wastes range from 5 to 70 per cent depending on the product group. The estimates for some important waste groups can be commented on as follows:

For floorings the achievable collection rate can be considerably high (about 80 per cent of waste arising) taking into account the experiences with similar separate collection systems. The material can be discarded separately in containers at the construction sites. However, only a limited percentage of that (in the order of 30 per cent) can be recycled to high-quality recyclates, due to contaminations like glue, sand or concrete. Thus the potential overall recycling rate is reduced to about 20–30 per cent.

For pipes and fittings the estimated potential recycling rate is 60–80 per cent. This figures refers to the quantity of used pipes which actually arises as waste. This is only a part of the total quantity of worn-out pipes (estimated 30 per cent), whilst the major part is left in the ground for cost reasons (unless the old pipes are replaced by new ones at the same location). Therefore, compared to the total quantity of used pipes the recycling rate would be 20–25 per cent only. It can be expected that the collection of PVC pipes will be organised together with other plastics pipes like PE, as it is practised in the existing recycling systems. Thus, a subsequent sorting process is necessary to separate the PVC from the other plastics. Due to the size of the collected parts and different colours the separation is comparatively easy to manage. This results in comparatively high potential recycling rates, taking also into account losses in collection (about 20 per cent), sorting (about 10 per cent) and mechanical treatment (about 10 per cent).

Also for window profiles comparatively high recycling rates (50–60 per cent) can be achieved. The conditions for collection are favourable like in the case of flooring, resulting in an estimated potential collection rate of 80 per cent. Old windows can be returned by the window producers when old windows are replaced by new ones (renovation) or dismantled and kept separately in demolition projects. Of the collected window profiles 60–70 per cent can be recycled to high-quality recyclates, taking into account that the mechanical recycling process requires extensive separation operations (rubber, metals, glass, sorting by colours, etc.).

Based upon the experiences with existing packaging recycling systems, the potential collection rate for PVC bottles can be in the order of 80 per cent (in mixed collection systems). Also based on the existing experiences we estimate that 50 per cent of this quantity can be recycled to high-quality recyclates, taking into account the sorting and mechanical treatment processes.

For PVC in households or commercial wastes, like book covers, bags, camping articles and shoes, only mixed collection systems are feasible with subsequent sorting and mechanical treatment. Therefore, for some products considerable collection rates at or above 50 per cent can be achieved. This applies for PVC in commercial wastes (office supply, printing films, sheets) especially, which arise in sufficiently large charges to make a separate collection feasible (e.g. together with other plastics wastes). Also for some consumer product wastes arising in considerable quantities (footwear especially) existing or new collection systems (e.g. textile collections) with subsequent separation of the PVC products are feasible. However the yield of high-quality recyclates in sorting and mechanical treatment processes will be comparatively low, resulting in overall potential recycling rates between 10 and 20 per cent only. In the remaining product/waste groups the recycling potentials are limited to a few candidates only. Table 13.5 includes also the estimated absorption rates which specify the maximum percentage of recyclates which can be used for the production of the related products as substitute for virgin PVC.

Correction by low-volume waste groups

As fixed above a quantity of about 20,000 tons per year (PVC output from the recycling systems) has been defined as a limit for a feasible PVC recycling. With the potential recycling rates and the projected development of those product/waste groups have been eliminated from the list of potential candidates for high-quality recycling where the achievable quantities are below 20,000 tons per year.

Technical potentials for 'low-quality' PVC recycling (downcycling)

'Low-quality recycling' refers to the mechanical recycling of PVC in mixed plastics waste fractions. It can be applied for those types of PVC wastes where high-quality recycling is not feasible but a collection and recycling together with other PVC or plastics materials is possible. This requires that the total waste arising in the related product groups reaches a minimum quantity to make a separate collection feasible or that the mixed plastics fraction can be collected in a ready-to-process quality, so that sorting or mechanical treatment is not necessary.

When checking this for the different PVC waste groups the following waste groups candidates have been identified for low-quality recycling:

The first group is constituted by the remaining preconsumer wastes which cannot be recycled in a high quality. For most of the preconsumer wastes, a separate collection is feasible, for example in the frame of general systems or business activities for plastics waste management and recycling for industry and commerce like they exist in some Member States. The achievable recycling rates depend on the origin of the preconsumer wastes. Production wastes are easier to collect reaching an estimated potential recycling rate of 70 per cent (taking into account collection, sorting and treatment efficiency). For installation wastes (e.g. cut-offs from the laying of floorings or cables) the separate collection requires more efforts, e.g. on construction sites where a number of different containers for the different waste fractions to be separated must be installed, the personal must be trained, etc. Therefore the willingness for waste separation is lower and a lower potential recycling rate of 50 per cent seems to be realistic for these wastes. Installation wastes in applications of PVC plastisols, organosols and coatings are not suitable for mechanical recycling, due to the composite structures. This concerns, e.g. a part of PVC floorings.

The estimated recycling rates for the preconsumer wastes are shown in Table 13.6. For those product groups where preconsumer wastes are composed of both production wastes and cut-offs, the average potential recycling rate is given, taking also into account that depending on the waste group a part of total preconsumer waste arising can be recycled to high-quality recyclates.

Table 13.6. Recycling Potentials of the major PVC product/waste groups suitable for low-quality mechanical recycling.

PVC product/waste group	Preconsumer wastes	Post-consumer wastes [a]
Construction products		
Cables (F)	60	70–90 (80)[c]
Flooring calandered (F)	55	–
Pipes (R)	45	–
Packaging products		
Rigid films (R)	25	15–25 (20)
Furniture components	60[b]	–
Other consumer and commercial products		
Credit cards (R)	90	–
Electric/electronics		
Cables (F)	70	30–50 (40)
Adhesive tapes (F)	15	30–50 (40)
Inject. moulding parts (F)	–	30–50 (40)
Automotive	60-90[b]	–
Other products	0–30[b]	–

F = Flexible PVC applications; R = Rigid PVC applications.
[a] in brackets = chosen rates.
[b] absolute quantities are low; range of recycling rate depending on the product group.
[c] This recycling rate refers to the quantity of used cables which actually arises as waste. It must be taken into account that the major part of used cables remain in the ground (approximately 70 per cent). This means that the recycling potential related to the total quantity of used cables is about 25 per cent only.

In the group of post-consumer wastes, the following PVC product groups are feasible for low-quality recycling:

Cables (domestic applications): The low-quality recycling of cables is one of the major areas of PVC recycling today. The recycling of post-consumer cable wastes is carried out together with cut-offs from the laying of cables. PVC is included in the mixed plastics fraction output from the cable recycling plants which can be readily used for the extrusion of piles, traffic control systems, etc. The estimated potential recycling rate is about 70–90 per cent. This recycling rate refers to the PVC in collected cables arising as wastes. Like in the case of pipes, the major part (about 70 per cent) of the worn-out underground cables are not dismantled or extracted from the ground for economic reasons. Therefore, if the recycling rate is related to the total amount of worn-out cables it results a figure of 20–25 per cent only. The estimate of the reycling rate takes account of material losses in the cable shredding and waste flows into alternative waste management options for the mixed plastics fraction (feedstock recycling, energy recovery).

Packaging wastes (rigid and soft films): PVC in packagings is or will be collected together with other packaging materials. From the experiences with the existing recycling systems in Germany and Austria especially, the achievable collection rates are in the order of 80 per cent. However, in the subsequent sorting and mechanical treatment processes considerable losses occur: Since the PVC content of the plastics collections is low the separation of PVC or PVC rich fractions in sorting plants is not feasible. Hence the PVC ends up in the sorting residues (which are landfilled or incinerated) or in mixed plastics fractions. Taking into account that a part of the mixed plastics fraction will go to energy recovery and feedstock recycling we estimate that at most 20–30 per cent of the PVC in packaging collections can be recovered for low-quality recycling. Hence, the estimated overall recycling potential for PVC in packaging wastes is about 15–25 per cent.

Other household and commerical wastes: The recycling of PVC in nonpackaging applications from household wastes will not be feasible, due to low quantities (compared, e.g. to packaging wastes) and/or the appearance of plastics in composite products (e.g. in the case of films for furniture). Only those PVC products which end up in commercial wastes and whose collection can thus be integrated into separate mixed plastics collections for commerce/industry are potential candidates for low-quality mechanical recycling. Hoses, technical applications of rigid films (office supply, printing films, others), and a part of the miscellaneous product groups are concerned here. The potential recycling rate has been estimated at 30 per cent. For applications which are used in both industry/commerce and households this rate has been reduced to 15 per cent.

Electric/electronic wastes: On an EU level and in several Member States regulations for enforcing/improving the recycling of electric/electronic wastes are in preparation. The recycling potentials are considerably high. The technological state-of-the-art for the recycling of electric/electronic products is the mechanical treatment, where the PVC components are recovered in a mixed plastics fraction—similar to cable recycling. Just like for cable recycling there is a potential to use this material for the extrusion of different products with low material specifications. Electronic waste are discussed in chapter 15 of this book. According to the experiences with a number of collection and recycling schemes for electric/electronic wastes and related pilot projects, the achievable collection rates (in relation to the estimated waste arising) are in the order of 50–60 per cent. Based upon these experiences and taking into account losses in the mechanical treatment process (in the order of 10 per cent) the maximum overall recycling rate for the PVC components is about 50 per cent.

For PVC in end-of-life vehicles we do not expect any significant mechanical recycling. As mentioned before, there may be a limited dismantling before shredding allowing for a limited high-quality PVC recycling but the major part of the PVC will still be contained in the shredder residues. According to the regulations in preparation, these residues will have to be recovered to a large extent. However, according to the present technological development, mechanical processes which allow for the material recovery of PVC and other plastics from the shredder residues will not be applied. In contrast to the plastics fraction from electronic waste shredding, the mechanical separation of plastics from car shredding residues requires a subsequent treatment for which technical solutions are not straightforward, lacking economic competitiveness too. Therefore, most probably energetic treatment processes will prevail, where the plastics content is used for energy recovery.

LIMITS TO THE PVC RECYCLING POTENTIALS

Environmental Limits

The general environmental limits of PVC recycling have been already discussed. Two major limits are relevant:

1. No life cycle improvements: If the recycling of the related PVC products provides no or no significant environmental savings (resource consumption, emissions into air and water, wastes) as compared to disposal, incineration or energetic/feedstock recovery, there is no justification for promoting PVC recycling (except in those cases where it provides economic advantages).
2. Potential toxicological risks: The dispersion of toxic heavy metal stabilisers in the PVC (cadmium and lead) into the products made of the recyclates and the possible contamination of cable scraps with toxic PCB involve potential health and ecological risks.

Based upon these limits the following conclusions can be drawn concerning the utilisation of the technical potentials of mechanical PVC recycling:

1. Based upon the results of life cycle assessment studies for plastics waste management the environmental benefits of 'low-quality' mechanical plastics recycling must be regarded as low. As a consequence, the technical potential for 'low-quality, PVC recycling is certain, since the environmental justification is weak. However, low-quality mechanical recycling can be justified for economic reasons in some cases. This applies for the 'low-quality' mechanical recycling of most preconsumer wastes as well as post-consumer cable insulation wastes which can be recycled at competitive costs. The same can be expected for the mixed PVC/plastics fraction from the mechanical treatment of electric/electronic scrap. In contrast, the recycling of mixed packaging wastes is not viable economically, for it involves excessive costs.
2. As already mentioned there are different points of view on the toxicological issue.

As a matter of fact, cadmium and lead are toxic and persistent substances. However, one part of experts argue that the actual risk of an exposure of humans and the environment to these substances is comparatively low: The lead and cadmium compounds are fixed in the PVC matrix. A release is not possible but potentially in landfills (eluation) and in case of accidents in waste handling areas, incineration plants or fires. Not only the likelihood of such occurrences is comparatively low but also are the released quantities, taking also into account that PVC stabilisers are not the major application of these heavy metals.

The other part of the experts argue that the control over the cadmium and lead flows could not be guaranteed. Thus in accordance with the precautionary principle cadmium and lead should be banned and removed from the technosphere. As a consequence of this point-of-view, the recycling of all PVC wastes containing cadmium and lead would have to be banned. This concerns major high-volume PVC construction products, thus reducing the utilisable potential for mechanical PVC recycling greatly. However, the ban would be effective for a transition period only, until the major part of the existing stocks of long-life products containing lead and cadmium will have been disposed off. Nevertheless, due to the long-lifetime of the affected products the transition period can be as long as several decades.

In contrast to heavy metals a possible contamination with PCB must be considered as more critical, since the PCB may be released out of the PVC matrix during use. In the past, recycling products with high PCB concentrations have been detected and taken from the market and that the flow of the recycling materials cannot be controlled since they are used in a variety of different products. Thus for those PVC wastes affected (cable insulations and electric/electronic products) further restrictions can be justified (e.g. control of PCB contents of input materials). As a consequence, it is likely that the related recycling potentials would be reduced, due to the cost increase associated with the required PCB measurements or other provisions to control the PCB content.

Economic Limits

Apart from the environmental limits, the realisation of the technical recycling potentials is also limited by the overall recycling costs. The recycling costs include:

Collection costs (e.g. container, transportation).
+ Costs of sorting and mechanical treatment.

= Gross recycling costs.
− Credits for recyclates.

= Net recycling costs.

An economic limit of PVC recycling is the economic profitability. If profitability cannot be reached the recycling potentials will not be made use of, unless there are legal regulations or voluntary measures promoting or enforcing the recycling. Today, most PVC recycling is carried out under 'free market' conditions, i.e. for economic reasons. Economic profitability of PVC recycling depends on three major factors:

1. Apart from the gross recycling costs which depend on the PVC waste or product groups respectively as well as the available recycling and collection technologies.
2. It is the cost level of the alternative waste management routes (landfilling and incineration especially) which are competing with mechanical recycling on the related markets for wastes.
3. The price level of virgin PVC which determines the achievable selling price for recyclates.

For rough estimates the economic profitability is reached when the not recycling costs are lower than the prices for alternative waste management routes for the related PVC wastes.

The available information on the costs of the existing PVC recycling systems is very limited. However, together with otherwise published cost data they should be sufficient for a rough classification of the economic profitability of the different PVC wastes:

High-quality PVC recycling

In general, the recycling of a major part of PVC preconsumer wastes is profitable. This is why the recycling of preconsumer wastes constitutes the overwhelming part of PVC recycling today. However, economic profitability is not achieved for all waste groups, like, e.g. for some installation wastes (like cut-offs from the laying of pipes or floorings). The profitability depends also on the changing conditions on the markets for virgin PVC and wastes. Today, in many EU Member States the prices for landfilling are considerably low, due to the upcoming changes in the related legal regulations (technical requirements for landfill sites, phase-out of the disposal of certain waste groups. At the same time the prices for virgin PVC are at a minimum level. Both developments have reduced the economic profitability. As a result, 'free market' PVC recycling has been reduced greatly, for example in Italy. However, it can be expected to rise again when market conditions will recover. With these market-related limitations in mind, it can be concluded that the recycling of most preconsumer PVC wastes is economic feasible in principle.

In contrast, the high-quality recycling of post-consumer wastes is not profitable, i.e. the net costs are well above the costs for landfilling or incineration at present:

1. Net recycling costs for the rigid PVC construction products pipes, window profiles and 'other profiles' (cable trays and shutters especially) are in the order of 200–300 Euro/ton, not including additional provisions for the separation of the wastes at the construction sites. The prices for landfilling vary greatly in the EU. 150 Euro/ton (including transportation) can be regarded as

an upper limit (only a few landfill sites have higher charges). The average level is much lower. Thus recycling cannot compete with landfilling if the net recycling costs are to be covered. However, in the next years the economic conditions for recycling are likely to improve. The phase-out of the direct landfilling of plastics in some Member States and the technical requirements imposed by the EU landfill directive will increase landfill costs or they will make incineration to the major waste disposal route. Incineration costs including transportation reaches 200 Euro/ton in the majority of the existing incineration plants which comply with stringent emissions standards. Therefore, the recycling of pipes, window profiles and 'other profiles' will come nearer to the economic threshold in the next years.

2. Net recycling costs for PVC flooring are in the order of 300–400 Euro/ton. We do not expect that economic profitability can be reached since the cost gap to landfilling or incineration is too large to be closed by the expected future cost changes.

3. The same conclusion holds for PVC bottle recycling. For the remaining PVC post-consumer products which are suitable for high-quality recycling (shoes/soles and sheets), there are no cost data available. However, due to the necessary considerable provisions for collection and sorting we expect that the net recycling costs for these products will also reach an order of magnitude which make competitiveness with landfilling or incineration impossible.

Low-quality PVC recycling

As in the case of high-quality recycling, the low-quality recycling of a major part of PVC preconsumer wastes is economic feasible in principle, taking into account the market-related limitations and the restriction that economic feasibility does not apply for 100 per cent of these wastes.

In the recycling of post-consumer wastes the situation varies greatly depending on the product groups:

1. Cable insulation is the only post-consumer waste which is recycled at competitive costs. This is due to the special situation that these materials arise as a waste fraction from cable recycling, free of collection and other costs, and in a ready-to-process quality for their use in the extrusion of new products. There is a similar situation for PVC in the mixed plastics fraction from the recycling of electric/electronic scrap. So, with electric/electronic scrap recycling increasing in the next years it can be expected that the recycling of the mixed plastics fraction (including PVC) will be economic feasible as well.

2. The costs for the recycling of PVC films in packaging wastes are considerably high. Generally the collection of PVC packaging films and other PVC products is included in the different packaging recycling schemes which are existing or are being established in the Member States. For the packaging recycling systems in Austria and Germany the costs for the plastics fraction are between 700 and more than 1000 Euro/ton. This is far from economic profitability.

3. For the remaining PVC wastes which are suitable for low-quality recycling (hoses, profiles, office supply, printing films, rigid profiles and sheets arising in commercial wastes) no specific cost information is available. We expect that these PVC wastes will be collected together with other plastics wastes at the company sites, with lower degrees of contamination than in the household collections and thus reducing the sorting requirements. As a consequence the costs will be lower than the household collections (packaging), although still far from reaching economic profitability.

Apart from the cost considerations low-quality PVC recycling can be limited due to market restrictions: Up to now, the low-quality recyclates have been applied for a limited number of products, a part of it have been especially designed for the use of the recyclates (e.g. traffic cones, back layers of industrial floorings). So, with expected increasing waste arisings in the future, the market of these 'recycling products' may be too small to absorb the resulting quantities of low-quality recyclates. However, the market for these products is not once-for-all given but it can be enlarged by the design of new products.

FUTURE PROSPECTS AND MEASURES TO IMPROVE MECHANICAL PVC RECYCLING

Today, PVC recycling is focused on those areas where economic profitability is achieved, that is preconsumer wastes and post-consumer cable insulation wastes. The majority of PVC post-consumer wastes is landfilled. Post-consumer recycling systems exist for packaging wastes — enforced by the packaging regulations of the EU and individual Member States — and for some construction wastes in a limited number of countries — all of them have been established voluntarily by the related PVC industry.

For the evaluation of the present situation two aspects must be taken into account:

1. PVC post-consumer waste arising is still comparatively low, due to the time lag to PVC consumption (which started to increase to the present level in the 1960s and early 1970s). Today, total PVC waste arising is not more than 40 per cent of PVC consumption. For the major product groups with high potentials for mechanical recycling the ratio is even smaller; 2 per cent for pipes, 5 per cent for windows and 20 per cent for other building profiles. For some of the important products a significant increase of the waste arising is not expected before 2015.
2. Not only the absolute recycling quantities are low but also the recycling rates: The recycled PVC post-consumer waste quantities are far from reaching their potentials. Today, only 3 per cent of PVC post-consumer wastes are recycled (whilst for preconsumer wastes recycling rates of more than 80 per cent are achieved).

Taking into account that there are no recycling regulations for the management of the major part of PVC-related wastes the main reason for the low recycling rates is that PVC recycling is too far from reaching economic competitiveness. This means that the (net) cost of the whole 'recycling chain', including the cost of the waste owners (e.g. for separation of PVC at construction sites or transport to collection points) and the costs for the operation of the recycling organisation (collection, sorting, treatment and marketing of the recyclates), is significantly higher than the cost of landfilling which is the major option for PVC waste disposal at present.

The cost disadvantage is mainly due to low PVC contents in the related waste streams, composite PVC applications or PVC in mixed or contaminated waste collections which require expensive collection (and sorting operations to separate PVC fractions of a suitable quality or in the case of low quality recycling, the achievable prices for the recycling materials are low.

Another short-term reason which is specific to the present situation is the 'double squeeze' of the profitability of plastics recycling by very low prices for virgin plastics on the one side and very low prices for landfilling on the other side, at least in countries like Germany where price dumping is practiced as a consequence of the coming ban of the landfilling of reactive wastes including plastics.

Consequently, unless there are no legal or administrative measures (e.g. the existing packaging regulations), voluntary agreements or public contracts (like, e.g. in the Netherlands or Denmark) the incentives for the recycling of PVC post-consumer wastes are low.

An important factor which determines the prospects of the existing voluntary recycling systems which have been established by the PVC industry to improve the environmental performance and/or the public image of PVC, is the type of financing. The prospects for 'subsidised systems' where the waste owner is charged with a fraction of the actual costs only (like, e.g. in existing plastics pipe recycling systems) are better than for 'non-subsidised' systems where the recycling fees cover total costs. With 'subsidised systems' we define recycling systems which are financed by voluntary contributions (in case of voluntary recycling systems established by industry) or non-voluntary fees (in case of legal obligations like the packaging regulations). Due to the comparatively high cost level of recycling the 'non-subsidised' fees (together with additional costs at the waste owners) would be much higher than the prices for alternative waste management routes (landfilling especially). The 'subsidisation' compensates the higher prices and the related economic disincentives for the waste holders to collect wastes for recycling.

There are additional factors which are responsible for the low level of PVC postconsumer recycling. Technical standards especially, which have excluded the use of plastics recyclates in important product groups like pipes by now, must be mentioned here.

However, the adaptation of many of the related standards is in progress. In some countries, like the Netherlands (for pipes) or Italy, industrial standards which permit the use of recyclates have been developed already. These changes will make an improved utilisation of the existing PVC recycling potentials possible.

Future Changes Due to Legal and Voluntary Measures in Force or in Preparation

Also as a consequence of the legal regulations and voluntary measures in force or in preparation an improvement of the conditions for PVC recycling can be expected for the future. The following legal regulations must be mentioned here:

1. The EU and national landfilling regulations impose technical and economic requirements for landfill sites which will involve an increase in costs or landfill fees respectively. In addition to this, some Member States like the UK, Denmark, Sweden and Finland have introduced landfill taxes, which aim at encouraging recovery and recycling to reduce landfilling. In some countries like Austria, the Netherlands or Germany landfilling of PVC and other plastics will be phased-out in the next years according to the national regulations. As a conclusion the competitive pressure to PVC recycling imposed by landfilling will be reduced.

2. As a consequence of the landfill regulations incineration will be the sole route for the final disposal of PVC wastes in some Member States, apart from recovery (mechanical recycling, feedstock recycling, energy recovery). On an average, incineration costs are higher than landfill costs, taking also into account that the operation of low-cost incineration plants which do not comply with state-of-the-art emission control technology is likely to be phased out, also as a result of the EU draft directive on waste incineration. This is an additional factor improving the conditions for PVC recovery and recycling.

3. For important waste streams (also including PVC) 'recycling regulations' have been adopted or are in preparation, fixing targets for recycled quantities or separate collection. They include the European and national packaging regulations and the draft directives on electric/electronic equipment and end-of-life vehicles. There is no European regulation for construction and demolition wastes which is the most important waste stream for PVC. Only in a few Member States related measures exist. For example, in the Netherlands, Sweden and Denmark there are national programs to increase recycling and recovery of these wastes, in Austria an ordinance requires the separation of plastics and other fractions at the construction sites and in Germany

there exist similar regulations on a regional level, being accompanied by local/regional landfill surcharges for commingled construction and demolition wastes.
4. In addition, the related voluntary agreements and contracts between Government and industry in some member states will encourage PVC recycling. They include, e.g. the existing commitments of industry to reduce PVC flows into incineration in Denmark, to establish recycling systems for window profiles and pipes in the Netherlands or the currently negotiated commitment to reduce PVC flows going to landfills in the UK. These agreements are the major basis for the existing recycling systems for PVC post-consumer wastes. However, many systems are not yet fully developed and they are limited to few Member States by now.

Further Measures to Improve PVC Recycling

Altogether, there exists already a lot of waste management measures in the EU whose implementation will improve the conditions for PVC recycling (including mechanical recycling, but feedstock recycling and energy recovery as well). As a consequence, measures to increase PVC recycling beyond this trend development must focus on:
1. Increasing the utilisation of the existing recycling systems and the further development of these systems.
2. Broadening the regional scope of the existing recycling systems aiming at an EU-wide coverage.
3. Establishing supplementary recycling systems for those PVC products/wastes which are suitable for mechanical recycling but have not yet included in the existing recycling activities; PVC profiles in construction wastes and some PVC products in household and commercial wastes (especially footwear, printing films and sheets) are concerned here.

With regard to the different PVC product groups the major focus of additional measures must be on PVC in construction and demolition wastes where the mechanical recycling potentials are high (both quantitatively and qualitatively) and the recycling activities are not reaching far enough with regard to both regional spread and PVC product groups covered.

With regard to the different steps in the recycling process the major focus of the additional measures must be on the collection of the wastes. The collection is the major bottleneck for mechanical PVC recycling in terms of costs and achievable recycling rates. To achieve sufficiently high collection rates:
1. The collection systems must include a sufficient number of collection points to make a broad regional accessibility possible.
2. The recycling costs to be paid by the waste owners must be lower than the fees for waste disposal (landfilling and incineration)—otherwise there are no incentives for recycling.

Potential political instruments to encourage separate collection of PVC range from:
1. Statutory orders or prohibitions.
2. Over 'economic' instruments.
3. To voluntary agreements.

Possible statutory regulations include an EU-wide requirement to separate mineral and non-mineral fractions (including PVC and other plastics) of construction and demolition wastes on the construction sites or EU-wide recycling quota for the relevant waste streams, i.e. construction and demolition waste, specific municipal solid waste fractions like footwear/textiles and commercial/industrial plastics wastes (including PVC printing films or sheets). Another potential statutory measure is the modification of the European landfill directive by inclusion of a definitive ban on the landfilling of plastics and other reactive components, following similar national regulations and being also valid for construction and demolition wastes.

Possible 'economic measures' are taxes or levies on landfilling or incineration setting financial incentives for recycling, while at the same time the income from the taxes can be used to subsidise recycling systems to make the separate collection at competitive costs possible. A specific measure in this context is the 'internalisation' of the PVC-specific external costs in waste incinerators (for flue gas treatment), e.g. by an allowance for the operators of incinerators paid by the PVC industry and re-financed over the PVC price. Another possibility is to enforce recycling organisations for construction and demolition wastes similar to those existing for packaging wastes, by imposing overall recycling goals and leaving implementation and financing up to industry. There are also 'soft' measures for supporting the acceptance of the existing and possible future recycling systems, like information campaigns.

Finally the related PVC industry can be committed to increase PVC recycling or the recycling of the relevant waste streams respectively by voluntary agreements, following the examples of some Member States. Such agreements may be made on the EU level or separately for the individual Member States.

Apart from the objective to increase the recycled quantities the environmental benefits and risks must be taken into account for a general improvement of PVC recycling. First of all, most of the potential measures to increase PVC recycling do not encourage specifically mechanical recycling but PVC recovery and recycling in general. High-quality mechanical recycling is the environmentally most favourable recovery option as a rule, however in most cases the costs are higher than for feedstock recycling or energy recovery.

Thus, for environmental reasons, flanking measures to 'protect' high-quality mechanical recycling against feedstock recycling or energy recovery may be necessary. In contrast, low-quality mechanical recycling should not be encouraged by specific measures unless significant environmental advantages over feedstock recycling or energy recovery can be proven.

Further measures may be necessary to reduce environmental risks associated with mechanical PVC recycling. Depending on the position on the assessment of these risk:

1. A ban on heavy metal stabilisers (cadmium and lead) in PVC recyclates may be justified (as a consequence, the recycling of end-of-life window profiles would have to be stopped then, for some of them were stabilised with cadmium, and products like pipes and building profiles which are stabilised with lead are affected too).

2. A phase-out of the mechanical recycling of PVC wastes from cable and electronic scrap recycling which may contain PCB may be reasonable for precautionary reasons.

Chapter 14

Tyre Recycling

INTRODUCTION

Tyre recycling or rubber recycling is the process of recycling vehicles tyres (or tyres) that are no longer suitable for use on vehicles due to wear or irreparable damage (such as punctures). These tyres are among the largest and most problematic sources of waste, due to the large volume produced and their durability. Those same characteristics which make waste tyres such a problem also make them one of the most reused waste materials, as the rubber is very resilient and can be reused in other products. Approximately one tyre is discarded per person per year. Tyres are also often recycled for use on basketball courts and new shoe products. However, material recovered from waste tyres, known as 'crumb', is generally only a cheap 'filler' material and is rarely used in high volumes.

TYRE LIFE CYCLE

The tyre life cycle can be identified by the following six steps:
1. Product developments and innovations such as improved compounds and camber tyre shaping increase tyre life, increments of replacement, consumer safety, and reduce tyre waste.
2. Proper manufacturing and quality of delivery reduces waste at production.
3. Direct distribution through retailers, reduces inventory time and ensures that the life span and the safety of the products are explained to customers.
4. Consumers use and maintenance choices like tyre rotation affect tyre wear and safety of operation.
5. Manufacturers and retailers set policies on return, re-tread, and replacement to reduce the waste generated from tyres and assume responsibility for taking the 'tyre to its grave' or to its reincarnation.
6. Recycling tyres by developing strategies that combust or process waste into new products, creates viable businesses, and fulfilling public policies.

LANDFILL DISPOSAL

Tyres are not desired at landfills, due to their large volumes and 75 per cent void space, which quickly consumes valuable space. Tyres can trap methane gases, causing them to become buoyant or 'bubble' to the surface. This 'bubbling' effect can damage landfill liners that have been installed to help keep landfill contaminants from polluting local surface and groundwater. Shredded tyres are now being used in landfills, replacing other construction materials, for a lightweight backfill in gas venting systems, leachate collection systems, and operational liners. Shredded tyre material may also be used to cap, close or daily cover landfill sites. Scrap tyres as a backfill and cover material are also more cost-effective, since tyres can be shredded on-site instead of hauling in other fill materials.

STOCKPILES AND ILLEGAL DUMPING

Tyre stockpiles create a great health and safety risk. Tyre fires can occur easily, burning for months, creating substantial pollution in the air and ground. Recycling helps to reduce the number of tyres in storage. An additional health risk, tyre piles provide harbourage for vermin and a breeding ground for mosquitoes that may carry diseases. Illegal dumping of scrap tyres pollutes ravines, woods, deserts, and empty lots; which has led many states to pass scrap tyre regulations requiring proper management. Tyre amnesty day events, in which community members can deposit a limited number of waste tyres free of charge, can be funded by state scrap tyre programs, helping decrease illegal dumping and improper storage of scrap tyres.

Uses

Tyres can be recycled into, among other things, the hot melt asphalt, typically as crumb rubber modifier-recycled asphalt pavement (CRM-RAP), and Portland Cement, Tyres can also be recycled into other tyres. Pyrolysis can be used to reprocess the tyres into fuel gas, oils, solid residue (char), and low-grade carbon black which cannot be used in tyre manufacture. A pyrolysis method which produces activated carbon and high-grade carbon black has been suggested.

Recent developments in devulcanisation enable dealing with substantial volumes, taking 40 mesh whole tyre crumb and converting it into value added compounds without degrading the polymer and without generating any pollution. This new generation in devulcanisation technologies operates with very high productivity while maintaining a low energy footprint. The compounds produced from processed tyre scrap can be blended with virgin rubber compounds, maintaining performance while substantially reducing the raw material cost. The substantial economies of scale and value addition now make it possible to make burning of tyres entirely unnecessary.

Tyre Pyrolysis

The pyrolysis method for recycling of used tyres is an innovation technique that uses a special mechanism to heat the used tyres in a closed, oxygen-free environment — a stove to melt down the tyres into the materials that they were made of. There are many different ways to achieve the melting procedure. For a long time, external heating methods were used. Recently an electro–magnetic field technology was developed by Coral group, in Dnepropetrovsk, Ukraine. This method produces carbon, metal, gas and artificial oil as by-products of the recycling process.

The quality of these by-products depends on the heating technique used, with simple outside heating techniques producing heavy oils (mazut); however, newer techniques that produce a 'softer' pyrolysis produce by-products such as benzene, kerosene and diesel.

Microwave Recycling

The process of remediation of tyre waste using microwaves to excite the rubber until it is in a gaseous state which will be condensed into its component parts including #3 diesel, syngas as well as carbon black and plated steel. No emissions are created in this process and all components can be reutilised.

Tyre Recycling Supply Chain

The tyre recycling supply chain is divided into three stages:

Tyre-derived products stage

Second stage of tyre recycling involves the production of alternate products for sale. New products derived from waste tyres generate more economic activity than combustion or other low multiplier production, while reducing waste stream without generating excessive pollution and emissions from recycling operations.

Tyre-derived products

Whole tyres can be reused in many different ways. One way is for a Steel mill to use the tyres as a carbon source, replacing coal or coke in steel manufacturing. Instead of mining coal from the ground and then burying tyres in landfills, the tyres are used directly. Tyres are also bound together and used as different types of barriers such as: collision reduction, erosion control, rainwater runoff, wave action—that protects piers and marshes, and sound barriers between roadways and residences. Entire homes can be built with whole tyres by ramming them full of earth and covering them with concrete, known as earthships.

Some Artificial reefs are built using tyres that are bonded together in groups, there is some controversy on how effective tyres are as an artificial reef system:

1. The process of stamping and cutting tyres is used in some apparel products, such as sandals and as a road sub-base, by connecting together the cut sidewalls to form a flexible net.
2. Chipped and shredded tyres are used as tyre derived fuel (TDF); this is not the same as recycling, but TDF helps to eliminate tyres from our waste stream and produces a fuel source. They are used in civil engineering applications such as sub grade fill and embankments, backfill for walls and bridge abutments, sub grade insulation for roads, landfill projects, and septic system drain fields.
3. Shredded tyres, known as tyre derived aggregate (TDA), have many civil engineering applications. TDA can be used as a backfill for retaining walls, fill for landfill gas trench collection wells, backfill for roadway landslide repair projects as well as a vibration damping material for railway lines.
4. Ground and crumb rubber, also known as size-reduced rubber, can be used in both paving type projects and in mouldable products. These types of paving are: Rubber modified asphalt (RMA), rubber modified concrete, and as a substitution for an aggregate. Examples of rubber-moulded products are carpet padding or underlay, flooring materials, dock bumpers, patio decks, railroad crossing blocks, livestock mats, sidewalks, rubber tiles and bricks, moveable speed bumps, and curbing/edging. The rubber can be moulded with plastic for products like pallets and railroad ties. Athletic and recreational areas can also be paved with the shock absorbing rubber-moulded material. Rubber from tyres is sometimes ground into medium-sized chunks and used as rubber mulch. Rubber crumb can also be used as an infill, alone or blended with coarse sand, as in infill for grass-like synthetic turf products such as FieldTurf.

Environmental Concerns

Due to heavy metals and other pollutants in tyres there is a potential risk for the leaching (leachate) of toxins into the groundwater when placed in wet soils. This impact on the environment varies according to the pH level and conditions of local water and soil. Research has shown that very little leaching occurs when shredded tyres are used as light fill material, however limitations have been put on use of this material; each site should be individually assessed determining if this product is appropriate for given conditions.

Ecotoxicity may be a bigger problem than first thought. Studies show that zinc, heavy metals, a host of vulcanisation and rubber chemicals leach into water from tyres. Shredded tyre pieces leach much more, creating a bigger concern, due to the increased surface area on the shredded pieces. Many organisms are sensitive, and without dilution, contaminated tyre water has been shown to kill some organisms.

Disposal of Tyres by Burning

Technically, it is quite feasible to destroy scrap tyres by incineration. Equipments are available to achieve this without infringement of the requirements of the Clean Air Acts. The problem is, however, cost. It costs 3–4 times more to incinerate than to dump tyres in landfill sites. The heating value of tyres varies between 12,000 to 15,000 Btu/lb. In comparison, normal garbage generates 3000 to 4000 Btu/lb. Scrap tyres are already a good source of fuel but the handling, storing, collecting of the tyres, the low demand for steam and high costs are against this method of disposal.

Fishing reefs

There is an increasing interest in the use of scrap tyres to construct artificial fishing reefs. The tyres are put together in small bundles of approximately 10 passenger tyres each, then five of these bundles are strapped together and taken out on barges where divers place the tyres on a sandy bottom and thus build up an artificial reef.

Fish like to inhabit an area where they can feed and be close to an area where they can feed and be close to an area where they can dart to safety if a large fish comes along. The artificial reef built of tyres provides the perfect safety as the large fish cannot fit into the nooks and crannies that bundles of tyres provide. The tyres do not rust away as car bodies did when they were first used to build artificial reefs. After a period of time, useful organisms attach themselves to the tyres and produce food for the fish. Artificial reefs built of tyres can be placed in a barren area, and fishermen are able to know the exact spot where the fish are biting. It should be understood that while using scrap tyres for artificial reefs is putting scrap tyres to a use, it is not the complete answer.

The tyres have to be collected, each tyre has to have a hole cut in it, and the tyres have to be bundled together and carried out on barges and handled by divers to be placed in position — a very high labour cost project — which is neither an economic nor a long-term answer to the problem.

Swamp reclamation

The reclamation of swampy ground lends itself as a method of scrap tyre disposal. The general complaint of tip operators, who use landfill with general garbage and tyres, is that the tyre will not consolidate in the tip because the inside cavity of the scrap tyre never fills with hard fill, hence the ground remains spongy and the tyres work their way to the top of the tip when traffic travels over it.

Swampy, useless ground that is on river level and the water level of which fluctuates with the rising and falling of the river, can be successfully reclaimed by using the tyres to raise the level of ground, so that the fluctuations of the normal river level would not affect the reclaimed land. Tyres will not pollute the water or land in any way and will remain there indefinitely.

Trenches are dug in the swamp and the earth removed is placed alongside the trench. A pump is used to keep the trench dry and then scrap tyres are stacked into the trench in a particular way. The earth previously removed is placed back on top of the tyres and another trench is dug parallel to the previous one. The type of earth being river slit oozes down through the tyres and into their cavities and consolidates into a solid mass. The rising and falling water level helps to move the silt into the cavities.

The site, when completed, is raised up above the high water level of the nearby river; it is solid and is able to be put to use.

Shedding

One of the great problems of scrap tyres is the volume they occupy and the difficulty of handling. If tyres could be shredded easily into a form where they could be handled by conveyor belts, hoppers and the general equipment available to reduce labour, they would be in a form that should be acceptable into normal landfill sites. Tyre engineers strive to produce a tyre that will not wear out, be abrasive and absorb impact. They use high tensile steel in the bead case and now there is a general swing to use steel cords in the radial belts in both passenger and truck radial ply tyres. All these qualities make for a better tyre but play havoc with any machinery used to try and shred tyres.

The machines that have seen so far for shredding tyres will not cut the beads up into sections and each tyre has to be handled individually into the machines and makes the use of such machines uneconomical. The Jacques shredder is reputed to be able to shred passenger and truck tyres in quantity. There is no reason why shredded tyres would not be acceptable into any landfill sites.

Novel uses of scrap tyres

Scrap tyres have been used for various purposes. One that springs to mind is that associated with the mini bike boom. Bike clubs have been using scrap tyres for edges of mini bike tracks.

Tyres have been used to form a barrier around plants like a normal pot, they have been stuck together and used as table legs in trendy furniture, placed on sides of wharves as buffers for ships, but this latter use is decreasing as rubber companies manufacture now specially-produced buffers.

RECYCLING OF POLYURETHANES

Polyurethanes are by far the most versatile group of polymers, because the products range from soft thermoplastic elastomers to hard thermoset rigid forms. Although polyurethane rubbers are speciality products, polyurethane foams are well known and widely used materials. While the use of plastics in automobile has increased steadily over the years, a major part of these plastics is polyurethane (PU), which is used for car upholstery; front, rear, and side coverings; as also for spoiler. In fact, about half of the weight of plastics in modern cars is accounted for by PU foams. Accordingly, in addition to production scrap, large quantities of used PU articles are now generated from automotive sources. Though most PU plastics are cross-linked polymers, they cannot be regarded as ordinary thermosetting plastics, owing to their chemical structure and physical domain structure. Thus in contrast to typical thermosetting plastics, various methods are available today for recycling PU scrap and used products.

There are basically two methods for recycling polyurethane scrap and used parts, namely, *material recycling* (primary, secondary, and tertiary recycling) and *energy recycling* (quaternary recycling). The former methods are preferred since in this way material resources are replenished. After multiple uses the material can finally be used for energy recovery by high-temperature combustion or gasification.

Processes of Recycling Polyurethanes

Among several processes described for PU material recycling, *thermopressing* and *kneader recycling* have attracted much attention. By the thermopressing process, granulated PU wastes can be converted into new moulded parts, while in the kneader recycling process a thermo-mechanical operation causes partial chemical breakdown of PU polymer chains that can be subsequently cross-linked by reacting with poly-isocyanates. Hydrolysis and glycolysis are important tertiary recycling processes for PU wastes.

Thermopressing process

Thermopressing or moulding by heat and compression, is a direct method of material recycling that is designed such that elastomeric, cross-linked polyurethanes can be recycled in much the same way as thermoplastic materials. The principle of thermopressing is based on the realisation that polyurethane and polyurea granules are capable of flowing into each other and building up new bonding forces under the influence of high temperature (185°–195°C), high pressure (300–800 bar), and strong shearing forces. The granules generally used for this purpose have a diameter of 0.5 to 3 mm. They completely fill the cavities of a mould meaning that mouldings with new geometries can also be manufactured. Unlike injection moulding of thermoplastics for which a cold mould is used, in the thermopressing process, the mould is kept constantly hot at a temperature of 190 ± 5°C and no release agent is used for demoulding. This relatively simple technique will permit 100 per cent recycling of polyurethane RIM and RRIM mouldings, particularly when the formulations of RIM systems to be used in future have been optimised for recycling. The steps in the thermopressing process are shown in Fig. 14.1.

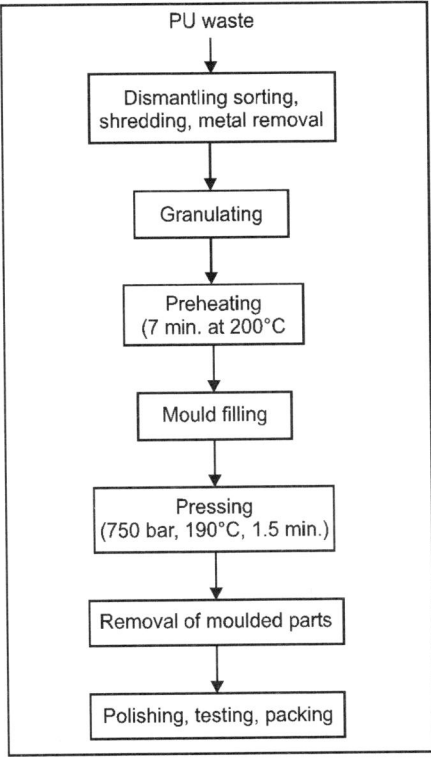

Fig. 14.1. Reprocessing of polyurethane waste by thermopressing.

The moulded parts obtained by thermopressing of granulated PU waste exhibit only slight reduction in hardness and impact strength but significant reduction in elongation at break. The last named property, for example, drops to about 10 per cent of the original value if painted PU wastes are used. Moreover, because of the use of granulated feed, the resulting moulded parts lack surface smoothness and thus should be used preferably in those areas where they are not visible. In a passenger car, there are many such parts that are not subjected to tensile stress but require dimensional and heat stability—properties fulfilled by PU recycled products. Examples of application are wheelboxes, reserve wheel covers and similar other covers, mudguard linings, glove boxes, and casings.

Kneader process

The basis of the kneader recycling process is a thermo-mechanical degradation of polymer chains to smaller-size segments. The hard elastic PU is thereby converted into a soft, plastic (unmolten) state, which is achieved with a kneader temperature of 150°C and additional frictional heating. This leads to temperatures above 200°C and causes thermal decomposition into a product that is soft at 150°–200°C but becomes brittle at room temperature, enabling it to be crushed to powder in a cold kneader or roller press. The resulting powder can be easily mixed with a powder form polyisocyanate and moulded into desired shapes by compression moulding at 150°C and 200 bar pressure. The scheme of the recycling process is shown in Fig. 14.2.

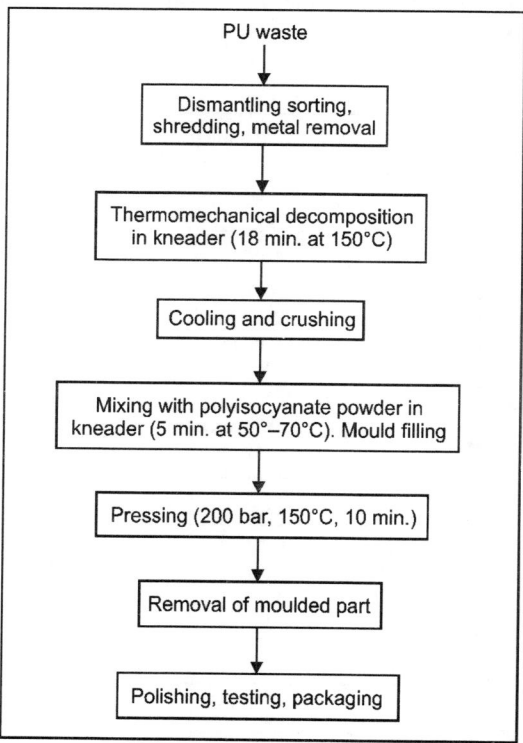

Fig. 14.2. Recycling of polyurethane waste via partial decomposition in kneader.

Partial breakdown of PU network in the kneader results in highly branched molecules with many functional groups necessitating addition of polyisocyanate in relatively high concentration for subsequent cross-linking to produce moulded articles. The process thus yields products of high hardness (with Shore hardness up to 80) and high tensile strength (30 MPa), but small elongation at break (6 to 8 per cent).

Hydrolysis

Hydrolysis of PU waste results in the formation of polyethers and polyamines that can be used as starting materials for producing foam. In this process, powdered PU waste is reacted with superheated steam at 160°–190°C and the polymer gets converted in about 15 min to a liquid heavier than water. The liquid is a mixture of toluene diamine and propylene oxide (polyether diol), the former accounting for 65–85 per cent of the theoretical yield.

The recovered polyether can be used in formulations for making PU foam, preferably in admixture with virgin polyether. A continuous hydrolysis reactor utilising a twin-screw extruder has been designed that can be heated to a temperature of 300°C and has a provision for injection of water into the extruder at a point where the scrap is almost in the pulp state. Polyurethane scrap in powder form is fed into the extruder and residence time is adjusted to 5–30 min. Separation of the two components, polyether and diamine, in the product may be effected by fractional distillation, by extraction with a suitable solvent or by chemical means. The PU foams made from these recycled products can be used in several applications, one example being protection boards for construction sites. Hydrolytic recycling has not, however, found much application, since virgin raw materials are cheaper than the regenerated products.

SECTION IV

E-Waste Management

Chapter 15

Electronic Wastes

INTRODUCTION

Electronic products have made our life easy by saving time and being efficient. Now it has become difficult for us to function without electronic equipments. Most of our household work is done by using electronic appliances. Communication systems have revolutionalised by way of mobile phones. Entertainment products like television and music system have added enjoyment to our life. Electronic products, which were once thought to be luxury, have presently become a need.

From villages to cities, all of them have using electronic products either in the form of radio or a high tech computer. There are places in India where people do not have access to electricity but they still have electronic products operated with battery. Increase in the use of electronic products have resulted in increase in production of these products and hence created a new waste, which is termed as Electronic waste or E-waste.

The E-waste is one of the fastest growing environmental problems of the world, as there is a lack of awareness among people about its treatment and serious impacts. E-waste needs to be treated as a hazardous waste.

E-waste is a popular, informal name for electronic products coming to the end of their 'useful life'. As per the Hazardous Wastes (management and handling) Rules, 2003, E-waste can be defined as waste electrical and electronic equipment including all components, subassemblies and their fractions except batteries falling under these rules.

COMPONENTS OF E-WASTE

E-waste comprises of wastes generated from used electronic devices and household appliances which are not fit for their original intended use and are destined for recovery, recycling or disposal. Computers, televisions, VCRs, fax machines are common electronic products. Electronic products are made up of a variety of components. Some of them contain toxic substances that have an adverse impact on human health and the environment, if not handled and disposed off properly.

For example, cathode ray tubes (CRTs) have content of barium, phosphorous and other heavy metals. Halogenated chlorides and bromides used as flame retardants is dangerous as they form persistent dioxins and furans on combustion at low temperature (600°–800°C), copper used in printed circuit board and cables, poly vinyl chloride (PVC) sheathing of wires also induces the formation of dioxins which is a known carcinogen.

HAZARDS ASSOCIATED WITH E-WASTE

E-waste contains significant quantities of toxic metals and chemicals. If these are left untreated and disposed off in landfills or not recycled by using proper methods of recycling, they leach into the surrounding soil, water and the atmosphere, and causes adverse effects on human health and environment. Many elements of this waste contain poisonous substances such as lead, tin, mercury, cadmium and barium, which cause severe diseases like cancer, birth defects, neurological and respiratory disorders. Some of the toxic substances of E-waste and their impact are listed in Table 15.1.

Table 15.1. Hazards associated with E-waste.

E-waste components	Process	Potential occupational hazard	Potential environmental hazard
Cathode ray tubes (CRTs)	Breaking and removal of copper yoke and dumping.	Silicosis Inhalation or contact with phosphor containing cadmium or other metals.	Lead, barium and other heavy metals leaching into groundwater, release of toxic phosphor.
Printed circuit boards	Disordering and removing computer chips.	Tin and lead inhalation. Possible brominated dioxin, beryllium, cadmium, mercury inhalation.	Air emission of same substances.
Dismantled printed circuit board processing.	Open burning of waste boards to remove inside metals.	Toxicity to workers and nearby residents from tin, lead, brominated dioxin, beryllium, cadmium and mercury inhalation. Respiratory irritation.	Tin and lead contamination of immediate environment including surface and groundwaters. Brominated dioxins, beryllium, cadmium and mercury emissions.
Chips and other gold plated components	Chemical stripping using nitric and hydrochloric acid along river banks.	Acid contact with eyes, skin may result in permanent injury. Inhalation of mists and fumes of acids, chlorine and sulphur dioxide gases can cause respiratory irritation to severe effects including pulmonary edema, circulatory failure and death.	Hydrocarbons, heavy metals, brominated substances, etc. discharged directly into river and banks. Acidifies the river destroying fish and flora.
Plastics from computer and peripherals, e.g. printers keyboards, etc.	–	–	Emissions of brominated dioxins and heavy metals and hydrocarbons.
Shredding and low temperature melting to be reutilised in poor grade plastics.	Probable hydrocarbon, brominated dioxin and heavy metal exposure.	Brominated and chlorinated dioxin, polycyclic aromatic hydrocarbons (PAH) are carcinogenic to workers living in the burning works area.	Hydrocarbon ashes including PAHs discharged to air, water and soil.

(Contd ...)

E-waste components	Process	Potential occupational hazard	Potential environmental hazard
Miscellaneous computer parts encased in rubber or plastic, e.g. steel rollers	Open burning to recover steel and other metals.	Hydrocarbon including PAHs and potential dioxin exposure.	Hydrocarbon ashes including PAHs discharged to air, water and soil.
Secondary steel or copper and precious metal smelting	Furnace recovers steel or copper from waste including organics	Exposure to dioxins and heavy metals.	Emission of dioxins and heavy metals.

EXISTING LEGISLATION

Factories Act 1948 (amended till 1987): There are several contaminants arising out from manufacturing or recycling of electronic components and are listed in this Act.

Environmental Protection Rules 1986 (amended till 2004): There is no direct standard, which can address pollutants from an electronics manufacturing or recycling industries. However certain PCB units fall in electroplating category and are therefore required to be abide by the effluent disposal norms as given in schedule 1 of this rule.

Hazardous waste (management and handling) rules 1989, amended in 2003:
1. Schedule 2 of this act can be applied for the disposal of E-waste.
2. Schedule 3 entry at Sl. no. A1180: Waste electrical and electronic assemblies (For EXIM, i.e. Export Import).
3. Schedule 3 entry at Sl. no. B1110: Electrical and electronic assemblies not valid for direct reuse but for recycling (For EXIM).

Hazardous waste (management, handling and transboundary movement) rules 08:
1. Part A of Schedule III (Basal no. 1180) consists of list of E-waste applicable for import with prior informed consent.
2. Part B of Schedule III (Basal no. 1110) deals with list of E-waste applicable for import and export not requiring prior informed consent.

Basal convention: The basal convention on the control of transboundary movements of hazardous wastes and their disposal, adopted by a conference in basal (Switzerland) in 1989, was developed under UNEP.

E-WASTE MANAGEMENT

International scenario:
1. 50 to 80 per cent E-wastes collected is exported for recycling by US Export is legal in US.
2. Exported E-waste recycling and disposal in China, India and Pakistan is highly polluting.

Indian scenario: The electronic waste management assumes greater significance in India not only due to the generation of our own waste but also dumping of E-waste particularly computer waste from the developed countries.
1. There are two small E-waste dismantling facilities functioning in Chennai and Bangalore.
2. Five E-waste recyclers around Chennai have been recognised by the Tamil Nadu Pollution Control Board — Thrishyiraya Recycling India Pvt Ltd, INAA Enterprises, AER World Wide (India) Private Ltd, TESAMM Recyclers India Pvt Ltd and Ultrust Solution (I) Pvt Ltd.

3. In Mumbai, Eco Reco Company that has been authorised by Maharashtra Pollution Control Board is involved in the management of E-waste. It collects E-waste across India and recycles it in an environment friendly manner. TCS, SBI, Castrol, M&M, Oberoi Groups of Hotels, Gati, Alfa Laval, Pfizer, HDFC, Aventis Pharma, GPEC, Tata Ficosa are recycling their E-waste with the help of Eco Reco. SIMS Recycling Ltd. a multinational company has submitted a proposal to Pune Municipal Corporation (PMC) to solve the problem of E-waste in Pune city. It will collect and treat the E-waste in their recycling plant outside India.

E-WASTE TREATMENT AND DISPOSAL METHODS

Landfilling

It is one of the most widely used methods for disposal of E-waste. In landfilling, trenches are made on the flat surfaces. Soil is excavated from the trenches and waste material is buried in it, which is covered by a thick layer of soil. Modern techniques like secure landfill are provided with some facilities like, impervious liner made up of plastic or clay, leachate collection basin that collects and transfer the leachate to waste-water treatment plant. The degradation processes in landfills are very complicated and run over a wide time span (Fig. 15.1).

Fig. 15.1. Landfilling.

The environmental risks from landfilling of E-waste cannot be neglected because the conditions in a landfill site are different from a native soil, particularly concerning the leaching behaviour of metals. Mercury, cadmium and lead are the most toxic leachates. Lead has been found to leach from broken lead-containing glass, such as the cone glass of cathode ray tubes from TVs and monitors. Cadmium also leaches into soil and groundwater. In addition, it is known that cadmium and mercury are emitted in diffuse form or via the landfill gas combustion plant. Landfills are also prone to uncontrolled fires, which can release toxic fumes. Therefore, landfilling does not appear to be an environmentally sound treatment method for substances, which are volatile and not biologically degradable (Cd, Hg), persistent (poly chlorinated biphenyls) or with unknown behaviour in a landfill site (brominated flame retardants).

Incineration

It is a controlled and complete combustion process, in which the waste material is burned in specially designed incinerators at a high temperature (900°–1000°C) (Fig. 15.2). Advantage of incineration of E-waste is the reduction of waste volume and the Utilisation of the energy content of combustible materials. Some plants remove iron from the slag for recycling. By incineration some environmentally hazardous organic substances are converted into less hazardous compounds.

Fig. 15.2. E-waste incinerator.

Disadvantage of incineration are the emission to air of substances escaping flue gas cleaning and the large amount of residues from gas cleaning and combustion. E-waste incineration plants contribute significantly to the annual emissions of cadmium and mercury. In addition, heavy metals not emitted into the atmosphere are transferred to slag and exhaust gas residues and can re-enter the environment on disposal. Therefore, E-waste incineration will increase these emissions, if no reduction measures like removal of heavy metals are taken.

Recycling of E-waste

Monitors and CRT, keyboards, laptops, modems, telephone boards, hard drives, floppy drives, compact disks, mobiles, fax machines, printers, CPUs, memory chips, connecting wires and cables can be recycled. Recycling involves dismantling i.e. removal of different parts of E-waste containing dangerous substances like PCB, Hg, separation of plastic, removal of CRT, segregation of ferrous and nonferrous metals and printed circuit boards. Recyclers use strong acids to remove precious metals such as copper, lead, gold. The value of recycling from the element could be much higher if appropriate technologies is used. The recyclers working in poorly—ventilated enclosed areas without mask and technical expertise results in exposure to dangerous and slow poisoning chemicals. The existing dumping grounds in India are full and overflowing beyond capacity and it is difficult to get new dumping sites due to scarcity of land. Therefore recycling is the best possible option for the management of E-waste.

Reuse

It constitutes direct second hand use or use after slight modifications to the original functioning equipment. It is commonly used for electronic equipments like computers, cell phones, etc. Inkjet cartridge is also used after refilling. This method also reduces the volume of E-waste generation. We can use above-mentioned methods for treatment and disposal of E-waste. The better option is to avoid its generation. To achieve this, buy back of old electronic equipments shall be made mandatory. Large companies should purchase the used equipments back from the customers and ensure proper treatment and disposal of E-waste by authorised processes. This can considerably reduce the volume of E-waste generation.

COMPUTER RECYCLING

Computer recycling or electronic recycling is the recycling or reuse of computers or other electronics. It includes both finding another use for materials (such as donation to charity), and having systems dismantled in a manner that allows for the safe extraction of the constituent materials for reuse in other products (Fig. 15.3).

Fig. 15.3. Computers for recycling.

Reasons for Recycling

Obsolete computers or other electronics are a valuable source for secondary raw materials, if treated properly; if not treated properly, they are a source of toxins and carcinogens. Rapid technology change, low initial cost, and even planned obsolescence have resulted in a fast-growing surplus of computer or other electronic components around the globe. Technical solutions are available, but in most cases a legal framework, a collection system, logistics, and other services need to be implemented before a technical solution can be applied. According to the US Environmental Protection Agency, an estimated 30 to 40 million surplus PCs, which it classifies under the term 'hazardous household waste', will be ready for end-of-life management in each of the next few years. The US National Safety Council estimates that 75 per cent of all personal computers ever sold are now surplus electronics.

In 2007, the United States Environmental Protection Agency (EPA) said that more than 63 million computers in the US were traded in for replacements or they simply were discarded. Today 15 per cent of electronic devices and equipment are recycled in the United States. Most electronic waste is sent to landfills or becomes incinerated, having a negative impact on the environment by releasing materials such as lead, mercury, or cadmium into the soil, groundwater, and atmosphere.

Many materials used in the construction of computer hardware can be recovered in the recycling process for use in future production. Reuse of tin, silicon, iron, aluminium, and a variety of plastics—

all present in bulk in computers or other electronics — can reduce the costs of constructing new systems. In addition, components frequently contain copper, gold, and other materials valuable enough to reclaim in their own right.

Computer components contain valuable elements and substances suitable for reclamation, including lead, copper, and gold. They also contain many toxic substances, such as dioxins, polychlorinated biphenyls (PCBs), cadmium, chromium, radioactive isotopes, and mercury. A typical computer monitor may contain more than 6 per cent lead by weight, much of which is in the lead glass of the cathode ray tube (CRT). A typical 15-inch computer monitor may contain 1.5 pounds of lead, but other monitors have been estimated as having up to 8 pounds of lead. Circuit boards contain considerable quantities of lead-tin solders and are even more likely to leach into groundwater or to create air pollution via incineration. Additionally, the processing required to reclaim the precious substances (including incineration and acid treatments) may release, generate, and synthesise further toxic by-products.

A major computer or electronic recycling concern is export of waste to countries with lower environmental standards. Companies may find it cost-effective in the short-term to sell outdated computers to less developed countries with lax regulations. It is commonly believed that a majority of surplus laptops are routed to developing nations as 'dumping grounds for E-waste'. The high value of working and reusable laptops, computers, and components (e.g. RAM) can help pay the cost of transportation for a large number of worthless 'commodities'. Broken monitors, obsolete circuit boards, and short-circuited transistors are difficult to spot in a container load of used electronics.

Regulations

Europe

In Switzerland, the first electronic waste recycling system was implemented in 1991, beginning with collection of old refrigerators; over the years, all other electric and electronic devices were gradually added to the system. The established producer responsibility organisation is SWICO, mainly handling information, communication, and organisation technology.

The European Union implemented a similar system in February 2003, under the Waste Electrical and Electronic Equipment Directive (WEEE Directive, 2002/96/EC).

United States

Federal

The United States Congress considers a number of electronic waste bills, including the National Computer Recycling Act introduced by Congressman Mike Thompson (D-CA). Meanwhile, the main federal law governing solid waste is the Resource Conservation and Recovery Act of 1976. It covers only CRTs, though state regulations may differ. There are also separate laws concerning battery disposal. On March 25, 2009, the House Science and Technology Committee approved funding for research on reducing electronic waste and mitigating environmental impact, regarded by sponsor Ralph Hall (R-TX) as the first federal bill to address electronic waste directly.

State

Many states have introduced legislation concerning recycling and reuse of computers or computer parts or other electronics. Most American computer recycling legislation addresses it from within the larger electronic waste issue.

In 2001, Arkansas enacted the Arkansas Computer and Electronic Solid Waste Management Act, which requires that state agencies manage and sell surplus computer equipment, establishes a computer and electronics recycling fund, and authorises the Department of Environmental Quality to regulate and/or ban the disposal of computer and electronic equipment in Arkansas landfills.

The recently passed Electronic Device Recycling Research and Development Act distributes grants to universities, government labs, and private industry for research in developing projects in line with E-waste recycling and refurbishment.

Asia

South Korea, Japan, and Taiwan require that sellers and manufacturers of electronics be responsible for recycling 75 per cent of them.

Recycling Methods

Consumer recycling

Consumer recycling options include sale, donating computers directly to organisations in need, sending devices directly back to their original manufacturers or getting components to a convenient recycler or refurbisher.

Corporate recycling

Businesses seeking a cost-effective way to recycle large amounts of computer equipment responsibly face a more complicated process. Businesses also have the options of sale or contacting the Original Equipment Manufacturers (OEMs) and arranging recycling options.

Some companies will pick up unwanted equipment from businesses, wipe the data clean from the systems, and provide an estimate of the product's remaining value. For unwanted items that still have value, these firms will buy the excess IT hardware and sell refurbished products to those seeking more affordable options than buying new.

Companies that specialise in data protection and green disposal processes dispose of both data and used equipment while at the same time employing strict procedures to help improve the environment. Professional IT Asset Disposition (ITAD) firms specialise in corporate computer disposal and recycling services in compliance with local laws and regulations and also offer secure data elimination services that comply with data erasure standards.

Corporations face risks both for incompletely destroyed data and for improperly disposed computers, and according to the Resource Conservation and Recovery Act, are liable for compliance with regulations even if the recycling process is outsourced. Companies can mitigate these risks by requiring waivers of liability, audit trails, certificates of data destruction, signed confidentiality agreements, and random audits of information security. The National Association of Information Destruction is an international trade association for data destruction providers.

Sale

Online auction at eBay is an alternative for consumers willing to resell for cash less fees, in a complicated, self-managed, competitive environment where paid listings might not sell. Craigslist can be similarly risky due to forgery scams and uncertainty.

Donation

A number of organisations, usually nonprofit organisations (NPOs), attempt to reuse computers. These NPOs usually refurbish usable computers for sale at discounted prices to the needy, to other nonprofit organisations, or to the general public.

Consumer recycling includes a variety of donation options, such as charitable NPOs [501(c)(3) organisations—for example, Free Geek] which may offer tax benefits in return.

For both corporations and consumers, NPOs (such as Nonprofit Technology Resources or Camara) will often accept and refurbish still-usable computers in return for tax benefits. The Computer Takeback Campaign and the TechSoup Donate Hardware List are resources for locating such refurbishers. Donated systems can also be directed to developing nations. However, in cases where the computer equipment comes from a wide variety of manufacturers, it may be more efficient to hire a third-party contractor to handle the recycling arrangements.

Takeback

When researching computer companies before a computer purchase, consumers can find out if they offer recycling services. Most major computer manufacturers offer some form of recycling. At the user's request they may mail in their old computers or arrange for pick-up from the manufacturer.

Hewlett-Packard also offers free recycling, but only one of its 'national' recycling programs is available nationally, rather than in one or two specific states. Hewlett-Packard also offers to pick up any computer product of any brand for a fee, and to offer a coupon against the purchase of future computers or components; it was the largest computer recycler in America in 2003, and it has recycled over 750 million pounds of electronic waste globally since 1995. It encourages the shared approach of collection points for consumers and recyclers to meet.

Exchange

Manufacturers often offer a free replacement service when purchasing a new PC. Dell Computers and Apple Inc. will take back old products when one buys a new one. Both refurbish and resell their own computers with a one-year warranty. Many companies purchase and recycle all brands of working and broken laptops and notebook computers, whether from individuals or corporations. Building a market for recycling of desktop computers has proven more difficult than exchange programs for laptops, smartphones, and other smaller electronics. A basic business model is to provide a seller an instant online quote based on laptop characteristics, then to send a shipping label and prepaid box to the seller, to erase, reformat, and process the laptop, and to pay rapidly by check. A majority of these companies are also generalised electronic waste recyclers as well; organisations that recycle computers exclusively include Cash For Laptops, a laptop refurbisher in Nevada that claims to be the first to buy laptops online, in 2001. Bulk laptops at a recycling affiliate, broken down into Dell, Gateway Computers, Hewlett-Packard, Sony, and other.

Scrapping/recycling

For systems which are obsolete or no longer useful to its user, recycling is often the only choice available. This is usually done by breaking down the equipment into its component parts, such as plastics and metals. These parts can then be recycled through various methods depending on the material. Recyclers typically charge a fee, but in return many have a zero-landfill policy and the sorted or shredded pieces are melted down to recover their component materials for reuse.

Early Pioneering Efforts to E-waste

The first major publication to report the recycling of computers and electronic waste was published on the front page of the *New York Times* on April 14, 1993 by columnist Steve Lohr.

Data Security

Data security is an important part of computer recycling. Federal regulations mandate that there are no information security leaks in the life-cycle of secure data; this includes its destruction and recycling. There are a number of federal laws and regulations, including HIPAA, Sarbanes-Oxley, FACTA, GLB, which govern the data life cycle and require that establishments with high- and low-profile data keep their data secure. Recycling computers can be dangerous when handling sensitive data, specifically to businesses storing tax records or employee information. While most people will try to wipe their hard drives clean before disposing of their old computers, only 5 per cent rely on an industry specialist or a third party to completely clean the system before it is disposed of according to an IBM survey. Industry standards recommend a 3X overwriting process for complete protection against retrieving confidential information. This means a hard drive must be wiped three times in order to ensure the data cannot be retrieved and possibly used by others.

Reasons to destroy and recycle securely

There are ways to ensure that not only hardware is destroyed but also the private data on the hard drive. Having customer data stolen, lost or misplaced contributes to the ever growing number of people who are affected by identity theft, which can cause corporations to lose more than just money. The image of a company that holds secure data, such as banks, pharmaceuticals, and credit corporations is also at risk. If a company's public image is hurt that could cause consumers to not use their services and could cost millions in business losses and positive public relation campaigns. The cost of data breaches varies widely ranging $90 to $305 per customer record, depending on whether the breach is 'low-profile' or 'high-profile' and the company is in a non-regulated or highly regulated area, such as banking. There is also a major backlash from the consumer if there is a data breach in a company that is supposed to be trusted to protect their private information.

Secure recycling

There are regulations that monitor the data security on end-of-life hardware. National Association for Information Destruction (NAID) 'is the international trade association for companies providing information destruction services. Suppliers of products, equipment and services to destruction companies are also eligible for membership. NAID's mission is to promote the information destruction industry and the standards and ethics of its member companies.' There are companies that follow the guidelines from NAID and also meet all Federal EPA and local DEP regulations.

The typical process for computer recycling aims to securely destroy hard drives while still recycling the by-product. A typical process for effective computer recycling accomplishes the following:
1. Receive hardware for destruction in locked and securely transported vehicles.
2. Shred hard drives.
3. Separate all aluminium from the waste metals with an electromagnet.
4. Collect and securely deliver the shredded remains to an aluminium recycling plant.
5. Mould the remaining hard drive parts into aluminium ingots.

SOLVING THE E-WASTE PROBLEM

Solving the E-waste problem (StEP) is an international initiative, created to develop solutions to address issues associated with waste electrical and electronic equipment (WEEE). Some of the most eminent players in the fields of production, reuse and recycling of electrical and electronic equipment (EEE), government agencies and NGOs as well as UN Organisations count themselves among its members. StEP encourages the collaboration of all stakeholders connected with E-waste, emphasising a holistic, scientific yet applicable approach to the problem.

Aims and Means

'One of the most important aims of the StEP Initiative is to elaborate a set of global guidelines for the treatment of E-waste and the promotion of sustainable material recycling.' Press communique of the initiative.

The initiative comprises five cooperating Task Forces, each addressing specific aspects of E-waste, while covering the entire life cycle of electric and electronic equipment. In all its activities, the initiative places emphasis on working with policy-making bodies to allow results from its research to impact current practices. StEP is being coordinated by the science and research body of the UN System, the United Nations University (UNU). The long-term goal of StEP 'is to develop—based on scientific analysis—a globally accepted standard for the refurbishment, recycling of E-waste. Herewith, StEP's aim is to reduce dangers to humans and the environment, which result from inadequate and irresponsible treatment practices, and advance resource efficiency.' To achieve this, StEP conceives and implements projects based on the results of multidisciplinary dialogues. The projects seek to develop sustainable solutions that reduce environmental risk and enhance development.

Organisation of the Initiative

The supreme body of the StEP Initiative is its General Assembly, which decides its general direction and development. This General Assembly is based on a Memorandum of Understanding, which is signed by all members and states the guiding principles of StEP. A Secretariat, hosted by the UNU in Bonn, is mandated with the accomplishment of the day-to-day managerial work of the initiative. A Steering Committee, composed of representatives from key stakeholders, monitors the progress of the Initiative. The core work is accomplished by the five Task Forces (TF): 'Policy', 'ReDesign', 'ReUse', 'Recycling' and 'Capacity Building'. These Task Forces conduct research and analysis in their respective domains and seek to implement innovative projects.

1. TF1—Policy: The aim of this Task Force is to assess and analyse current governmental approaches and regulations related to WEEE. Starting from this analysis, recommendations for future regulating activities shall be formulated.
2. TF2—ReDesign: This Task Force works on the design of EEE, focusing on the reduction of negative consequences of electrical and electronic appliances throughout their entire life cycle. The Task Force especially takes heed of the situation in developing countries.
3. TF3—ReUse: The focus of this Task Force lies in the development of sustainable, transmissible principles and standards for the reuse of EEE.
4. TF4—Recycling: The objective of this Task Force is to improve infrastructures, systems and technologies to realise a sustainable recycling on a global level.
5. TF5—Capacity building: The aim of this Task Force is to draw attention to the problems connected to WEEE. This aim shall be achieved by making the results of the research of the

Task Forces and other stakeholders publicly available. In doing so, the Task Force relies on personal networks, the internet, collaborative working tools, etc.

Guiding principles

1. StEP's work is founded on scientific assessments and incorporates a comprehensive view of the social, environmental and economic aspects of E-waste.
2. StEP conducts research on the entire life cycle of electronic and electrical equipment and their corresponding global supply, process and material flows.
3. StEP's research and pilot projects are meant to contribute to the solution of E-waste problems.
4. StEP condemns all illegal activities related to E-waste including illegal shipments and reuse/recycling practices that are harmful to the environment and human health.
5. StEP seeks to foster safe and eco/energy-efficient reuse and recycling practices around the globe in a socially responsible manner.

Chapter 16

Environmentally Sound Options for E-Waste Management

INTRODUCTION

E-waste is a popular, informal name for electronic products nearing the end of their useful life. E-wastes are considered dangerous, as certain components of some electronic products contain materials that are hazardous, depending on their condition and density. The hazardous content of these materials pose a threat to human health and environment. Discarded computers, televisions, VCRs, stereos, copiers, fax machines, electric lamps, cell phones, audio equipment and batteries if improperly disposed can leach lead and other substances into soil and groundwater. Many of these products can be reused, refurbished or recycled in an environmentally sound manner so that they are less harmful to the ecosystem. This chapter highlights the hazards of E-wastes, the need for its appropriate management and options that can be implemented.

Industrial revolution followed by the advances in information technology during the last century has radically changed people's lifestyle. Although this development has helped the human race, mismanagement has led to new problems of contamination and pollution. The technical prowess acquired during the last century has posed a new challenge in the management of wastes. For example, personal computers (PCs) contain certain components, which are highly toxic, such as chlorinated and brominated substances, toxic gases, toxic metals, biologically active materials, acids, plastics and plastic additives. The hazardous content of these materials pose an environmental and health threat. Thus proper management is necessary while disposing or recycling E-wastes.

These days computer has become most common and widely used gadget in all kinds of activities ranging from schools, residences, offices to manufacturing industries. E-toxic components in computers could be summarised as circuit boards containing heavy metals like lead and cadmium; batteries containing cadmium; cathode ray tubes with lead oxide and barium; brominated flame-retardants used on printed circuit boards, cables and plastic casing; polyvinyl chloride (PVC) coated copper cables and plastic computer casings that release highly toxic dioxins and furans when burnt to recover valuable metals; mercury switches; mercury in flat screens; polychlorinated biphenyl's (PCB's) present in older capacitors; transformers, etc.

Basel Action Network (BAN) estimates that the 500 million computers in the world contain 2.87 billion kgs of plastics, 716.7 million kgs of lead and 2,86,700 kgs of mercury. The average 14-inch monitor uses a tube that contains an estimated 2.5 to 4 kgs of lead. The lead can seep into the ground water from landfills thereby contaminating it. If the tube is crushed and burned, it emits toxic fumes into the air.

EFFECTS ON ENVIRONMENT AND HUMAN HEALTH

Disposal of E-wastes is a particular problem faced in many regions across the globe. Computer wastes that are landfilled produces contaminated leachates which eventually pollute the groundwater. Acids and sludge obtained from melting computer chips, if disposed on the ground causes acidification of soil. For example, Guiyu, Hong Kong a thriving area of illegal E-waste recycling is facing acute water shortages due to the contamination of water resources.

This is due to disposal of recycling wastes such as acids, sludges, etc. in rivers. Now water is being transported from faraway towns to cater to the demands of the population. Incineration of E-wastes can emit toxic fumes and gases, thereby polluting the surrounding air. Improperly monitored landfills can cause environmental hazards. Mercury will leach when certain electronic devices, such as circuit breakers are destroyed. The same is true for polychlorinated biphenyls (PCBs) from condensers. When brominated flame retardant plastic or cadmium containing plastics are landfilled, both polybrominated diphenyl ethers (PBDE) and cadmium may leach into the soil and groundwater. It has been found that significant amounts of lead ion are dissolved from broken lead containing glass, such as the cone glass of cathode ray tubes, gets mixed with acid waters and are a common occurrence in landfills.

Not only does the leaching of mercury poses specific problems, the vapourisation of metallic mercury and dimethylene mercury, both part of waste electrical and electronic equipment (WEEE) is also of concern. In addition, uncontrolled fires may arise at landfills and this could be a frequent occurrence in many countries. When exposed to fire, metals and other chemical substances, such as the extremely toxic dioxins and furans (TCDD tetrachloro dibenzo-dioxin, PCDDs-polychlorinated dibenzo-dioxins. PBDDs-polybrominated dibenzo-dioxin and PCDFs-poly chlorinated dibenzo furans) from halogenated flame retardant products and PCB containing condensers can be emitted. The most dangerous form of burning E-waste is the open-air burning of plastics in order to recover copper and other metals. The toxic fallout from open air burning affects both the local environment and broader global air currents, depositing highly toxic by-products in many places throughout the world.

Table 16.1 summarises the health effects of certain constituents in E-wastes. If these electronic items are discarded with other household garbage, the toxics pose a threat to both health and vital components of the ecosystem. In view of the ill-effects of hazardous wastes to both environment and health, several countries exhorted the need for a global agreement to address the problems and challenges posed by hazardous waste. Also, in the late 1980s, a tightening of environmental regulations in industrialised countries led to a dramatic rise in the cost of hazardous waste disposal. Searching for cheaper ways to get rid of the wastes, 'toxic traders' began shipping hazardous waste to developing countries. International outrage following these irresponsible activities led to the drafting and adoption of strategic plans and regulations at the Basel Convention. The Convention secretariat, in Geneva, Switzerland, facilitates and implementation of the Convention and related agreements. It also provides assistance and guidelines on legal and technical issues, gathers statistical data, and conducts training on the proper management of hazardous waste.

BASEL CONVENTION

The fundamental aims of the basel convention are the control and reduction of transboundary movements of hazardous and other wastes including the prevention and minimisation of their generation, the environmentally sound management of such wastes and the active promotion of the transfer and use of technologies.

A draft strategic plan has been proposed for the implementation of the basel convention. The draft strategic plan takes into account existing regional plans, programs or strategies, the decisions of the conference of the parties and its subsidiary bodies, ongoing project activities and process of international environmental governance and sustainable development. The draft requires action at all levels of society: training, information, communication, methodological tools, capacity building with financial support, transfer of know-how, knowledge and sound, proven cleaner technologies and processes to assist in the concrete implementation of the basel declaration. It also calls for the effective involvement and coordination by all concerned stakeholders as essential for achieving the aims of the basel declaration within the approach of common but differentiated responsibility.

Table 16.1. Effects of E-waste constituent on health.

Source of E-wastes	Constituent	Health effects
Solder in printed circuit boards, glass panels and gaskets in computer monitors	Lead (PB)	Damage to central and peripheral nervous systems, blood systems and kidney damage. Affects brain development of children.
Chip resistors and semiconductors	Cadmium (CD)	Toxic irreversible effects on human health. Accumulates in kidney and liver. Causes neural damage. Teratogenic.
Relays and switches, printed circuit boards	Mercury (Hg)	Chronic damage to the brain. Respiratory and skin disorders due to bioaccumulation in fishes.
Corrosion protection of untreated and galvanised steel plates, decorator or hardner for steel housings	Hexavalent chromium (Cr) VI	Asthmatic bronchitis. DNA damage.
Cabling and computer housing	Plastics including PVC	Burning produces dioxin. It causes: Reproductive and developmental problems. Immune system damage. Interfere with regulatory hormones.
Plastic housing of electronic equipments and circuit boards	Brominated flame retardants (BFR)	Disrupts endocrine system functions
Front panel of CRTs	Barium (Ba)	Short-term exposure causes: Muscle weakness. Damage to heart, liver and spleen.
Motherboard	Beryllium (Be)	Carcinogenic (lung cancer) Inhalation of fumes and dust. Causes chronic beryllium disease or beryllicosis. Skin diseases such as warts.

A set of interrelated and mutually supportive strategies are proposed to support the concrete implementation of the activities as indicated in the website (www.basel.int/DraftstrateKJcpian4Seot.pdf) is described below:

1. To involve experts in designing communication tools for creating awareness at the highest level to promote the aims of the basel declaration on environmentally sound management and the

ratification and implementation of the basel convention, its amendments and protocol with the emphasis on the short-term activities.

2. To engage and stimulate a group of interested parties to assist the secretariat in exploring fund raising strategies including the preparation of projects and in making full use of expertise in non-governmental organisations and other institutions in joint projects.

3. To motivate selective partners among various stakeholders to bring added value to making progress in the short-term.

4. To disseminate and make information easily accessible through the internet and other electronic and printed materials on the transfer of know-how, in particular through Basel Convention Regional Centres (BCRCs).

5. To undertake periodic review of activities in relation to the agreed indicators.

6. To collaborate with existing institutions and programs to promote better use of cleaner technology and its transfer, methodology, economic instruments or policy to facilitate or support capacity-building for the environmentally sound management of hazardous and other wastes.

The basel convention brought about a respite to the transboundary movement of hazardous waste. India and other countries have ratified the convention. However United States (US) is not a party to the ban and is responsible for disposing hazardous waste, such as, E-waste to Asian countries even today. Developed countries such as US should enforce stricter legislations in their own country for the prevention of this horrifying act.

In the European Union where the annual quantity of electronic waste is likely to double in the next 12 years, the European Parliament recently passed legislation that will require manufacturers to take back their electronic products when consumers discard them. This is called Extended Producer Responsibility. It also mandates a timetable for phasing out most toxic substances in electronic products.

MANAGEMENT OF E-WASTES

It is estimated that 75 per cent of electronic items are stored due to uncertainty of how to manage it. These electronic junks lie unattended in houses, offices, warehouses, etc. and normally mixed with household wastes, which are finally disposed off at landfills. This necessitates implementable management measures. In industries management of E-waste should begin at the point of generation. This can be done by waste minimisation techniques and by sustainable product design. Waste minimisation in industries involves adopting:

1. Inventory management.
2. Production-process modification.
3. Volume reduction.
4. Recovery and reuse.

Inventory Management

Proper control over the materials used in the manufacturing process is an important way to reduce waste generation. By reducing both the quantity of hazardous materials used in the process and the amount of excess raw materials in stock, the quantity of waste generated can be reduced. This can be done in two ways, i.e. establishing material-purchase review and control procedures and inventory tracking system.

Developing review procedures for all material purchased is the first step in establishing an inventory management program. Procedures should require that all materials be approved prior to purchase. In the approval process all production materials are evaluated to examine if they contain hazardous constituents and whether alternative nonhazardous materials are available.

Another inventory management procedure for waste reduction is to ensure that only the needed quantity of a material is ordered. This will require the establishment of a strict inventory tracking system. Purchase procedures must be implemented which ensure that materials are ordered only on an as-needed basis and that only the amount needed for a specific period of time is ordered.

Production Process Modification

Changes can be made in the production process, which will reduce waste generation. This reduction can be accomplished by changing the materials used to make the product or by the more efficient use of input materials in the production process or both. Potential waste minimisation techniques can be broken down into three categories:
1. Improved operating and maintenance procedures.
2. Material change.
3. Process-equipment modification.

Improvements in the operation and maintenance of process equipment can result in significant waste reduction. This can be accomplished by reviewing current operational procedures or lack of procedures and examination of the production process for ways to improve its efficiency. Instituting standard operation procedures can optimise the use of raw materials in the production process and reduce the potential for materials to be lost through leaks and spills.

A strict maintenance program, which stresses corrective maintenance, can reduce waste generation caused by equipment failure. An employee-training program is a key element of any waste reduction program. Training should include correct operating and handling procedures, proper equipment use, recommended maintenance and inspection schedules, correct process control specifications and proper management of waste materials.

Hazardous materials used in either a product formulation or a production process may be replaced with a less hazardous or nonhazardous material. This is a very widely used technique and is applicable to most manufacturing processes. Implementation of this waste-reduction technique may require only some minor process adjustments or it may require extensive new process equipment. For example, a circuit board manufacturer can replace solvent-based product with water-based flux and simultaneously replace solvent vapour degreaser with detergent parts washer.

Installing more efficient process equipment or modifying existing equipment to take advantage of better production techniques can significantly reduce waste generation. New or updated equipment can use process materials more efficiently producing less waste. Additionally such efficiency reduces the number of rejected or off-specification products, thereby reducing the amount of material which has to be reworked or disposed of.

Modifying existing process equipment can be a very cost-effective method of reducing waste generation. In many cases the modification can just be relatively simple changes in the way the materials are handled within the process to ensure that they are not wasted. For example, in many electronic manufacturing operations, which involve coating a product, such as electroplating or painting, chemicals are used to strip off coating from rejected products so that they can be recoated.

These chemicals, which can include acids, caustics, cyanides, etc. are often a hazardous waste and must be properly managed. By reducing the number of parts that have to be reworked, the quantity of waste can be significantly reduced.

Volume Reduction

Volume reduction includes those techniques that remove the hazardous portion of a waste from a nonhazardous portion. These techniques are usually to reduce the volume, and thus the cost of disposing of a waste material. The techniques that can be used to reduce waste-stream volume can be divided into two general categories: source segregation and waste concentration. Segregation of wastes is in many cases a simple and economical technique for waste reduction.

Wastes containing different types of metals can be treated separately so that the metal value in the sludge can be recovered. Concentration of a waste stream may increase the likelihood that the material can be recycled or reused. Methods include gravity and vacuum filtration, ultra filtration, reverse osmosis, freeze vapourisation, etc.

For example, an electronic component manufacturer can use compaction equipments to reduce volume of waste cathode ray-tube.

Recovery and Reuse

This technique could eliminate waste disposal costs, reduce raw material costs and provide income from a saleable waste. Waste can be recovered on-site, or at an off-site recovery facility, or through inter industry exchange. A number of physical and chemical techniques are available to reclaim a waste material such as reverse osmosis, electrolysis, condensation, electrolytic recovery, filtration, centrifugation, etc. For example, a printed-circuit board manufacturer can use electrolytic recovery to reclaim metals from copper and tin-lead plating bath.

However recycling of hazardous products has little environmental benefit if it simply moves the hazards into secondary products that eventually have to be disposed of. Unless the goal is to redesign the product to use non-hazardous materials, such recycling is a false solution.

Sustainable Product Design

Minimisation of hazardous wastes should be at product design stage itself keeping in mind the following factors:

1. Rethink the product design: Efforts should be made to design a product with fewer amounts of hazardous materials. For example, the efforts to reduce material use are reflected in some new computer designs that are flatter, lighter and more integrated. Other companies propose centralised networks similar to the telephone system.
2. Use of renewable materials and energy: Biobased plastics are plastics made with plant-based chemicals or plant-produced polymers rather than from petrochemicals. Biobased toners, glues and inks are used more frequently. Solar computers also exist but they are currently very expensive.
3. Use of non-renewable materials that are safer: Because many of the materials used are non-renewable, designers could ensure the product is built for reuse, repair and/or upgradability. Some computer manufacturers such as Dell and Gateway lease out their products thereby ensuring they get them back to further upgrade and lease out again.

THE INDIAN SCENARIO

While the world is marvelling at the technological revolution, countries like India are facing an imminent danger. E-waste of developed countries, such as the US, dispose their wastes to India and other Asian countries. A recent investigation revealed that much of the electronics turned over for recycling in the United States ends up in Asia, where they are either disposed of or recycled with little or no regard for environmental or worker health and safety. Major reasons for exports are cheap labour and lack of environmental and occupational standards in Asia and in this way the toxic effluent of the developed nations would flood towards the world's poorest nations. The magnitude of these problems is yet to be documented. However, groups like Toxic Links India are already working on collating data that could be a step towards controlling this hazardous trade.

It is imperative that developing countries and India in particular wake up to the monopoly of the developed countries and set up appropriate management measures to prevent the hazards and mishaps due to mismanagement of E-wastes.

MANAGEMENT OPTIONS

Considering the severity of the problem, it is imperative that certain management options be adopted to handle the bulk E-wastes. Following are some of the management options suggested for the government, industries and the public.

Responsibilities of the Government

1. Governments should set up regulatory agencies in each district, which are vested with the responsibility of co-ordinating and consolidating the regulatory functions of the various government authorities regarding hazardous substances.
2. Governments should be responsible for providing an adequate system of laws, controls and administrative procedures for hazardous waste management. Existing laws concerning E-waste disposal be reviewed and revamped. A comprehensive law that provides E-waste regulation and management and proper disposal of hazardous wastes is required. Such a law should empower the agency to control, supervise and regulate the relevant activities of government departments. Under this law, the agency concerned should:
   (a) Collect basic information on the materials from manufacturers, processors and importers and to maintain an inventory of these materials. The information should include toxicity and potential harmful effects.
   (b) Identify potentially harmful substances and require the industry to test them for adverse health and environmental effects.
   (c) Control risks from manufacture, processing, distribution, use and disposal of electronic wastes.
   (d) Encourage beneficial reuse of 'E-waste' and encouraging business activities that use waste. Set up programs so as to promote recycling among citizens and businesses.
   (e) Educate E-waste generators on reuse/recycling options.
3. Governments must encourage research into the development and standard of hazardous waste management, environmental monitoring and the regulation of hazardous waste-disposal.
4. Governments should enforce strict regulations against dumping E-waste in the country by outsiders. Where the laws are flouted, stringent penalties must be imposed. In particular, custodial sentences should be preferred to paltry fines, which these outsiders/foreign nationals can pay.
5. Governments should enforce strict regulations and heavy fines levied on industries, which do not practice waste prevention and recovery in the production facilities.

6. Polluter pays principle and extended producer responsibility should be adopted.
7. Governments should encourage and support NGOs and other organisations to involve actively in solving the nation's E-waste problems.
8. Uncontrolled dumping is an unsatisfactory method for disposal of hazardous waste and should be phased out.
9. Governments should explore opportunities to partner with manufacturers and retailers to provide recycling services.

Responsibility and Role of Industries

1. Generators of wastes should take responsibility to determine the output characteristics of wastes and if hazardous, should provide management options.
2. All personnel involved in handling E-waste in industries including those at the policy, management, control and operational levels, should be properly qualified and trained. Companies can adopt their own policies while handling E-wastes. Some are given below:
   (a) Use label materials to assist in recycling (particularly plastics).
   (b) Standardise components for easy disassembly.
   (c) Re-evaluate 'cheap products' use, make product cycle 'cheap' and so that it has no inherent value that would encourage a recycling infrastructure.
   (d) Create computer components and peripherals of biodegradable materials.
   (e) Utilise technology sharing particularly for manufacturing and de-manufacturing.
   (f) Encourage/promote/require green procurement for corporate buyers.
   (g) Look at green packaging options.
3. Companies can and should adopt waste minimisation techniques, which will make a significant reduction in the quantity of E-waste generated and thereby lessening the impact on the environment. It is a 'reverse production' system that designs infrastructure to recover and reuse every material contained within E-wastes metals such as lead, copper, aluminium and gold, and various plastics, glass and wire. Such a 'closed loop' manufacturing and recovery system offers a win-win situation for everyone, less of the earth will be mined for raw materials, and groundwater will be protected, researchers explain.
4. Manufacturers, distributors, and retailers should undertake the responsibility of recycling/disposal of their own products.
5. Manufacturers of computer monitors, television sets and other electronic devices containing hazardous materials must be responsible for educating consumers and the general public regarding the potential threat to public health and the environment posed by their products. At minimum, all computer monitors, television sets and other electronic devices containing hazardous materials must be clearly labelled to identify environmental hazards and proper materials management.

Responsibilities of the Citizen

Waste prevention is perhaps more preferred to any other waste management option including recycling. Donating electronics for reuse extends the lives of valuable products and keeps them out of the waste management system for a longer time. But care should be taken while donating such items i.e. the items should be in working condition.

Reuse, in addition to being an environmentally preferable alternative, also benefits society. By donating used electronics, schools, non-profit organisations, and lower-income families can afford to use equipment that they otherwise could not afford. E-wastes should never be disposed with garbage and other household wastes. This should be segregated at the site and sold or donated to various organisations. While buying electronic products opt for those that:

1. Are made with fewer toxic constituents.
2. Use recycled content.
3. Are energy efficient.
4. Are designed for easy upgrading or disassembly.
5. Utilise minimal packaging.
6. Offer leasing or take back options.
7. Have been certified by regulatory authorities. Customers should opt for upgrading their computers or other electronic items to the latest versions rather than buying new equipments.

NGOs should adopt a participatory approach in management of E-wastes.

Chapter 17

Electronic Waste Management in India

INTRODUCTION

The current practices of E-waste management in India suffer from a number of drawbacks like the difficulty in inventorisation, unhealthy conditions of informal recycling, inadequate legislation, poor awareness and reluctance on part of the corporate to address the critical issues. The consequences are that: (i) toxic materials enter the waste stream with no special precautions to avoid the known adverse effects on the environment and human health, and (ii) resources are wasted when economically valuable materials are dumped or unhealthy conditions are developed during the informal recycling. This chapter highlights the associated issues and strategies to address this emerging problem, in the light of initiatives in India. The chapter presents a waste management system with shared responsibility for the collection and recycling of electronic wastes amongst the manufacturers/assemblers, importers, recyclers, regulatory bodies and the consumers.

The electronic industry is the world's largest and fastest growing manufacturing industry. During the last decade, it has assumed the role of providing a forceful leverage to the socio-economic and technological growth of a developing society. The consequence of its consumer oriented growth combined with rapid product obsolescence and technological advances are a new environmental challenge — the growing menace of 'Electronics Waste' or 'E-waste' that consists of obsolete electronic devices. It is an emerging problem as well as a business opportunity of increasing significance, given the volumes of E-waste being generated and the content of both toxic and valuable materials in them. The fraction including iron, copper, aluminium, gold and other metals in E-waste is over 60 per cent, while plastics account for about 30 per cent and the hazardous pollutants comprise only about 2.70 per cent.

Solid waste management, which is already a mammoth task in India, is becoming more complicated by the invasion of E-waste, particularly computer waste. E-waste from developed countries find an easy way into developing countries in the name of free trade is further complicating the problems associated with waste management. The chapter also highlights the associated issues and strategies to address this emerging problem, in the light of initiatives in India.

E-WASTE IN INDIA

As there is no separate collection of E-waste in India, there is no clear data on the quantity generated and disposed of each year and the resulting extent of environmental risk. The preferred practice to get rid of obsolete electronic items in India is to get them in exchange from retailers when purchasing a new item. The business sector is estimated to account for 78 per cent of all installed computers in India.

Obsolete computers from the business sector are sold by auctions. Sometimes educational institutes or charitable institutions receive old computers for reuse. It is estimated that the total number of obsolete personal computers emanating each year from business and individual households in India will be around 1.38 million. According to a report of Confederation of Indian Industries, the total waste generated by obsolete or broken down electronic and electrical equipment in India has been estimated to be 1,46,000 tons per year.

The results of a field survey conducted in the Chennai, a metroplolitan city of India to assess the average usage and life of the personal computers (PCs), television (TV) and mobile phone showed that the average household usage of the PC ranges from 0.39 to 1.70 depending on the income class. In the case of TV it varied from 1.07 to 1.78 and for mobile phones it varied from 0.88 to 1.70. The low-income households use the PC for 5.94 years, TV for 8.16 years and the mobile phones for 2.34 years while, the upper income class uses the PC for 3.21 years, TV for 5.13 years and mobile phones for 1.63 years. Although the per-capita waste production in India is still relatively small, the total absolute volume of wastes generated will be huge. Further, it is growing at a faster rate. The growth rate of the mobile phones (80 per cent) is very high compared to that of PC (20 per cent) and TV (18 per cent). The public awareness on E-wastes and the willingness of the public to pay for E-waste management as assessed during the study based on an organised questionnaire revealed that about 50 per cent of the public are aware of environmental and health impacts of the electronic items. The willingness of public to pay for E-waste management ranges from 3.57 to 5.92 per cent of the product cost for PC, 3.94 to 5.95 per cent for TV and 3.4 per cent to 5 per cent for the mobile phones.

Additionally considerable quantities of E-waste are reported to be imported. However, no confirmed figures available on how substantial are these transboundary E-waste streams, as most of such trade in E-waste is camouflaged and conducted under the pretext of obtaining 'reusable' equipment or 'donations' from developed nations. The government trade data does not distinguish between imports of new and old computers and peripheral parts and so it is difficult to track what share of imports is used in electronic goods.

IMPACTS OF E-WASTES

Electronic wastes can cause widespread environmental damage due to the use of toxic materials in the manufacture of electronic goods. Hazardous materials such as lead, mercury and hexavalent chromium in one form or the other are present in such wastes primarily consisting of cathode ray tubes (CRTs), printed board assemblies, capacitors, mercury switches and relays, batteries, liquid crystal displays (LCDs), cartridges from photocopying machines, selenium drums (photocopier) and electrolytes. Although it is hardly known, E-waste contains toxic substances such as lead and cadmium in circuit boards; lead oxide and cadmium in monitor cathode ray tubes (CRTs); mercury in switches and flat screen monitors; cadmium in computer batteries; polychlorinated biphenyls (PCBs) in older capacitors and transformers; and brominated flame retardants on printed circuit boards, plastic casings, cables and polyvinyl chloride (PVC) cable insulation that releases highly toxic dioxins and furans when burned to retrieve copper from the wires. All electronic equipments contain printed circuit boards which are hazardous because of their content of lead (in solder), brominated flame retardants (typically 5–10 per cent by weight) and antimony oxide, which is also present as a flame retardant (typically 1–2 per cent by weight).

Landfilling of E-wastes can lead to the leaching of lead into the groundwater. If the CRT is crushed and burned, it emits toxic fumes into the air. These products contain several rechargeable battery types, all of which contain toxic substances that can contaminate the environment when burned in incinerators or disposed of in landfills. The cadmium from one mobile phone battery is enough to pollute 600 m^3 of water. The quantity of cadmium in landfill sites is significant, and considerable toxic contamination is caused by the inevitable medium and long-term effects of cadmium leaking into the surrounding soil. Because plastics are highly flammable, the printed wiring board and housings of electronic products contain brominated flame retardants, a number of which are clearly damaging to human health and the environment.

Impacts of Informal Recycling

The accrued electronic and electric waste in India is dismantled and sorted manually to fractions such as printed wiring boards, cathode ray tubes (CRT), cables, plastics, metals, condensers and other, nowadays invaluable materials like batteries. It is a livelihood for unorganised recyclers and due to lack of awareness, they are risking their health and the environment as well. The valuable fractions are processed to directly reusable components and to secondary raw materials in a variety of refining and conditioning processes. No sophisticated machinery or personal protective equipment is used for the extraction of different materials. All the work is done by bare hands and only with the help of hammers and screwdrivers. Children and women are routinely involved in the operations. Waste components which does not have any resale or reuse value are openly burnt or disposed off in open dumps. Pollution problems associated with such backyard smelting using crude processes are resulting in fugitive emissions and slag containing heavy metals of health concern.

CRT breaking operations result in injuries from cuts and acids used for removal of heavy metals and respiratory problems due to shredding, burning, etc. They use strong acids to retrieve precious metals such as gold. Working in poorly ventilated enclosed areas without masks and technical expertise results in exposure to dangerous and slow poisoning chemicals. Polychlorinated biphenyls (PCBs) in older capacitors and transformers; and brominated flame retardants on printed circuit boards, plastic casings, cables and polyvinyl chloride (PVC) cable insulation can release highly toxic dioxins and furans when burned to retrieve copper from the wires.

On a broader scale, analysing the environmental and societal impacts of E-waste reveals a mosaic of benefits and costs. Proponents of E-waste recycling claim that greater employment, new access to raw materials and electronics, and improved infrastructure will result. These will further boost the region's advance towards prosperity. Yet the reality is that the new wealth and benefits are unequally distributed, and the contribution of electronics to societal growth is sometimes illusory.

Most E-waste 'recycling' involve small enterprises that are numerous, widespread, and difficult to regulate. They take advantage of low labour costs due to high unemployment rates, internal migration of poor peasants, and the lack of protest or political mobilisation by affected villagers who believe that E-wastes provide the only viable source of income or entry into modern development pathways. They are largely invisible to state scrutiny because they border on the informal economy and are therefore not included in official statistics.

STATUS OF E-WASTE MANAGEMENT IN INDIA

Despite a wide range of environmental legislation in India there are no specific laws or guidelines for electronic waste or computer waste. As per the hazardous waste rules, E-waste is not treated as hazardous unless proved to have higher concentration of certain substances. Though PCBs and CRTs would always exceed these parameters, there are several grey areas that need to be addressed. Basel convention has waste electronic assemblies in A1180 and mirror entry in B1110, mainly on concerns of mercury, lead and cadmium. Electronic waste is included under List-A and List-B of Schedule-3 of the hazardous wastes. The import of this waste therefore requires specific permission of the Ministry of Environment and Forests.

As the collection and recycling of electronic wastes is being done by the informal sector in the country at present, the Government has taken the following action/steps to enhance awareness about environmentally sound management of electronic waste:

1. Several workshops on electronic waste management was organised by the Central Pollution Control Board (CPCB) in collaboration with Toxics Link, CII, etc.
2. Action has been initiated by CPCB for rapid assessment of the E-waste generated in major cities of the country.
3. A National Working Group has been constituted for formulating a strategy for E-waste management.
4. A comprehensive technical guide on 'Environmental Management for Information Technology Industry in India' has been published and circulated widely by the Department of Information Technology (DIT), Ministry of Communication and Information Technology.
5. Demonstration projects has also been set up by the DIT at the Indian Telephone Industries for recovery of copper from Printed Circuit Boards.

Although awareness and readiness for implementing improvements is increasing rapidly, the major obstacles to manage the E-wastes safely and effectively remain. These include:

1. The lack of reliable data that poses a challenge to policy makers wishing to design an E-waste management strategy and to an industry wishing to make rational investment decisions.
2. Only a fraction of the E-waste (estimated 10 per cent) finds its way to recyclers due to absence of an efficient take back scheme for consumers.
3. The lack of a safe E-waste recycling infrastructure in the formal sector and thus reliance on the capacities of the informal sector pose severe risks to the environment and human health.
4. The existing E-waste recycling systems are purely business-driven that have come about without any government intervention. Any development in these E-waste sectors will have to be built on the existing setup as the waste collection and preprocessing can be handled efficiently by the informal sector, at the same time offer numerous job opportunities.

The Swiss State Secretariat for Economic Affairs mandated the Swiss Federal Laboratories for Materials Testing and Research (EMPA) to implement the program 'knowledge partnerships in E-waste recycling' and India is one of the partner countries. The program aims at improving E-waste management systems through knowledge management and capacity building. It has analysed E-waste recycling frameworks and processes in different parts of the world in its first phase and all results of the project are documented on the website http://www.ewaste.ch/.

E-waste Policy and Regulation

The policy shall address all issues ranging from production and trade to final disposal, including technology transfers for the recycling of electronic waste. Clear regulatory instruments, adequate to control both legal and illegal exports and imports of E-wastes and ensuring their environmentally sound management should be in place. There is also a need to address the loop holes in the prevailing legal framework to ensure that E-wastes from developed countries are not reaching the country for disposal. The Port and the Custom authorities need to monitor these aspects. The regulations should prohibit the disposal of E-wastes in municipal landfills and encourage owners and generators of E-wastes to properly recycle the wastes. Manufactures of products must be made financially, physically and legally responsible for their products. Policies and regulations that cover design for environment (DfE) and better management of restricted substances may be implemented through measures such as:

1. Specific product take-back obligations for industry.
2. Financial responsibility for actions and schemes.
3. Greater attention to the role of new product design.
4. Material and/or substance bans including stringent restrictions on certain substances.
5. Greater scrutiny of cross-border movements of electrical and electronic products and E-waste.
6. Increasing public awareness by labelling products as 'environmental hazard'.

The key questions about the effectiveness of legislation would include:

1. What is to be covered by the term electronic waste?
2. Who pays for disposal?
3. Is producer responsibility the answer?
4. What would be the benefits of voluntary commitments?
5. How can sufficient recovery of material be achieved to guarantee recycling firms a reliable and adequate flow of secondary material?

A complete national level inventory, covering all the cities and all the sectors must be initiated. A public-private participatory forum (E-waste agency) of decision-making and problem resolution in E-waste management must be developed. This could be a working group comprising regulatory agencies, NGOs, Industry Associations, experts, etc. to keep pace with the temporal and spatial changes in structure and content of E-waste. This Working Group can be the feedback providing mechanism to the Government that will periodically review the existing rules, plans and strategies for E-waste management. Mandatory labelling of all computer monitors, television sets and other household/industrial electronic devices may be implemented for declaration of hazardous material contents with a view to identifying environmental hazards and ensuring proper material management and E-waste disposal.

The efforts to improve the situation through regulations, though an important step; are usually only modestly effective because of the lack of enforcement. While there has been some progress made in this direction with the support of agencies such as GTZ, enforcement of regulations is often weak due to lack of resources and underdeveloped legal systems. Penalties for noncompliance and targets for collection or recycling are often used to ensure compliance.

Extended Producer Responsibility

Extended producer responsibility (EPR) is an environmental policy approach in which a producer's responsibility for a product is extended to the post consumer stage of the product's life cycle, including its final disposal. In principle, all the actors along the product chain share responsibility for the life-cycle environmental impacts of the whole product system. The greater the ability of the actor to influence the environmental impacts of the product system, the greater the share of responsibility for addressing those impacts should be. These actors are the consumers, the suppliers, and the product manufacturers. Consumers can affect the environmental impacts of products in a number of ways: via purchase choices (choosing environmentally friendly products), via maintenance and the environmentally conscious operation of products, and via careful disposal (e.g. separated disposal of appliances for recycling). Suppliers may have a significant influence by providing manufacturers with environmentally friendly materials and components. Manufacturers can reduce the life-cycle environmental impacts of their products through their influence on product design, material choices, manufacturing processes, product delivery, and product system support. The system design needs to be such that there are checks and balances, especially to prevent free riders. The goals of the product designer could include reducing toxicity, reducing energy use, streamlining product weight and materials, identifying opportunities for easier reuse, and more. Manufacturers have to improve the design by: (i) the substitution of hazardous substances such as lead, mercury, cadmium, hexavalent chromium and certain brominated flame retardants, (ii) measures to facilitate identification and reuse of components and materials, particularly plastics, and (iii) measures to promote the use of recycled plastics in new products.

Manufacturers should give incentives to their customers for product return through a 'buy back approach' whereby old electronic goods are collected and a discount could be given on new products purchased by the consumer. All vendors of electronic devices shall provide take-back and management services for their products at the end of life of those products. The old electronic product should then be sent back to be carefully dismantled for its parts to be either recycled or reused, either in a separate recycling division at the manufacturing unit or in a common facility.

Collection systems are to be established so that E-waste is collected from the right places ensuring that this directly comes to the recycling unit. Collection can be accomplished through collection centres. Each electronic equipment manufacturer shall work cooperatively with collection centres to ensure implementation of a practical and feasible financing system. Collection Centres may only ship wastes to dismantlers and recyclers that are having authorisation for handling, processing, refurbishment, and recycling meeting environmentally sound management guidelines.

E-waste Recycling

Many discarded machines contain usable parts which could be salvaged and combined with other used equipment to create a working unit. It is labour intensive to remove, inspect and test components and then reassemble them into complete working machines. Institutional infrastructures, including E-waste collection, transportation, treatment, storage, recovery and disposal, need to be established, at national and/or regional levels for the environmentally sound management of E-wastes. These facilities should be approved by the regulatory authorities and if required provided with appropriate incentives. Establishment of E-waste collection, exchange and recycling centers should be encouraged in partnership with governments, NGOs and manufacturers.

Environmentally sound recycling of E-waste requires sophisticated technology and processes, which are not only very expensive, but also need specific skills and training for the operation. Proper recycling of complex materials requires the expertise to recognise or determine the presence of hazardous or potentially hazardous constituents as well as desirable constituents (i.e. those with recoverable value), and then be able to apply the company's capabilities and process systems to properly recycle both of these streams. Appropriate air pollution control devices for the fugitive and point source emissions are required. Guidelines are to be developed for environmentally sound recycling of E-wastes. Private Sector are coming forward to invest in the E-waste projects once they are sure of the returns.

Capacity Building, Training and Awareness Programs

The future of E-waste management depends not only on the effectiveness of local government, the operator of recycling services, but also on the attitude of citizens, and on the key role of manufactures and bulk consumers to shape and develop community participation. Lack of civic sense and awareness among city residents will be a major hurdle to keep E-waste out of municipal waste stream. Collaborative campaigns are required to sensitise the users and consumers should pay for recycling of electronic goods. Consumers are to be informed of their role in the system through a labelling requirement for items. Consumers to be educated to buy only necessary products that utilise some of the emerging technologies (i.e. lead-free, halogen-free, recycled plastics and from manufacturers or retailers that will 'take-back' their product) to be identified through ecolabelling.

Awareness raising programs and activities on issues related to the environmentally sound management (ESM), health and safety aspects of E-wastes in order to encourage better management practices should be implemented for different target groups. Technical guidelines for the ESM of E-wastes should be developed as soon as possible.

WASTE MANAGEMENT STRATEGIES

The best option for dealing with E-wastes is to reduce the volume. Designers should ensure that the product is built for reuse, repair and/or upgradability. Stress should be laid on use of less toxic, easily recoverable and recyclable materials which can be taken back for refurbishment, remanufacturing, disassembly and reuse. Recycling and reuse of material are the next level of potential options to reduce E-waste. Recovery of metals, plastic, glass and other materials reduces the magnitude of E-waste. These options have a potential to conserve the energy and keep the environment free of toxic material that would otherwise have been released.

It is high time the manufactures, consumers, regulators, municipal authorities, state governments, and policy makers take up the matter seriously so that the different critical elements depicted in Fig. 17.1 are addressed in an integrated manner. It is the need of the hour to have an 'E-waste-policy' and national regulatory framework for promotion of such activities. An E-waste policy is best created by those who understand the issues. So it is best for industry to initiate policy formation collectively, but with user involvement. Sustainability of E-waste management systems has to be ensured by improving the effectiveness of collection and recycling systems (e.g. public–private-partnerships in setting up buy-back or drop-off centers) and by designing-in additional funding, e.g. advance recycling fees.

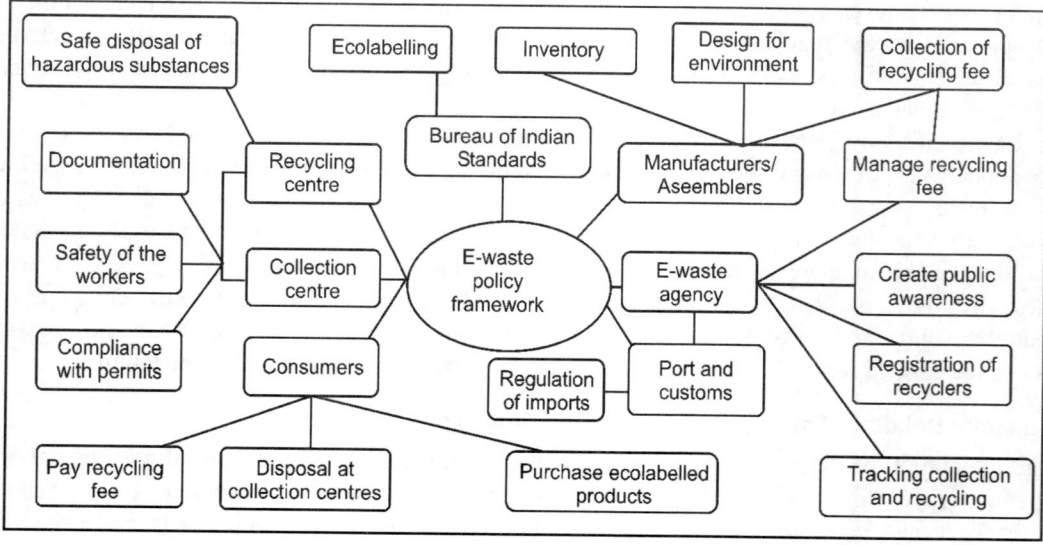

Fig. 17.1. Elements of E-waste management system for India.

To sum up, solid waste management, which is already a mammoth task in India, is becoming more complicated by the invasion of E-waste, particularly computer waste. There exists an urgent need for a detailed assessment of the current and future scenario including quantification, characteristics, existing disposal practices, environmental impacts, etc. Institutional infrastructures, including E-waste collection, transportation, treatment, storage, recovery and disposal, need to be established, at national and/or regional levels for the environmentally sound management of E-wastes. Establishment of E-waste collection, exchange and recycling centers should be encouraged in partnership with private entrepreneurs and manufacturers.

Model facilities employing environmentally sound technologies and methods for recycling and recovery are to be established. Criteria are to be developed for recovery and disposal of E-wastes. Policy level interventions should include development of E-waste regulation, control of import and export of E-wastes and facilitation in development of infrastructure. An effective take-back program providing incentives for producers to design products that are less wasteful, contain fewer toxic components, and are easier to disassemble, reuse, and recycle may help in reducing the wastes. It should set targets for collection and reuse/recycling, impose reporting requirements and include enforcement mechanisms and deposit/refund schemes to encourage consumers to return electronic devices for collection and reuse/recycling. End of life management should be made a priority in the design of new electronic products.

SECTION V

Waste to Energy

Chapter 18

Pyrolysis, Gasification and Combined Pyrolysis/Gasification Systems

INTRODUCTION

Pyrolysis is a thermochemical decomposition of organic material at elevated temperatures in the absence of oxygen. Pyrolysis typically occurs under pressure and at operating temperatures above 430°C (800°F). The word is coined from the Greek-derived elements pyr 'fire' and lysis 'separating'. Pyrolysis is a special case of thermolysis, and is most commonly used for organic materials, being, therefore, one of the processes involved in charring. The pyrolysis of wood, which starts at 200°–300°C (390°–570°F), occurs for example in fires or when vegetation comes into contact with lava in volcanic eruptions. In general, pyrolysis of organic substances produces gas and liquid products and leaves a solid residue richer in carbon content. Extreme pyrolysis, which leaves mostly carbon as the residue, is called carbonisation. The process is used heavily in the chemical industry, for example, to produce charcoal, activated carbon, methanol, and other chemicals from wood, to convert ethylene dichloride into vinyl chloride to make PVC, to produce coke from coal, to convert biomass into syngas, to turn waste into safely disposable substances, and for transforming medium-weight hydrocarbons from oil into lighter ones like gasoline. These specialised uses of pyrolysis may be called various names, such as dry distillation, destructive distillation or cracking. Pyrolysis also plays an important role in several cooking procedures, such as baking, frying, grilling, and caramelising. And it is a tool of chemical analysis, for example, in mass spectrometry and in carbon-14 dating. Indeed, many important chemical substances, such as phosphorus and sulphuric acid, were first obtained by this process. Pyrolysis has been assumed to take place during catagenesis, the conversion of buried organic matter to fossil fuels. It is also the basis of pyrography. In their embalming process, the ancient Egyptians used a mixture of substances, including methanol, which they obtained from the pyrolysis of wood. Pyrolysis differs from other high-temperature processes like combustion and hydrolysis in that it does not involve reactions with oxygen, water, or any other reagents. In practice, it is not possible to achieve a completely oxygen-free atmosphere. Because some oxygen is present in any pyrolysis system, a small amount of oxidation occurs.

The term has also been applied to the decomposition of organic material in the presence of superheated water or steam (hydrous pyrolysis), for example, in the steam cracking of oil.

OCCURRENCE AND USES

Fire

Pyrolysis is usually the first chemical reaction that occurs in the burning of many solid organic fuels, like wood, cloth, and paper, and also of some kinds of plastic. In a wood fire, the visible flames are not due to combustion of the wood itself, but rather of the gases released by its pyrolysis, whereas the flameless burning of embers is the combustion of the solid residue (charcoal) left behind by it. Thus, the pyrolysis of common materials like wood, plastic, and clothing is extremely important for fire safety and fire-fighting.

Cooking

Pyrolysis occurs whenever food is exposed to high enough temperatures in a dry environment, such as roasting, baking, toasting, grilling, etc. It is the chemical process responsible for the formation of the golden-brown crust in foods prepared by those methods. In normal cooking, the main food components that undergo pyrolysis are carbohydrates (including sugars, starch, and fibre) and proteins. Pyrolysis of fats requires a much higher temperature, and, since it produces toxic and flammable products (such as acrolein), it is, in general, avoided in normal cooking. It may occur, however, when barbecuing fatty meats over hot coals.

Even though cooking is normally carried out in air, the temperatures and environmental conditions are such that there is little or no combustion of the original substances or their decomposition products. In particular, the pyrolysis of proteins and carbohydrates begins at temperatures much lower than the ignition temperature of the solid residue, and the volatile subproducts are too diluted in air to ignite. (In flambé dishes, the flame is due mostly to combustion of the alcohol, while the crust is formed by pyrolysis as in baking.)

Pyrolysis of carbohydrates and proteins requires temperatures substantially higher than 100°C (212°F), so pyrolysis does not occur as long as free water is present, e.g. in boiling food—not even in a pressure cooker. When heated in the presence of water, carbohydrates and proteins suffer gradual hydrolysis rather than pyrolysis. Indeed, for most foods, pyrolysis is usually confined to the outer layers of food, and begins only after those layers have dried out. Food pyrolysis temperatures are, however, lower than the boiling point of lipids, so pyrolysis occurs when frying in vegetable oil or suet, or basting meat in its own fat.

Pyrolysis also plays an essential role in the production of barley tea, coffee, and roasted nuts such as peanuts and almonds. As these consist mostly of dry materials, the process of pyrolysis is not limited to the outermost layers but extends throughout the materials. In all these cases, pyrolysis creates or releases many of the substances that contribute to the flavour, colour, and biological properties of the final product. It may also destroy some substances that are toxic, unpleasant in taste or those that may contribute to spoilage. Controlled pyrolysis of sugars starting at 170°C (338°F) produces caramel, a beige to brown water-soluble product widely used in confectionery and (in the form of caramel colouring) as a colouring agent for soft drinks and other industrialised food products.

Solid residue from the pyrolysis of spilled and splattered food creates the brown-black encrustation often seen on cooking vessels, stove tops, and the interior surfaces of ovens.

Charcoal

Pyrolysis has been used since ancient times for turning wood into charcoal on an industrial scale. Besides wood, the process can also use sawdust and other wood waste products. Charcoal is obtained by heating wood until its complete pyrolysis (carbonisation) occurs, leaving only carbon and inorganic ash. In many parts of the world, charcoal is still produced semi-industrially, by burning a pile of wood that has been mostly covered with mud or bricks. The heat generated by burning part of the wood and the volatile by-products pyrolyses the rest of the pile. The limited supply of oxygen prevents the charcoal from burning. A more modern alternative is to heat the wood in an airtight metal vessel, which is much less polluting and allows the volatile products to be condensed.

The original vascular structure of the wood and the pores created by escaping gases combine to produce a light and porous material. By starting with a dense wood-like material, such as nutshells or peach stones, one obtains a form of charcoal with particularly fine pores (and hence a much larger pore surface area), called activated carbon, which is used as an adsorbent for a wide range of chemical substances.

Biochar

Residues of incomplete organic pyrolysis, e.g. from cooking fires, are thought to be the key component of the terra preta soils associated with ancient indigenous communities of the Amazon basin. Terra preta is much sought by local farmers for its superior fertility compared to the natural red soil of the region. Efforts are underway to recreate these soils through biochar, the solid residue of pyrolysis of various materials, mostly organic waste.

Biochar improves the soil texture and ecology, increasing its ability to retain fertilisers and release them slowly. It naturally contains many of the micronutrients needed by plants, such as selenium. It is also safer than other 'natural' fertilisers such as manure or sewage, since it has been disinfected at high temperature. And, since it releases its nutrients at a slow rate, it greatly reduces the risk of water table contamination. Biochar is also being considered for carbon sequestration, with the aim of mitigation of global warming. Because pyrolysis burns the volatile gases, biochar emits only water vapour. By burning the harmful gases, a stabile form of carbon can be sequestered into the ground, where it will remain for thousands of years.

Coke

Pyrolysis is used on a massive scale to turn coal into coke for metallurgy, especially steelmaking. Coke can also be produced from the solid residue left from petroleum refining.

Those starting materials typically contain hydrogen, nitrogen, or oxygen atoms combined with carbon into molecules of medium to high molecular weight. The coke-making or 'coking' process consists of heating the material in closed vessels to very high temperatures (up to 2000°C or 3600°F) so that those molecules are broken down into lighter volatile substances, which leave the vessel and a porous but hard residue that is mostly carbon and inorganic ash. The amount of volatiles varies with the source material, but is typically 25–30 per cent of it by weight.

Carbon Fibre

Carbon fibres are filaments of carbon that can be used to make very strong yarns and textiles. Carbon fibre items are often produced by spinning and weaving the desired item from fibres of a suitable polymer, and then pyrolysing the material at a high temperature (from 1500°–3000°C or 2730°–5430°F).

The first carbon fibres were made from rayon, but polyacrylonitrile has become the most common starting material. For their first workable electric lamps, Joseph Wilson Swan and Thomas Edison used carbon filaments made by pyrolysis of cotton yarns and bamboo splinters, respectively.

Biofuel

Pyrolysis is the basis of several methods that are being developed for producing fuel from biomass, which may include either crops grown for the purpose or biological waste products from other industries.

Although synthetic diesel fuel cannot yet be produced directly by pyrolysis of organic materials, there is a way to produce similar liquid (bio-oil) that can be used as a fuel, after the removal of valuable bio-chemicals that can be used as food additives or pharmaceuticals. Higher efficiency is achieved by the so-called flash pyrolysis, in which finely divided feedstock is quickly heated to between 350° and 500°C (660° and 930°F) for less than 2 seconds.

Fuel bio-oil resembling light crude oil can also be produced by hydrous pyrolysis from many kinds of feedstock, including waste from pig and turkey farming, by a process called thermal depolymerisation (which may, however, include other reactions besides pyrolysis).

Plastic waste disposal

Anhydrous pyrolysis can also be used to produce liquid fuel similar to diesel from plastic waste.

Processes

In many industrial applications, the process is done under pressure and at operating temperatures above 430°C (806°F). For agricultural waste, for example, typical temperatures are 450° to 550°C (840° to 1000°F).

Vacuum Pyrolysis

In vacuum pyrolysis, organic material is heated in a vacuum in order to decrease boiling point and avoid adverse chemical reactions. It is used in organic chemistry as a synthetic tool. In flash vacuum thermolysis (FVT) the residence time of the substrate at the working temperature is limited as much as possible, again in order to minimise secondary reactions.

Processes for biomass pyrolysis

Since pyrolysis is endothermic, various methods to provide heat to the reacting biomass particles have been proposed:
1. Partial combustion of the biomass products through air injection. This results in poor-quality products.
2. Direct heat transfer with a hot gas, the ideal one being product gas that is reheated and recycled. The problem is to provide enough heat with reasonable gas flow-rates.
3. Indirect heat transfer with exchange surfaces (wall, tubes). It is difficult to achieve good heat transfer on both sides of the heat exchange surface.
4. Direct heat transfer with circulating solids: Solids transfer heat between a burner and a pyrolysis reactor. This is an effective but complex technology.

For flash pyrolysis, the biomass must be ground into fine particles and the insulating char layer that forms at the surface of the reacting particles must be continuously removed. The following technologies have been proposed for biomass pyrolysis:
1. Fixed beds used for the traditional production of charcoal. Poor, slow heat transfer result in very low liquid yields.
2. Augers: This technology is adapted from a Lurgi process for coal gasification. Hot sand and biomass particles are fed at one end of a screw. The screw mixes the sand and biomass and conveys them along. It provides a good control of the biomass residence time. It does not dilute the pyrolysis products with a carrier or fluidising gas. However, sand must be reheated in a separate vessel, and mechanical reliability is a concern. There is no large-scale commercial implementation.

3. Ablative processes: Biomass particles are moved at high speed against a hot metal surface. Ablation of any char forming at the particles surface maintains a high rate of heat transfer. This can be achieved by using a metal surface spinning at high speed within a bed of biomass particles, which may present mechanical reliability problems but prevents any dilution of the products. As an alternative, the particles may be suspended in a carrier gas and introduced at high speed through a cyclone whose wall is heated; the products are diluted with the carrier gas. A problem shared with all ablative processes is that scale-up is made difficult, since the ratio of the wall surface to the reactor volume decreases as the reactor size is increased. There is no large-scale commercial implementation.

4. Rotating cone: Preheated hot sand and biomass particles are introduced into a rotating cone. Due to the rotation of the cone, the mixture of sand and biomass is transported across the cone surface by centrifugal force. Like other shallow transported-bed reactors relatively fine particles are required to obtain a good liquid yield. There is no large-scale commercial implementation.

5. Fluidised beds: Biomass particles are introduced into a bed of hot sand fluidised by a gas, which is usually a recirculated product gas. High heat transfer rates from fluidised sand result in rapid heating of biomass particles. There is some ablation by attrition with the sand particles, but it is not as effective as in the ablative processes. Heat is usually provided by heat exchanger tubes through which hot combustion gas flows. There is some dilution of the products, which makes it more difficult to condense and then remove the bio-oil mist from the gas exiting the condensers. This process has been scaled up by companies such as dynamotive and Agri-Therm. The main challenges are in improving the quality and consistency of the bio-oil.

6. Circulating fluidised beds: Biomass particles are introduced into a circulating fluidised bed of hot sand. Gas, sand, and biomass particles move together, with the transport gas usually being a recirculated product gas, although it may also be a combustion gas. High heat transfer rates from sand ensure rapid heating of biomass particles and ablation stronger than with regular fluidised beds. A fast separator separates the product gases and vapours from the sand and char particles. The sand particles are reheated in fluidised burner vessel and recycled to the reactor. Although this process can be easily scaled up, it is rather complex and the products are much diluted, which greatly complicates the recovery of the liquid products.

Industrial Sources

Many sources of organic matter can be used as feedstock for pyrolysis. Suitable plant material includes greenwaste, sawdust, waste wood, woody weeds; and agricultural sources including nut shells, straw, cotton trash, rice hulls, switch grass; and animal waste including poultry litter, dairy manure, and potentially other manures. Pyrolysis is used as a form of thermal treatment to reduce waste volumes of domestic refuse. Some industrial by-products are also suitable feedstock including paper sludge and distillers grain.

There is also the possibility of integrating with other processes such as mechanical biological treatment and anaerobic digestion.

The thermal treatment options of pyrolysis, gasification and combined pyrolysis/gasification systems, are generating increasing interest as viable alternative environmental and economic options for waste processing. These options have a number of advantages over conventional incineration or landfilling of waste. Depending on the technology, the waste can be processed to produce not only energy, but also gas or oil products for use as petrochemical feedstocks and/or a carbonaceous char for use in applications such as effluent treatment or for gasification feedstock. The production of storable end products such as a gas, oil or char, enables the possibility of decoupling the end use of that product, either for energy production or petrochemical use from the waste treatment process. The EC Waste Incineration Directive, regulates the emissions to air, land and water from incineration and also details the operational, emissions monitoring, process conditions, etc. of the incineration plant. Incineration is defined in the Directive as any thermal process dedicated to the thermal treatment of wastes, with or without energy recovery. In addition, the directive specifically includes the thermal treatment processes of pyrolysis and gasification processes insofar as the substances resulting from the treatment are subsequently incinerated.

Figure 18.1 characterises the main differences between pyrolysis, gasification and incineration. The key difference is the amount of oxygen supplied to the thermal reactor. For pyrolysis there is an absence of oxygen, and for gasification there is a limited supply of oxygen, such that complete combustion does not take place, instead the combustible gases; carbon monoxide and hydrogen are produced. The oxygen for gasification is supplied in the form of air, steam or pure oxygen. Incineration involves the complete oxidation of the waste in an excess supply of oxygen to produce carbon dioxide, water and ash, plus some other products such as metals, trace hydrocarbons, acid gases, etc.

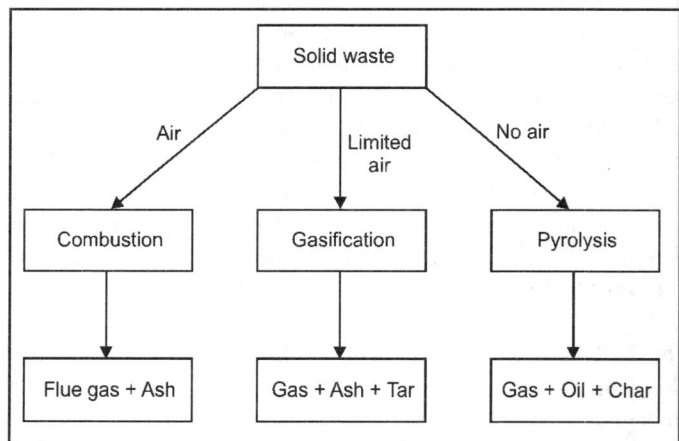

Fig. 18.1. Process characterisation of incineration, gasification and pyrolysis.

Waste materials are composed of complex chemical compounds, for example, municipal solid waste contains paper and cardboard which are composed of large, complex polymeric, organic molecular chains such as cellulose, hemicellulose and lignin. Similarly, wastes such as forestry wastes and biomass are also mainly composed of cellulose, hemicellulose and lignin polymeric molecules. Plastics are also composed of large polymer chains. The process of thermal degradation or pyrolysis of such materials, in the absence of oxygen, results in the long polymer chains breaking to produce shorter molecular weight chains and molecules. These shorter molecules result in the formation of the oils and gases characteristic of pyrolysis of waste. The exact mechanisms of thermal degradation of waste are not clear. Figures 18.2 and 18.3 show examples of the large polymer chains of cellulose, hemicellulose, lignin, and several plastics and rubbers found in waste materials.

Relatively low temperatures are used for pyrolysis, in the range 400°–800°C. The application of pyrolysis to waste materials is a relatively recent development. In particular, the production of oils from the pyrolysis of waste has been investigated, with the aim of using the oils directly in fuel applications or, after upgrading, to produce refined fuels. The pyrolysis oils derived from a variety of wastes have also been shown to be complex in composition and contain a wide variety of chemicals which may be used as chemical feedstock. The oil has a higher energy density, that is a higher energy content per unit weight, than the raw waste. The solid char can be used as a solid fuel or as a char-oil, char-water slurry for fuel. Alternatively the char can be used as carbon black or upgraded to activated carbon. The gases generated have medium to high calorific values and may contain sufficient energy to supply the energy requirements of a pyrolysis plant.

The process conditions are altered to produce the desired char, gas or oil end product, with the pyrolysis temperature and heating rate having the most influence on the product distribution. The heat is supplied by indirect heating, such as the combustion of the gases or oil, or directly by hot gas transfer. Pyrolysis has the advantage that the gases or oil product derived from the waste can be used to provide the fuel for the pyrolysis process itself. Pyrolysis systems for municipal solid waste, tyres, plastics, composite plastics, sewage sludge, textile waste and biomass have been investigated.

Very slow heating rates coupled with a low final maximum temperature, maximises the yield of char, because the production of char from wood in the form of charcoal involves a very slow heating rate to moderate temperatures. The process of carbonisation of waste results in reduced concentrations of oil/tar and gas product and these are regarded as by-products of the main charcoal-forming process. Moderate heating rates in the range of about 20°C/min. to 100°C/min. and maximum temperatures of 600°C give an approximately equal distribution of oils, char and gases. This is referred to as conventional pyrolysis or slow pyrolysis. Because of the slow heating rates and generally slow removal of the products of pyrolysis from the hot pyrolysis reactor, secondary reactions of the products can take place. Generally, a more complex product slate is found.

Very high heating rates of about 100°C/s to 1000°C/s at temperatures below 650°C and with rapid quenching, lead to the formation of a mainly liquid product, which is referred to as fast or flash pyrolysis. Liquid yields up to 70 per cent have been reported for biomass feedstocks using flash pyrolysis. In addition, the carbonaceous char and gas production are minimised. The primary liquid products of pyrolysis are rapidly quenched and this prevents breakdown of the products to gases in the hot reactor. The high reaction rates also cause char-forming reactions from the oil products to be minimised.

At high heating rates and high temperatures the oil products quickly breakdown to yield a mainly gas product. The typical yield of gas from the original feedstock hydrocarbon is 70 per cent. This process differs from gasification which is a series of reactions involving carbon and oxygen in the form of oxygen gas, air or steam, to produce a gas product consisting mainly of CO, CO_2, H_2 and CH_4. Table 18.1 shows the typical characteristics of different types of pyrolysis.

Pyrolysis process conditions can be optimised to produce either a solid char, gas or liquid/oil product. Table 18.2 shows the yields of char, oil/liquid and gas from various waste feedstocks.

The solid char product from carbonisation or slow pyrolysis of wood has been used for centuries as the process to produce charcoal for use as fuel and charcoal product yields of between 30 and 40 per cent are common. Pyrolysis of waste materials also produces a char product, the parentage production depending on process conditions. Pyrolysis of municipal solid waste produces a 35 per cent char product, which has a high ash content of up to 37 per cent, and tyre pyrolysis under slow heating rate conditions produces a char of up to 50 per cent with an ash content of about 10 per cent. The chars may be used directly as fuels, briquetted to produce solid fuels, used as adsorptive materials such as activated carbon, upgraded to produce a higher grade activated carbon or crushed and mixed with the pyrolysis oil product to produce a slurry for combustion.

Table 18.1. Typical characteristics of different types of pyrolysis.

Pyrolysis	Residence time	Heating rate	Reaction environment	Pressure (bar)	Temperature (°C)	Major product
Carbonisation	hrs–days	Very low	Combustion products	1	400	Charcoal
Conventional	10 s–10 min	Low-moderate	Primary/ secondary products	1	<600	Gas, char liquid
Flash–liquid	<1 s	High	Primary products	1	<600	Liquid
Flash–gas	<1 s	High	Primary products	1	>700	Gas
Ultra	<0.5 s	Very high	Primary products	1	1000	Gas, chemicals
Other pyrolysis types						
Vacuum	2–30 s	Medium	Vacuum	<0.1	400	Liquid
Hydropyrolysis	<10 s	High	H_2 + primary	~20	<500	Liquid, chemicals
Methanolysis	0.5–1.5 s	High	CH_4 + primary	~3	1050	Benzene, toluene, xylene + alkenes

Table 18.2. Product yields from the pyrolysis of waste.

Waste	Pyrolysis process	Temperature (°C)	Heating rate	Char (%)	Liquid (%)	Gas (%)
Wood	Moderate (batch)	600	20°C/min.	22.6	50.4	27.0
Wood	Fast (fluidised bed)	550	–300°C/s	17.3	67.0	14.9
Tyre	Moderate (batch)	600	20°C/min.	39.2	54.0	6.8
Tyre	Slow/moderate (batch)	850	–5°C/min.	49.5	32.5	18.0
Tyre	Fast (fluidised bed)	640	–	38	40	18
RDF*	Moderate (batch)	600	20°C/min.	35.2	49.2	18.8
RDF	Moderate (batch)	700	–	30	49	22
Plastic (mixed)	Moderate (batch)	700	25°C/min.	2.9	75.1	9.6
Textile flax	Slow (batch)	450	2°C/min.	25.0	52.5	22.5

*Refuse derived fuel from municipal solid waste.

The calorific value of the chars are relatively high, for example, char derived from municipal solid waste has a calorific value of about 19 MJ/kg, tyre char about 29 MJ/kg and wood waste produces a char of calorific value about 33 MJ/kg. These figures compare with a typical bituminous coal of calorific value 30 MJ/kg. As such, the chars could be used as a medium grade solid fuel.

The significance of a high ash content in the chars means that the value of the char as a fuel is reduced. In addition, the use of pyrolysis chars as substitutes for activated carbon are greatly diminished if they have a high ash content. Chars from wood have very low ash contents, typically less than 2 per cent, whereas the ash content of tyre-derived pyrolysis chars are over 10 per cent. The upgrading of pyrolysis chars to activated carbon for biomass-derived pyrolysis chars, has been achieved using steam activation.

However, the upgrading of tyre chars to activated carbon requires an additional processing step of de-ashing to make the product acceptable to the activated carbon industry. In addition, the specifications of activated carbon derived from traditional routes such as coconut shell are well established, and as with most new products, it is difficult for an alternative product to break into an established market. Even though the waste-derived chars may be cheaper, the specifications, quality and maintenance of quality have to be guaranteed. Commercially used activated carbons have surface areas typically in the range of 500–2000 m^2/g and pore sizes which can be manipulated by the process conditions or source feedstock to produce the desired pore-size distribution for a particular application. Activated carbons may be produced by either physical or chemical activation. Chemical activation involves impregnation with a chemical, such as zinc chloride, followed by carbonisation using the pyrolysis process. Physical activation involves pyrolysis of the source material to produce a char, followed by steam of carbon dioxide gasification. Such techniques have been applied to waste materials to enhance the properties of the derived char in order to produce an activated carbon with properties similar to those produced commercially. For example, char derived from tyres has an initial surface area of about 60 m^2/g, but activation of the carbonaceous char with steam at temperatures above 800°C produces an activated carbon with a surface area of over 650 m^2/g. The action of the steam is to react with the carbon to produce carbon monoxide, carbon dioxide and hydrogen, opening up pores and increasing the surface area. Surface areas of flax textile waste are less than 5 m^2/g, but steam activation can produce activated carbons of over 900 m^2/g. The chemical activation of waste, i.e. where an absorbent waste, such as textile flax and hemp biomass waste absorbs a reacting chemical, such as zinc chloride or potassium hydroxide, followed by pyrolysis at 500°C has produced very high surface area activated carbons of over 2000 m^2/g.

The product oil from pyrolysis of waste has the advantage of being able to be used in conventional electricity-generating systems, such as diesel engines and gas turbines. However, the properties of the pyrolysis oil fuel may not match the specifications of a petroleum-derived fuel and may require modifications to the power plant or upgrading of the fuel. In some cases the oil product is described as a liquid but, depending on the feedstock and the pyrolysis process conditions, it may represent either a true oil, an oil/aqueous phase, separated oil and aqueous phases or, for some waste feedstocks, a waxy material. The advantages of producing an oil product from waste are that the oil can be transported away from the pyrolysis process plant and therefore decouples the processing of the waste from the product utilisation.

The oil may be used directly as a fuel, added to petroleum refinery stocks, upgraded using catalysts to a premium grade fuel or used as a chemical feedstock. The composition of the oil is dependent on the chemical composition of the feedstock and the processing conditions. For example, oils derived from biomass have a high oxygen content, of the order of 35 per cent by weight, due to the content of cellulose, hemicellulose and lignin in the biomass. These are large polymeric structures containing mainly carbon, hydrogen and oxygen. Similarly, oils derived from municipal solid waste have a high oxygen content due to the presence of cellulosic components in the waste such as paper, cardboard and wood. Biomass and municipal solid waste pyrolysis oils derived from flash pyrolysis processes, tend to have a lower viscosity and consist of a single water/oil phase. The oils are, therefore, high in water, which markedly reduces their calorific value. Slow pyrolysis produces liquid products with higher viscosities which tend to have two phases due to the more extensive degree of secondary reactions which occur. Oils derived from scrap tyre pyrolysis and plastics, on the other hand, are composed of mainly carbon and hydrogen.

The oils have significant calorific values ranging from 25 MK/kg for oils derived from municipal solid waste to 42 MJ/kg for oils derived from scrap tyres, compared with a typical petroleum-derived fuel oil at 46 MJ/kg (Table 18.3). Table 18.3 shows the properties of oils derived from the pyrolysis of tyres, municipal solid waste and wood. Comparison with petroleum-derived diesel fuel shows that in many respects the oils derived from waste are quite similar. However, the direct use of such fuels in combustion systems designed and optimised on fuels refined from petroleum, may be difficult. For example, biomass and municipal solid waste pyrolysis oils can be viscous, highly acidic, due to the organic acids present in the oils, and can readily polymerise. In addition, pyrolysis oils may contain solid char particles due to carry-over from the pyrolysis reactor. Consequently, their use in liquid spray or atomisation combustion systems such as diesel engines, furnaces and boilers, may result in the spray or atomisation system becoming blocked and/or corroded.

Table 18.3. Typical fuel properties of waste derived pyrolysis oils.

Parameter	Tyre oil	MSW oil	Biomass oil	Diesel oil
Carbon residue (%)	0.7	–	–	<0.35
Mid B.Pt. (°C)	230	–	–	300
Viscosity (cSt)	2.12 (60°C)	–	17 (100°C)	1.3 (60°C)
	3.50(40°C)	–	90 (50°C)	3.3 (40°C)
Density (kg/m^3)	0.91	1.3	1.2	0.78
API gravity	20.41	–	–	31
Flash point (DC)	24	56	110–120	75
Hydrogen (%)	9.98	7.6	7–8	12.8
Carbon (%)	87.0	57.5	50–67	–
Nitrogen (%)	0.4	0.9	0.8-1	–
Oxygen (%)	0.7	33.4	15–25	–
Initial B.Pt. (°C)	80	–	–	180
10% B.Pt. (°C)	140	–	–	–
50% B.Pt. (°C)	230	–	–	300
90% B.Pt. (°C)	340	–	–	–
CV (MJ/kg)	42.0	24.4	24.7 (lower)	46.0
Sulphur (%)	1.5	0.1–0.3	<0.01	0.9

Performance guarantees for the use of nonstandard fuels in combustion systems may invalidate the manufacturers warranties, which would be based around standard, i.e. petroleum-refined fuels. Emission limits from the combustion system, set at National and European level, would also have to be met irrespective of the fuel being used. However, the fuels derived from waste materials such as tyres, wood and municipal solid waste, have been successfully combusted in a variety of systems.

The oils derived from the pyrolysis of waste materials tend to be chemically very complex, due to the polymeric nature of the wastes and the range of potential primary and secondary reactions. Biomass and municipal solid waste pyrolysis oils contain hundreds of different chemical compounds including organic acids phenols, alcohols aldehydes ketones, furans, etc. Tyre pyrolysis oils consist mainly of alkanes, alkenes and monoaromatic and polycyclic aromatic compounds. Oils derived from mixed plastic waste at typical pyrolysis temperatures of 500°C are highly viscous and consist largely of alkanes, alkenes and aromatic compounds. Where single plastics are pyrolysed, the wax or oil is similar to the basic structure from which the plastic was formed. Consequently, for example, polyethylene and polypropylene will produce mainly alkane and alkene waxes, whilst polystyrene will produce an oil consisting of alkane, alkene and aromatic product. Polyvinyl chloride also produces an aromatic oil product when pyrolysed, in addition to hydrocarbon gases and hydrogen chloride gas.

Because of the range of compounds found in pyrolysis oils, there is some interest in using the oils as chemical feedstocks for speciality chemicals. For example, wood pyrolysis oils contain oxygenated compounds such as methylphenols (cresol), methyoxyphenol (guaiacol), furaldehyde (fufural) and methoxypropenylphenol (isoeugenol) which have applications in the pharmaceutical, food and paint industries. Tyre oil contains dl-limonene used in the formulation of industrial solvents, resins and adhesives and as a replacement for chlorofluorocarbon for cleaning electronic circuit boards. The oils also contain significant concentrations of benzene, xylenes, styrene and toluene used extensively in the chemical and pharmaceutical industries. The wax/oil-like product derived from the pyrolysis of mixed plastic waste has been successfully reprocessed in petroleum refineries, using catalyst cracking to produce gasoline or plastics.

To overcome some of the problems of high oxygen content, high viscosity, acidity and polymerisation associated with the oils derived from waste materials containing high oxygen contents, e.g. biomass and municipal solid waste, research has been undertaken to upgrade the oils. The research has concentrated on the use of catalysts to produce a premium quality fuel or high value chemical feedstock. Two main routes to catalytic upgrading have been investigated: high pressure catalytic hydrotreatment, and low pressure catalysis using shape-selective catalysts of the zeolite type. Catalytic hydrotreatment of the oils with hydrogen or hydrogen and carbon monoxide under high pressure and/or in the presence of hydrogen donor solvents using transition metal catalysts, has produced oils similar in composition to gasoline and diesel. The upgrading takes place through deoxygenation and hydrocracking of the heavy fractions in the oil. Zeolite ZSM-5 catalysts have a strong acidity, high activities and shape selectivities, which convert the oxygenated oil to a light hydrocarbon mixture in the C_1–C_{10} range by dehydration and deoxygenation reactions. The oxygen in the oxygenated compounds of biomass pyrolysis oils is converted largely to CO, CO_2 and H_2O, and the resultant oil is highly aromatic with a dominance of single-ring aromatic compounds, and is similar in composition to gasoline.

Catalytic upgrading of pyrolysis oils has also been undertaken for tyres and plastics, also using Zeolite-type catalysts. The application of catalysts to the processing of scrap tyres and waste plastics is to enhance the oil product from the pyrolysis of such wastes to produce a premium-grade fuel or chemical feedstock. The derived oils from pyrolysis–catalysis of waste plastics have been shown to be very aromatic. Combined pyrolysis–catalysis of tyres with Zeolite-type catalysts produces an oil very high in concentrations of the high-value chemicals benzene, xylenes and toluene, such that it has the potential to be used as a chemical feedstock rather than a liquid fuel.

The gases produced from municipal solid waste and biomass waste pyrolysis are mainly carbon dioxide, carbon monoxide, hydrogen, methane and lower concentrations of other hydrocarbon gases. The high concentration of carbon dioxide and carbon monoxide is derived from the oxygenated structures in the original material, such as cellulose, hemicellulose and lignin. In addition, the gas contains a significant proportion of uncondensed pyrolysis oils. The pyrolysis of scrap tyre and mixed plastics waste produces higher concentrations of hydrogen, methane and other hydrocarbon gases, since the waste material is high in carbon and hydrogen compounds and has less oxygenated compounds. The gases have a significant calorific value, for example, the gas produced from the conventional pyrolysis of municipal solid waste has a calorific value of the order of 18 MJ/m³ and wood waste produces a gas of calorific value 16 MJ/m³. Tyre pyrolysis produces a gas of much higher calorific value, of about 40 MJ/m³ depending on the process conditions. The high calorific value is due to the high concentrations of hydrogen and other hydrocarbons.

By comparison, the calorific value of natural gas is about 37 MJ/m³. The high calorific value of pyrolysis gases means that the gas could be used to provide the energy requirements for the pyrolysis process plant. The gases are produced from the thermal degradation reactions of the waste constituents as they breakdown, and also through secondary cracking reactions of the primary products. Consequently, higher gas yields are found where the products of pyrolysis spend a relatively longer time in the hot zone of the reactor rather than rapid quenching which produces higher oil yields and lower char yields. Also, higher gas yields are found at pyrolysis temperatures above about 750°C, where accelerated cracking of the pyrolysis products occurs. At such high-temperature conditions, the main product is gas and some process descriptions term this process as gasification. However, pyrolysis implies the absence of any oxygen, whereas gasification implies that limited oxygen is supplied to the process as air, steam or pure oxygen, to gasify the waste. A wide variety of pyrolysis technologies have been investigated for the pyrolysis of waste materials. Examples of technologies which have been used for waste pyrolysis include fluidised beds, fixed-bed reactors, ablative pyrolysis at hot surfaces, rotary kilns, entrained flow reactors and vacuum pyrolysis. The design is dictated by the type of pyrolysis being undertaken, for example, fast or slow heating rates to produce the targeted end product. Many are still at the pilot-scale stage, whilst others are at the commercial or near commercial stage.

GASIFICATION

Gasification differs from pyrolysis in that oxygen in the form of air, steam or pure oxygen is reacted at high temperature with the available carbon in the waste to produce a gas product ash and a tar product. Partial combustion occurs to produce heat and the reaction proceeds exothermically to produce a low to medium calorific value fuel gas. The operating temperatures are relatively high compared to pyrolysis, at 800°–1100°C with air gasification, and 1000°–1400°C with oxygen. Calorific values of the product gas are low for air gasification in the region of 4–6 MJ/m³, and medium, about 10–15 MJ/m³ for oxygen gasification. Steam gasification is endothermic for the main char-steam reaction and consequently steam is usually added as a supplement to oxygen gasification to control the temperature.

Steam gasification under pressure is, however, exothermic and steam gasification at pressures up to 20 bar and temperatures of between 700° and 900°C produces a fuel gas of medium calorific value, approximately 15–20 MJ/m³. The product calorific values can be compared with natural gas at about 37 MJ/m³.

The principle reactions occurring during gasification of waste in air are:

$$C + O_2 \Rightarrow CO_2 \qquad \text{Oxidation – exothermic}$$
$$C + CO_2 \Rightarrow 2CO \qquad \text{Boudouard reaction – endothermic}$$

Overall:

$$2C + O_2 \Rightarrow 2CO \qquad \text{Exothermic}$$

Steam gasification:

$$C + H_2O \Rightarrow CO + H_2 \qquad \text{Carbon-steam reaction – endothermic}$$
$$C + 2H_2O \Rightarrow CO_2 + 2H_2 \qquad \text{Carbon-steam reaction – endothermic}$$
$$CO + H_2O \Rightarrow CO_2 + H_2 \qquad \text{Water-gas shift reaction – exothermic}$$
$$C + 2H_2 \Rightarrow CH_4 \qquad \text{Hydrogenation – exothermic}$$

In high pressure steam gasification, additional reactions include:

$$CO + 3H_2 \Rightarrow CH_4 + H_2O \quad \text{Hydrogenation – exothermic}$$

$$CO_2 + 4H_2 \Rightarrow CH_4 + 2H_2O \quad \text{Hydrogenation – exothermic}$$

In practice there is usually some moisture present with the air which produces some hydrogen. In addition, the beating of the waste produces pyrolytic reactions and methane, and higher molecular weight hydrocarbons or tar are formed. When air is used, the noncombustible nitrogen in the air inevitably reduces the calorific value of the product gas by dilution. Therefore, the major components of the product gas from waste gasification are carbon monoxide, carbon dioxide, hydrogen and methane and, where air gasification is used, nitrogen will also occur as a major component.

For wastes and biomass, the development of gasification technologies has been via air or oxygen/steam gasification. Table 18.4 shows examples of the systems used for waste gasification. The characteristics of the gasifier system, the waste composition and operational conditions can give rise to tars, hydrocarbon gases and char; these are products of the incomplete gasification of the waste. The characteristics of the gasifier have most influence on the quality of the product gas, for example, downdraft gasifiers have all the products of gasification passing through a high temperature zone and with high turbulence.

Table 18.4. The main types of waste gasifier reactor systems.

Updraft gasification

Air flows up from the base of the reactor with the waste flowing down countercurrent to the air flow. Gasification takes place in a slowing moving 'fixed' bed. Because the moisture, tar and gases generated do not pass through a hot bed of char there is less thermal breakdown of the tars and heavy hydrocarbons, therefore the product gas is relatively high in tar. The tars may be condensed and recycled to increase thermal breakdown of the tars.

Downdraft gasification

The air and the waste flow co-currently down the reactor. Gasification takes place in a slowing moving 'fixed' bed. There is an increased level of thermal breakdown of the tars and heavy hydrocarbons as they are drawn through the high-temperature oxidation zone, producing increased concentrations of hydrogen and light hydrocarbons. The air/steam or oxygen is introduced just above a 'throat' or narrow section in the reactor, which influences the degree of tar cracking.

Fluidised bed gasification

Waste is fed into the fluidised bed at high temperature. The fluidised bed may be a bubbling bed where the solids are retained in the bed through the gasification process. Alternatively, circulating beds may be used with high fluidising velocities; the solids are elutriated, separated and recycled to the reactor in a high solids/gas ratio resulting in increased reaction. Twin fluidised-bed reactors may be used where the first bed is used to gasify the waste, and the char is passed to a separation unit and then to a second fluidised bed where combustion of the char occurs to provide heat for the gasifier reactor.

Entrained flow gasification

A widely used technology, where the gasification reactions take place in suspension in an entrained flow of gas. The waste feedstock is introduced into a vertical reactor with steam and oxygen. The residence time is very short and the gasification takes place at high temperature and pressure. The waste can be in liquid or solid form, but where solids are used, the particle size must be small. The entrained flow gasifier tends to produce a high conversion of the waste to produce a low tar content gas.

Rotary kiln gasification

Rotary kilns involve a slowly rotating, inclined, ceramic-lined cylinder, which slowly moves the waste down the cylinder, whilst the waste is gasified. Gasification is with air, steam or oxygen. The residence time is much longer than for fluidised bed and entrained flow gasification reactors.

This arrangement results in a high conversion of the pyrolysis intermediates and a gas with a low tar content, whereas, the up-draft gasifier produces gas which is hot and, when passing up through the down-flowing waste, produces pyrolysis reactions and a higher concentration of tar in the final product gas. Fluidised bed reactors produce intermediate pyrolysis tar products which are passed out of the fluid bed into the freeboard by the fluidising gas. As the tars pass up through the hot freeboard of the fluidised bed, some thermal cracking of the tars to gases may occur, but an overall gas tar content similar to that of an up-draft gasifier designs require the waste to be entrained flow gasifier and fluidised bed gasifier designs require the waste to be processed to produce fine granules to enable efficient feeding of the waste to the gasifier.

For wastes such as sewage sludge, this is not such a problem, but for municipal solid waste and forestry residues, pretreatment of the waste is required. For rotary kilns, no such pretreatment and size reduction of the waste is required. Table 18.5 shows the characteristics of gasification product gas from different gasifier types.

Table 18.5. Product gas characteristics from different gasifier types.

Gasifier type	Calorific value of the product gas (MJ/m³)	Gas quality[2]	Efficiency (%)
Downdraft-air	4.0–6.0	****	70–90
Downdraft-O_2	9–11	****	60–80[1]
Updraft-air	4.0–6.0	***	75–95
Updraft-O_2	8–14	***	65–85[1]
Fluidised bed-air	4–6	***	70–90
Fluidised bed-O_2	8–14	***	60–75[1]
Fluidised bed-steam	12–18	***	70–80
Circulating fluidised bed-air	5–6.5	**	75–95
Circulating fluidised bed-O_2	10–13	***	70–80[1]
Twin fluid bed	13–20	***	65–75
Cross flow-air	4.0–6.0	*	75–95
Horizontal moving bed-air	4.0–6.0	**	60–70
Rotary kiln-air	4.0–6.0	**	70–85
Multiple hearth	4.0–6.0	**	60–80

[1]Oxygen system efficiencies include a notional energy used for oxygen production.

[2]Gas quality is a relative assessment in terms of tars and particulates in raw gas: *, worst; ******, best.

Utilisation of the gaseous product is often by direct combustion in a boiler or furnace. The heat energy is used for process heat or to produce steam for electricity generation. However, the raw gas will contain tar, char and hydrocarbon gases and therefore the boiler or furnace burner system must be able to tolerate these contaminants and not be susceptible to fouling or clogging. In addition, gasification of heterogeneous waste, such as municipal solid waste, produces a gas which can vary in composition, and consequently, the burner system of the boiler or furnace should be able to handle a range of gas compositions and calorific values. The advantages of direct combustion systems are that the gas does not have to be cleaned to any great extent before combustion and that the gases are used hot, maintaining the sensible heat in the system.

Where the utilisation of the product gas is into gas turbines or internal combustion engines to generate power or electricity, then the gas has to be cleaned to a higher specification than in direct combustion systems. Piping of the gas to die combustion unit requires that it be cooled and cleaned before utilisation to prevent pipe corrosion and deposition of tars and water.

Removal of particulate material is by cyclones and bag filters, and tar removal is by secondary cracking at high temperature or catalyst cracking at lower temperatures. Gas turbines have been suggested as a suitable utilisation system for electricity generation, particularly for pressurised waste gasifiers. However, the fuel gas specifications for gas turbines are very stringent.

COMBINED PYROLYSIS-GASIFICATION

Some modern developments in thermochemical processing of waste have utilised both pyrolysis and gasification in combined technologies, which may then involve a further combustion step to combust the gases produced in the first two stages. Such pyrolysis/gasification/combustion technologies are, in effect, incinerators, but each step is separated into a separate temperature and pressure controlled reactor rather than in an incinerator, where the three thermal degradation steps are combined in a one-step grate combustion system. The decoupling of the thermal degradation steps has the advantage of flexibility in determining which targeted end product is best suited to each application. Further advantages include the option that the product gas may be cleaned to remove acid gases prior to the combustion of the gas for energy recovery.

This results in reduced high-temperature corrosion within the energy recovery system. Also, pyrolysis/ gasification systems produce significantly reduced gas volumes for clean-up compared with a conventional waste incinerator, resulting in scale-down of the gas cleaning system and a consequent reduction in cost. In addition, as an alternative, the gas product may be further cleaned and used as a chemical feedstock.

More than 100 pyrolysis/gasification process technologies have been identified worldwide, of which 60 have been technologically and economically evaluated in detail. Figure 18.2 shows the different combinations of different processes involved under the category of pyrolysis, gasification and combined pyrolysis/gasification systems. Several pyrolysis, gasification and combined systems for processing wastes have been described.

The commercial development of the various systems ranges from pilot scale through to commercialisation. Clearly, the range and complexity of such systems and the lack of a proven long-term track record, inhibits the full-scale commercial development of pyrolysis—gasification systems as an alternative to mass burn incineration. It has been suggested that the increasing complexity of combined pyrolysis-gasification systems, which may then also include combustion and ash melting, have developed because of the need to handle more complex and less homogeneous waste streams. The processes that include ash melting, aim to ensure the stability of the solid residue output and thereby maximise the recycling potential of the waste in the process.

Systems where the second stage consists of combustion, such as pyrolysis-combustion or gasification-combustion aim to maximise the energy recovery without the need to clean the product gas to any extent. The complex systems such as pyrolysis-gasification-ash melting, followed by full gas cleaning, aim at producing a clean product gas with low levels of particulate, tars and acid gases, suitable for use in combined-cycle gas turbine power generation.

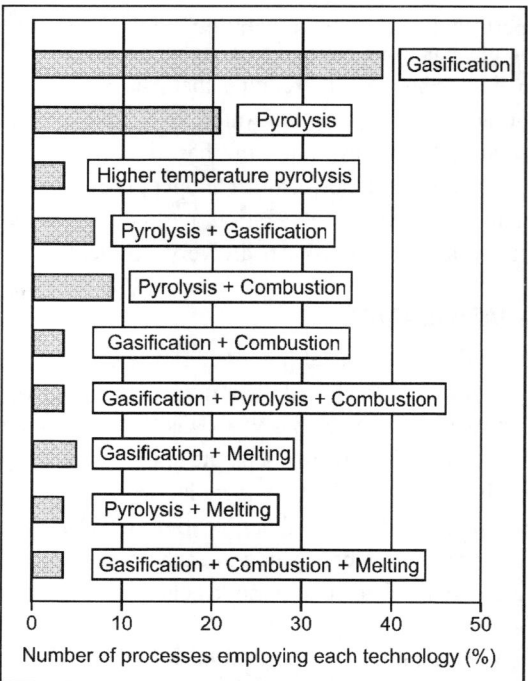

Fig. 18.2. Types of technology combinations employed in combined thermal processing systems (60 types analysed).

Schematic of inputs and outputs of a typical pyrolysis process are shown in Fig. 18.3.

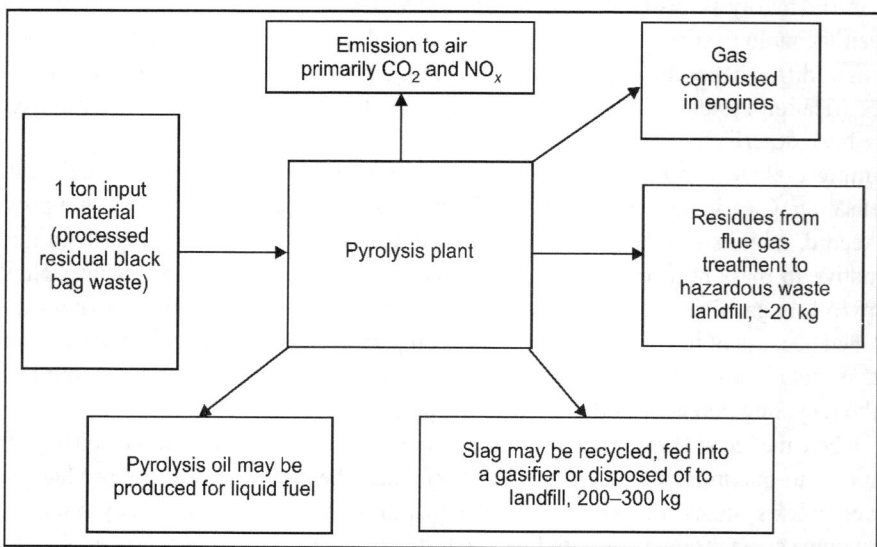

Fig. 18.3. Inputs and outputs of a typical pyrolysis process.

Gasification usually operates at a higher temperature range to pyrolysis, with the addition of an oxidant (either air or oxygen) and the output from a pyrolysis plant may be fed into this process. Gasification of organic derived wastes will produce a gas which can be combusted to generate electricity and a char which usually requires disposal if no markets are available. Schematic of inputs and outputs of a typical gasification process are shown in Fig. 18.4.

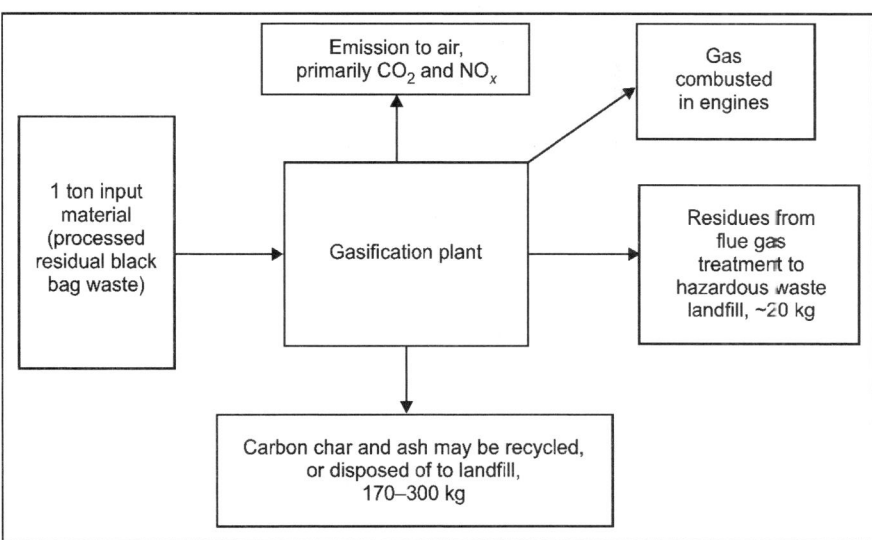

Fig. 18.4. Inputs and outputs of a typical gasification process.

Table 18.6 gives strengths and weaknesses of pyrolysis and gasification processes.

Table 18.6. Strengths and weaknesses of pyrolysis and gasification processes.

Strengths	Weaknesses
Not incineration	May suffer from the same negative perception as incineration, some evidence of this overseas, yet to be tested in the UK
Qualifies for the renewables obligation for a substantial proportion of the feedstock processed	Requires extensive pretreatment to be able to handle MSW
Efficient electricity generation through combustion of gas through engines	Many processes will still have residues to be disposed of, some of which (from flue gas treatment) will be hazardous in nature
Potential to recycle a large proportion of residues depending on the process	Unproven on a commercial scale on MSW in the UK, patchy experience overseas
High temperatures may make the system more flexible for other waste streams such as clinical	More sensitive system than moving grate incineration technology
Smaller units more acceptable and part of an integrated system	More expensive (in terms of gate fee) than energy from waste
Capable of being integrated with other processes such as the output from MBT/refuse derived fuel (RDF) production	

Chapter 19

Incineration

INTRODUCTION

Incineration is a waste treatment process that involves the combustion of organic substances contained in waste materials. Incineration and other high temperature waste treatment systems are described as 'thermal treatment'. Incineration of waste materials converts the waste into ash, flue gas, and heat. The ash is mostly formed by the inorganic constituents of the waste, and may take the form of solid lumps or particulates carried by the flue gas. The flue gases must be cleaned of gaseous and particulate pollutants before they are dispersed into the atmosphere. In some cases, the heat generated by incineration can be used to generate electric power. Incineration with energy recovery is one of several waste-to-energy (WtE) technologies such as gasification, plasma arc gasification, pyrolysis and anaerobic digestion. Incineration may also be implemented without energy and materials recovery.

In several countries, there are still concerns from experts and local communities about the environmental impact of incinerators. In some countries, incinerators built just a few decades ago often did not include a materials separation to remove hazardous, bulky or recyclable materials before combustion. These facilities tended to risk the health of the plant workers and the local environment due to inadequate levels of gas cleaning and combustion process control. Most of these facilities did not generate electricity. Incinerators reduce the solid mass of the original waste by 80–85 per cent and the volume (already compressed somewhat in garbage trucks) by 95–96 per cent, depending on composition and degree of recovery of materials such as metals from the ash for recycling. This means that while incineration does not completely replace landfilling, it significantly reduces the necessary volume for disposal. Garbage trucks often reduce the volume of waste in a built-in compressor before delivery to the incinerator. Alternatively, at landfills, the volume of the uncompressed garbage can be reduced by approximately 70 per cent by using a stationary steel compressor, albeit with a significant energy cost. In many countries, simpler waste compaction is a common practice for compaction at landfills.

Incineration has particularly strong benefits for the treatment of certain waste types in niche areas such as clinical wastes and certain hazardous wastes where pathogens and toxins can be destroyed by high temperatures. Examples include chemical multi-product plants with diverse toxic or very toxic waste-water streams, which cannot be routed to a conventional waste-water treatment plant.

INCINERATION SYSTEMS

The modern incinerator is an efficient combustion system with sophisticated gas clean-up which produces energy and reduces the waste to an inert residue with minimum pollution. Incineration plants may be classified on a variety of criteria, for example, their capacity, the nature of the waste to be combusted, the type of system, etc. However, a broad classification may be made between mass burn incineration and other types.

Mass burn incineration: Large-scale incineration of municipal solid waste is a single-stage chamber unit in which complete combustion or oxidation occurs. Typical throughputs of waste are between 10 and 50 tons per hour.

Other types of incineration: Other types of incineration involves smaller scale throughputs of between 1 and 2 tons per hour of wastes such as clinical waste, sewage sludge and hazardous waste. Typical examples of such systems include fluidised bed, cyclonic, starved air or pyrolytic, rotary kiln, rocking kiln, cement kiln, and liquid and gaseous incinerators.

Mass Burn Incineration

Mass burn incineration is used for the treatment and disposal of municipal solid waste throughout the world. Within Europe, the amount of municipal solid waste incineration undertaken varies between countries. In addition, the number of incinerator plants and their capacities also varies. For example, the average waste incinerator plant size in the Netherlands is more than 4,80,000 tons per year throughput, whereas in Italy and Norway, the average size is less than 1,00,000 tons per year throughput. The economic viability of incineration as a waste treatment and disposal route for municipal solid waste, depends on the recovery of energy from the process to offset the high costs involved in incineration.

The composition and characteristics of the waste will influence the combustion properties and emissions produced from the combustion system. A typical calorific value for municipal solid waste is approximately 9000 kJ kg^{-1}. Ash and moisture contents tend to be high and thus, in terms of a fuel, the waste would compare poorly with coal, for example. Of particular importance are the 'fuel' properties of the waste, the proximate analysis (ash, moisture, volatile contents) and the ultimate (elemental) analysis which can be used to assess how the waste will burn in the incinerator and the emissions which are likely to result. Moisture content is obviously important since ignition will not occur if the material is wet and moisture also diminishes the gross calorific value of a fuel. Volatile matter contains the combustible fraction of the waste and consists of gases such as hydrogen, carbon monoxide, methane, ethane, etc. a more complex organic hydrocarbon fraction, and an aqueous phase derived by decomposition of water-bound compounds. The ash content is important since a high ash percentage will lower the calorific value of the waste and must be removed and disposed of after combustion. Waste ash is highly heterogeneous and contains inert non-combusted material such as glass and metal cans. The municipal solid waste also contains significant concentrations of heavy metals such as cadmium, lead, zinc and chromium and will influence the emissions of such metals. Similarly, the sulphur and chlorine content will produce emissions of sulphur dioxide and hydrogen chloride.

In most cases, waste incinerator operators have limited control of the precise composition of the incoming waste. Consequently, mass burn incinerators are designed to be sufficiently flexible to cope with the wide range of waste compositions that they may receive. The composition of waste may be generally represented in a ternary diagram, shown in Fig. 19.1 and shows the range of analyses acceptable to the combustion system. The shaded area represents the typical composition of municipal solid waste which can sustain combustion without the requirement for auxiliary fuel. The area encloses the minimum acceptable calorific value and the maximum permissible moisture content. In addition, the influences of pretreatment of the waste, prior to arrival at the waste incinerator, may influence the composition and properties such as the metal content and calorific value. For example, removal of glass and metals for recycling would increase the calorific value of the waste and reduce the emission of metals to either the flue gases or bottom ash. Recovery of paper, card and plastic would decrease the calorific value of the incoming waste. Recycling of organic food and garden waste, for example to composting, would reduce the moisture content of the municipal solid waste and thereby increase the net calorific value.

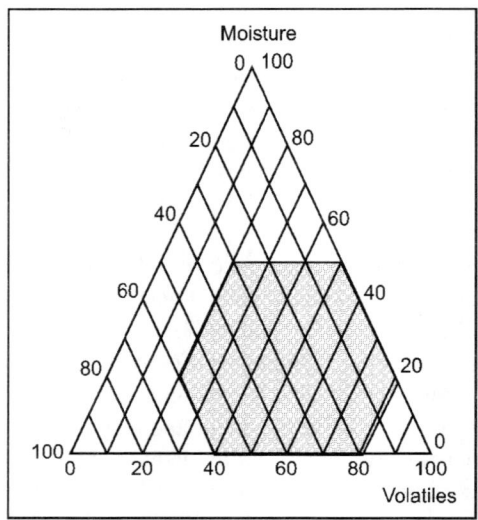

Fig. 19.1. Suitability of municipal solid waste composition for incineration.

A typical modern municipal waste incineration plant with energy recovery is shown in Fig. 19.2. The incinerator may be divided into five main areas:

1. Waste delivery, bunker and feeding system.
2. Furnace.
3. Heat recovery.
4. Emissions control.
5. Energy recovery via district heating and electricity generation.

Waste delivery, bunker and feeding system

The waste is usually delivered by collection vehicles, although in some European incinerators, barges or trains may be used. The collection vehicles are weighed on arrival and departure to provide accurate weights of the waste throughput for determining the fees to be charged for disposal and for incinerator operational control. The incinerator may handle a variety of wastes from households, commercial sites and industry and these would be monitored not only to differentiate the fees charged, but also since they may have very different combustion properties which would influence incinerator performance. Odour may result from the waste due to biodegradation and handling, and therefore plants are normally kept under a slight negative pressure, because the combustion air is taken from the waste storage area, which prevents escape of odour. The EC Waste Incineration Directive sets out the requirements for the handling of waste at the incinerator to minimise the environmental impact of waste incineration and hazard to human health.

The bunker is large enough to allow for storing the waste to ensure a balance between the uneven delivery of the waste and the continuous operation of the plant. Therefore, the bunker would be designed to hold about 2–3 days equivalent of weight of waste which would be typically 1000–3000 tons of waste. Longer periods of storage are undesirable due to the rotting of the waste and consequent bad odours. The waste is delivered to the bunker which may be divided into different sections in separate unloading bays to allow for the mixing of the waste of different calorific values and combustion properties by the crane operator.

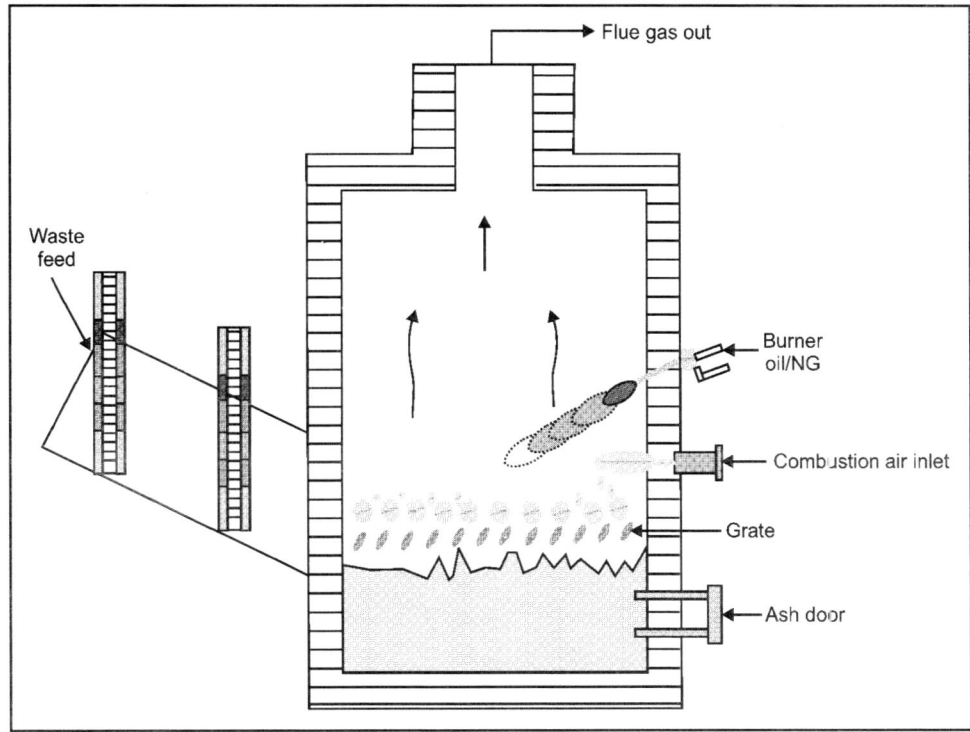

Fig. 19.2. Schematic diagram of a typical mass burn municipal solid waste incinerator.

The crane is of a travelling type and the crane operator will not only mix the wastes, but will also extract any bulky or dangerous items from the refuse for separate treatment. The operator then loads the waste to the feeding system. The crane grab can hold up to 6 m^3 of waste.

The feeding system is a steel hopper where the waste is allowed to flow into the incinerator under its own weight and is fed into the grate system by a hydraulic ram or other conveying system without bridging or blocking. The hoppers are kept partly filled with waste to minimise air leakage into the furnace and to ensure there is no interruption of feed to the grate. Monitors are used to measure the level of waste in the hopper. To prevent the fire in the furnace from burning back-up into the feeding hopper, hydraulic shutters are used to seal the hopper at the furnace entrance. Also, the feed chute may be water-cooled or refractory-lined to prevent fire.

Furnace

Figure 19.3 shows a schematic diagram of a typical furnace system for a mass burn municipal solid waste incinerator. Each incinerator may have several furnaces fed by the operator from the waste bunker.

For example, a typical 50 ton/hr incinerator might have five separate 10 ton/hr furnaces. The use of multiple furnaces allows for down time of the furnace for repair and regular maintenance. During the start-up of the incinerator, auxiliary burners are used to raise the temperature of the gases to initiate waste combustion.

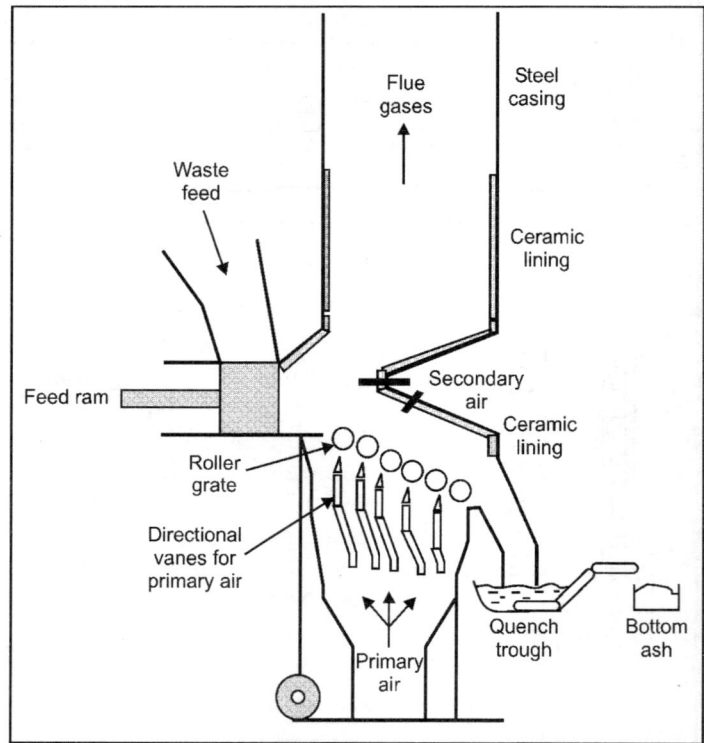

Fig. 19.3. Schematic diagram of the furnace of a mass burn municipal solid waste incinerator.

The waste is fed into the furnace usually by an independently controlled ram. In the furnace the waste undergoes three stages of incineration:

1. Drying and devolatilisation.
2. Combustion of volatiles and soot.
3. Combustion of the solid carbonaceous residue.

In practice the various stages merge, since the components of the waste stream differ in moisture content, thermal degradation temperature, volatile composition and ignition temperature and carbon (fixed) content.

As the waste enters the hot furnace, the waste is heated up via contact with hot combustion gases, preheated air or radiated heat from the incinerator walls, and initially moisture is driven off in the temperature range 50°–100°C. The water content of waste is very important since heat is required to evaporate the moisture, and therefore more of the available calorific value of the waste is lost in heating up the wet waste and so less energy is available. In addition, the rate of heating up of the waste, and therefore the rate of thermal decomposition, will also be affected by the water content of the waste.

Water contents of municipal solid waste can vary between 25 and 50 per cent. After moisture release, the waste then undergoes thermal decomposition and pyrolysis of the organic material such as paper, plastics, food waste, textiles, etc. in the waste which generates the volatile matter, the combustible gases and vapours. The volatile components of organic material in municipal solid waste comprise typically between 70 and 90 per cent and are produced in the form of hydrogen, carbon monoxide, methane, ethane and other higher molecular weight hydrocarbons. Devolatilisation takes place over a wide range of temperatures from about 200°–750°C with the main release of volatiles between 425° and 550°C. Thermal decomposition of the waste and volatile release will also be dependent on the different components present in the waste. For example, polystyrene decomposes over the temperature range 450°–500°C and yields almost 99 per cent volatiles, whereas wood decomposes over the temperature range 280°–500°C and produces about 70 per cent volatiles. In addition to composition, the physical state of the waste will influence the rate of thermal decomposition. For example, cellulosic material in thin form such as paper will decompose in a few seconds, whereas in the form of a large piece of wood may take several minutes to decompose totally.

The combustion of volatiles to produce the flames of the fire takes place immediately above the surface of the waste on the grate and in the combustion chamber above the grate. Complete combustion of the gases and vapours requires sufficiently high temperature, adequate residence time and excess turbulent air to ensure good mixing. The EC Waste Incineration Directive stipulates that the gases derived from incineration of the waste are to be raised to a temperature of 850°C for 2 s to ensure complete burnout of the volatile hydrocarbons. The volatile gases and vapours released, immediately ignite in the furnace since the furnace gas temperature will be typically between 750° and 1000°C, but can occur at temperatures up to 1600°C. The ignition temperature of the volatiles derived from the waste are well below these temperatures. Combustion chamber temperatures above about 1200°C are avoided, as above this temperature ash fusion is likely to occur leading to a build-up of slag on refractory material. Typical mean residence times of the gases and vapours in the combustion chamber are 2–4 s, to comply with the EC Waste Incineration Directive which compares with typical burnout times for volatile hydrocarbons of the order of milliseconds. Secondary air is blown in through nozzles above the grate to ensure excess air for combustion and to provide turbulence. Excess secondary air is required to avoid areas of zero oxygen levels, which serve to pyrolyse rather than combust the hydrocarbons as this can produce potentially hazardous high molecular weight hydrocarbons and soot. Therefore, the distribution and turbulence characteristics of the secondary air are important factors in minimising the formation of pollutants in the combustion chamber.

After the drying and devolatilisation stages, the residue consists of a carbonaceous char and the inert material. The carbonaceous char which is defined as the fixed carbon rather than the volatile carbon contained in the volatile gases such as methane, ethane and other hydrocarbons, combusts on the grate and may take between 30 and 60 min for complete burnout. The ash and metals residue is discharged continuously at the end of the last grate section into a water trough and quenched or air-cooled. The handling equipment is subject to heavy wear, due to the moist and abrasive nature of the material. The ash residue should be completely burnt out and biologically sterile. The EC Waste Incineration Directive sets a TOC content for the ash, of less than 3 per cent, to ensure complete burnout of the waste. The ash, known as bottom ash, is removed continuously or periodically via a conveyor and is disposed of in landfill sites or used for recycling as secondary aggregate for construction projects such as road building and concrete production. The bottom ash comprises about 30 per cent of the total mass of waste input. The lighter fly-ash is transported through the system as particulate material which will also adsorb metals and organic material as it cools through the incinerator heat recovery and gas clean-up system. The fly-ash is collected in the cyclones, electrostatic precipitators and bag filters of the gas clean-up system. Because the fly-ash contains heavy metals, polycyclic aromatic hydrocarbons and dioxins and furans it has no recyclable value, is regarded as hazardous waste and is therefore landfilled. Fly-ash comprises only a few per cent of the waste mass input.

At the heart of the incinerator is the grate, and a number of different types of furnace grate exist for municipal waste incineration, for example, the roller system (Fig. 19.3) and the rocker, stoker, forward reciprocating systems and reverse reciprocating systems (Fig. 19.4).

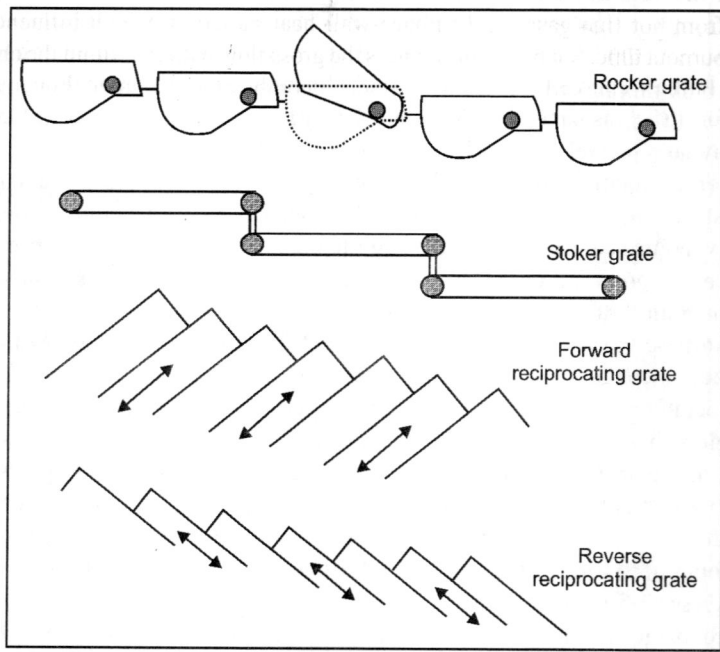

Fig. 19.4. Types of municipal solid waste incinerator grate.

The grates are automatic and serve to move the waste from the inlet hopper end to the discharge end, whilst providing agitation or tumbling of the waste to stoke the furnace fire and loosen the combusting materials. The grate has a variable speed drive to adjust the residence time of the waste in the combustion zone to allow for changes in composition. For example, the roller grate system (Fig. 19.3) has a roller diameter of 1.5 m and a circumferential speed of between 5 and 15 m/hr. The grates in the roller system are inclined at about 30°, which assists in the movement of the burning waste through the furnace. Rocker grates have alternative rows of mechanical rockers which are pivoted or rocked to produce an upward and forward motion, advancing and agitating the waste. Horizontal stoker-type or travelling grates are generally arranged in sections of drying, ignition and burnout, and also assist the distribution and control of primary air. Reciprocating grates consist of three or more sections with a step of about 0.5–1 m between sections. Each section consists of a series of fixed bars and moveable bars in a staircase-like arrangement (Fig. 19.4). The movement of the mobile bars serves to agitate and move the waste down the grate. Most grates are cooled, most often with air. Control of the air supply to the furnace and combustion chamber is essential for efficient combustion. Primary air is blown evenly through the waste bed via the underside of the grate through slits in the grate which assists in combustion and cooling of the grate. Overgrate or secondary air is introduced through nozzles above the fuel bed, and in some plants tertiary air is added to cool the flue gases before gas cleaning treatment. Some fine material, referred to as riddlings, fall through the grate and may be recycled back to the incinerator or removed for disposal.

The size and shape of the combustion chamber itself are both important in determining optimum combustion efficiency and a number of different designs exist (Fig. 19.5). The size determines the mean residence time of the volatiles affecting their burnout. The shape affects the heating pattern of the incoming waste from hot flue gases and furnace-wall heat radiation, which influences drying time, ignition time and burnout time. Shape also influences the gross flow patterns within the chamber including recirculation and bulk mixing, which in turn influence combustion. Counter flow, medium flow and unidirectional flows of gases can be produced depending on design (Fig. 19.5). The initial drying and devolatilisation of municipal solid waste can be very malodorous and the flow pattern of the gases and vapours should be through the hottest part of the furnace to completely combust and therefore destroy the odorous organic compounds.

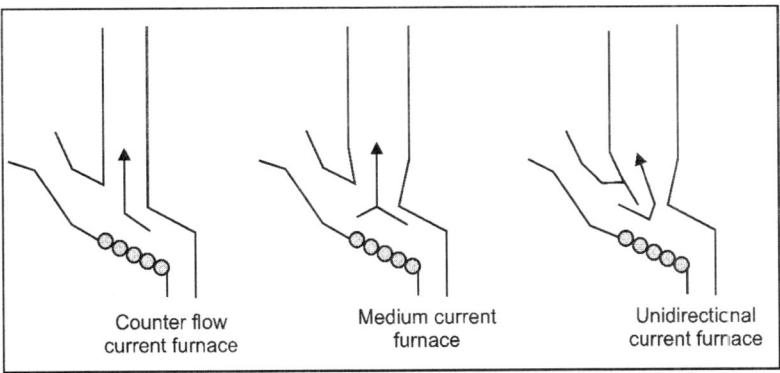

Fig. 19.5. Types of incinerator furnace design.

The furnace and combustion chamber are lined throughout with refractory materials within a steel outer casing. Between the outer casing and the refractory of the combustion chamber may also be contained the water tubes of the boiler, which generate the steam for energy recovery. The main boiler tubes are located in the main boiler chamber above the combustion chamber, through which the hot flue gases flow. The refractory material of the furnace and combustion chamber essentially contains the combustion process in an area which will not fail due to thermal stress or degradation from high-temperature corrosion and abrasion. Refractories also re-radiate heat to accelerate drying, ignition and combustion of the incoming waste. The type of refractories used vary with the different parts of the furnace and combustion chamber. This is because at each stage the temperatures and fluctuations in temperature, oxidation and reduction conditions, abrasion from hard objects, erosion from dust laden flue gases, and corrosion from gases and slags, will require different properties from the refractories. For example, the alumino-silicates with high alumina refractories or silicon carbide bricks are used in the hotter, grate-level part of the furnace, and the upper walls of the combustion chamber would be lower specification alumino-silicate firebricks.

Heat recovery

The combustion of waste is an exothermic or heat-generating process and the majority of the heat generated is transferred to the flue gases. The potential for heat recovery from the incineration process is due to the fact that the combustion gases must be cooled before they can be discharged through the flue gas cleaning system. The temperature of the gases leaving the combustion zone, at typically 750°–1000°C is too high for direct discharge since gas temperatures below 250°–300°C are required for the gas cleaning equipment such as electrostatic precipitators, scrubbers and bag filters. Cooling is by the integral boiler and boiler chamber system in the modern municipal waste incinerator, although older incinerators where heat recovery was not practised, used water injection and air cooling. The heat of the flue gases is transferred to the water in the boiler tubes to produce steam. The boiler consists of banks of steel tubes through which water flows. The integral type or water-wall boilers are constructed around, and integrated with, the combustion chamber. This stage is usually an empty shaft since the flue gases are corrosive and high in particulate matter with typical temperatures of between 650° and 700°C and fusion of hot fly-ash at the boiler tube surface may occur. The main bank of boiler-tube bundles is located in separate boiler chambers, the first pass of hot flue gases being across superheater boiler tubes which allow for higher temperature heat transfer, followed by evaporator-tube bundles which operate at lower temperatures. After the boiler there may be an economiser, which is a heat-exchange system, to heat water in a tube bank in order to produce further hot water from the flue gases before they enter the flue gas cleaning system. The boiler water/steam flow arrangement through the boiler tubes is from the economiser, to the evaporator, to the superheater, gradually producing hotter and hotter steam. The generated steam may be used for electrical power generation, district heating or may also be used within the plant to provide power and space heating.

In an incineration plant, waste is burned at a more or less constant rate, generally near to the design capacity, and therefore the output of energy cannot be varied to meet fluctuating demand. For space or district heating utilisation of this energy may be a problem, whereas electricity generated may be sold to the mains grid. Therefore if continuous output of heat is required, a back-up furnace is required with consequent additional investment and maintenance costs. Similarly, if heat demand is reduced, for example, in the summer months, alternative uses for the heat or a system of flue gas cooling is required so that waste incineration may be continued. The boiler is designed to ensure good heat transfer with the optimum circulation of the water without the occurrence of excessive fouling, allow for the cleaning of the boiler surfaces and to be mechanically stable under operating conditions. Figure 19.6 shows a typical boiler configuration with four passes of the hot gases over the boiler tube bundles. Progressively, the temperature of the flue gases is reduced from about 1000°–1200°C just above the grate, until eventually the gases leave the fourth boiler/economiser stage at about 250°C.

A major factor in the efficient operation of the boiler is the fouling of the tubes with deposits from the flue gases which contain fly-ash, soot, volatilised metal compounds, etc.

The deposits stick to the boiler tubes and, therefore, reduce the transfer of heat from the hot flue gases to the water in the steel tubes, and hence the generation of steam and recovery of energy. The rate at which tube fouling deposits build-up depends on the dust loading of the flue gases, the stickiness of the fly-ash, which in turn depends on temperature, flue gas velocity and tube bank geometry. The boiler tubes should be arranged parallel to the gas flow to minimise fouling and corrosion. The adherence of fly-ash to boiler tubes is mainly determined by the presence of molten salts such as calcium, magnesium and sodium, sulphates, oxides, bisulphates, chlorides, pyrosulphates, etc. in the fly-ash, and the presence of SO_3 and HCl. Scale deposits can be partially removed by means of soot blowers (using superheated steam), shot cleaning (dropping cast iron shot on the tubes to knock off the deposits), or by rapping the tubes (rapping the tube banks to knock off the deposits). Soot blowers are the most common and are usually operated once per operational shift. When the outlet temperature of the flue gases reaches a predetermined maximum value, the operation has to be halted for a thorough mechanical or wet cleaning of the boiler, approximately every 4000 hr of operation.

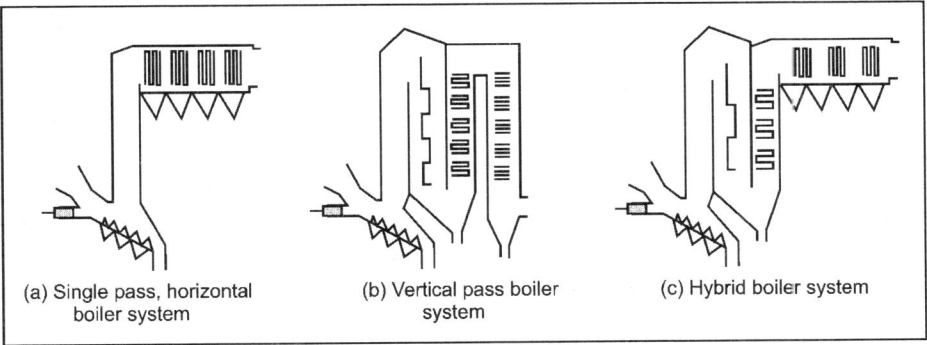

(a) Single pass, horizontal boiler system

(b) Vertical pass boiler system

(c) Hybrid boiler system

Fig. 19.6. Typical boiler configurations for energy recovery.

Corrosion is another primary consideration in the design and operation of incinerator boilers. The formation of HCl by the combustion of chlorine-containing wastes such as paper and board and plastics such as PVC, may cause serious corrosion of tubes due to low-temperature acid corrosion. Critical control of temperature is required to prevent high-temperature and low-temperature corrosion of the boiler. High-temperature corrosion involves superheater boiler tubes in the boiler chamber at temperatures above 450°C, and involves a series of chemical interactions between tube metal, tube scale deposits, slag deposits and flue gases. The rate of corrosion is influenced by temperature, the presence of low melting phases such as alkali bisulphates and pyrosulphates, acid gases such as HCl and SO_3, the nature of the tube metal and the periodic occurrence of reducing conditions. Low-temperature corrosion is due to condensation of acid gases such as HCl and H_2SO_4 formed as the temperature falls below the dew-point. The dew-point for H_2SO_4 is between about 40° and 155°C and for HCl it is between 27° and 60°C depending on gas concentration and water content in the flue gas. Therefore, gas temperatures of more than 200°C are required to minimise downstream dew-point corrosion.

Erosion of boiler tubes may also be a consideration. Flue gases from waste incineration are high in particulate or dust concentration and the dust particles can be very hard, causing erosion of the tubes by abrasion.

Emissions control

The emissions to air and water and the solid residues arising from the incineration of waste are highly regulated by the EC Waste Incineration Directive. In particular, the gas clean-up system required to meet the requirements of the directive now constitutes a major proportion of the cost, technological sophistication and space requirement of an incinerator.

Of the pollutant emissions arising from the incineration of waste, those emitted to the atmosphere have received most attention from environmentalists and legislators. There are a wide variety of emissions limits, but it is clear that the emissions of most concern are total particulate or dust, acidic gases such as hydrogen chloride, hydrogen fluoride and sulphur dioxide, and heavy metals such as mercury, cadmium and lead. In addition, the combustion efficiency is controlled by limits on the emission of carbon monoxide and organic carbon. There are also limits on the emission of dioxins of 0.1 ng TEQ/m^3 (TEQ = toxic equivalent). All legislative emission limits and plant emissions data across the world are related to a set of reference conditions, such as 7 per cent O_2, 9 per cent O_2, 9 per cent CO_2, 11 per cent O_2, 11 per cent CO_2, etc. so that emissions can be compared from different plants which may be actually operated at very different conditions. In fact this makes the emission limit data for most countries very similar. Municipal solid waste incinerators generally produce flue gas volumes typically between 4500 and 6000 m^3 per ton at 11 per cent O_2, reference conditions.

Table 19.1 shows typical concentration ranges for emissions before any gas clean-up treatment for a range of European municipal solid waste incineration plant. The emissions are very much higher than are legally permitted under the EC Waste Incineration Directive which emphasises the need for efficient and sophisticated gas clean-up to reduce the emissions to below legislated values. The layout of a hypothetical gas clean-up system for a municipal waste incinerator is shown in Fig. 19.7. The particulate material is first removed by an electrostatic precipitator, and precollector, then the acid gases are removed by a lime scrubber which may be of the dry-lime or wet-lime type. After the lime scrubber, and addition of an additive, such as activated carbon and lime, to adsorb mercury and dioxins and furans, there is a fabric filter to remove the fine particulate and activated carbon with the adsorbed pollutants. Finally, the oxides of nitrogen are removed by addition of ammonia to form inert nitrogen.

Table 19.1. Typical concentration ranges of emissions from municipal solid waste mass burn incineration after the boiler and before gas clean-up.

Emission	Units	Range
Total dust	mg/m^3	1000-5000
TOC	mg/m^3	1–10
Hydrogen chloride	mg/m^3	500–2000
Hydrogen fluoride	mg/m^3	5–20
Carbon monoxide	mg/m^3	5–50
Sulphur oxides	mg/m^3	200–1000
Nitrogen oxides	mg/m^3	250–500
Cadmium + thallium	mg/m^3	<3
Mercury	mg/m^3	0.05–0.50
Other heavy metals Pb, Sb, As, Cr, Co, Mn, Ni, V, Sn	mg/m^3	<50
Dioxins and furans (PCDD/PCDF)	$ngTEQ/m^3$	0.5–10

Formation and control of emissions

The emissions to the environment of most concern in relation to mass burn, municipal waste incinerators are those covered by the legislation and are listed below:

1. Dust (particulate matter).
2. Heavy metals such as mercury, cadmium, lead, arsenic, zinc, chromium, copper, nickel, etc.
3. Acidic and corrosive gases such as hydrogen chloride, hydrogen fluoride, sulphur dioxide and nitrogen oxides.

4. Products of incomplete combustion such as polycyclic aromatic hydrocarbons, dioxins and furans.
5. Waste-water.
6. Ash residue.

Each will be discussed in turn in relation to their formation, environmental impact and control.

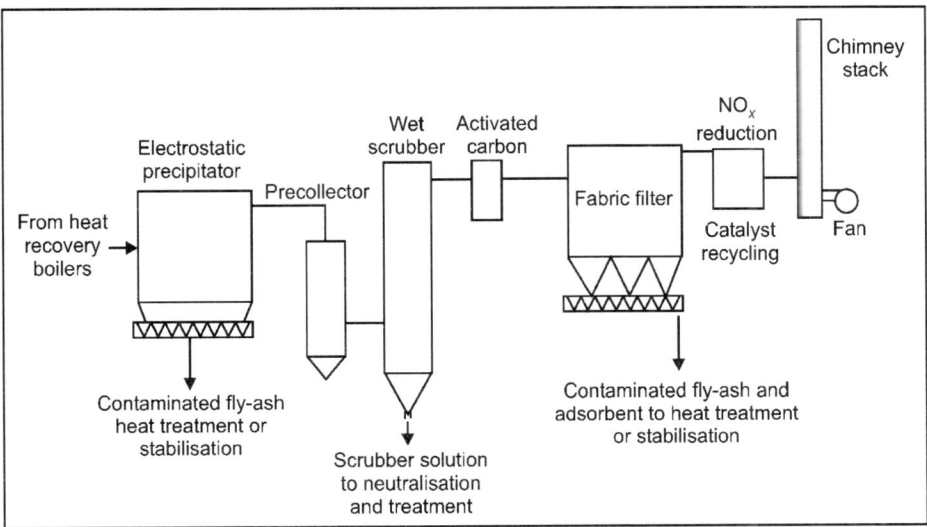

Fig. 19.7. Hypothetical advanced gas clean-up system for a municipal solid waste incinerator.

Dust or particulate matter

The combustion of waste is a very dusty process. The agitation of the waste as it tumbles down the grate, the blowing of primary air through the bed, the high ash content of the waste, and the heterogeneous nature of the waste, all serve to produce a high particulate loading in the flue gases. The design of the incinerator also influences the particulate loading in the flue gases. Such design factors include the size of the incinerator, the grate type, and the combustion chamber design. Particulate emissions from incinerators are the most visual to the public and require efficient and high levels of removal so that complaints do not arise. The particulate is largely composed of ash. However, in addition, pollutants of a more toxic nature such as heavy metals and dioxins and furans are associated with particulate matter, either as individual solid particles or adsorbed on the surface of the particles.

The particulates may also contain carbon and adsorbed acidic gases such as hydrochloric, sulphuric or even hydrofluoric acid to produce corrosive acid 'smuts'. Soot is formed when carbon-containing wastes are combusted in conditions of high temperature and low oxygen content. Polycyclic aromatic hydrocarbons (PAHs) have been cited as chemical intermediaries in soot formation, an alternative proposed mechanism is via acetylene radicals which build to form large soot molecules. The control of soot formation is via adequate residence time for the combustion process to completely burn out any soot being formed, with good mixing of the primary and secondary combustion air.

The emission of untreated flue gases would give rise to a dark plume and the deposition of dust downwind of the incinerator stack. The size range of incinerator particulates found in the flue gases is from <1 μm to 75 μm, the larger particles tending to settle out prior to the flue. It is the ultrafine particles that are of particular concern in assessment of health effects since they contain ash and adsorbed acid gases, heavy metals and organic micropollutants which, because of their size, can pass deep into the respiratory system of humans. There is currently some concern that the important factor in determining the deleterious effects of fine particles is not particularly their composition but their ultrafine nature and the fact that they can penetrate deep into the lungs. A separate size category of particulate matter of environmental concern, PM to (particulate matter) has been designated for particulate matter of less than 10 μm in size. Exposure to PM_{10} particles is associated with both acute and chronic health effects. These include increased rates of bronchitis, reduced lung function, respiratory symptoms and cancer. A large fraction of municipal waste incinerator particulates are of such a small size. The human health concern relates to the interaction of the respiratory system with the high surface area, ultrafine particles and their associated adsorbed heavy metals and organic pollutants. In addition, the small size of waste incinerator particulate emissions, promotes both short and long-range dispersion from the chimney stack into the environment. However, Rabl and Spadaro have assessed the risk to human health from the particulates emitted from municipal solid waste incinerators and have concluded that the health risks are insignificant. The environmental and human health impact of particulates from municipal solid waste incinerators were calculated to be insignificant compared with other sources of particulates in the atmosphere.

The selection of gas cleaning equipment used for the control of particulates in municipal solid waste incinerator flue gases is determined by a range of parameters. These include: the particle load in the gas stream; the average particle size; the particle size distribution; the flow-rate of gas; the flue gas temperature; the required outlet gas concentration and the other components of the overall clean-up system used. Particulate emissions from mass burn incinerators are controlled by a range of possible equipment which are effective in removing particulate material. These include cyclones, electrostatic precipitators and fabric filters. The most common initial particulate removal apparatus is an electrostatic precipitator, since fabric filters will be found down stream in most incinerators because of the need to trap heavy metals and organic micropollutants which are in the solid phase or adsorbed on fine ash particles. Electrostatic precipitators can reach low exit gas particulate concentrations, typically in the range 15–25 mg/m^3 and even lower exit gas values of less than 5 mg/m^3 are possible, where additional electrostatic fields or larger volume equipment is used. Cyclones are not common in municipal solid waste incinerator systems because of their lower capture efficiency compared with electrostatic precipitators and fabric filters for the particular size range of incinerator flue gas dust emissions. However, they may be used as an initial stage to reduce the particulate load for other devices. Cyclones typically achieve exit gas concentrations of only between 200 and 300 mg/m^3. Cyclones also have the advantage that they may be used at higher temperatures of above 500°C. Wet scrubbers are also effective in reducing the particulate load of flue gases with about 50 per cent capture efficiency. Fabric filters can achieve very low exit gas particulate concentrations, of typically less than 5 mg/m, and are increasingly used in most municipal solid waste incinerators. There are a range of materials available from which the fabric filters are made, the choice depending on the operational temperature required and their resistance to acid or alkali gas attack. Materials include polypropylene which has a maximum operational temperature of 95°C, wool at 100°C, polyester at 135°C, nylon at 205°C and fibreglass at 260°C.

Heavy metals

Metals and metal compounds are present in the components of raw waste. For example, municipal refuse may contain lead from lead-based paints, mercury and cadmium from batteries, aluminium foil, lead plumbing, zinc sheets, volatile metal compounds, etc. High levels occur and the concentrations are very variable. The behaviour of metals during the incineration process are illustrated schematically in Fig. 19.8. Within the environment of the furnace of the incinerator, the release of heavy metals from the waste and incorporation into the flue gases is a function of many factors, including volatility, combustion conditions and ash entrainment. Metals and metal compounds may evaporate in the furnace to condense eventually in the colder parts of the flues and generate an aerosol of submicron particles or they may become adsorbed onto fly-ash particles through a range of processes. The extent of evaporation of these metals and metal compounds in the furnace depends on complex and interrelated factors such as operating temperature, oxidative or reductive conditions and the presence of scavengers, mainly halogens such as chlorine. The volatility of these metals and salts is low, for example: Cd, 765°C; Hg, 357°C; As, 130°C; $PbCl_2$, 950°C; and $HgCl_2$, 302°C. However, for some compounds the volatility temperatures are not known. As these metals enter the incinerator in the waste they are subject to combustion temperatures of anything between 800° and 1400°C, well above the boiling points of metals such as mercury and cadmium and metal compounds such as lead chloride. The metals therefore enter the gas phase. The metals may also react with hydrogen chloride or oxygen to form compounds which are more volatile than the metal. Zhang have shown that there is a direct relationship between temperature and the amount of volatilisation of a whole range of heavy metals as the temperature of municipal solid waste is raised from 500° to 1000°C.

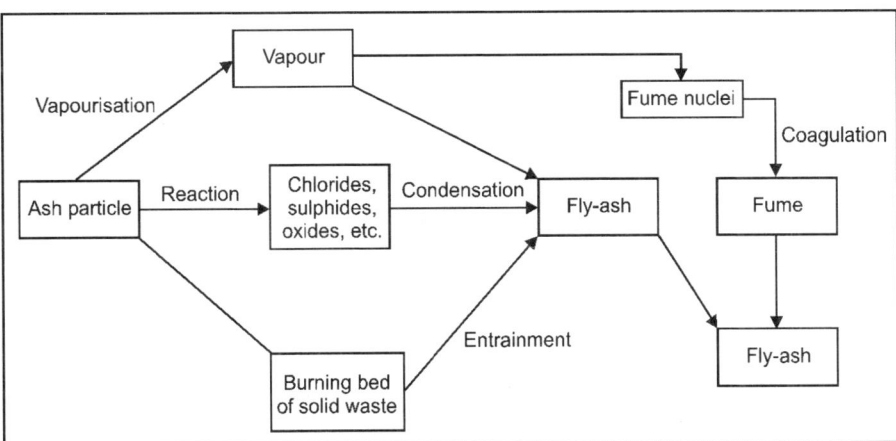

Fig. 19.8. Metals behaviour during waste incineration.

The partitioning of the heavy metals in the incinerator system is a function of their physico-chemical properties. For example, cadmium and mercury being the more volatile of the heavy metals, with high vapour pressures and low boiling points, are most likely to be found in the flue gas. Other metals with low vapour pressure and high boiling points, such as iron and copper, are almost completely trapped in the bottom ash. In addition to the physical properties of volatilisation, simultaneous chemical processes such as decomposition, chlorination, oxidation and reduction may take place. For example, it has been shown that chlorine influences the volatility of heavy metals via the formation of chlorides. Nickel, because of its low vapour pressure and high boiling point, will not vapourise under the conditions of incinerator furnaces, but will do so in the presence of chlorine. Also, cadmium is easily volatilised during incineration and is oxidised in the presence of hydrogen chloride to form mainly cadmium chloride. A further route for the heavy metals to enter the flue gas stream is via entrainment of fine ash particles containing the metal either as the metal itself or as metal compounds. Entrainment is a function of the size, shape and density of the ash particles as well as the incinerator operating conditions. Changes in the oxidising and reducing conditions within the incinerator can also influence the volatilisation of heavy metals. Low-volatility metal compounds may also react under reducing conditions to form metal compounds which are more readily volatilised.

As the furnace off-gases cool on passing through the flue gas system, the heavy metals are subject to a series of condensation reactions involving homogeneous nucleation to form a fine fume of metal particles and heterogeneous deposition onto fly-ash. Homogeneous nucleation occurs, for example, when the partial pressure of an inorganic vapour species exceeds a certain critical value. The incineration gases may become supersaturated as a result of rapid cooling of the gas or rapid formation of a new metal species of lower volatility. Heterogeneous deposition involves fly-ash particles in the flue gases providing sites for condensation of the cooling metal vapour. It has been shown that the relative rates of homogeneous nucleation and heterogeneous deposition also depend on the time/temperature gradient experienced by the metal-containing flue gas. Following homogeneous nucleation and heterogeneous deposition, particles will subsequently grow by coagulation.

The distribution of the metals in the various outputs from municipal waste incinerators has been investigated by a number of workers. Figure 19.9 shows the percentage distribution of heavy metals as a mass balance into and out of an incinerator, equipped with an electrostatic precipitator, as the only gas clean-up measure, in terms of that fraction either emitted to the flue gas, or captured in the electrostatic precipitator (fly-ash) or the bottom ash from the furnace.

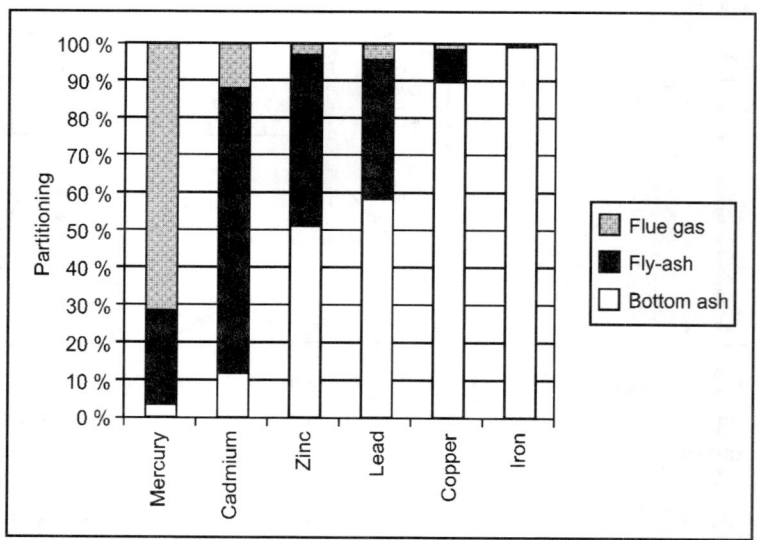

Fig. 19.9. Partitioning of heavy metals during municipal solid waste incineration.

It is suggested that the partitioning is a function of the physico-chemical properties of the elements and their derived compounds, such that volatile mercury and cadmium compounds with high vapour pressures and low boiling points are most likely to be found in the flue gas. Metals with a medium vapour pressure and boiling points such as lead and zinc are retained better in the slag and are less concentrated in the electrostatic precipitator dust. Other metals with low vapour pressure and high boiling points, such as iron and copper, are almost completely trapped in the slag.

The speciation of these metals, whether metal or metal compounds, in the incinerator off-gas is strongly influenced by the presence of compounds of chlorine, sulphur, carbon, nitrogen, fluorine and others during combustion and gas cooling. The form of the metal in the flue gases is important in determining the extent to which it will be captured by the gas-cleaning system. For example, mercury chloride is more easily captured than mercury metal is by wet scrubber systems. The off-gases containing metals and chlorine species, particularly hydrogen chloride, leads to the formation of metal chlorides, many of which have lower boiling points than the parent metal. For example, cadmium is easily volatilised during incineration and is oxidised in the presence of hydrogen chloride to give cadmium chloride as the main product. Mercury has also been shown to be present largely in the halogenated form, predominantly mercury(II)chloride and to a lesser extent mercury(I)chloride. Whilst initially mercury is vapourised as the metal in the furnace, it quickly becomes oxidised to the halogenated form and only a small percentage is present as metal vapour.

Bergstrom also reported on theoretical calculations of equilibrium and mass transfer data and concluded that mercury initially exists as mercury metal (Hg°), but during the cooling of the combustion gases an increasing fraction reacts to form $HgCl_2$. Experiments on a pilot plant and an incinerator also suggested that only a small proportion of mercury is present in the flue gases as mercury metal, most of it is oxidised and is present as gaseous salts, with $HgCl_2$ being the most likely since its high vapour pressure permits existence at temperatures down to 140°C. Home and Williams reported thermochemical equilibrium studies of mercury with oxygen, HCl and SO_2. At high temperatures, above 700°C, HgO was the stable species, below 700°C, the presence of $HgCl_2$ and HgO becomes significant and below 500°C, $HgCl_2$ is the dominant species under simulated flue gas conditions.

Organomercury compounds are particularly toxic and their formation in waste incineration flue gases would be of concern. However, Lindquist suggests that organomercury compounds decompose during combustion. It is unclear whether recombination and/or formation of organomercury compounds takes place in the flue. Sampling of flue gases from coal combustion indicated that there were no measurable amounts of organomercury compounds present. However, Chow have reported that methyl mercury can be 10 per cent of the total mercury species emitted from the combustion of bituminous coal. The form of the metal species for other heavy metals is also important. Uberoi and Shadman have reported that, based on thermodynamic equilibrium calculations, when chlorine containing wastes are incinerated, lead compounds exist predominantly as PbO, $PbCl_2$, and $PbCl_4$ and in the presence of sulphur $PbSO_4$ may also be formed. Wey, using thermochemical equilibrium analysis, suggested that Cr_2O_3, $PbCl_2$ and $CdCl_2$ were the dominant species of Cr, Pb and Cd respectively. The thermodynamic analysis assumed that equilibrium had been attained in the reaction system. Seeker has suggested that the principal species for arsenic is As_2O_3.

The gas clean-up systems required to control Heavy metals are dependent on the metal's volatility. Measures used to control total particulate emissions, such as electrostatic precipitators and fabric filters, will collect the associated heavy metals which are in the fly-ash, are adsorbed to the surface or are discrete heavy metal particles. The heavy metals are associated with the particulate, because of the volatilisation of metals during the combustion of the waste, subsequent condensation at lower temperatures and adsorption onto the fine particulates in the flue gas. The heavy metals tend to concentrate in the finer grained size fraction of the particulate. For example, Walsh found heavy metal enrichment factors, including cadmium, zinc and lead, of more than ten-fold for fine particles emitted from waste incinerators compared with those larger particles retained on the air pollution control system as fly-ash. Zhang, have also reported that the emitted heavy metals are concentrated in the fine particulate submicron size fraction from municipal solid waste incinerators.

Fabric filters are particularly effective for the removal of heavy metals, since they are operated at temperatures typically below 250°C where the metals have condensed to form particulate material which are effectively trapped by the fabric filter. Exit gas concentrations of particulate from fabric filters are typically less than 5 mg/m^3 and therefore fabric filters efficiently and effectively remove heavy metals adsorbed to the particulate material.

Mercury is present in the flue gases mainly as mercury and mercury chloride. Because mercury has a boiling point of 357°C and mercury chloride has a boiling point of 302°C, all the mercury will be in the gas phase as it exits the furnace and boiler system (Fig. 19.9). Even at the operating temperatures of the fabric filter (and also electrostatic precipitator) mercury and mercury chloride, with their high vapour pressure and low boiling points, will pass through the particulate trapping system. Therefore, since the emission-limit value for total mercury set by the EC Waste Incineration Directive is 0.05 mg/m^3, most municipal solid waste incinerators require the addition of special gas-cleaning measures specifically for the capture of mercury, in order to meet the required emission limit.

Scrubber systems, used to remove acid gases, develop an acidic solution with a low pH which are effective at trapping mercury chloride at the level of about 95 per cent, but for mercury metal the level is only 0–10 per cent. Addition of sulphur additives to the scrubber solution enables an improvement in the trapping of mercury metal but other methods are required to enable the emission limits for mercury to be attained. Systems used to trap total mercury include addition of reagent upstream of the scrubber system or fabric filter. Additives which have proven effective are sodium sulphide, TMT 15 (trimercapto-s-triazine) and activated carbon. The additives add to the cost of gas clean-up with sodium sulphide being the most cost-effective, followed by activated carbon, which is approximately three times the cost, and TMT 15 at seven times the cost, of sodium sulphide.

Activated carbon is the most commonly used additive to trap mercury, coupled with a downstream fabric filter, and it also has the advantage of removing dioxins and furans from the gas stream. Activated carbon is a high surface area material produced as a fine carbon powder, typically with surface areas in the range of 1500–2000 m^2/g. The high surface area of the carbon surface adsorbs the mercury vapour by physical adsorption on the surface active sites or, where the activated carbon is impregnated with sulphur, by chemical adsorption as mercury sulphide. The activated carbon additive is added at concentration levels of between 0.1 and 0.5 g/m^3 of waste gas, and high removal efficiencies have been reported. For example, capture efficiencies for total mercury of more than 95 per cent have been achieved, resulting in emission levels of less than 0.03 mg/m^3. The activated carbon traps the mercury vapour in the flue gas stream and also on the surface of the fabric filter, where it deposits and acts as a further adsorbing filter deposit. There is a risk of fire with the use of fine-sized activated carbon powder and therefore the carbon is generally mixed with other reagents, such as calcium hydroxide, at a ratio of 10 per cent activated carbon to 90 per cent calcium hydroxide. The calcium hydroxide also acts to remove acid gases.

An alternative to gas clean-up of heavy metals is to eliminate them from the raw waste material, the recycling of batteries for the removal of cadmium and mercury has been shown to be effective in reducing the emissions of these metals.

The human health effects of heavy metals reported in the literature, relate to occupational health exposure studies and to accidental exposure or to animal studies. Heavy metals exert a range of toxic health effects including carcinogenic, neurological, hepatic and renal effects. For example, cadmium represents a health risk via accumulation in living tissue and has been associated with an increased risk of lung cancer, emphysema and kidney damage and in extreme circumstances, damage to bones and joints. Mercury and mercury compounds give rise to toxic effects associated with the central nervous system, the major areas affected being associated with the sensory, visual and auditory functions as well as those concerned with co-ordination. Lead exposure has been associated with disfunction in the haematological system and central nervous system.

Decreases in intelligence and behaviour have been reported in children subject to exposure of increased levels of lead. The primary route for human exposure to heavy metals released by incineration is the food chain. Of the heavy metals, cadmium, mercury and lead are deemed of most importance in relation to municipal waste incinerators since, whilst other metals do occur, their toxicities or emission levels are much lower. The health effects of heavy metals arising from incineration is increased because they are readily available to the body as they are concentrated on the finer size fraction, tend to be adsorbed to the surface of particles, and their fine size means they are more easily ingested. Whilst the effects of heavy metals on human health are significant, the gas clean-up measures required to meet legislative limits of heavy metal emissions from incinerators do reduce emissions of a range of heavy metals to very low levels. Investigations into the influence on human health of the emissions of heavy metals from waste incinerators suggest that they do not pose a significant problem to the environment. Reports on the effects of heavy metals and human health studies, in relation to municipal solid waste incinerators, have concluded that no effects on health have been linked to the release of heavy metals from incineration plants. A later review of epidemiological studies on the impact of a range of pollutants, including heavy metals, emitted from waste incinerators, on workers in the waste incineration industry and surrounding population, also concluded that no consistent pattern of ill health had emerged from such studies. This was attributed to the very low level of concentration and the effects of compounding and modifying factors, such as smoking and socio-economic effects.

Hu and Shy reviewed several epidemiological studies investigating the health effects of waste incinerator emissions on incinerator workers and community residents. They reported that, whilst some studies showed increased body levels of some organic chemicals and heavy metals, there were no effects on respiratory symptoms or pulmonary function and the findings for cancer and reproduction were inconsistent. Rabl and Spadaro undertook a risk assessment of the emissions from waste incinerators, including heavy metals, and concluded that the health impacts of municipal solid waste incinerators were insignificant compared with the background air quality, provided that the emissions complied with the emission limits of the EC Waste Incineration Directive.

Acidic and corrosive gases

Municipal waste contains a range of compounds which contain chlorine, fluorine, sulphur, nitrogen and other elements which may result in the generation of acidic, toxic or corrosive gases. Nitrogen oxides also result from the nitrogen in the combustion air formed by reaction with oxygen at the high temperatures of the combustion zone and from nitrogen in the waste. Typical waste contains about 4000 mg/kg chlorine, 100–350 mg/kg fluorine, 2000 mg/kg sulphur and 5000 mg/kg nitrogen. The waste chloride and fluoride are in the form of waste plastics, for example, PVC and PTFE (polytetrafluoroethylene), chlorides are also found in paper and board, rubber, leather and as sodium chloride, for example, from street sweepings after road slating in icy conditions. The sulphur content of municipal solid waste is low compared with coal sulphur contents.

Normally, because the combustion of the hydrocarbon volatile fraction in an incinerator is almost complete, the flue gases consist mainly of nitrogen, oxygen, water vapour and carbon dioxide. However, the combustible waste compounds which contain the chlorine, fluorine, sulphur or nitrogen during combustion, generate gaseous contaminants such as hydrogen chloride, hydrogen fluoride, sulphur oxides and nitrogen oxides:

$$C, HCl, F, S, N + O_2 \Rightarrow CO_2 + H_2O + HCl + HF + SO_2 + NO_x$$

At the high temperatures of the combustion zone, carbon dioxide and water vapour may partially dissociate, but the resulting carbon monoxide, hydrogen and oxygen recombine when the temperature decreases. Many of the other products, including hydrogen chloride, hydrogen fluoride and sulphur dioxide, are stable. Normally chlorine is not detectable in the furnace emissions since it is reduced by numerous gases or solid reducing agents to hydrogen chloride.

The origin of HCl in incinerator flue gas has been the subject of much research, due to the corrosive nature of HCl at low temperature, i.e. dew-point corrosion and high temperature corrosion and its implication in dioxin formation. One of the major sources of HCl is regarded as being PVC plastic at more than 50 per cent. However, other sources such as metal chlorides like NaCl or $CaCl_2$ from paper and board, rubber, leather and vegetable matter are regarded as significant sources of HCl. PVC emits HCl by a gradual process of thermal decomposition, which takes place between 180° and 600°C. Chlorides are also implicated in the formation of dioxins and furans. Hydrogen fluoride is even more reactive and corrosive than HCl and arises from combustion of fluorinated hydrocarbons, such as plastics like PTFE.

Municipal waste incinerators are regarded as only a minor source of sulphur dioxide (SO_2) emission when compared to power plants and industrial boilers firing heavy fuel oil or coal. About 1 per cent of the SO_2 may be further oxidised to sulphur trioxide, SO_3, which reacts with water vapour to form highly corrosive sulphuric acid, H_2SO_4, in the flue gas.

Nitrogen oxide (which includes all the nitrogen oxides, but particularly nitric oxide, NO, and nitrogen dioxide, NO_2) from waste incineration arises mainly from the nitrogen in the waste (fuel NO_x) and by direct combination of the nitrogen and oxygen present in the combustion air, which occurs more rapidly at high temperatures (thermal NO_x). In practice, thermal NO is formed almost exclusively at high temperatures in the flame, particularly under oxidising conditions. In reducing conditions, little NO is formed. Fuel NO_x is formed from the nitrogen compounds in the waste, but may also form nitrogen gas instead of NO_x.

The potential problem of acid gases in the back-end of the incinerator plant is due to their low dew-point, which results in corrosive damage to metals. The emission of the pollutant gases to the atmosphere produces the well-documented acid rain with its associated environmental damage, whilst NO, after atmospheric oxidation to NO_2, is active in the generation of photochemical smog. HCl, HF, SO_2 and NO_x all produce acids in the atmosphere, which contribute to acid rain, forming hydrochloric, hydrofluoric, sulphuric and nitric acids, respectively. Increased acidification of the atmosphere has resulted in damage to buildings, acidification of lakes, respiratory problems, die-back of forests, etc. The health effects associated with acid gases and NO_x include respiratory and bronchial problems and contribute to the general environmental deterioration of the urban atmosphere.

The EC Waste Incineration Directive sets emission limits for hydrogen chloride, hydrogen fluoride, sulphur dioxide and nitrogen oxides (EC Waste Incineration Directive). Wet, dry and semi-dry processes are used to remove the acid gases produced by waste combustion. Wet scrubbing systems use slurries and solutions at relatively low temperatures and produce a liquid or wet solid/sludge reaction product. Generally, wet scrubbers operate in two stages, a first stage which uses water to trap the highly soluble acid gases, hydrogen chloride and hydrogen fluoride, and a second stage which uses added alkali absorbents to trap the less soluble sulphur dioxide. The absorbents used include calcium hydroxide and sodium hydroxide. That is, alkali solutions which neutralise and scrub out the acid gases. Gas–liquid absorption of acid gases is very efficient, but the liquid or sludge product is highly polluted and difficult and expensive to treat. In addition, the slurry or solution used, requires to be made up to certain specifications which again adds to the cost. A further disadvantage is that the flue gases produced are high in moisture content and may result in acid condensation and corrosion.

Consequently, developments have centred on new methods of control which generate a solid residue which is easier to handle. Dry systems use a dry powder such as calcium oxide (lime) or sodium bicarbonate and possibly upstream humidification to improve gas/sorbent reaction. The dosage rate of the dry powder is approximately three times the stoichiometric amount, that is, three times more than the amount required to exactly trap the acidic component of the acid gases. The reaction products are solid particles and the process increases the dust load of the flue gas stream and would have to be trapped out of the flue gases, usually by a fabric filter. The build-up of the scrubber powder on the fabric filter produces a layer of adsorbent, which also aids the acid gas removal process.

Gas–solid adsorption tends to be less efficient than gas-liquid absorption. Therefore, a development is the semi-dry system which is also called the semi-wet system or spray absorption. Semi-dry processes use an alkaline sorbent slurry or solution which is atomised into fine droplets into the flue gas, the droplets react and dry in the hot flue gases to produce a dry powder. Absorption of the acid gases takes place efficiently at the gas–liquid interface, but then the heat of the flue gases causes evaporation of the water to produce a dry powder which is easier to handle. The flue gases have an increased particulate content and have to be cleaned by a downstream fabric filter. For the semi-dry systems, as was the case for the dry system, adsorption is improved by the use of a downstream fabric filter which increases contact time between the gases and the alkaline filter cake formed on the filter by the adsorbent.

Nitrogen oxides (NO_x) may be reduced by primary measures such as the temperature of combustion, recirculation of the flue gases, staged combustion or by secondary downstream clean-up measures such as ammonia addition. Since the formation of NO_x is directly related to increasing temperature, the use of primary and secondary air to ensure good gas mixing minimises high temperature excursions, which thereby results in lower formation of NO_x.

The use of flue gas recirculation, where the flue gases are fed back into the combustion chamber as secondary air, results in the reduction of NO_x formation, since the recirculated flue gases have a lower nitrogen content and therefore produce lower thermal NO_x. A further method to reduce the nitrogen content of the air supply which is required for combustion, is to use either pure oxygen or enriched oxygen-air. Again, the reduced nitrogen in the incoming gas has less nitrogen available to be converted to thermal NO_x. Other combustion operational measures to reduce NO_x include staged combustion and re-burn techniques which are common in the fossil fuel combustion power generation industry. Staged combustion involves the supply of a reduced amount of combustion air, and therefore oxygen, to the combustion zone which reduces the amount of NO_x formed because of the lower levels of nitrogen. The combustion of the waste is less complete and volatile unburnt hydrocarbons are formed which are then burnt-out in a second stage where excess air is added.

Reburning is a three-step process involving injection of a reburning fuel such as natural gas to a combustion zone above the burning bed of waste. NO_x formed in the primary combustion zone is reduced to molecular nitrogen by reaction with hydrocarbon fragments formed in the re-burn zone from the combustion of natural gas.

Nitrogen oxide, the main oxide of nitrogen found in flue gases, cannot be reduced by scrubbing because of its low solubility in scrubber systems. Therefore, secondary techniques to control NO, emission rely on addition of a reactant, such as ammonia (NH_x) or urea (a derivative of ammonia – H_2NCONH_2) to convert the NO_x to nitrogen and water by reaction. The ammonia or urea are added in aqueous solution. The secondary selective reduction takes place at either high temperature, known as the selective non-catalytic reduction (SNCR) process or at lower temperature using a catalyst, known as the selective catalytic reduction (SCR) process. SNCR involves addition of the ammonia or urea into the incinerator furnace at temperatures typically between 850° and 1000°C, with the optimum temperature range between 850° and 950°C, where the reaction with NO_x is maximised. Higher levels of NO_x removal require higher inputs of ammonia or urea, but can result in 'ammonia slip' where some of the added ammonia or urea is not utilised and passes into the flue gases unreacted. SCR operates with the addition of ammonia or urea as an additive to form nitrogen and water from NO_x, but operates at temperatures, typically between 250° and 400°C, with the use of a catalyst. The catalyst is usually in mesh form and generally consists of TiO_2 impregnated with V_2O_5 and WO_3 catalyst and lasts for approximately three to five years. High NO_x reduction rates of over 90 per cent can be produced with SCR. For waste incineration emissions control systems, the SCR process is usually applied after the scrubber and particulate removal system and the flue gases may need to be reheated to the temperature of the SCR unit (250°–400°C), resulting in consumption of energy. Odours from incineration of waste may not pose a health hazard as such, since the concentration levels of compounds where the nose can detect odour are extremely low. However, odours cause nuisance to the local population and are often the most common cause of complaint. Odours associated with incineration of waste are usually complex mixtures of organic compounds and result from incomplete combustion. Threshold limit values, above which the odour can be detected, can be very low, for example, many organic compounds which have been associated with waste incinerators have a threshold limit value between 0.001 and 10 ppm. The odours released into the atmosphere will be influenced by the efficiency of the combustion process and the gas clean-up system and peaks of emission may last only a few seconds. Emission to the atmosphere is influenced by dilution in the surrounding atmosphere and the dispersion of the stack plume, which in turn will be influenced by meteorological conditions. Brunner has reviewed the odours from incineration of waste, the threshold limit values of common odours and their control.

Products of incomplete combustion: polycyclic aromatic hydrocarbons (PAHs), dioxins and furans

The volatile matter arising from the thermal degradation of waste is normally completely combusted by providing adequate residence time, post combustion temperature and turbulent mixing.

Polycyclic aromatic hydrocarbons

PAHs are compounds based on aromatic benzene rings which are fused to form two or more polycyclic rings. PAHs are known to occur naturally in the environment, for example, in sediments, fossil fuels and by natural combustion in forest fires. The major sources of PAHs, however, are anthropogenic; examples include oil and coal-fired power generation plant, coke production, residential furnaces, diesel and gasoline engines and in waste combustion. Concern over the emission of PAHs to the environment is centred on the associated health hazard, because PAHs comprise the largest group of carcinogens among the environmental chemical groups.

The PAHs reported from municipal waste incineration include some species known to be biologically active in human and bacterial cell tests, for example, benzo[a]pyrene, benzo[e]pyrene, phenanthrene, methylphenanthenes, fluoranthene and the methylfluorenes.

Dioxins and furans

Polychlorinated dibenzo-ρ-dioxins (PCDD) or 'dioxins' and the closely related polychlorinated dibenzofurans (PCDF) or 'furans' constitute a group of chemicals that have been demonstrated to occur ubiquitously in the environment. They have been detected in soils and sediments, rivers and lakes, chemical formulations and wastes, herbicides, hazardous waste site samples, landfill sludges and leachates.

Sampling and analysis of emissions to air

The compliance of an incinerator with the legislated emission limits requires sampling and analysis of the emissions. The EC Waste Incineration Directive stipulates which emissions are required to be analysed on a continuous or periodic basis. Continuous measurement of nitrogen oxides, carbon monoxide, total dust (particulate), TOC, hydrogen chloride, hydrogen fluoride and sulphur dioxide, is required. Combustion parameters such as oxygen, temperature, and water vapour monitoring, is also stipulated to be carried out on a continuous basis. At least two measurements per year of heavy metals and PCDD and PCDF are required, but more frequent monitoring is required in the first twelve months of operation. There are variations in the requirements, subject to the authorisation of the regulatory authority.

The point of analysis is within the stack by means of *in situ* measurement or else a sample of the stack gas is withdrawn for analysis, for example, for heavy metals and PCDD and PCDF. Where a sample is withdrawn from the stack, the sample is usually conditioned, i.e. water vapour, particulates, etc. are removed and the gas is cooled. Some analyses require a heated sampling system. Acidic gases such as HCl should be maintained at temperatures above the dew-point to prevent acid corrosion in the instrument. Water in the sample may be removed with a drying tube, alternatively, chilling the sample may be appropriate.

Particulate analysis is required on a continuous basis and the main method of analysis is usually by infrared absorption, measured by a reduction of light intensity in relation to concentration of dust. The particles absorb the light and the reduction in intensity, compared to clean flue gas, is proportional to the concentration of particles in the gas stream. The continuous measurement system is normally calibrated by a periodic method, such as a gravimetric system.

The analysis of hydrogen chloride is required on a continuous basis. Instruments which fulfil this criteria are infrared and electrochemical analysers. Hydrogen fluoride analysis may be made on a periodic basis, provided it can be demonstrated that control measures for hydrogen chloride will not be exceeded. Sulphur dioxide and carbon monoxide may be analysed on a continuous basis using infrared analysis as described for hydrogen chloride. The preferred method of analysis for TOC is the flame ionisation detector. The sample is burnt in a hydrogen/air flame and the organic carbon molecules produce ions. An electric potential difference across the ionised molecules causes a current to flow between the electrodes and is proportional to the mass flow of common atoms.

Nitrogen oxides, for most purposes, can be regarded as NO and NO_2. Infrared and electrochemical systems can be used, but perhaps the preferred method is chemiluminescence. Nitric oxide (NO) is reacted with ozone produced in the analyser in a reaction chamber, and is oxidised to nitrogen dioxide (NO_2) in a chemically excited state. The excited molecule loses energy and reverts to the ground state by emission of energy as light. The emitted light energy is detected and measured by a photomultiplier tube. The chemiluminescence reaction is specific to NO and NO plus NO_2 is detected and reported as NO_x by passing the sample gas through a catalytic converter which reduces the NO_2 to NO prior to the reaction chamber.

Heavy metals are monitored by periodic gravimetric analysers followed by laboratory analysis. Isokinetic sampling traps the coarse and fine particulates on a glass fibre filter paper held in an oven. Mercury may largely be in the vapour phase and a vapour-trapping system should also be used in conjunction with the filter system. The filter paper is weighed before and after sampling at constant humidity and temperature to determine the mass of particulate, per metre cubed, of flue gas. The weighed filter is digested in hot concentrated acids in either an open or closed vessel in order to take the metals into solution tor analysis. The resulting solution is made up to a known volume and analysed, usually by atomic absorption spectrometry or inductively couped plasma spectrometry.

Monitoring of PCDD and PCDF is a time consuming, expensive and meticulous process requiring great analytical skill. The sampling of PCDDs and PCDFs is required at a minimum time period of 6 hr and maximum of 8 hr using standardised sampling protocols which trap the particulate and vapour phase PCDD and PCDF. At each stage of the sampling and analysis procedure, radio labelled standard PCDD and PCDF are added to the sample to measure the recovery of the sampling, sample clean-up and analysis efficiency. The analysis of PCDD and PCDF is difficult, since they occur at very low concentrations, in a sample matrix which contains other chlorinated hydrocarbons, and there are many congeners of PCDD and PCDF. Analysis is performed in three stages. Extraction of the PCDD and PCDF from the sample which may be ash, particulate adsorbed on glass fibre filter paper or condensed water, each requiring a different extraction procedure. The extract is analysed in a series of steps to eliminate compounds which interfere with the analysis and to remove the PCDD and PCDF as a separate fraction from the extracted sample which may contain many chemical groups in high concentration, in addition to the PCDD and PCDF. Final analysis of the fraction containing PCDD and PCDF is most frequently carried out by coupled capillary gas chromatography/mass spectrometry which separates the individual PCDD and PCDF congeners and detects them using mass spectrometry.

Waste-water

Water pollution from incinerators is not generally regarded as an important problem, because of the limited amount of waste-water generated. A typical European municipal solid waste incinerator would generate approximately 0.15–0.3 m³/ton of waste-water depending on the type of gas cleaning system used. The main sources of waste-water from incinerators are from flue gas treatment as flue gas scrubber water, and alkaline scrubbing of the gases to remove acid gases and the quenching of incinerator ash. Other minor sources include, for example, scrubber water pretreatment and the purification of boiler feedwater where a boiler plant is installed. Such water is contained in a closed system and would not come into contact with the pollutant flue gases.

Where the flue gases are scrubbed or cooled with water, the absorbed acid gases will make the water very acidic and will also consequently contain significant quantities of heavy metals which are soluble in the acidic solution. Where the flue gases are scrubbed with an alkaline solution, such as sodium hydroxide or calcium hydroxide to remove acid gases, the scrubber water will be very alkaline. Recirculation of waste-water in the wet scrubbing system can result in a substantial reduction in the amount of waste-water.

Ash residue

If the incinerator is operating correctly, the residue or ash should be completely burnt out and biologically sterile. The EC Waste Incineration Directive specifies that the carbon content of the bottom ash should be less than 3 wt% as a measure of the burnout efficiency of the incinerator. Bottom ash from the furnace grate represents the bulk of total ash and is composed mainly of mineral oxides. Fly-ash comprises only a few per cent of the waste mass input. Three types of solid residue or ash may be distinguished as the solid residues of a municipal solid waste incinerator, these are bottom ash collected at the bottom of the grate, boiler ash collected in the heat recovery boiler system of the incinerator, and fly-ash collected from the air pollution control system.

Bottom ash from the furnace grate represents the bulk of total ash and is a heterogeneous mixture of slag, ferrous and nonferrous metals, ceramics, glass, other noncombustible material and uncombusted organic material. The bottom ash consists mainly of silicates, oxides and carbonates. Various compounds and mineral species have been identified in bottom ash, for example, $Ca_2Al_2SiO_7$, $Ca_2MgSi_2O_7$, SiO_2, Fe_3O_4, Fe_2O_3, $CaCO_3$, $MgCO_3$, $Ca(OH)_2$, $CaSO_4$, NaCl and KCl and elemental Fe, Al and Cu. Bottom ash contains metals which are less volatile than those released during the high temperatures of the furnace. For example, iron and nickel concentrations tend to be higher in the bottom ash than the fly-ash. More volatile metals such as cadmium, mercury, zinc and lead tend to be higher in concentration in the fly-ash compared with the bottom ash. The bottom ash PCDD and PCDF concentrations are much lower than the fly-ash. When the TEQ of the PCDD and PCDF congeners is calculated, the bottom ash TEQ is very low compared with the fly-ash. PCDD and PCDF tend to be much lower in bottom ashes since they are quickly quenched before significant PCDD and PCDF production can take place.

Bottom ash recycling, for aggregate use in the construction industry and road building is common in Europe. For example, in Germany, approximately 80 per cent of bottom ash is utilised, in the Netherlands more than 90 per cent, Denmark 90 per cent, France more than 70 per cent and in the UK more than 50 per cent. Other countries, such as Austria, Switzerland, Portugal, Italy and Norway recycle less than 10 per cent of the bottom ash, the large majority going to waste landfill. The main use in Denmark is for development of a granular sub-base for car parking, bicycle paths and paved and un-paved roads, etc. In Germany, utilisation has been through, for example, sub-base paving applications and, in the Netherlands, bottom ash has been used in construction for granular base, or in-fill road base, embankments and noise and wind barriers. In the Netherlands, bottom ash has also been used as aggregate in asphalt and concrete.

Dispersion of emissions from the chimney stack

A serious consideration in relation to incinerators of all types is the height of the chimney stack and the dispersion of the plume of exhaust gases. The dispersion of the plume involves an initial rise, followed by a horizontal spreading about the plume centre line (Fig. 19.10).

The dispersion is assumed to proceed from an imaginary point source. The concentration profiles in the vertical and horizontal directions are assumed to approach a Gaussian distribution form. The dispersion of the plume will influence the downwind ground concentrations of pollutants. Such considerations are vitally important in the assessment of the impact of an incinerator on the local environment. The rate of dispersion, and hence ground concentrations, are influenced by meteorological considerations, wind speed and rate of plume emission. Unstable conditions promote dispersion, whilst stable conditions, such as fog, result in very low spreading plumes. The ground level concentration close to the stack is zero. However, eventually the plume spreads out and reaches ground level a point known as the radius of maximum effect. For incinerators, this point could be several kilometres from the incinerator. At further distances from the stack, the ground concentration becomes reduced as the plume becomes more diluted in the atmosphere. Figure 19.10 shows plume dispersal in stable, unstable and neutral meteorological conditions. Unstable conditions give the highest ground level concentration closer to the stack than for stable conditions. There is a range of computer software to calculate the dispersion from the chimney stack. Such calculations involve input data in relation to a range of different parameters, for example, the rate of pollutant emissions, stack height, volumetric emissions rate, gas temperature, surrounding geographical terrain, surrounding building heights, wind speed, wind direction, atmospheric stability, any atmospheric chemical reactions, etc.

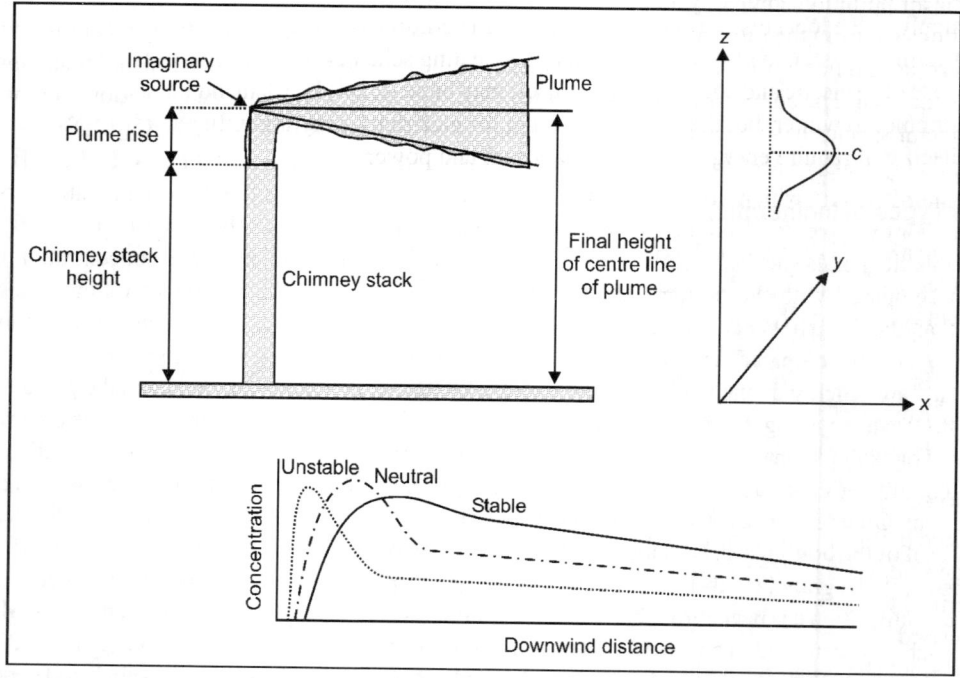

Fig. 19.10. Plume dispersion from chimney stacks and ground level concentration in relation to atmospheric stability.

Dispersion is also influenced by the height of the stack, as higher stacks promote increased rates of dispersion. The use of dispersion models and guidelines give recommended stack heights for incinerators.

Energy recovery via district heating, electricity generation and combined heat and power

The modern municipal waste incinerator relies on the production of steam for electricity generation or district heating to ensure the cost-effectiveness of the process. Some schemes may incorporate both electricity generation and district heating as combined heat and power (CHP) systems. Reported thermal efficiencies for electricity generation—only systems are between 25 and 30 per cent, district heating systems are between 80 and 90 per cent and combined heat and power systems are between 70 and 85 per cent. Electricity is generated from the steam produced in the boilers via a steam-condensing turbine. The high-pressure, high-temperature steam enters the turbine and passes through the various stages of the turbine and, as it does so, it expands and reaches high velocity, turning the blades of the turbine and hence the turbine shaft, which generates the electricity. In general, about 0.3–0.7 MWh of electricity can be generated in a municipal solid waste incinerator from one ton of municipal solid waste, depending on the plant size, steam parameters, steam utilisation efficiency and the calorific value of the waste. Where district heating is the objective, the high-temperature, high-pressure steam passes through heat exchangers which generate hot water under pressure for distribution to homes, offices and institutions. The water is often superheated. CHP systems would use a different type of steam turbine which would generate a lower amount of electricity, but the steam effluent from the turbine would be at a higher temperature, enabling district heating to be incorporated.

Evaluation of the best option is very much a site-specific issue. Contracts to sell the electricity or heat supply to the local district should be secured. In addition, contracting for waste to fuel the plant on a long-term basis should also be secured. District heating schemes rely on a market for the heat which, in the case of domestic and commercial premises, may be seasonal. The demand for heat will be different from summer to winter, but this is not a problem for electricity generation. In addition, the incinerator plant itself will require energy for hot water, steam and power.

Other Types of Incineration

There are a wide variety of incineration types used to incinerate a wide variety of wastes. In this section a number of different design types will be discussed in relation to their technologies and application to different types of waste:
1. Fluidised bed incinerators.
2. Starved air incinerators.
3. Rotary kiln incinerators and cement kilns.
4. Liquid and gaseous waste incinerators.

Fluidised bed incinerators

Fluidised bed incinerators have been used for a wide variety of wastes including municipal solid waste, sewage sludge, hazardous waste, liquid and gaseous wastes and those wastes with difficult combustion properties. Fluidised beds are mainly of the bubbling, turbulent or circulating bed type, although some pressurised fluidised beds have been built for coal combustion for power generation. Figure 19.11 shows schematic diagrams of bubbling bed, turbulent and circulating fluidised beds.

Fluidised beds consist of a bed of sand particles contained in a vertical refractory-lined chamber through which the primary combustion air is blown from below; the sand particles are hence fluidised by adjusting the air flow. Increasing the air flow produces a turbulent flow of solids, and to prevent elutriation of the bed material out of the freeboard, cyclones are placed within the freeboard to recirculate the solids back into the bed. Further increase in air flow produces a circulating fluidised bed where, intentionally, the solids are elutriated out of the bed into a cyclone and the material is recirculated back to the bed. However, in the circulating fluidised bed, combustion also takes place in the cyclone. Such beds are much longer and produce longer residence times of the solid particles of waste in the hot zone, resulting in higher burnout of the products of combustion and reduced organic emissions.

The bed of sand is heated by preheated air or gas or oil burners to raise the temperature, such that incoming waste will ignite and combust efficiently. The processed waste, in the form of shredded municipal solid waste or refuse-derived fuel pelts or indeed other wastes such as sewage sludge or industrial waste, is fed continuously into the hot sand bed. The waste combusts and start-up fuel is then no longer required. The fluidised bed reactor promotes the dispersion of incoming waste, with rapid heating to ignition temperature and promotes sufficient residence time in the reactor for their complete combustion. In the fluidised bed, drying, devolatilisation, ignition and combustion all take place within the bed. Secondary functions include the uniform heating of excess air, good heat transfer for heat exchange surfaces within the bed, and the ability to reduce gaseous emissions by control of temperature or the addition of additives directly to the bed to adsorb pollutants, for example, the addition of lime to reduce sulphur dioxide emissions. A further feature of fluidised beds is their lower operating temperatures, typically around 850°–950°C maximum combustion temperatures, which therefore produce lower levels of thermal NO_x.

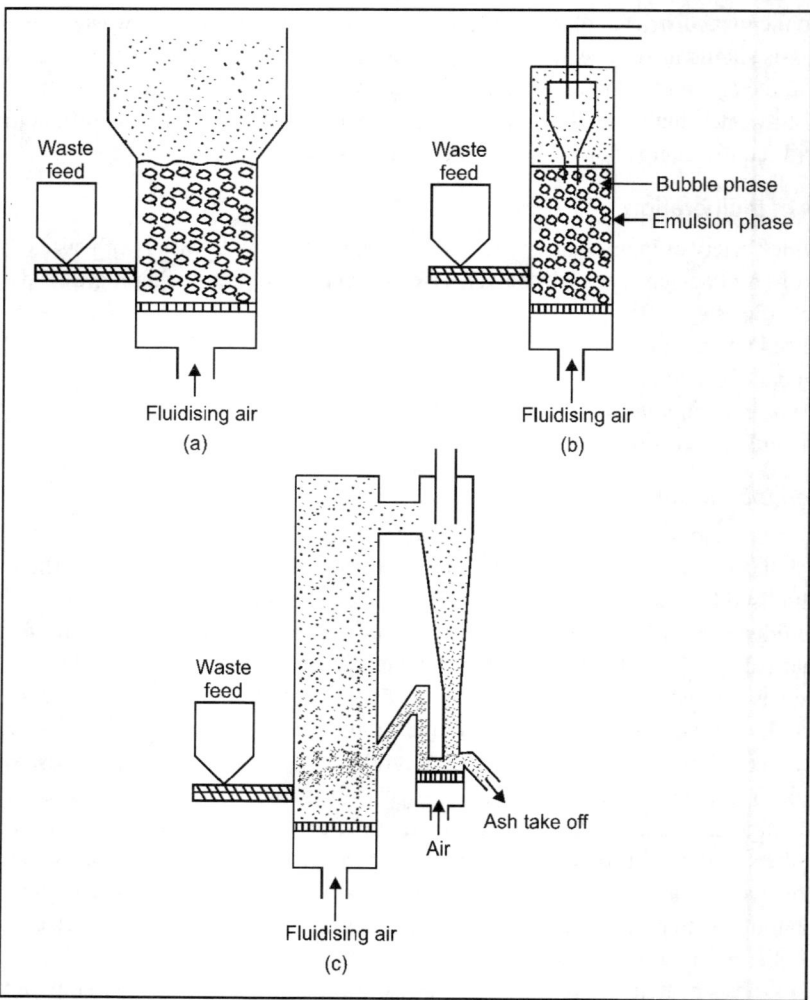

Fig. 19.11. Schematic diagram of: (a) a bubbling bed, (b) a turbulent bed, and (c) a circulating fluidised bed.

The fluidised bed reactor greatly increases the burning rate of waste, since the rate of pyrolysis of the solid waste material is increased by direct contact with the hot, inert bed material. Gases in the bed are continuously mixed by the bed material, thus enhancing the flow of gases to and from the burning solid surface and enhancing the completeness and rate of the gas-phase combustion reaction, this factor becomes more pronounced with circulating fluidised beds which have longer residence times in the hot zone. In addition, the charred surface of the burning solid material is continuously abraded by the bed material, enhancing the rate of new char formation and the rate of char oxidation. Fluidised beds are compact and have high heat storage with fast dynamic response to throughput or demand. They also have high heat transfer rates and thus enable faster ignition of low combustible waste. Because of the high heat transfer rates found in fluidised beds, they are very good for heat-recovery processes and the heat-transfer surfaces may be placed within the bed. The ash residue from waste combustion is usually removed from the bottom of the furnace.

Fluidised beds, by the nature of their combustion, are able to incinerate a range of wastes. Municipal solid waste may be incinerated in a fluidised bed incinerator but is best achieved by some form of prescreening and shredding or the production of refuse derived fuel (RDF) pellets. The solid waste is then fed by screw gravity feeding or pneumatically. Processing routes have involved the production of refuse-derived fuel in shredded or pelletised form. Large particles in fluidised beds can cause problems due to agglomeration, which prevents fluidisation, and the bed then consequently slumps. The agglomerated masses will then have to be removed. Fusion of ash particles in very hot zones in the bed can also cause agglomerates to form. Fluidised beds for municipal solid waste tend to be in the range 12,000–2,00,000 tons per year, whereas large-scale municipal solid waste incinerators range from 50,000 to 8,00,000 tons per year. Fluidised bed incinerators for processed municipal solid waste are common in the USA, Japan and Sweden. Fluidised beds in the USA are often linked to integrated materials recycling and energy recovery facilities, which pre-sort and recover recyclable materials and then shred the waste to produce a combustible fraction for the fluidised bed. Japan also has a large fluidised bed industry to incinerate MSW, where the incinerators are fed by waste which has already been pre-sorted by the householder to produce a combustible fraction. Consequently, there is a lower requirement for preprocessing, only shredding being generally required.

Fluidised beds have been successfully developed for the incineration of sewage sludge. Sewage sludge has a high water content, typically 96 per cent water. The dried solids have a relatively high calorific value of about 20–24 MJ/kg but a high ash content of between 20 and 50 per cent. Dewatering the sludge is expensive so that a balance is struck between dewatering and raising the calorific value of the wet sludge sufficiently to enable combustion to take place. Lower levels of dewatering can be allowed if supplementary fuel is used with sludge.

Normally, the sludge entering the fluidised bed would have been thickened and dewatered to some extent, using mechanical dewatering. The limiting dry solids content of a sludge which, if fed to the furnace, would require no supplementary fuel, is termed the 'autothermic solids content'. A typical bubbling fluidised bed for sewage sludge incineration is shown in Fig. 19.12. Figure 19.13 shows two sewage sludge fluidised bed incinerator designs. Figure 19.13(a) shows a simple mechanical dewatering system to produce 24 per cent of dry solids in the sludge. Such a sludge would not be autothermic and therefore supplementary fuel is required. Figure 19.13(b) shows that, after the mechanical dewatering stage, flue gas heat from the combustion process is used to dry the sludge and increase the solids content to 45 per cent. Supplementary fuel is therefore not required, except in the initial start-up of the process. The hot flue gases may be passed to a boiler system for energy recovery. Typical throughputs of sewage sludge are between 15,000 and 25,000 tons of dry solids per year.

Sewage sludge incinerators of the multiple-hearth type have also been used extensively. The incinerator has between 5 and 12 hearths and is designed to handle wastes with high moisture contents. The sludge, which requires dewatering to at least 15 per cent dry solids content, is fed to the top of the incinerator. The sludge moves down through the hearths by movement of the rabble arms which move the sludge alternatively through the centre and edge of the hearths. Flue gases pass up through the furnace. The upper hearths act as drying hearths utilising the hot flue gases from the burning sludge in the middle and lower hearths. Furnace temperatures are up to 900°C. Ash residues exit at the base of the incinerator.

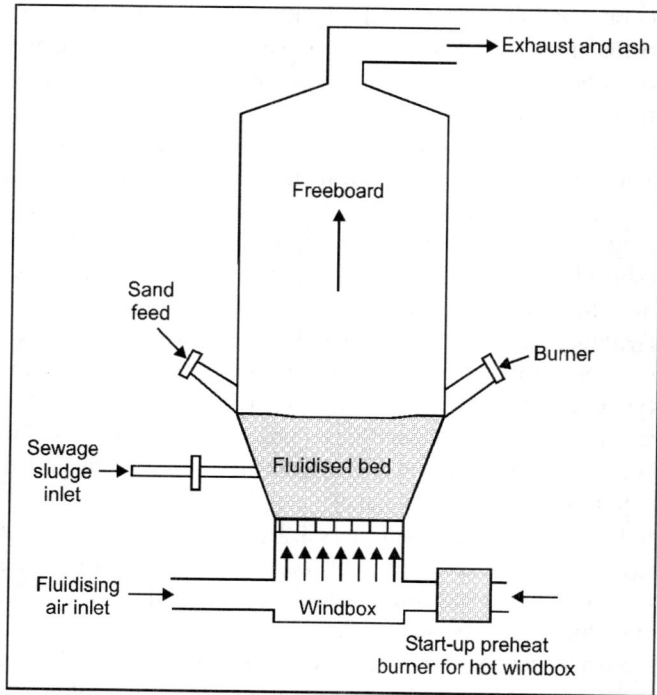

Fig. 19.12. Schematic diagram of a typical bubbling fluidised bed incinerator for sewage sludge.

Starved air incinerators

Starved air or pyrolytic incinerators are two-stage combustion-type incinerators which are widely used for clinical waste incineration and also for some industrial wastes. The two stages, consist of a pyrolytic stage and a combustion stage. Typical throughputs of waste are between 400 and 25,000 tons per year. The system is used mainly for solid waste. A typical two-stage solid waste incinerator with waste heat recovery system is shown in Fig. 19.14.

The advantages of the starved air incinerator are a more controlled combustion process leading to lower releases of volatile organic compounds and carbon monoxide. In addition, the low combustion air flow results in low entrainment of particulate in the flue gases, which also reduces other particulate-borne pollutants, such as heavy meals, dioxins and furans.

Pyrolysis is defined as the chemical decomposition of the waste by the action of heat. Heating the waste in an inert atmosphere produces a gas which, when ignited, is self-supporting in air. In practice the two-stage combustor relies on semi-pyrolysis, where the heat for the thermal decomposition or gasification of the waste is produced by substoichiometric combustion of the waste. The waste is combusted under substoichiometric conditions, i.e. where there is insufficient air to provide complete combustion and therefore there is a high proportion of the products of incomplete combustion which pass through to the second stage.

The two-stage process ensures that gas velocities are relatively low and particulate matter is largely retained in the first stage. However, with the stringent legislative requirements on emissions, the full range of gas clean-up systems are required to control other emissions such as acid gases, heavy metals and dioxins and furans.

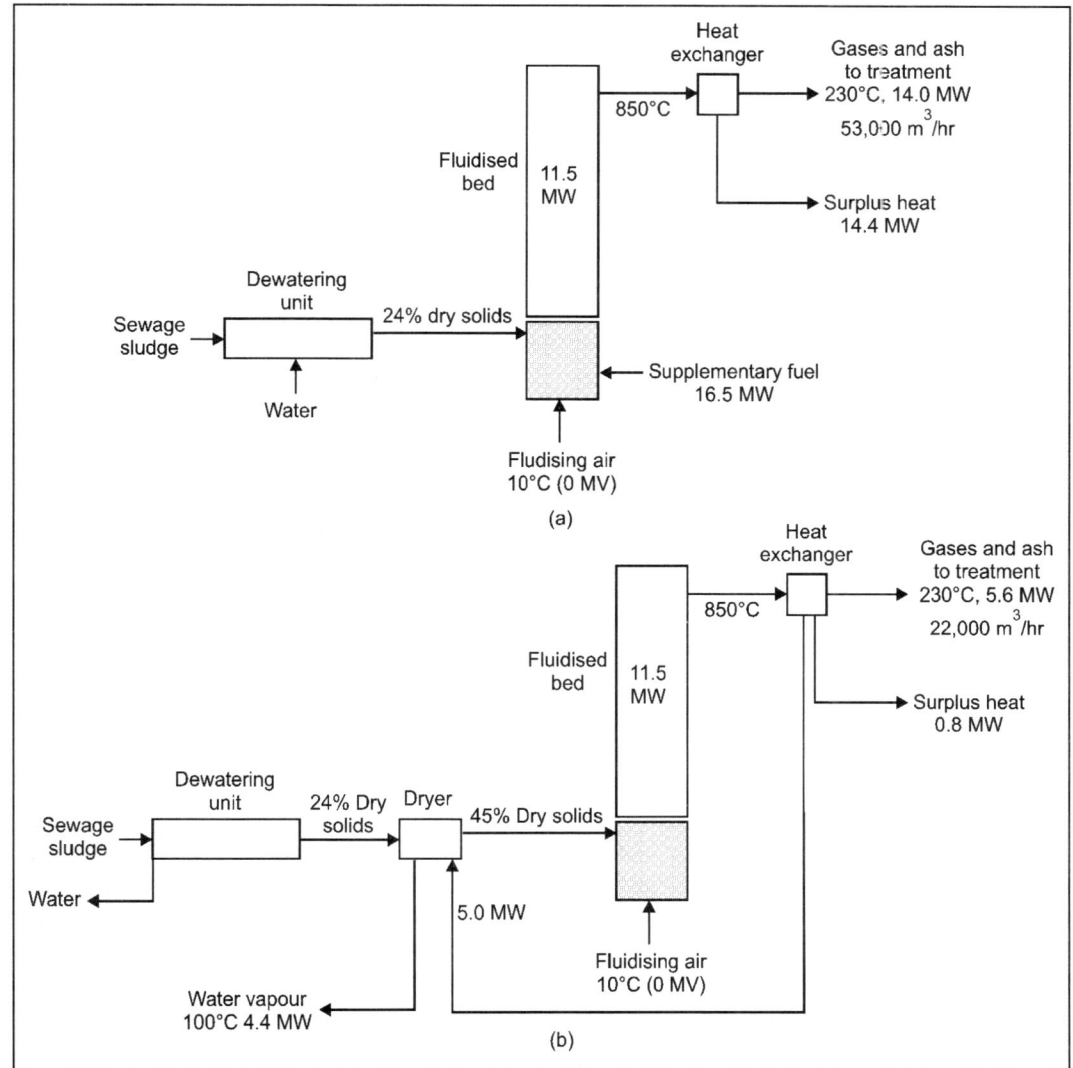

Fig. 19.13. Schematic diagrams of bubbling fluidised bed designs for incineration of sewage sludge.

The pyrolytic/gasification reactions that take place are numerous and complex. Figure 19.15 shows the reactions for a typical hydrocarbon of chemical composition $(CH_2)_n$. The substoichiometric conditions produce a reducing atmosphere within the primary chamber, and the heat generated breaks down the hydrocarbon in the pyrolysis zone into carbon and hydrogen. The carbon reacts with the CO_2 and H_2O, generated earlier, to give CO and H_2 which passes to the second-stage combustion zone where complete combustion takes place.

The temperature of the gases leaving the pyrolytic section are of the order of 700°–800°C since a high proportion of the heat generated is used in the endothermic pyrolytic process. These gases will then pass to the secondary section, where secondary excess air, approximately 200 per cent stoichiometric, is added to give a temperature of 1000°–1200°C, which completes the combustion process, combusting the hydrogen, carbon monoxide and hydrocarbons. The two-stage combustion process inhibits the formation of NO_x.

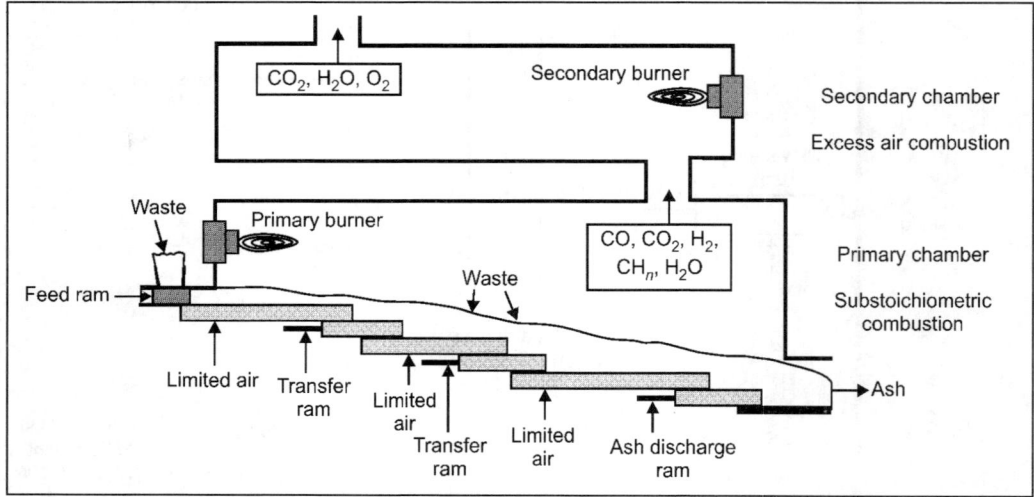

Fig. 19.14. Schematic diagram of a starved air two-stage incinerator.

The relatively long residence time, within the secondary chamber, plus the high temperature of over 1000°C, will destroy any dioxins and PCBs that are contained in the secondary gases. This does not preclude any *de novo* formation of dioxins and consequently clean-up procedures of, for example, additive activated carbon and/or lime plus a fabric filter, are also required. In practical terms, the smaller two-stage incinerator of up to 0.75 tons/hr capacity tend to be vertical units operating on a batch basis. The larger units 0.75–5 tons/hr are designed as horizontal units and have automatic feeding and deashing.

In most cases it is not necessary to pretreat the waste prior to loading onto the furnace. Charging is normally via a hydraulic ram system which pushes the waste, via a refractory-lined guillotine door, to the pyrolysis chamber. The primary chamber process consists of combustion, partial combustion, pyrolysis and drying. Primary air is fed through the grate to give even combustion at the base of the waste. The residence time of the waste in the primary chamber will be dependent on the hearth area and the characteristics of the waste, and should produce an almost carbon-free ash, the time may be anything from 6–12 hr. The ash is finally discharged at the rear of the incinerator into a water trough.

The gases entering the secondary chamber have a sufficient calorific value to be self-sustaining in combustion. Secondary air is introduced to provide the excess air conditions with a high degree of turbulence to create sufficient mixing in order to sustain combustion without the use of support fuel at typical operating temperatures of between 1000 and 1200°C. Long residence times in the secondary chamber also enable complete burnout of the combustible gases, vapours, tars and soot. Auxiliary burners are also employed for the initial start-up and then combustion is sustained by the gases from the pyrolytic stage, when the burner may be switched off.

Rotary kiln incinerators

The rotary kiln is a two-stage incineration type, but the first stage is usually operated in the oxidative mode, i.e. with about 50–200 per cent excess air, rather than the semi-pyrolytic mode found in starved air incinerators. Typical throughputs of waste in rotary kiln incinerators are of the order of 4000–50,000 tons per year.

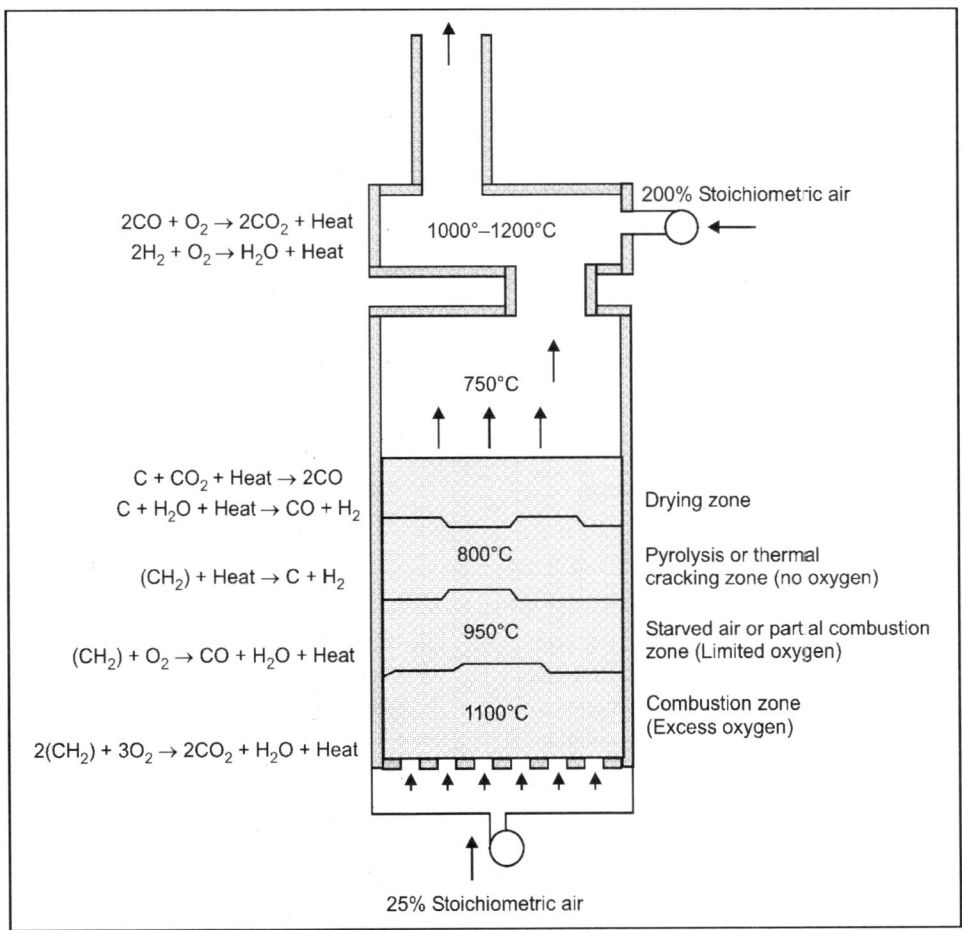

Fig. 19.15. Combustion reactions for a typical hydrocarbon in a starved air, two-stage incinerator.

Rotary kilns have been used for a wide variety of wastes, including municipal solid waste, sewage sludge, industrial waste and hazardous waste and for clean-up of contaminated soils. However, they are most common for the treatment of hazardous, clinical and industrial wastes, where in some cases whole drums of waste are fed to the rotary kiln to be completely destroyed. Figure 19.16 shows a schematic diagram of a rotary kiln incinerator.

The rotary kiln is the primary chamber, consisting of an inclined cylinder lined with ceramic material which rotates on rollers at rates which can vary between two revolutions per minute to six revolutions per hour, depending on the type of waste and type of rotary kiln.

The size of the rotary kiln can be 1–6 m in diameter and 4–20 m in length. The kiln is rotated by a series of rollers on which the kiln is located. The waste is fed to the front end and ignited by a burner, the combusting wastes are tumbled and agitated by the rotation of the kiln and move down the kiln to reach the end as ash. Residence times of the waste in the rotary kiln are generally more than 30 minutes. Internal baffles may be used to increase the mixing and turning of the waste. The kiln typically operates at temperatures around 1200°C when incinerating hazardous wastes. The 'slagging type' of rotary kiln operates at temperatures up to 1500°C.

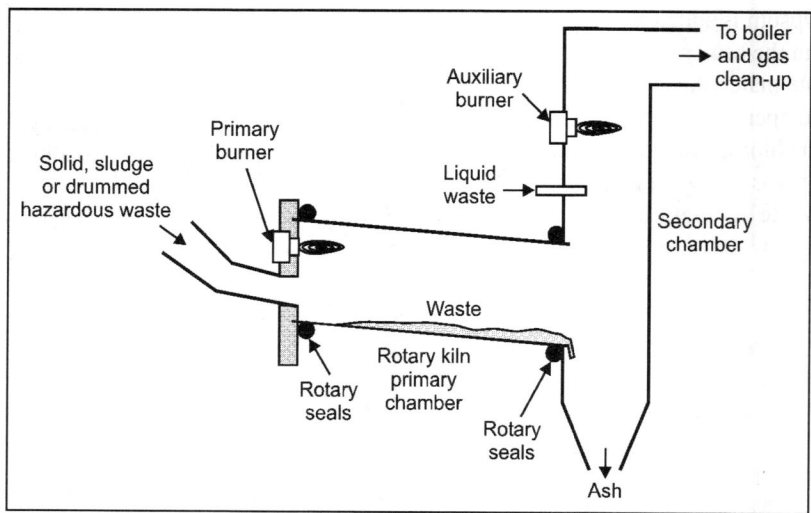

Fig. 19.16. Schematic diagram of a rotary kiln incinerator.

The high temperature of the 'slagging type' rotary kiln allows the formation of a molten slag of the ash and whole drums of waste can be incinerated as the metal drums will melt. The presence of the molten slag absorbs particulate matter, including heavy metals. The ash or molten slag exits the kiln into a quench pit. The molten ash forms a glasslike material, which is less susceptible to leaching of the dissolved metals.

The gases from the primary rotary kiln pass to the secondary chamber where excess air conditions with auxiliary burners serve to completely burn out the combustible gases, vapours, tars and soot. A secondary chamber is particularly necessary for hazardous wastes where the time, temperature and turbulence may be insufficient to guarantee the complete combustion of all the organic components of the waste in the primary chamber. Typical temperatures in the secondary chamber would be up to 1400°C, with residence times of between 1 and 3 s and up to 200 per cent excess air levels. The rotary kiln combines a long residence time plus a high temperature, which enables the complete combustion of complex hazardous wastes. To protect the rotary kiln and secondary chamber from the high temperatures of combustion, the walls are lined with refractory ceramic material.

Cement kilns used for the production of cement are used in some countries to dispose of a variety of wastes including municipal solid waste, industrial waste, tyres and hazardous wastes. Substitute liquid fuel or secondary liquid fuel (SLF) used in cement kilns is produced by blending organic wastes. The main constituents of SLF include solvents, working fluids (oils, lubricants, etc.) contaminated fuels, organic sludge (e.g. food industry wastes) and other organic chemical products. The wastes are derived as solvent wastes produced by the chemical industry, but also include some aqueous wastes and wastes containing high concentrations of halogen or metal contents. The waste is blended with other fuels in cement kilns which utilise rotary kiln technology. The length of the rotary kiln cylinder for cement manufacture is exceedingly long, typically up to 250 m in length and 4 m diameter and is lined with alumina bricks. Normally coal or oil is used the fuel but has been supplemented w th waste. Chalk or limestone, plus clay or shale are mixed with water to form a slurry which is passed through the high-temperature furnace to form cement clinker. After processing through the kiln, the cement clinker is ground and gypsum is added to produce cement. The combustion temperatures within the cement kiln are very high in the cement kiln, typically more than 1400°C. The process is very energy intensive and the use of waste material offsets the costs of fuel.

The high temperatures and long residence times used in the process serve to destroy the waste. In addition, when chlorinated or fluorinated wastes are combusted, the large mass of alkaline clinker from the process absorbs and neutralises the acidic stack gases.

The EC Waste Incineration Directive covers not only the incineration of waste but also the coincineration of waste. A co-incineration plant is any plant whose main purpose is the generation of energy or production of material products and which uses wastes as a regular or additional fuel, or in which waste is thermally treated for the purpose of disposal. The emission-limit values to air for co-incineration of waste are set down in the directive. The directive states that the co-incineration of waste in plants not primarily intended to incinerate waste should not be allowed to cause higher emissions of polluting substances in that part of the exhaust-gas volume resulting from such co-incineration, than those plants permitted for dedicated incineration. For the co-incineration of waste, the air emission-limit value is determined by a mixing rule formula.

The use of tyres in cement kilns has been shown to reduce the emissions of NO_x, through the formation of reducing zones when the tyres are being burned. It has also been proposed that the use of secondary liquid fuel (SLF), at a fuel input level of 40 per cent, reduced NO_x emission levels by 50 per cent. Similar reductions in NO_x have been reported when plastic waste has been used in cement kilns. The influence on the emissions of other pollutants from cement kilns using waste has shown, for example, no significant difference in SO_2 emissions whether coal or coal plus SLF was combusted in the cement kiln. The influence of using waste in cement kilns on heavy metal emissions is not clear. It has been suggested that the efficiency of the cement kiln in retaining the heavy metals present in the waste, binds the metals to the cement or cement kiln dust. Consequently, no significant increase in heavy metal emissions may be expected. However, Sarofim, and Guo and Eckert report that, when waste-derived fuels are used in cement kilns, there can be an increase in the emissions of certain metals. Because of the high operational temperature and long residence times of cement kilns, their destruction efficiency for organic compounds including PCDD, PCDF and PCBs present in the fuel, waste or raw feed material is high, of the order of >99.995 per cent.

The emissions of PCDD/PCDF from cement kilns using conventional fuel and waste fuels such as tyres, refuse-derived fuel and solvent derived fuels, have been reviewed. In general it was concluded that the ranges of PCDD/PCDF emission concentration resulting from the use of conventional fuel, such as coal and petroleum coke, overlap with the ranges obtained with the use of secondary waste-derived fuels and raw materials, regardless of the type of secondary fuel. It was also shown that, irrespective of which fuel was used in the cement kiln, emissions were below the target emission standard of 0.1 ng/m^3 generally applied throughout Europe, to regulate emissions of dioxins and furans from incinerators. Examination of cement kiln emissions, when using waste tyres as fuel, in relation to health-impact assessment, ambient monitoring, soil sampling and air-quality modelling, have concluded that, in general, the use of tyres as substitute fuel does not increase environmental impacts from the cement-making process.

Liquid and gaseous waste incinerators

Liquid and gaseous waste incinerators pass the waste into a burner which mixes the (combustible) waste with air to form a flame zone which burns the waste. Figure 19.17 shows a typical liquid waste burner. Supplementary fuel may be required, depending on the calorific value of the waste or else the liquid waste is pumped directly into a flame generated by the burner, fired by a conventional fuel.

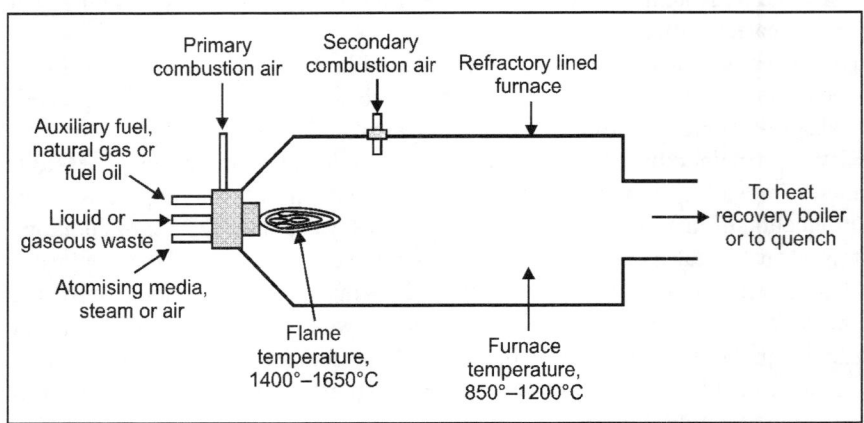

Fig. 19.17. Typical liquid and gaseous waste incinerator.

The conventional fuel, such as natural gas or fuel oil, helps to maintain steady combustion conditions. The flame is fired into a ceramically lined combustion chamber which radiates heat back into the exhaust gases, thus providing an extended hot zone to completely burnout the products from the combustion of the waste. Very high temperatures occur in the flame, of the order of 1400°–1650°C and furnace chamber temperatures are between 820° and 1200°C. The chamber may be horizontal or vertical. Liquid waste incinerators are used extensively for the combustion of hazardous wastes. The key section of the incinerator is the burner, which essentially serves to atomise the waste to form a fine spray of droplets, and vapour which ignites to form the flame. Several different designs of burner nozzle exist to cope with the wide range of properties found with liquid wastes and sludges. The liquid waste is pumped under high pressure through the burner nozzle, which produces a fine spray of atomised droplets of size typically between 10 and 150 μm. The smaller the droplet size, the easier vapourisation becomes and consequently the burnout of each droplet takes place in a much shorter time.

Gaseous waste incinerators operate on a similar system to liquid waste incinerators but the difficulties of producing a fine spray or vapour for combustion are already overcome. Gaseous wastes usually consist of organic hydrocarbons which are combustible. The gases or vapours may be of low concentration and consequently are not autothermic, and therefore the gases or vapours are passed into a burner with either the supplementary fuel gas or combustion air, or may be passed directly into the flame zone. The combustion chamber provides a long residence time for complete burnout of the gaseous waste.

SECTION VI

Composting

Chapter 20

Pretreatment of Municipal Solid Waste by Windrow Composting

INTRODUCTION

Cost of incineration and landfill are soaring with energy and land shortages and the present way of dumping the municipal refuse without proper pretreatment will stem many environmental problems. Management of solid wastes is a problem of increasing concern throughout the world. The organic portion of the solid waste however could be utilised in a very profitable way by composting or by using vermicomposting. *Vermes* is Latin word for worms and vermicomposting is essentially composting with worms. So in such case vermicomposting aids in the disposal by improving the physical quantities of waste. Vermistabilisation represents a technology that is environmentally sound and relatively new technology that can be classified as an innovative and alternative technology. Earthworms have been used for waste stabilisation for many years, especially in Southeast Asia and some third world countries mainly, Canada, United States, Australia and France. Earthworms, they are the unheralded soldiers of the soil. Among the primary benefits of having earthworms in the soil are that they aerate it, break it up for easier access by plant roots, help the soil hold more water, clean up dead organic matter by eating it and turning it into the world's best plant food, (whether natural or chemical). They contain 60 per cent protein and are raised as a very high-grade animal feed and much more. Earthworms are one of the major soil macroinvertebrates and are known for their contributions to soil formation and turnover with their widespread global distribution. Darwin first highlighted the role of earthworm. Later in the passage of time, different experimental studies are carried out in studying the role of earthworm in maintaining the soil fertility and in the degradation of the organic matter present in the soil. Different laboratories have tried the possibility of utilising earthworms to break down the organic waste, which has been the major organic pollution.

During the composting process, micro-organisms decompose organic compounds, which consist of carbohydrates, sugar, proteins, fats, cellulose and lignin. Carbohydrates are more easily decomposed whereas lignin is more resistance to decomposition. Many factors affect the composting process. Aerobic micro-organisms need oxygen, water and nutrients for their metabolism and cell synthesis. As a result of microbial activity heat is liberated and, if contained within the composting mass, the temperature rises. Temperature increases through the mesophilic phase into a thermophilic phase and then back into the mesophilic phase. During the course of these transitions, the microbial population changes, thereby affecting the rate of organic matter decomposition.

Earthworms can stabilise biodegradable organic matter, such as animal and vegetable and municipal sludge like the conventional composting. The worms maintain aerobic conditions in the mixture, ingest solids, and convert a portion of the organic into worm biomass and to respiration products, and expel the remaining partially stabilised matter as discrete material (castings). The worms and the micro-organisms act symbiotically to accelerate and enhance the decomposition of the organic matter.

This chapter is focused to investigate:

1. The most conducive method towards the pretreatment of municipal solid waste by aerobic composting and vermicomposting.
3. The degradation of organic portion in a mixed municipal solid waste by windrow composting and vermicomposting.
4. The best method involved in the separation of the organic parts from the commingled waste using both windrow composting and the vermicomposting process under different conditions.
5. The optimum operating condition and other parameters for best post-sorting of the pretreated municipal solid waste.
6. To find an optimum condition in the sorting efficiency of the mechanical shaker.

Vermicomposting is the conversion of biodegradable garbage into a high quality chemical free biofertiliser with the aid of earthworms. Whereas, the composting is the other way round where the organic part of the refuse is consumed by a series of successive bacteria according to the heat of the system.

Earthworms have from time immemorial played a key role in soil biology by serving as versatile natural bioreactors to harness and destroy soil pathogens, thus converting organic wastes into valuable biofertilisers, enzymes, growth hormones and proteinaceous worm biomass. The worms do it by feeding voraciously on all biodegradable refuse such as leaves, paper (nonaromatic), kitchen waste, vegetable refuse. It then burrows deep into the soil, positioning its castings towards the surface of the soil thereby enriching the soil with a predigested, easy to assimilate biofertiliser that is now rich with NPK. So when looking for a fertiliser for a farm or garden it would do well if people would consider the revolutionary vermicompost as an option. Certain types of earthworms ingest, digest, and excrete vermicompost with excellent nutrient content. Ingestion ensures the sorting out of only organic matter while the digestion accelerates the maturing process.

Excretion ensures the grading of the vermicompost as opposed to any inorganic matter, which may be existing in the waste and not concerned with the biological activity in the earthworm gut.

During the composting process, micro-organisms decompose organic compounds, which consist of carbohydrates, sugar, proteins, fats, cellulose and lignin. Carbohydrates are more easily decomposed whereas lignin is more resistance to decomposition. Many factors affect the composting process. Aerobic micro-organisms need oxygen, water and nutrients for their metabolism and cell synthesis. As a result of microbial activity heat is liberated and, if contained within the composting mass, the temperature rises. Temperature increases through the mesophilic phase into a thermophilic phase and then back in to the mesophilic phase. During the course of these transitions, the microbial population changes, thereby affecting the rate of organic matter decomposition.

IMPORTANT DEFINITIONS

Vermicomposting

The method of employing earthworms in reducing the organic matter present in the waste is called as the vermicomposting. Vermicomposting, also known as worm composting, is simply the way redworms transform decaying organic matter into worm castings. Vermicomposting is the process involved in the degradation of organic waste into useful components by using earthworms. It is all-together a natural system in which the earthworms play their major roles in degrading the organic portion of the waste. The use of earthworm in sludge management is called as vermicomposting or vermistabilisation.

Composting

Composting is the biological decomposition of organic matter under controlled aerobic condition. Composting in a way is one such method by which we can practically and economically use those waste streams dominated by organic refuse. As stated by Roger that there is no universally accepted definition of composting. According to him, it is the biological decomposition and stabilisation of organic substrates, under conditions that allow development of thermophilic temperatures as a result of biologically produced heat, to produce a final product that is stable, free of pathogens and plant seeds, and can be beneficially applied to land. Or simply it can be defined as the biological reduction of the organic waste to humus. In brief we can consider composting as a way of stabilising the waste.

Types of Vermicomposting Systems

Windrows

This system takes into account the availability of large land and other appropriate technology for operating the whole system. These systems are extensively being used for both in the open and under cover.

Wedge system

This is a modified type of windrow system where one can easily harvest the vermicompost without disturbing the earthworms. In this system organic materials are applied in layers against a finished windrow at a 450 angle.

Bed and bin system

Here in this systems bins are used to breed and harvest the vermicompost and also in some case beds are made on the ground for the same purpose. This method is labour intensive but is much easier to handle and is widely used.

Reactor system

Reactor systems have raised beds with mesh bottoms. Feedstocks are added daily in layers on top of the mesh or grate. Finished vermicompost is harvested by scraping a thin layer from just above the grate, and then it falls into a chamber below. These systems can be relatively simple and manually operated or fully automated with temperature and moisture controls. For maximum efficiency, they should be under cover.

Factors that may be considered for selecting the appropriate vermicomposting technology for a project include: Amount of feedstock to be processed; funding available; site and space restrictions; climate and weather; state and local regulatory restrictions; facilities and equipment on hand; and availability of low-cost labours, etc.

Breeding of Earthworms

Worms will be breeded by setting up a vermibed in a suitable container or a site under a shade, in an area on upland or an elevated level to prevent water stagnation in the pit.

Bedding preparation

For the preparation of the bedding we can have various choice regarding the availability of the bedding materials. Figure 20.1 shows the basic requirements for a vermibed. In the setting up of an ideal vermibed one should have the following layers, basal layer comprising of broken bricks or pebbles to a small extent followed by coarse sand to a thickness of 6–7.5 cm, this layer is to ensure proper drainage. This is topped by a layer of loamy soil up to a height of not less than 15 cm after it is moistened.

Now we can inoculate worms here in this layer. Over this small lump of cattle dung are scattered over the soil and this is covered by layer of hay up to a height of 10 cm. Broad leaves finally cover the unit and a net can be used to prevent the intrusion of any unwanted worms or other predators.

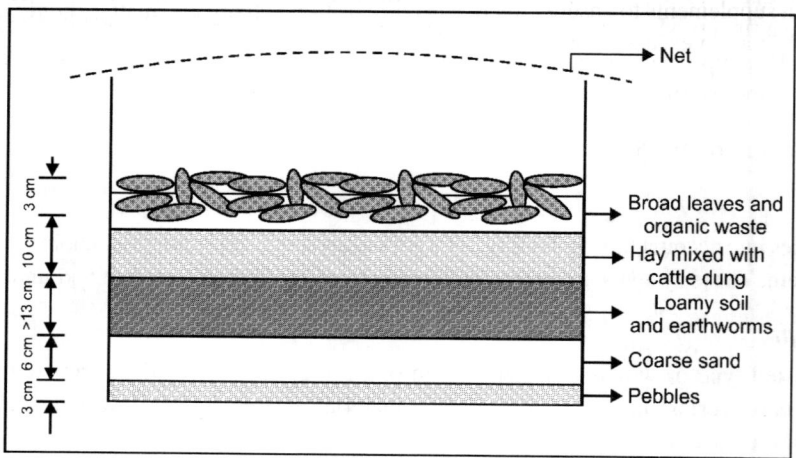

Fig. 20.1. Setting up of a vermibed.

An ideal environment for earthworms

The following are the environmental conditions, which are vital and may affect the breeding, cocoon production and hatching of young earthworms. They are lots of literature describing the various limiting parameters towards a successful breeding.

Temperature

In vermicomposting, temperatures are generally below 35°C. Most worm species used in vermicomposting require moderate temperatures from 10°–35°C. While tolerances and preferences vary from species to species, temperature requirements are generally pretty similar. The majority of vermicomposting worms can tolerate temperatures ranging from 50° to 85°F but decrease activity as temperatures move toward the extremes. Most species prefer temperatures within roughly ten degrees of 70°F. Earthworms tolerate cold and moist conditions far better than they can hot and dry conditions.

Moisture

Earthworm requires plenty of moisture for growth and survival, they need generally moisture at the range from 60–75 per cent. The soil should not be too wet else it may create an anaerobic condition which may drive the earthworms from the bed. It is very important to moisten the dry bedding material before putting them in the bin, so that the overall moisture level is well balanced.

pH

Although studies have suggested that worms perform best in neutral pH. It has been recorded by Edward that different species of earthworms have their own pH sensitivity and generally most of them can survive at the pH range between 4.5–9 per cent. The alteration of pH in the bedding is due to the fragmentation of the organic matter under series of chemical reaction.

Feed

The first step in starting a vermicomposting unit is to arrange for regular input of feed materials for the earthworms. These can be in the form of a nitrogen rich material like goat manure cattle dung and pig manure. When the material with high carbon content is used with C/N ratio exceeding 40:1, it is advisable to add nitrogen supplements to ensure effective decomposition. All organic matter should be added only as a limited layer as an excess of the former may generate heat. From the waste eaten up by the worms 5–10 per cent are being assimilated in their body and the rest are being excreted in the form of a nutrient rich cast.

Stimulants

They are no known stimulants which will force the earthworms to breed but fairly fresh manure or other nitrogen rich green organic matter seems to be the best stimulant to rapid breeding.

Biology of Earthworm

The earthworm is a tube shaped, segmented, invertebrate. Lacking bones or cartilage, its body holds its shape because it's full of a thick mucous-like liquid called coelomic fluid. If one were to view a cross section of the worm body it would resemble a target, with the center representing the internal organs and the outer circle representing the skin or dermal layer. The cavity between the internal organs and dermal layer is filled with the coelomic fluid. The pressure of this fluid against the dermal layer gives the worm its shape.

Classification of earthworm

According to their feeding habits, earthworms are classified into detritivores and goephages. Detritivores feed near the soil surface. They feed mainly on the plant litter or dead roots and other plant debris in the soil. These worms comprise the epigeic and the anecic forms. Geophagous worms, feeding deeper beneath the surface ingest large quantities of organically rich soil. These are generally called as humus feeders and comprise of endogeic earthworms.

Epigeic are surface dwellers serving as efficient agents in fragmentation of organic matters on the soil surface. Whereas the anecics feed on the organic matter mixed with soil. Endogeic earthworms live deep within the soil and derive their nutrition from the organically rich soil they ingest. The distribution of earthworm in the soil is influenced by several factors of which are soil texture and aeration, temperature, moisture, pH, inorganic salts and the organic matter.

Types of earthworms

In general we have six common types of earthworms:
1. The native night crawler or *Lumbricus terrestris*.
2. The common field worm or *Helodrilus caliginosus*.
3. The green worm or *Helodrilus chloroticus*.
4. The manure worm or *Eisenia foetida*.

5. The slim earthworm or *Diplocardia verrucusa*.
6. The red worm or *Lumbricus rubellus*.

Reproduction

Earthworms are hermaphrodite, which means each individual has its own male and female reproductive organs. In sexually matured earthworms the body wall of the forward segment is thickened by gland cells, forming a more or less conspicuous girdle known as clitellum. During reproduction they exchange sperm at a point just above the clitellum, the swollen band encircling the worms body. After exchanging sperm the worms move apart and secrete a thick mucous around the clitellum, which forms a jellylike band. Once the band slips off the worm's body, the ends close, forming a cocoon with sperm and eggs inside where fertilisation takes place. Both worms continue to generate cocoons until all of the sperm received from the mate is used. The number of young worms inside the cocoon and gestation time varies with worm species and environmental factors. While most species reproduce with the help of a mate, they can become self-fertile under stressed conditions.

Life cycle

In their natural habitat, earthworms follow a well-defined yearly cycle. This cycle is considered to starting at the autumn season. Life cycle of an earthworm is divided into four major phases (Fig. 20.2).

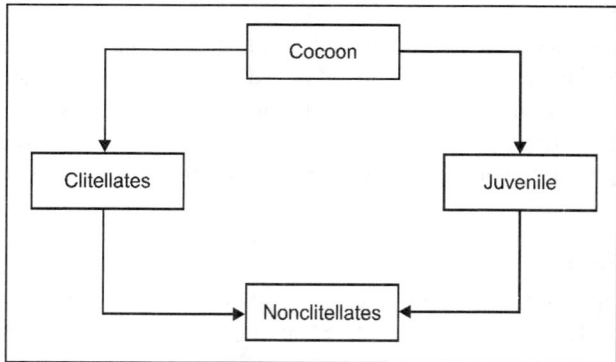

Fig. 20.2. Schematic diagram representing the life cycle of an earthworm.

Cocoon phase

Many cocoons were produced when the temperature rises up and the greatest number occurred between May and July. Thereafter, the numbers of cocoons produced decreased quite rapidly with falling temperature. Lofty reported that fewest cocoons were produced during winter and there was a temperature threshold of about 3°C, below which no cocoons were produced.

Juvenile phase

On hatching the worms measure to an average up to 0.8–1.5 mm in length and weigh around 7 mg. Their length gradually increases to about 4 cm and may latter weigh up to 150 mg.

Nonclitellates

Young earthworms whose *clitellum* are yet to develop are grouped into this nonclitellates here the young worms are very active at this stage and will weigh up to from 150 mg to 450 mg.

Clitellates

Clitellates are the mature and adult worms. Clitellates have the potentials for reproduction, the worms at this stage will appear bit darker in their colour due to the pigmentation of the epithelial cells. Here in this stage of life the body wall of the forward cell is thickened by gland cells, forming a conspicuous girdle known as *clitellum*.

At the time when many earthworms are young with favourable climatic condition, they become very active. This high level of physical activeness normally continues throughout the year except during a hot summer where they become inactive.

Thus, summer is the period when there is a sharp decline in the activity of the earthworm. This particular cycle might be considered as ending by the late summer as shown in the Fig 20.3. Saayman reported that the constant temperature and controlled temperature favours the development of clitella and cocoon production. He further noticed that the cocoon also hatched better in the controlled temperature (25°C) as compared to the fluctuating temperature 25°–37°C. Amongst the epegeic species the *E. foetida* is the one with most adaptive nature. It can survive to a wide range of temperature from 5° to 43°C. It is also noted by May, that the growth rate of the earthworm can also be accelerated by either addition of a new bedding materials and also by making some change in the worm diet, so that it allows the worm to eat more.

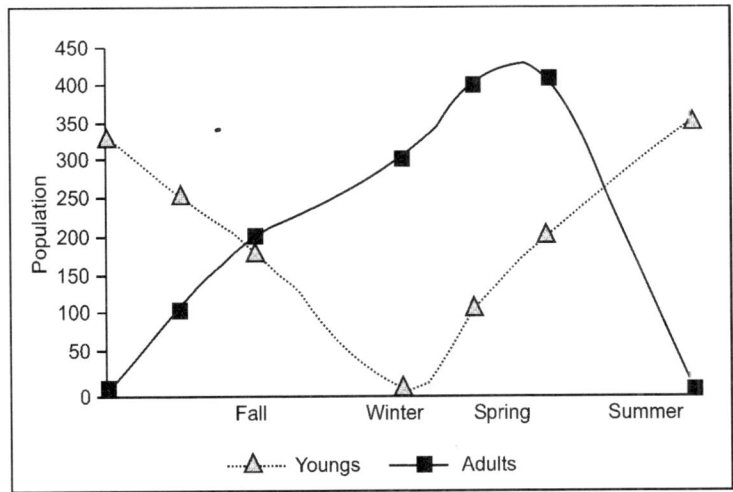

Fig. 20.3. Annual cycle of earthworm in natural habitat.

Role of Earthworm in Organic Matter Recycling

The rate of decomposition depends on the type of the litter. If physical conditions are suitable then the number of earthworms increase until the food becomes a limiting factor. Even when the organic matter such as dung is present in sufficient or are freely available still then the worms ingest large quantity of mineral soils. The increased interest in diverting organic wastes from landfill has raised concerns about the need to ensure that the end products are safe to use and meet minimum quality standards. One such concern is that strategies are adopted that minimises the risk of spreading pests, plant diseases (pathogens) and weeds in recycling organic wastes.

Recycling of organic waste

Fragmentation and breakdown

The rate of organic matter breakdown depends mainly on the type of litter. Very soft plant and animal residues may be decomposed by the microflora but much organic matter, particularly the tougher plant leaves, stems and root material does not break down without first being disintegrated by the soil animals, mainly earthworm. Earthworms have an important role in this initial process of the cycling of organic matter. Soils with only few earthworms have a well-developed layer of undecomposed organic matter lying on the soil surface. Many sorts of leaves are not acceptable to the earthworms when they first fall on the ground, but require a period of weathering before they become palatable. It is believed that this weathering leaches the water-soluble polyphenols from the leaves. These tiny creatures are responsible for translocating the accumulated organic debris from the soil surface the subsurface layers and during this process much of the organic materials is ingested, macerated and excreted. Earthworms also contribute several kinds of nutrients in the form of nitrogenous waste.

Consumption and Humification

Earthworms can consume more organic matter from the soil surface than all of the other small soil animals together. Earthworms seem to consume very large amounts of litter, and the amount they turn over seems to be more dependent on the total amount of suitable organic matter available. If the physical soil conditions are suitable than the number of earthworms increase until the food becomes a limiting factor. They pass a mixture of organic and an inorganic matter through their guts when feeding or burrowing. Edward reported that the smaller earthworms that feed on the litters on the woodland produce cast that are almost entirely fragmented litter.

Whereas the larger species consume large proportion of soil, and there is less organic matter in their casts. Crossley and others calculated the rate of through put of soil by *Octolasim* sp. in a culture of soil tagged with radioactive caesium (^{137}Cs). They found that the soil passed through the worms gut at the rate of about 86 mg per day per worm, equivalent to the 28.8 per cent of live weight of the earthworm. The final process in organic matter decomposition is the humification, in which the large organic particles breakdown into a complex amorphous colloid containing phenolic materials. Only about one fourth of the organic matter becomes converted to humus. The major contribution of the earthworms seems to be in breaking up organic matter, combining it with soil particles and enhancing microbial activity when humification is well advanced. Nevertheless, earthworms are also important in mixing the humified material into soil.

Nitrogen mineralisation

Earth worms greatly increases the soil fertility, and at least part of this must be due to the increased amounts of mineralised nitrogen that they make available for the plant growth. There have been reports of increase in the amount of nitrogen in which the earthworms are reared. This may be due to the decay of the bodies of dead earthworms. Since the body of the earthworm are rich in proteins. Govindan reported that earthworm body contains 65 per cent protein, 14 per cent fats, 14 per cent carbohydrates and 3 per cent ash. Similarly Ronald reported that 72 per cent of the dry weight of an earthworm is protein and that the death of an earthworm will releases up to 0.01 gram of nitrate in the soil. Earthworms consume large amount of plant organic matter that contains considerable quantities of nitrogen, and much of this is returned to the soil in their excretions. Smith reported that nitrogen mineralisation was greater in the presence of earthworms, and this mineral nitrogen was retained in nitrate form.

Effects on the C/N ratio

Plants root in general cannot assimilate the mineral nitrogen unless the ratio is in the order of 20:1 or lower. Therefore the ratio of carbon to nitrogen is important for the proper growth of any plant. Earthworms help to lower the carbon to nitrogen ratio of fresh organic matter by consuming the matter, breaking it down and using the carbon for energy during respiration. To assess the role of earthworm in lowering the C/N ratio, the consumption of the carbon must be measured, and this can be done approximately, by measuring the respiration. So there is always the disadvantages to this method, since the laboratory test will not always reflect the actual situation. Daniel and others have done an experiment in vermicomposting of selected leaf litter and cowdung mixtures (1:1) and shown a substantial variation in the Electrical conductivity, NPK, organic carbon and C/N ratio than worm unworked compost. The C/N ratio also showed here a remarkable reduction in the worm worked vermicompost than the worm unworked compost. Such type of reduction has been brought about by the respiratory activity and microflora present in the system.

Composition of Municipal Solid Waste

With the dawn of the consumerist culture, the composition of the waste will change drastically and pose an even bigger problem. The solid waste so generated can be of two types; biodegradable or organic and nonbiodegradable or the inorganic. The organic waste includes the mainly kitchen waste, straw, hay, paper and animal excreta. The inorganic portion of the waste is generally dominated by the ash, stone, cinders, plastics, rubber and ferrous and nonferrous metals. The residential and the commercial portion make up to about 50 to 70 per cent of the total MSW generated in a community. The actual percentage distribution will depend on the:

1. Extent of the construction and demolition activities.
2. Extent of the municipal service provided.
3. Types of water and waste-water treatment facilities that are used.

The wide variation in the special waste category is due to the fact that in many communities yard waste are collected separately. Typical values for paper in the residential MSW range for about 20 to 40 per cent. For the remaining components of the waste can vary to about 40 to 100 per cent. It is being noted that percentage of food waste is high because most vegetables are not pretrimmed.

Municipal waste

Municipal waste is distinguished into municipal refuse and sewage sludge. The municipal refuse also called as the municipal garbage, is compose of discarded material by the people in the home and in industry. It is composed of paper, plastics, food and paints. An average composition of municipal refuse is given in the Table 20.1.

Table 20.1. Components of municipal refuse.

Components	%
Paper	58.5
Food residue	9.2
Garden refuse	10.1
Metals	7.5
Glass, ceramics and ash	8.5
Miscellaneous	5.9

Here the paper makes the largest amount of this refuse, whereas the plastic comprises only very few and is considered as a miscellaneous item. The composition of MSW of a developing country and of industrialised countries is different.

Physical properties

The important physical properties of MSW include density (sometimes referred to as specific weight), moisture content, particle size and distribution, field capacity, and porosity. Although talking about MSW, it is important to note that the same fundamentals apply to all types of solid wastes.

Density

This is the weight per unit volume and is expressed as kg/m^3. Density varies because of the large variety of waste constituents, the degree of compaction, the state of decomposition, and in landfills because of the amount of daily cover and the total depth of waste. Inert wastes such as construction and demolition materials may have higher densities, and density can change as in landfills where the formation of landfill gas and decomposition may bring about significant mass loss. Density is important because it is needed to assess the total mass and volume of waste, which must be managed. Density varies not only because of the type of treatment it gets (collection vs compaction, etc.) but also because of geographic location, season, and length of time in storage.

Moisture content

The most commonly used method of expressing moisture content is as a percentage of the wet weight of material. Moisture content is important in regards to density, compaction, the role moisture plays in decomposition processes, the flushing of inorganic components, and the use of MSW in incinerators. Pretreatment of waste to ensure uniform moisture content can be carried out prior to landfill disposal (Table 20.2).

Table 20.2. Typical moisture contents of wastes.

Type of waste	Moisture content range (%)	Moisture content typical (%)
Residential		
Food wastes (mixed)	50–80	70
Paper	4–10	6
Plastics	1–4	2
Yard wastes	30–80	60
Glass	1–4	2
Commercial		
Food wastes	50–80	70
Rubbish (mixed)	10–25	15
Construction and demolition		
Mixed demolition combustibles	4–15	8
Mixed construction combustibles	4–15	8
Industrial		
Chemical sludge (wet)	75–99	80
Sawdust	10–40	20
Wood (mixed)	30–60	35

Particle size and distribution

The size and distribution of the components of wastes are important for the recovery of materials, especially when mechanical means are used, such as trommel screens and magnetic separators. For example, ferrous items, which are of a large size, may be to heavy to be separated by a magnetic belt or drum system.

Field capacity

The field capacity of MSW is the total amount of moisture which can be retained in a waste sample subject to gravitational pull. It is a critical measure because water in excess of field capacity will form leachate, and leachate can be a major problem in landfills as we will discuss next week. Field capacity varies with the degree of applied pressure and the state of decomposition of the wastes, but typical values for uncompacted commingled wastes from residential and commercial sources are in the range of 50–60 per cent.

Permeability of compacted wastes

The hydraulic conductivity of compacted wastes is an important physical property because it governs the movement of liquids and gases in a landfill. Permeability depends on the other properties of the solid material include pore size distribution, surface area and porosity.

Chemical properties

Knowledge of the chemical composition of waste is important to help evaluate alternative processing and recovery options. This is especially important where wastes are burned for energy recovery, in which case the four most important properties are proximate analysis, fusing point of ash, elemental analysis, and energy content. Elemental analysis is also important in determining nutrient availability.

Proximate analysis

Proximate analysis includes four tests—loss of moisture when heated to 105°C for 1 hour; volatile combustible matter (loss on ignition); fixed carbon; and ash (weight of residue after combustion).

Transformation of waste

Transformations of waste can occur through the intervention of people or by natural phenomena.

Physical transformation

These include component separation, mechanical volume reduction, and mechanical size reduction. Component separation is used to describe the separation processes (manual and/or mechanical) in commingled waste. It can include such things as magnetic separation. The usual materials recovered include separation of recyclable, the removal of hazardous wastes, and the recovery of energy products. Volume reduction refers to the processes whereby waste volumes are reduced, usually by force or pressure.

Chemical transformation

This usually involves a change of phase, e.g. solid to liquid, solid to gas, etc. The main processes are combustion, pyrolysis, and gasification. Combustion is the chemical reaction with oxygen of organic materials accompanied by the emission of light and heat.

Biological transformation

The biological transformation of the organic fraction bother to reduces the volume and weight of material but also produces compost. When carried out anaerobically methane is produced—a typical component of landfill gas.

Importance of waste transformation

Typically waste transformations (Table 20.3) are used:

1. To improve the efficiency of solid waste management systems.
2. To recover reusable and recyclable materials.
3. To recover conversion products and energy.

Table 20.3. Transformation processes in solid waste management.

Process	Methods	Principal conversion products
Physical		
Separation	Manual and/or mechanical	Individual components found in commingled MSW
Volume reduction	Force or pressure	Original waste reduced in volume
Size reduction	Shredding, grinding or milling	Altered in form and reduced in size
Biological		
Aerobic compost	Aerobic biological conversion	Compost
Anaerobic digestion	Anaerobic biological conversion	Methane, CO_2, trace gases, humus
Anaerobic composting (in landfills)	Anaerobic biological conversion	Methane, CO_2, digested waste

Environmental problem in disposal of the municipal solid waste

Most of the municipal refuse is still discarded in the sanitary landfills. One of the problems it will create in it is the generation of the methane gas due to the anaerobic decomposition of the organic materials compacted inside. It further can cause explosion and a great potential hazard to the public safety.

Sewage sludge: Being the product of waste-water treatment plant it is usually of liquid mixture composed of solid and dissolve organic and inorganic material. This solid part being biologically unstable consisting of organic and inorganic materials is usually referred to as the sewage sludge.

Heavy metal: Heavy metal in the sewage sludge is of great concern to the public health. Heavy metals are those metals that have density, >5–6 gm/cm^3. The heavy metal content the sewage sludge depends greatly on the type of industry. In general municipal sludge is high in Al, Fe, Zn, Cu and Cr content.

Importance of recycling

A significant challenge confronting engineers and scientist in the developing countries is the search for appropriate solutions to the treatment and disposal of the domestic waste. The treatment and recycling of organic waste can be most effectively accomplished by biological process employing the activities of micro-organisms and higher life forms. The main objective behind this recycling is to treat the waste and to reclaim the valuable substance present in the waste for possible reuse. Here, in vermicomposting we are collecting the waste as input raw material for our reactors. Latter these wastes are converted into useful compost by the action of the earthworms within a given period of time.

General fate of organic waste

When the domestic waste and other household waste enter the municipal waste steam without any pretreatment or source segregation, it either ends up in a landfill or in an open dumping site. This, then become a major sanitation issue and also a big problem in the landfill leachate management. The organic portion of the waste is the major contributor to the landfill leachate.

Earthworm in Waste Stabilisation and Processing

Potential for sewage sludge processing

Earthworms are these days intensively used in the degradation of sewage sludge. Research done by Neuhauser has shown that aerobic sewage sludge can supply the nutrients necessary for the growth and reproduction of the earthworm.

Degree of stabilisation

If earthworms are useful in stabilising sludge they must increase the stabilisation rate. This can be shown if the presence of earthworm in the sludge causes an increase in the rate of volatile solids reduction. Maximum reduction of the volatile solids is a goal of any sludge stabilisation system. *Esenia foetida* can increase the rate of volatile solid sludge destruction, when present in aerobic sludge. This increase in the sludge solids destruction rate reduces the probability of putrefaction occurring in the sludge due to anaerobic conditions.

The more rapid degradation of organic matter was probably due to increased aeration and other factors brought about by the earthworms (Fig. 20.4).

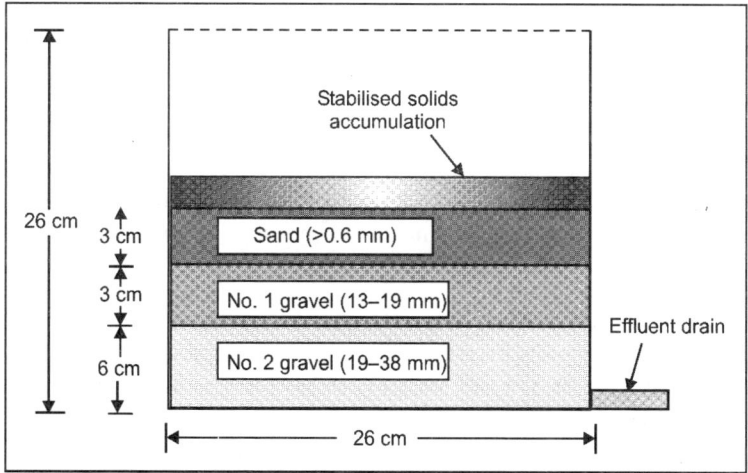

Fig. 20.4. Schematic of reactor used in the liquid vermistabilisation studies.

Bhiday reported that the aerobic and the anaerobic stages of the sludge help to convert the organic matter into the right type for rapid consumption and digestion by the earthworms. The role of earthworm in vermistabilisation is linked to the aerobic condition of the sludge and the moisture content. In sludge's that are aerobic, the worms have a strong and positive effect. Earthworms cannot exist in anaerobic sludges and will not have a positive effect until aerobic condition occurs.

Sludge age

It is also very important to relate the rate of earthworm growth to the age of the sludge, i.e. the time after the sludge was removed from the aerobic reactor and dewatered. Neuhauser reported that as the sludge ages, its nutritive value to the earthworm decreases rapidly after about twelve weeks removal from the digester, whereas the ash content of the sludge increases with the time, an indication of sludge stabilisation (Fig. 20.5).

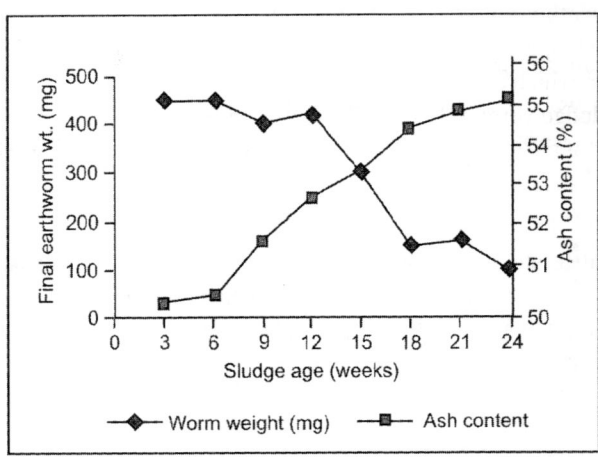

Fig. 20.5. Effect of ageing aerobically digested sludge on the growth of *E. foetida*.

Moisture content of the sludge

Both excessive and insufficient moisture can adversely impact earthworm growth. A series of experiments were conducted to determine the moisture content of the media that will exist in vermistabilisation units. In an experiment conducted by Neuhauser, an aerobically digested sludge was dewatered to a moisture content of 75 per cent and exposed to a temperature of 25°C. Here in this media four earthworms of one species (*E. foetida*) were placed and their growth was recorded accordingly for four weeks. Latter they found out that optimum worm growth occurred when the total solid content of the media is from 9–16 per cent. Very high and very low moisture concentration will also inhibit the earthworms.

Choice of earthworm

There are many earthworm species that have the potential to be used in sludge stabilisation systems. Since earthworm growth and reproductive rates are the way of indicating the potential for use in sludge management systems, therefore the proper choice of earthworm is an important factor that might effect the rate of sludge stabilisation. Neuhauser and others have used five species of earthworm to determine the optimum temperature for growth and reproduction in dewatered (10–12 per cent solids) aerobically digested sludge for twenty weeks. He estimated the overall reproductive capability of each species by using the total number of cocoons produced over the study period and fund out that amongst the other four species (*D. veneta, E. eugenia, P. excavatus* and *P. hawayana*), *E. foetida* appears to be the appropriate species to use in vermistabilisation studies. Ismail, reported that the local worms can also be used effectively in the combined process of litter and soil management. Since the introduction of foreign species may create a complex chain of interaction amongst the soil organisms which may lead to the competition among the species for the food.

Stabilisation of liquid municipal sludge

Earthworms are also used effectively in the stabilisation of the municipal sludges, the worms maintain aerobic conditions in the mixture, ingesting the solid content and also converting a portion of the organic matter into its biomass and to respiration products, and discard the remaining partially stabilised matter as casting. As reported by Neuhauser the degree to which the organic matter is degraded is a function of, the portion of the waste that is biodegradable, maintenance of aerobic condition and avoidance of toxic conditions. The role of earthworm in stabilisation of municipal sewage sludge is greatly linked to the aerobic condition of the sludge, ash content or the sludge age, the moisture content and the loading rate. The study done by Loehr, which clearly indicated the stabilisation of a primary sludge or the removal of volatile solid in liquid sludge vermistabilisation (LSVS) units. The liquid that drained from the reactor contains the by-products of the stabilisation that occurred in the reactor. It was observed that the loading rate as high as 1000 grams of volatile solids/m²/week may result in satisfactory operation of LSVS reactors involved in stabilising the primary sludge. Similarly the optimum loading rate for the waste activated sludge and the aerobically digested sludge were recorded to be 1000 grams/m²/week and 1200 grams/m²/week respectively.

Failure occurred in those reactors where the liquid no longer flow through the accumulated sludge solids, resulting in ponding of the media and latter development of anaerobic condition. Other possibilities to make these waste more palatable to the worms is to aerate the sludge in rotor drums to encourage the aerobic microbes before feeding them to earthworms (Fig. 20.6).

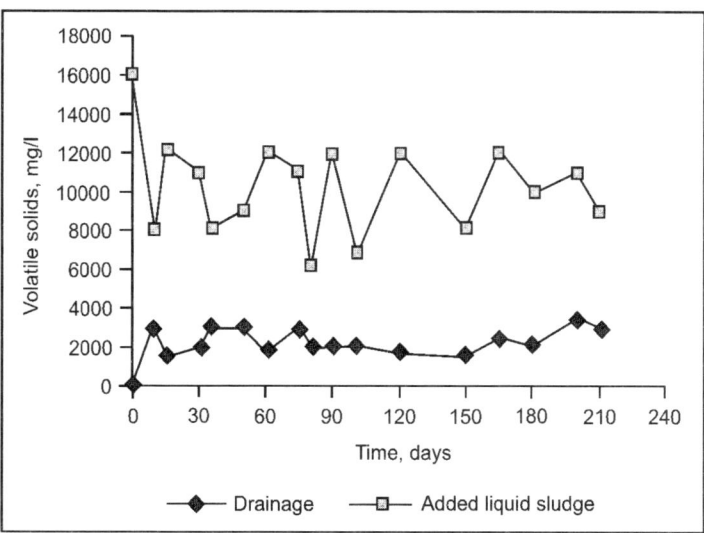

Fig. 20.6. Volatile solid removal pattern for primary sludge.

Break down of organic, animal and industrial waste

The role of earthworm in humification and breakdown of plant litter in natural soil has been known since the time of Darwin, but the potential use of these humble species in stabilising the organic refuse into useful components has been known recently. There has been an extensive research done by using earthworms to break down the various animal manure like pig, cattle solids, slurries and also waste from the poultry farms. Amongst all, Edward, reported five earthworm species (*D. veneta, E. eugenia, P. excavatus* and *P. hawayana* and *E. foetida*) to be the most potential earthworms to breakdown the organic refuse.

Generally most organic wastes can be broken down but in some case we have to do a pretreatment in various ways to make them acceptable to the worms. The earthworms are highly adaptable to different types of organic waste, provided the physical structure, pH and the salt concentration are not above the tolerance level.

Types of refused feed

Animal manures

Cattle solids

They are the easiest animal wastes in which the earthworms will grow. They usually contain no materials that are unfavourable to the earthworms. Cow slurry is a suitable substrate for vermicomposting, both when mixed with solid materials and when applied to the surface of bedding materials containing earthworms. Hand, reported that the mixture of slurry with paper tissue waste produced greater earthworm growth and cocoon production per unit of slurry consumed.

Horse manure

Horse manure is very suitable for the growth of earthworms. It also doesn't need any pretreatment and thus can be applied directly as a feed. Ronald reported that horse manure contains 0.7 per cent of nitrogen, 4.38 per cent of protein and 60 per cent of organic matter and trace amounts of phosphoric acid and potassium oxide. Another reason for recommending the horse manure is that it doesn't require addition of other material for moisture retention, ageing, porosity and above all it doesn't require to check for the acidity of the bedding.

Pig solids

Waste from the piggeries is probably regarded as the most productive refuse for growing the earthworms. This waste if it comes in slurry, the solids must be separated either by sedimentation or other mechanical means. Edward reported the presence of some inorganic salts and some ammonia, which have to be washed out can be composted for about two weeks or longer prior to inoculation with earthworms.

Poultry waste

These manures are higher in protein content, nitrogen and in terms of phosphoric acid than any other animal manure. These wastes have to be pretreated by composting, washing or simply by ageing process to reduce the inorganic salt and to reduce the heating potential. The fresh waste generated from the poultry farms contains significant amount of inorganic salts, and if used directly may threatened the survival of the worms.

Paper pulp and card board solids

The papers and the cardboard are primarily made up of cellulose; therefore these can become an excellent material both for feeding and for the bedding. Ronald reported that earthworms can convert cellulose into its food value faster than the proteins and carbohydrates. In a nutshell, these wastes are excellent materials for the growth of earthworms. Most of these wastes doesn't need any special pretreatment and can be applied directly as a feed.

Compost and waste products

Spent mushroom compost

Spent mushroom compost is also a good medium to grow the worms. According to Edward it is low in plant nutrients.

Brewery waste

Brewery waste needs no modification, in terms of moisture and the worms too can process it quickly and will grow rapidly.

Urban waste

Here there is too much waste-type materials available for feeding. Waste such as from the canning plant, potato chip or corn chip manufacturers are excellent food for worms. Wastes generated from vegetable oil factory (flowers and plants) are also considered to be suitable for the feed. The food waste from domestic households and restaurants and other yard waste are used as the feed and are all good growth media for the worms. Waste from the mining industries that contains sulphur causes a major disposal problem as well as nuisance to the public.

Smith and others reported the use of this sulphur waste residue in a vermicomposting system by mixing it with organic matter. They reported that not only the organic matter serves as food but also the mineral constituents are subjected to digestive enzymes and to a grinding action within the animals. The optimum-mixing ratio of the sulphur waste residue to the organic matter is 4 per cent at which they observe the maximum numbers of young earthworms.

Similarly, Glick and others have stated that the waste from logging and carpentry industries and sugar factory waste can also be used as a substrate to feed earthworms. They reported that, when the earthworms are reared in the mixture ratio 1:1 of the sawdust and pressmud, the cast so generated shows 1.2 times more CFU (Colony Forming Units) and 1.6 times more than the pressmud and 1.7 times more than the soil. It is also said that the earthworms can also partially detoxify the waste. The fly-ash waste generated from the thermal power plants is creating a major disposal problem due to its heavy metal content, although it is supposed to be very rich in microbial biomass. It was found out that the organic waste, sisal green pulp, parthenium and green grass cuttings admixed with 25 per cent of fly-ash proved to be a potential valuable material for the *E. foetida* biomass. The vermicompost so produced from it contains higher NPK content than the rest available commercial manures.

In some cases, earthworms are also used in the management of distillery waste containing waste of malt, spent grain wash, yeast and molasses settled at the bottom of the lagoon. Smith observed that the total volume of cowdung leaf litter should be proportional to the total volume of distillery waste and the pressmud to have positive impact on the growth and production of worm biomass. Collin reported that the filter pressmud from the sugar factory can be used as a feed in the vermicomposting units. It is seen that this pressmud is converted to nutrient rich manure and that the macro and micronutrients as well as the physico-chemical features increased after vermicomposting.

Composting

With the advancement of human civilisation and its never-ending tertiary consumers, our planet is deliberately heading towards a fateful direction. With the dawn of a theoretic new century, we are knowingly and unknowingly contributing thousands of tons of solid waste per day. Which if not regulated or channelled properly might bury us within it, in the long run just to get composted, freely.

The solid waste generated from a locality or from a larger domain can be utilised in a healthier way depending on its source and the nature of the waste. It would be best if we all produce least waste and practice the source separation in the first hand before dumping them in the municipal stream, which is truly not impossible. So the only solution for this unavoidable crossroad is to pretreat the waste utilise it at the most and to dispose the rest safely in the most hygienic and ecofriendly means.

During the composting process, micro-organisms decompose organic compounds, which consist of carbohydrates, sugar, proteins, fats, cellulose and lignin. Carbohydrates are more easily decomposed whereas lignin is more resistance to decomposition. Many factors affect the composting process. Aerobic micro-organisms need oxygen, water and nutrients for their metabolism and cell synthesis. As a result of microbial activity heat is liberated and, if contained within the composting mass, the temperature rises. Temperature increases through the mesophilic phase into a thermophilic phase and then back into the mesophilic phase. During the course of these transitions, the microbial population changes, thereby affecting the rate of organic matter decomposition.

Composting has long been considered an interesting option to reduce the amounts of waste to be transported and disposed of in landfills. One of the most pressing problems today is what to do with various waste products. Recent legislation mandated that certain wastes can no longer be deposited in landfills, and tipping fees are on the rise. Composting is a very acceptable and viable option for handling most organic wastes presently being landfilled. Windrowing is the most efficient and economical composting method and may be accomplished within a building, out-of-doors or under a simple roof.

Fundamentals of composting

Modern composting is an aerobic, thermophilic, biochemical process that, with the assistance of mechanical equipment and controls. Compost systems can be classified on three general bases, which are oxygen usage, temperature, and technological approach respectively. Oxygen usage is divided into aerobic and anaerobic. When temperature serves as the basis, the division becomes mesophilic and thermophilic. Finally, using technology as the key, the classification is divided into static pile or windrow, and mechanical or 'enclosed' composting. Aerobic composting involves the activity of aerobic microbes, and hence the provision of oxygen during the composting process. Aerobic composting generally is characterised by high temperatures, the absence of foul odours, and is more rapid than anaerobic composting. Anaerobic composting is characterised by low temperatures, the production of odorous intermediate products, and generally proceeds at a slower rate than does aerobic composting.

In mesophilic composting the temperatures are kept at intermediate temperatures 15° to 40°C, which in most cases is the ambient temperature. Thermophilic composting is conducted at temperatures from 45° to 65°C. One of the most important factors in producing quality compost is the exclusion or removal of undesirable materials prior to composting. Undesirable materials include plastics, metals, glass, paper products, construction wastes with nails, and woody plant materials, which have not been reduced in size. The best methods for avoiding contaminants include knowing the source of materials, inspecting materials carefully before processing, and paying attention to composting conditions. Kitchen wastes, manures, biosolids (waste-water residuals) and other organic solid wastes are compostable.

Briefly stated, essential factors are those features of the physical, chemical, and biological background that are necessary to the establishment and proliferation of the micro-organisms specific to the desired process. Five essential factors that have become key design features in recent compost technology are suitable microbial population or populations, aeration (oxygen availability), temperature, moisture content, and carbon availability.

Feedstock preparation

Feedstock preparation is an important first step to ensure that the composting process is optimised. Moisture content, carbon, nitrogen, and micronutrients are the raw materials that will determine the formulation of the mix. In general, a feedstock of organic materials high in carbon, such as brown leaves, ground branches, and twigs, should be mixed with nitrogen-rich materials, such as grass, weeds, and green leafy materials. A mixture ratio of 2 to 3 parts brown material (carbonaceous) to 1 part green material (nitrogen) is ideal.

Amendments and bulking agents

An amendment is a material added to the other substrate to condition the feed mixture. Two types of amendments can be defined as follows:

1. Structural drying amendments, which refer to an organic or inorganic material, added to reduce the bulk weight and increase air voids allowing for proper aeration.
2. Energy or fuel amendments, which refers to an organic material added to increase the quantity of the biodegradable organics in the mixture, and thereby, increase the energy content of the mixture.

Amendments that have been used to condition wet substrates such as sludge cake includes sawdust, straw, rice hulls, cotton gin trash, manure and variety of other waste materials. According to Polprasert the materials such as organic amendments and bulking agents are added to these waste to raise the C/N ratio, provide structural support for the composting pile, and would create void spaces in case of aerobic composting. Organic amendments are those materials added to the composting feed to increase the quantity of the degradable organic carbon, reduce bulk weight, and to increase air voids of the compost mixture. Whereas the bulking materials can be either organic or inorganic of sufficient size, bulking agent will provide the structural support and maintain air space in the composting mixture. Bulking agent is a material, organic or inorganic, of sufficient size to provide structural support and maintain air spaces within the composting matrix.

Methods of Composting

Composting system can be classified according to the reactor type, solid flow mechanism, and bed conditions in the reactor and in the manner of air supply. These systems are:

1. Windrow composting.
2. High rate composting.

Windrow composting

The windrow system is the most popular example of a non-reactor, agitated solid bed system. Height, width and shape of windrows vary depending on the nature of the feed material and the type of equipment used for turning. In the forced aeration windrow system, the oxygen transfer into the windrow is aided by forced or induced aeration from blowers.

Composting using the windrow method requires some management to setup windrow piles, turn them periodically, and monitor the composting process. Compost forms as microbes in the manure decompose the organic wastes. Since this method requires a moist environment containing oxygen, one must aerate the material by turning the piles two to four times during the process.

A basic understanding of the compost process can help produce a high quality product, while preventing many common problems. The micro-organisms that do the work in composting have a few basic requirements, which need to be provided. Air, water, the right food and temperature combine to create a good composting environment. Composting is an aerobic process, which means it occurs in the presence of oxygen. Oxygen is provided in two ways:

1. By turning the compost.
2. By building the pile correctly.

Micro-organisms need water, just like any other organism. Ideally, the moisture content should be between 40 and 60 per cent. Too wet and anaerobic condition result. Too dry and the decomposition process will slow down. Bacteria, fungi, and other micro-organisms get their energy from carbon sources, such as leaves, paper or wood chips. Nitrogen is required for population growth, but excess nitrogen can generate ammonia and other odours, and can pollute runoff water. If high nitrogen materials such as grass clippings are used, they must be thoroughly mixed with a carbon source. Surface area is also important in this relationship, as the carbon in leaves is much more available than the carbon in a large wood chip.

As the micro-organisms are working away, decomposing waste, they generate heat. When temperatures rise above 60°C, the organisms start to die. Turning the pile when temperatures reach this point will prevent overheating, which can result in drastic population fluctuations and odours.

Eventually, the micro-organisms will use up most of the readily decomposable waste, and the composting process will slow. Temperature will drop, and the compost takes on a dark, granular texture. At this point, the compost can be placed in large stockpiles to cure, and will continue to improve until it is ready for use.

Carbon to nitrogen ratio

MSW waste consists of a variety of different materials, each of which has its own characteristics and requirements. When combining different materials such as leaves and clippings to make compost, the concept of carbon to nitrogen ratios (C/N) is critical. The ideal proportion of these two elements is about 30:1 or 30 parts carbon to 1 part nitrogen by weight although this ratio may need to be adjusted based on the bioavailability of carbon and nitrogen. If carbon and nitrogen are too far out of balance, the microbial system will suffer. When there is little nitrogen, the microbial population will not grow to its optimum size, and composting will slow down. In contrast, too much nitrogen allows rapid microbial growth and accelerates decomposition, but this can create serious odour problems as oxygen is used up and anaerobic conditions occur. In addition, some of this excess nitrogen will be given off as ammonia gas that generates odours while allowing valuable nitrogen to escape. The Table 20.4 presents estimation of the C/N ratios of various compostable materials.

High nitrogen materials

Nitrogen contents in compostable materials are given in Table 20.4.

Table 20.4. Nitrogen contents in compostable materials.

High nitrogen materials	C/N
Grass clippings	19:1
Sewage sludge (digested)	16:1
Food wastes	15:1
Cow manure	20:1
Horse manure	25:1

High carbon materials

Carbon contents in compostable materials are given in Table 20.5.

Table 20.5. Carbon contents in compostable materials.

High carbon materials	C/N
Leaves and foliage	40–80:1
Bark	100–130:1
Paper	170:1
Wood and sawdust	300–700:1

Moisture content

Active micro-organisms need a moist environment. Ideally, composting materials should be between 40 and 60 per cent. When conditions are too wet, water will fill the pore space needed for air movement, and anaerobic conditions can result. If conditions are too dry, the decomposition rate will slow down. Leaves are often quite dry when collected in the fall, and water may need to be added. Some materials, like grass clippings, may seem dry to the touch but contain a great deal of water in their cell structure. As that structure breaks down, the water is released, turning the grass into a slimy mess. If compost becomes too wet, it may be necessary to add some drier material, such as partially decomposed leaves or wood chips. Coarse material is especially helpful in this situation, as it increases the porosity allowing water to drain out and air to flow in. The shape of a compost pile has an important effect on moisture content. Scooping out the top of the pile to create a concave shape will maximise water absorption. However, if the pile is over saturated, anaerobic odours and leachate will be produced. Water can be added to the compost pile in various ways.

Oxygen

MSW waste composting is an aerobic process, which means it occurs in the presence of oxygen. The air we breathe is about 21 per cent oxygen. During composting if the oxygen level falls in the large pores, parts of the compost pile can become anaerobic producing methane gas and loss of Nitrogen from the system. Also during the anaerobic phase the odour problem is unavoidable. Because odour complaints are the most common problem at MSW composting sites, maintaining an adequate oxygen supply is critical. Successful composting depends on the availability of a sufficient supply of air to the materials in the compost pile. Oxygen is essential for microbial activities and for adequate decomposition to take place. While composting with adequate aeration proceeds more rapidly, decomposition of plant materials can proceed, but more slowly, anaerobically.

Air can be supplied by either passive or active means. If pile size remains moderate, fresh air can flow in from the outside of the pile. The passive processes supplying air in this way include diffusion and natural convection. Natural convection is driven by a chimney effect, with warm air from the middle rising out of the top of the pile, and cool fresh air sucked in at the bottom sides. Materials that decompose more quickly, such as a mixture of MSW and carbon source, must be placed in smaller piles or oxygen will be depleted. If the pile is too large, oxygen will not penetrate to the middle of the pile, resulting in an anaerobic core. Moisture content and the size of composting particles will also affect the effectiveness of natural convection.

Additional oxygen can be provided mechanically, by turning the compost with a front-end loader or manually in case of a small scale. Although the oxygen added by turning only lasts a few hours, turning also loosens the piles so that air can flow more easily by natural convection. In some compost operations additional oxygen is supplied by a system of blowers and perforated pipes. These forced aeration systems are somewhat more expensive, but the cost may be justified if the high nitrogen source is causing consistent odour problems or if the MSW is being composted with other materials such as sludge.

Temperature

Proper temperature is an important factor, particular in the aerobic composting process. Considerable amounts of heat are released by aerobic decomposition. Since composting material has relatively good insulation properties, a sufficiently large composting mass will retain the heat of the exothermal-biological reaction and high temperatures will develop. High temperatures are essential for destruction of pathogenic organisms and undesirable weed seeds. Decomposition also proceeds much more rapidly in the thermophilic temperature range. The optimum temperature range is $57°$–$71°C$, around $65°C$ usually being the best. In normal practice composting begins at ambient temperature (mesophilic range) and progresses to and through a thermophilic phase, followed by a descent to the mesophilic level.

During the decomposition organic waste, they generate heat and the decomposition is most rapid when the temperature is between $32°$–$60°C$. Below $32°C$, the process slows considerably, while above $60°C$ most micro-organisms cannot survive.

In windrow composting the pile temperature depends on how the heat produced by micro-organisms is offset by the heat lost through aeration or surface cooling. During periods of extremely cold weather, piles may need to be larger than usual to minimise surface heat loss.

After an initial high temperature period of a few days to several weeks, compost pile temperatures will gradually drop. By measuring temperatures regularly, we can tell how fast material is composting, and whether there are hot or cold spots in the pile.

A drop in temperature in the compost pile before material is stabilised indicates that the pile is becoming anaerobic and should be aerated. High temperatures do not persist when the pile becomes anaerobic. The temperature curve for different parts of the pile varies somewhat with the size of the pile, the ambient (surrounding) temperature, the moisture content, the degree of aeration, and the character of the composting material. The provision of aerobic conditions, however, is the important factor in maintaining high temperatures during decomposition. To destroy the pathogens, the compost needs to maintain temperature above $55°C$.

Particle size and surface area

Particle size is critical to the size and structure of the windrow because of its effect on pile aeration. Large, brushy yard trash should be reduced in volume for uniformity of size and to obtain a particle size that will promote airflow in the composting pile. The ground material provides a greater surface area for the micro-organisms (e.g. fungi, bacteria, and actinomycetes) which will promote decomposition. The pile should be constructed with a variety of particle structures that permits the chimney effect to occur within the pile. The chimney effect allows cool air to be drawn into the bottom of the pile and heated air to be vented through the top, providing the necessary oxygen for microbial activity. In a properly constructed pile, this venting is apparent when the area along the ridge of the pile is giving off visible vapours.

Micro-organisms

Solid waste such as municipal refuse and other mixed vegetable waste contain many types of bacteria, actinomycetes and other fungi. Temperature, moisture content, aeration, pH and the nutrient content should be within a suitable range to provide the best environment for growth of micro-organisms. The hundreds of types of micro-organisms involved with composting are generally classified into three categories according to temperatures most favourable to their metabolism and growth (Table 20.6).

Table 20.6. Ideal temperature range for the composting micro-organisms.

Psychrophilic	Less than 25°C
Mesophilic	25°–45°C
Thermophilic	>45°C

As the micro-organisms decompose (oxidise) organic matter, heat is generated and the temperature of the compost is raised a few degrees as a result. In composting, as in the decomposition of any complex substance, the breakdown is a dynamic process accomplished by a succession of micro-organisms with each group reaching its peak population when conditions have become optimum for its activity. One group of micro-organisms dies and another group thrive until the next incremental change in nutrition and temperature occurs, etc.

Composting rate is generally measured by rate of carbon dioxide production. The maximum rate occurs when compost temperature range from 43°–65°C. As the temperature exceeds 65°C, the composting rate drops rapidly and becomes negligible at temperatures higher than 65°C. Most composting should include temperatures in the thermophilic range 37°–65°C. At these temperatures the rates of organic matter decomposition are maximum, and weed seeds and most microbes of pathogenic significance cannot survive. It is important that piles are turned frequently to ensure that all parts are exposed to high temperatures.

The organisms responsible for composting are facultative and obligate aerobic bacteria, actinomycetes and other organisms. Bacteria are characteristically predominant throughout the process, with fungi appearing after few days, and actinomycetes appearing only in the final stages.

Micro-organisms have also been found to grow at very high temperatures. Although theoretically it has been stated that the composting ceases or is reduced at temperatures exceeding 60°C. Temperature becomes an important parameter affecting the number and types of micro-organisms in a composting pile. As the temperature increase, the growth of organisms accelerates. While many composting organisms thrive at 50°C, numerous organisms grow at temperatures exceeding 50°C.

Building Windrows

The first stages of composting are in many ways the most important and proper windrow construction is the key to getting the process off to a good start. The two aspects of windrow building are:

1. Mixing materials.
2. Forming and shaping the windrow.

If several different types of waste are going to be composted together they must first be thoroughly blended. Mixing is required to balance the carbon and nitrogen ratio and distribute moisture throughout the pile, and also to insure an even distribution of large pores so that oxygen can move freely. If MSW or other high nitrogen materials are being composted, this blending process is particularly critical. Mixing can be accomplished with a front-end loader, although other equipment such as tub grinders or specialised windrow turning machines are commonly used when mixing MSW. The size and shape of the windrow are designed to allow oxygen to flow throughout the pile while maintaining temperatures in the proper range. If windrows are too large, oxygen cannot penetrate to the middle, while if they are too small they will not heat up properly. The optimum size varies both with the type of material and with the time of year. The sides of the windrow can be as steep as the material will naturally pile up, which typically leads to a windrow about twice as wide as it is high. Windrows can be as long as is convenient for the site.

Turning windrows

There are two goals to keep in mind when turning a compost windrow. The first is to move material from the outside of the pile to the middle, where it can decompose more quickly. The second goal is to loosen and fluff the material, so it will be more porous and air can move freely. Specialised windrow turners are designed to accomplish both of these goals. A front-end loader can do the job as well. First flip the top of the windrow over just beyond the existing windrow. Second, take the compost from the bottom of the old windrow and place it on top of the new windrow. Let the compost cascade out of the loader, to keep it as loose as possible.

Turning frequency should normally be based on temperature, and should occur whenever temperatures exceed to 60°C, or drop below 30°C. If the compost is staying in this range on its own, regular turning can accelerate decomposition by mixing the material and exposing new surface. The carbon source will only need to be turned a few times a year but will benefit from turning as often as every two weeks. On the other hand, MSW even when properly mixed with carbon sources may initially need turning 3–4 times a day. As decomposition proceeds and the compost becomes more stable, frequent turning becomes less important.

If the compost has become anaerobic and smells, turning will temporarily add oxygen but may also release VOCs (bad odours). Schedule compost turnings to minimise any negative impacts by considering such factors as wind direction, when people are home, and whether they are likely to be outside or have their windows open.

Common impediments and their solutions

The common problems and their solutions during composting is given in Table 20.7.

Table 20.7. The common problems and their solutions during composting.

Problem	Cause	Solution
Anaerobic odour	Excess moisture	Turn windrow
	Window too large	Make windrow smaller
	Temperature greater than 60°C	Turn windrow
	Leaf compaction	Turn or reduce windrow size
	Surface ponding	Eliminate ponding
Low windrow	Windrow too small	Combine windrows
Temperature	Insufficient moisture	Add water while turning windrow
	Poor aeration	Turn windrow
High windrow temperature	Windrow too large	Reduce windrow size
	Carbon compaction	Turn windrow
Surface ponding	Depression or ruts	Fill depression and/or regrade
	Inadequate slope	Grade site to recommended slope design

METHODOLOGY

Overview of the General Methodology and Experimental Set up

The main objective behind this experimental study is to pretreat the municipal solid waste by composting and vermicomposting on a pilot scale before sending them for landfill or any other final disposal. The conceptual lay out of this experimental study is given in Fig. 20.7. Within this methodology framework, the most favourable biological and physical conditions to the degradation of the organic portion of the waste was also determined. Along with that, the best operating condition and other physical parameters to achieve the best sorting result of the commingled waste stream was also obtained.

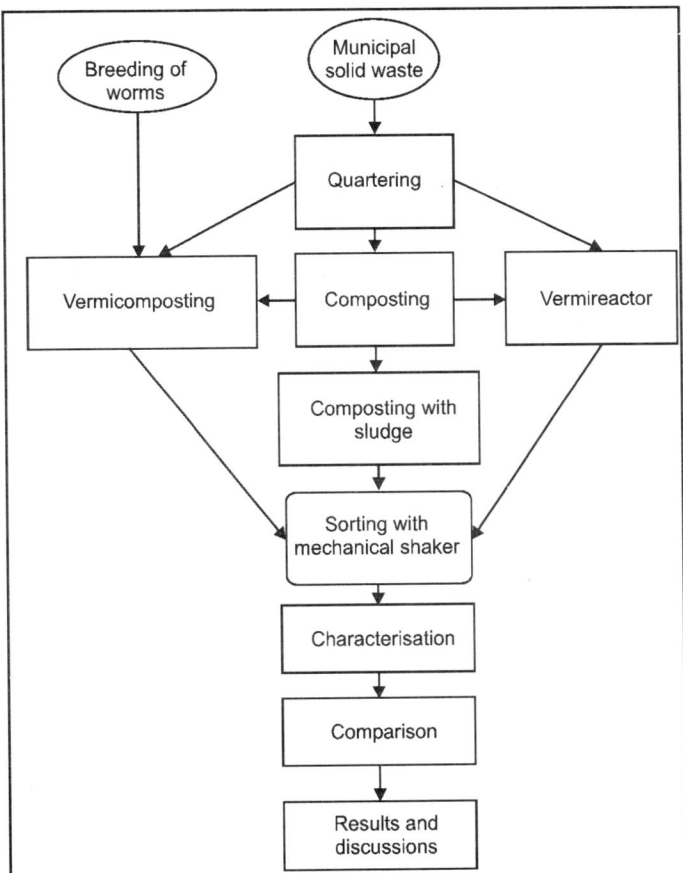

Fig. 20.7. Conceptual layout of this experiment.

Breeding of worms

The initial phase of this experimental study was involved in the breeding of local worms in plastic bins. For the bedding material, pebbles are used as the first layer. This layer ensures the proper drainage in the system. A layer of loamy soil or a garden soil was used as the next layer. Moistened coconut dust and shredded paper are used as the third layer of the bedding. Finally, on top of that some dry leaves and crushed eggshells are sprayed as shown in Fig. 20.8. To protect the worms from predators and to avoid any direct contact with sunlight, these vermibins are constantly monitored and kept at a higher ground with proper shading. To avoid the water stagnation and for better aeration, some holes are drilled on the bottom of the bin as a precautionary method. Watering and feeding of the breeding unit was done on biweekly basis. During the first few weeks, the worms were fed with vegetable peels and corn flour and gradually shifted towards the waste stream as their feed except the breeding units. The frequency of the feed thereafter depends on their consumption rate and their reproductive potentials and most of all, the avoidance of any unfavourable conditions.

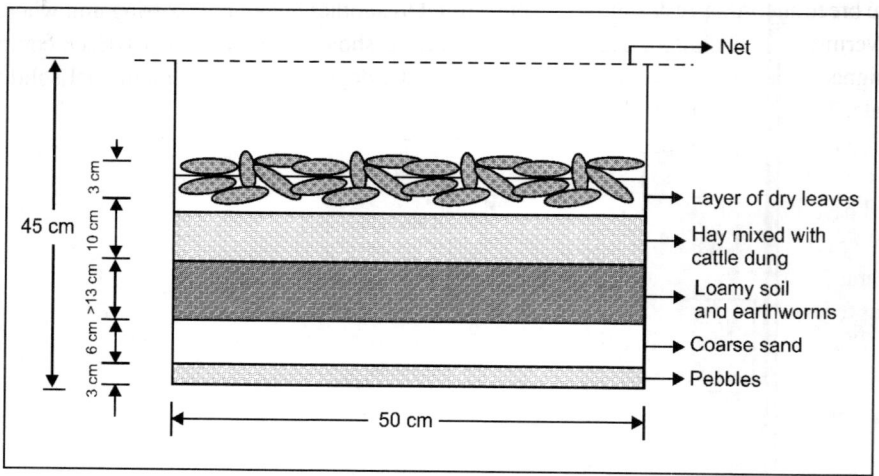

Fig. 20.8. Setting up a breeding vermibin.

The vermibins are fed twice a week and all the important parameters are maintained for almost three months. Latter when time came to harvest the worms it is seen that the pebbles used in the bottom layer creates a bit problem when separating the worms from the earth but this complication is overcome by using the light retraction method as discussed already. Smith found that the population of the local worms have not increased that much as per expected.

Breeding of *Esenia foetida* or more commonly known as the red worm has started from the third week of February from a few samples of above mentioned species. The initial breeding of these worms are done in an open bucket with a black shade. There is no special change in the bedding material other than a thin layer of chicken manure, which was obtained initially along with the worms. It is seen that these worms grow very fast and have much higher reproductive rate as compared with the local worms. Similarly here also the worms are first bred in a closed system but due to its high reproduction rate, after couple of weeks these worms are shifted to an open vermibed at the composting site. The pH of the open vermibed are maintained in the range of (7 ±1). These worms are latter used in the vermireactor study to determine the total mass reduction and the stabilisation of the same waste.

Important parameters for breeding

As discussed already, the reproduction of earthworm depends on several important parameters, which we should always monitor them whenever necessary. The most important of all parameters are temperature, moisture and the pH. The following parameters shown in the Table 20.8 are measured twice a week for a period of four weeks and the average values are considered.

Table 20.8. Regulated parameters during the breeding of earthworms.

Parameters	Optimum range	Average measured value
Temperature	10°–35°C	30°C
Moisture	60–75%	65% (wet basis)
pH	4.5–9	7.5

Harvesting and inoculation of earthworms

After the breeding part, the earthworms were shifted to another bigger composting unit where they were bred in vermibed. The composting site was designed as shown in the Fig. 20.10. Each composting bed was designed in a way that, it will further enhance the degradation of the municipal solid waste on a pilot scale. The dimensions of each bed was (3 × 2.5 × 0.5) metres.

The bottom layer of the concrete bed was designed in such a way that any excess water will be drained off within and will further promote the free passage of air. Overall side view of the composting unit used here is shown in the Fig. 20.9. Before introducing the worms, the bedding was first laid with a thin layer of garden soil and the worms were fed with the mixture of fruit waste and the MSW. This feed alteration is done slowly in order to give some time for the worms to adjust themselves to the changing feed in the latter stage and as a precautionary method.

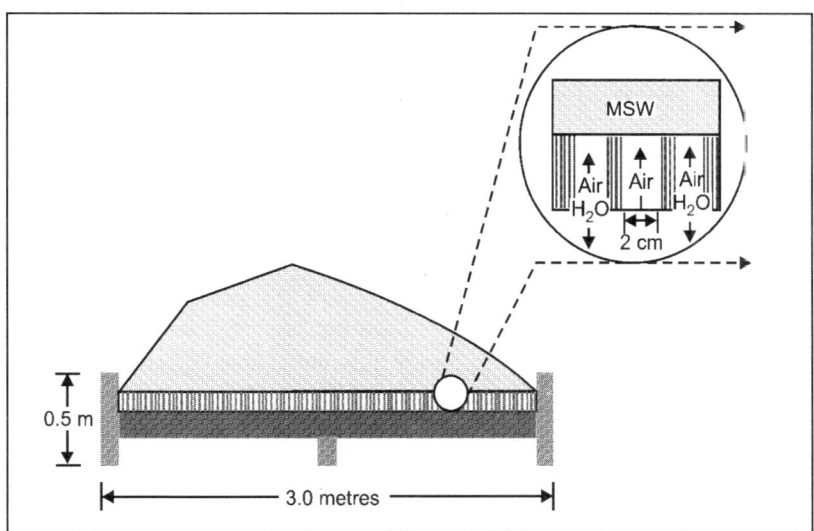

Fig. 20.9. Side view of a composting bed.

During the harvesting of the worms from the vermibins it is seen that the worms tend to bury themselves into the pebbles and it further makes our job more tedious in transferring them to the new bedding. So for the new bedding the pebbles are replaced by a thicker layer of coarse sand and loamy soil. The pH of the beddings' are checked twice a week to make sure whether it falls in the favourable range. In some case we had to spray a little lime in order to adjust the pH. As for the feed, the worm beds are fed twice a week with cornflour, vegetable peels and kitchen scraps as before we have also used the food scraps and the fruit peels from our university cafeteria, which otherwise will be in the trash bin.

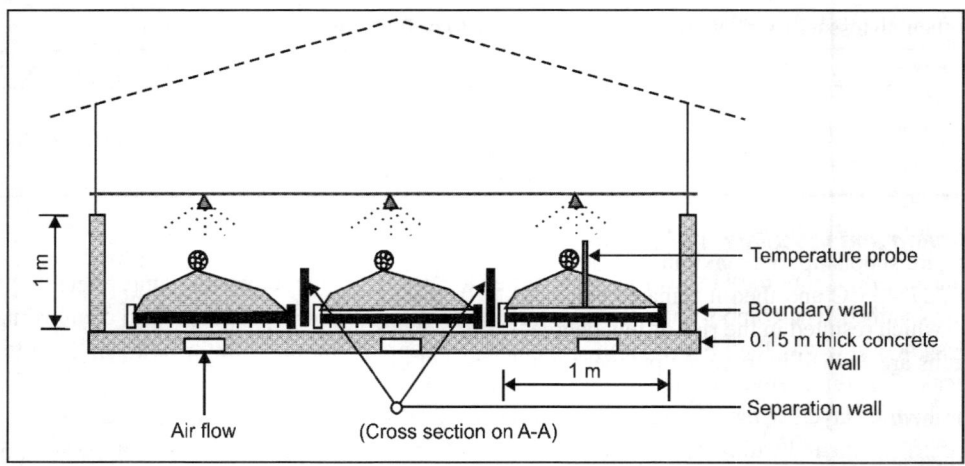

Fig. 20.10. Composting unit at the site.

Growth rate determination

Growth rate of the earthworms in different waste was determined by comparing the rate of the worms to the breeding unit setup in the first part of the study. The growth rate here is defined as the quotient of the difference obtained from the initial total count of worms and the total number of living worms at the end of the study divided by the experimental time period. This is further simplified by this relationship,

$$R = (N_2 - N_1)/ T \qquad \qquad \dots (20.1)$$

where,

    R = Growth rate.

    N_1 = Total number of initial worms.

    N_2 = Total number of living worms by the end of time T.

    T = Time period of the experiment in days.

By recording the feed fed per turn (twice a week) we are here able to relate the consumption rate of the food to the growth rate of the earthworms during their initial breeding phase in the vermibins.

Collection of Waste

For this study the solid waste was collected from Rangsit fruit market. The waste as received was introduced in the composting unit without any source separation. About one ton of raw wastes are being collected per trip from Rangsit vegetable market. This waste was mainly dominated by the organic part consisting of generally vegetable refuse and certain amount of fruit peels, major part of the non-biodegradable fraction of 17–21 per cent mainly consist of polyethene bags, plastic particles and expanded polystyrene (EPS). The organic fraction of the waste is very high to about 75 to 77 per cent of the total waste on wet weight basis. The cumulative average moisture content of the MSW as collected from the site is 67 per cent (wet basis).

The collection frequency of the solid waste was once a month. These waste were studied for a maximum of three weeks only since our main emphasis is not in the composting but the pretreatment of the raw waste and for the optimum separation of the refractory substance from the mixed waste. These wastes are collected from the site in the AIT pick-up truck. All the waste collected are unloaded at the environmental research station behind the Biotechnology building. At the composting site the waste collected are piled up in three cells. Before making up the pile these wastes are properly mixed using pitchforks and spade. Daily temperatures of the piles are recorded and maintained at the composting site using compost thermometer. The raw waste that we receive, although doesn't contain any hazardous materials but since it was handled manually during its composting period therefore gloves and other precautionary measures were also practiced.

The initial temperature of the waste in the heap is measured and is recorded as 43°–52°C. The temperature shoots up to a maximum of 60°–62°C during the first week and latter it decreased rapidly to about 37°–35°C and then it remains constant for couple of days. This is then followed by a brief turning, which resulted in the rise of few degrees but falls back to its same old profile within few days. Three runs are studied in the same manner within a period of three weeks.

Experimental Set up

A weighed sample of the then solid waste, was introduced in the composting cells and windrow piles are constructed to a height of approximately one metre to a surface area of about 1.3 m². The initial moisture content, temperature, pH, carbon content, volatile solid content and most of all the nitrogen content of the waste were all analysed according to the methods mentioned in Table 20.10.

A representative sample of the raw waste was passed through the sorting machine and its degree of separation was recorded as per the data sheet given in the Table 20.9. After three days of precomposting, another sorting was done and the above mentioned parameters were again analysed. Except for the temperature, which is recorded daily at the site by using a compost thermometer inserted within the pile. After each experimental period of three weeks, a comparative parametric study and discussion was done regarding the analytical data so obtained during the composting. And also its influence, in the sorting percentage of the MSW.

Pretreatment study

Here a concise pretreatment study was investigated by developing various indicators for a pretreated waste:

1. Volatile solid.
2. Biochemical oxygen demand.
3. Carbon to nitrogen ratio.
4. pH.

These parameters were compared amongst, composting of:

1. Raw waste only.
2. Raw waste without plastics.
3. Raw waste with sludge.

Process involved within

Pretreatment by composting and vermicomposting process

As the Fig. 20.11 describes clearly the methods that was adopted to compare and to study the biological pretreatability potential between composting and the vermicomposting. The initial and the final sample from each system was analysed in the laboratory and are being compared.

Pretreatment by composting process

Unlike the above method, here the raw waste is mixed with digested sludge and then piled up in a windrow. The pile was disturbed every four-day interval to record the sorting output and to see whether the addition of the sludge has increased the sorting percentage of the waste. Composting of the waste with sludge in different ratio was also tried and the results were also recorded down. Figures 20.11 and 20.12 briefly describe these two methods.

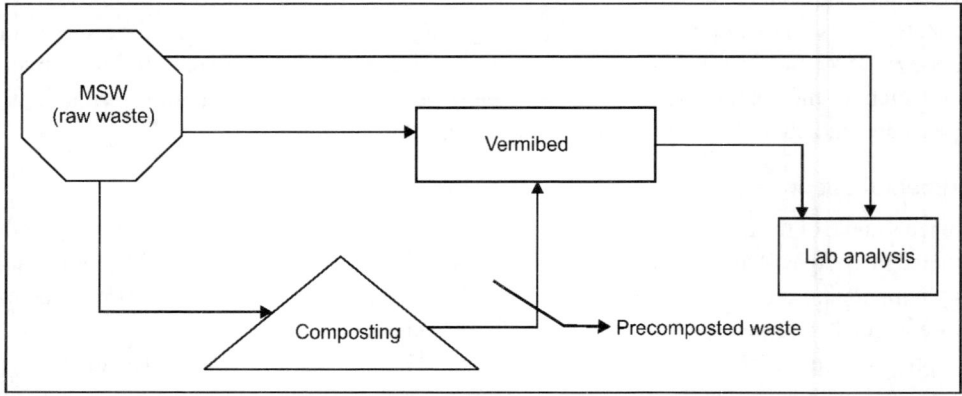

Fig. 20.11. Vermicomposting study.

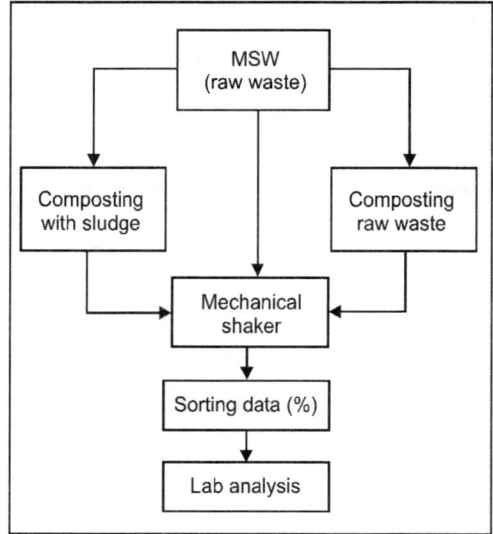

Fig. 20.12. Pretreatment study process.

The samples that we collected for the lab analysis will be used to check the parameters listed in the Table 20.10. In all the case except for pH and the moisture the samples are dried at 60°C for 24 hours and then grounded to a size of less than 2 mm using the shredder machine. The alteration in parameter with time is recorded and was used to compare the reliability of the then methods employed.

Characterisation

Characterisation of the raw waste was done before and after a brief period of windrow composting, and the result so obtained was compared amongst each other (Fig. 20.13).

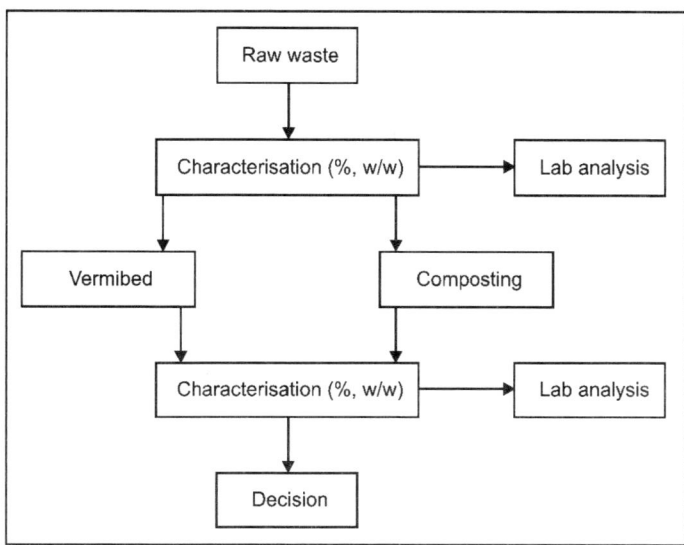

Fig. 20.13. Characterisation study.

Characterisation of the waste reflects the change in the waste components, which might have occurred due to the further degradation of the waste. The waste which are passed through the vermibeds are seen more easier to handle, if compared to the ones that are being composted aerobically. As it is clear from the above Fig. 20.13 that the lab analysis was also done after the characterisation to check the chemical parameters listed in the Table 20.10.

Vermireactor Design and Set Up

Another study was carried out parallel to the vermibed, but in a smaller scale using a vermireactor. Using this vermireactor we have investigated the actual mass reduction in the municipal solid waste after vermicomposting (Fig. 20.14).

This volume and mass reduction of the waste depends lot on certain specific parameters specially pH and the loading rate. By using the data obtained from the vermireactor we can calculate the amount of waste consumed by the worms, which is given by the difference in the mass of the initial and final waste. This vermireactor was also used to study the stabilisation of the waste and also to run a comparative study with composting pile but in a smaller scale.

Parameters mentioned in the Table 20.10 were also analysed from the waste collected from this vermireactor.

Mechanical Sorter

The mechanical sorter used here is like an inclined vibrating screen. The skeletal structure of this sorter comprises of four frames including a pan at the bottom. The surface area of each frame is 0.63 m^2. These frames are mounted alternatively on a firm metallic case holding the motor and the pulley inside. The topmost frame has a square opening of 60 mm, where the larger particles are retained. The wastes which passes through these openings falls on the next screening deck with yet another square opening of 30 mm. Then from here to next screening unit of 10 mm and finally the smallest particles are collected in the bottom pan. From this unit four products of different sizes and range are produced. The detail dimension and the photograph of the sorter are given in Appendix A at the end of this chapter.

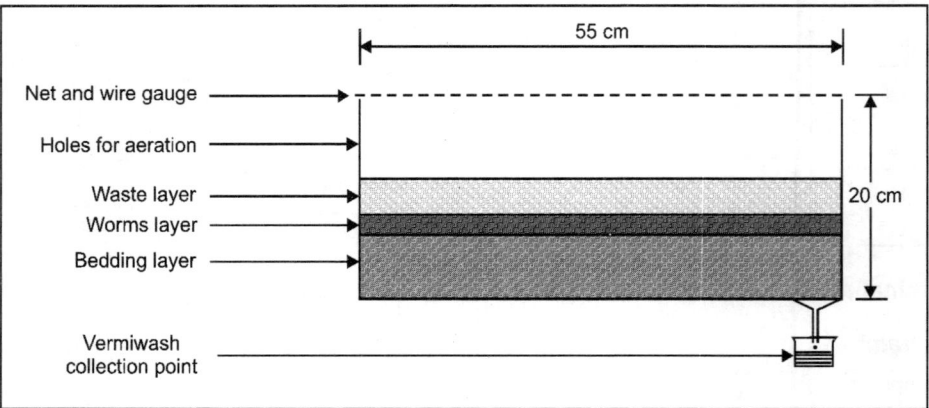

Fig. 20.14. Vermireactor applied.

By using this mechanical shaker it was able to get the sorting percentage of different pretreated waste. The main idea behind this mechanical sorter is to separate the inorganic parts and those matters that cannot be easily degraded during composting. Those wastes that were retained on the topmost screen are analysed for the stabilisation indicator. Those wastes, which are collected in the lower part, are put back in the composting system.

Following conditions were set during the operation of the mechanical shaker:
1. Operating time of three minutes.
2. Weight of the sample to be sorted was set to 10 kg.

From this sorting data we have investigated the following things:
1. Minimum period to composting of the waste for the best sorting.
2. Optimum physical variables for the best sorting.
3. Comparative sorting data between different waste mixture.
4. The convenience of plastic separation from the precomposted and the raw waste.

Following weighed samples were operated on the mechanical shaker and their data output is maintained for further comparison.
1. Raw waste as received.
2. Precomposted waste.
3. Raw waste with sludge.

Along with the above samples a representative waste sample from different timeframe of the composting was also operated on the sorting device. The data so obtained are recorded in the format shown in Table 20.9.

Table 20.9. Sorting data output format from the mechanical shaker.

Run no.

Date

Sample (identification)

Operating time

Weight of the sample

Sieve no.	Alternative (in mesh)	Weight retained	% Retained
60 mm	60 mm		
30 mm	30 mm		
10 mm	10 mm		
In the pan	>10 mm		
Total			

Determination of Physical Variables

Temperature

The temperature in each of the pile is recorded daily throughout the composting period of three weeks. A compost thermometer was inserted in the pile and the temperature was recorded when it reaches to equilibrium. Turning of the windrow pile was done without any fixed time frame. The turning of the pile in some case was done particularly when the pile temperature drops below 40°C during its first week and in some case when the incoming waste have moisture content higher than 75 per cent (wet basis).

Bulk density

The incoming raw waste were weighed in a one metre cubical cage to determine the bulk density of the, as received raw waste. We have seen that the bulk density of the market waste changes with seasonal variations. In the later stage we have used this cubical cage to study the pile settlement of the raw waste, with and without sludge, 10 per cent (w/w).

Bulk density of the raw waste was given by this simple relationship:

$$\text{Bulk density} = \frac{\text{Weight of the sample (kg)}}{\text{Volume of the cubical cage (1 m}^3\text{)}} \qquad \dots (20.2)$$

$$\text{Dry bulk density} = \frac{\text{Weight of the dry sample (kg)}}{\text{Volume of the cubical cage (1 m}^3\text{)}} \qquad \dots (20.3)$$

Despite the seasonal variation of the waste, the moisture content of the raw waste also affected the bulk density.

Pile settlement

Pile height reduction was studied using the same cage, which was used to measure the bulk density. The settlements of the wastes were recorded daily at 12 hours at the site. This cubical cage was kept at the composting site away from direct sunlight and the uncertain rainfall. A scaling system was attached to one side of the cage, the raw waste was dumped in without any external pressure or any kind of compaction as per done in the windrow system. After this run, another run was made in the same manner with a measured amount of sludge and the reductions were recorded. The detail design of the cubical cage used here is given in Appendix (A).

Analytical Methods

Sampling for chemical analysis

Sample from the compost pile is taken from three different levels to obtain a representative sample. Raw waste sample as received from the generation source will be only used to measure the pH of the waste. Other portion of the sample was dried in the oven for 24 hours at 60°C till a constant weight was obtained. The sample was then grounded to obtain particle size of less than 2 mm and following parameters was determined as given in Table 20.10.

Table 20.10. Experimental parameters and analytical methods.

Parameters	Principle	Unit
Volatile solids	Ignition at 550°C	% (Dry basis)
Moisture %	Oven heating at 105°C	%
Temperature	Compost thermometer	°C
pH	Glass electrode	–
Total organic carbon %	Calculation	% (Dry basis)
TKN or N %	Macro–Kjedahl method	% (Dry basis)
COD	Closed reflux method	mg of O_2/l
BOD	Oxitop bottle	mg of O_2/l

Analytical methods

The analytical methods used here are strictly followed according to the what is recommended in the standard method and in ASTM. These parameters were monitored periodically.

RESULTS AND DISCUSSIONS

Characteristic of Waste

About one ton of raw wastes are being collected per trip from Rangsit vegetable market (Sim Mung Mung). Initial moisture content of the raw waste lies in the range of 60 to 75 per cent. This is due to the domination of the major part of the waste by fruit peel and vegetable scraps, which further increases the field capacity of the waste stream. A brief characteristic of the waste is given in Table 20.11. Similarly the high initial temperature of the waste might be due to the heat liberated due to the then biochemical degradation of the waste.

Nitrogen content of the waste was higher than the general MSW stream due to presence green leafy vegetables and fruit peels. The presence of these vegetables and fruit parts was the main reason behind the high moisture content of the waste from the source. As shown in the Table 20.10 the volatile solid content of the waste was above 70 per cent this might be due to presence of high lignin content in the waste and the plastic particles. The pH of raw waste lies in the acidic range, this might be due to the mineralisation of organic acids that might have taken place at the source and during transportation. According to the BOD and COD value we can say that the raw waste has a high degradability potential.

Katarina, reported that the BOD/COD ratio, if lies below the range of 0.5 indicates that the sample contains less organic matter that are degradable and vice versa. The BOD/COD of the raw waste lies in the range of 0.94–0.71, which clearly indicates that the waste can be further degraded.

Table 20.11. Characteristics of the raw waste.

Content (parameters)	Measuring unit	Raw waste	
		Range	Average
Moisture	(%)	60–75	67
Carbon	(%)	44–47	45.5
TKN	(%)	1–1.05	1.025
Volatile solid	(%)	75–85	80
Temperature	°C	45–51	48
pH	–	5.5–6	5.75
BOD	mg O_2/l	4000–5000	4500
COD	mg O_2/l	4700–7000	5850
C/N Ratio	–	40–47	43.5
BOD/COD	–	0.94-0.71	0.86

The waste used for our study was mainly dominated by the organic part consisting of generally vegetable refuse and certain amount of fruit peels. Major part of the non-biodegradable fraction of 17–21 per cent mainly consists of polyethene bags, rubber tubes, plastic particles and expanded polystyrene (EPS). The organic fraction of the waste, if compared to the general MSW stream was very high since it was mainly dominated by vegetable refuse and fruit peels. This organic fraction of the waste collected from the generation source was not always the same they also differs during each collection period due to the seasonal variation of the vegetable and their shelf life. Since our main emphasis was not on composting the waste but the pretreatment of the waste, therefore source separation was not done. About 250 to 300 kg (wet weight) of the solid waste was piled up in each of the three composting cells at the site and were monitored as Pile 1, 2 and 3 respectively. During each collection of fresh waste from the source it was referred to as another run.

The moisture content of the incoming waste from the source varies during each collection according to the seasonal demand of the fruit and vegetable. The initial moisture content of the raw municipal solid waste used for the composting pile, was between 60–73 per cent (wet basis). Loss in the moisture content and the organic matter during the decomposition depends on the magnitude of the temperature and are indirectly proportional to each other. The moisture content rapidly dropped in the composting pile during the first week due to the biodegradation of the readily available organic matter, which resulted in the temperature rise. Which further promotes the evapotranspiration of the water molecules. Due to this high temperature the whole system is shifted towards the thermophilic phase, where the fast degradation of the organic matter resulted in the generation of extra heat and the loss of moisture from the pile through bacterial respiration.

Development of Pretreatment Indicators

The main objective behind the pretreatment of MSW was to reduce the overall emission from the landfills. Here, under this scope of study the main focus was channelled to reduce the organic content of the waste and to see the efficiency in sorting out the inert parts from the waste stream within the shortest possible time. Therefore, certain pretreatment indicators were developed and tested for their applicability on aerobic composting and vermicomposting, should they be checked for stabilisation and inactivation of the secondary pathogens.

Aerobic Composting

Volatile solid

It was found that the average volatile solid content of the raw was 80 per cent, when the samples are burnt at 550°C for 1 hour in and the oven. The volatile solids, representing the organic matter, were steadily decomposed throughout the experimental period. The maximum decrease in the volatile solids was achieved during the second week and third week of the composting period. This decrease in the volatile solid and subsequent increase in the ash content of the sample indicates that the biological constituents in the MSW were reduced during the composting. The reduction of volatile solids of the refuse was rather fast from the beginning as shown in Fig. 20.15. This indicates that the degradation of organic matter is taking place at the expense of carbon, since Epstein, stated that the volatile solid represents the total carbon content of that substance. In other words, greater the organic degradation, greater is the production of CO_2, which directly reduces the volatile solid and increases the temperature of the pile. The highest reduction in the volatile solid was obtained during the second run, with the reduction of 10.3, 28 and 30 per cent during the first three weeks of composting.

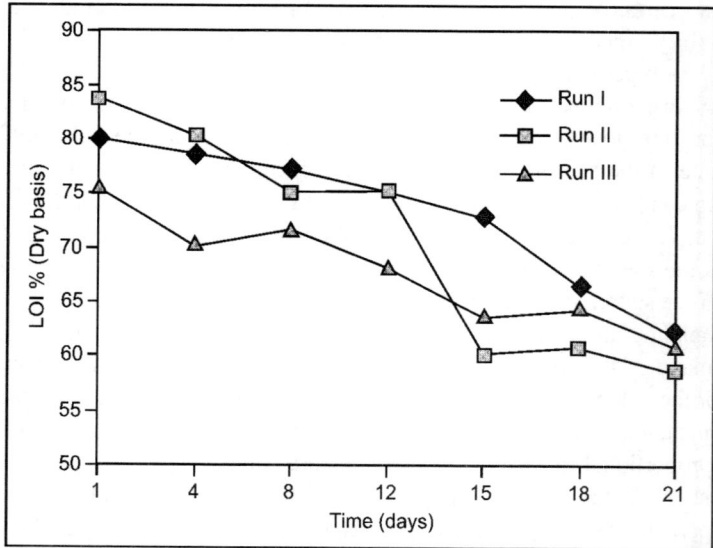

Fig. 20.15. Typical volatile solid profile during the composting, Run I, II and III.

The carbon content was obtained from the volatile solid data. The difference is in a constant factor used for the conversion from volatile solids to carbon. For practical composting, the carbon may be determined by an empirical equation as reported by Gottas:

$$\%\text{Carbon} = (\%\text{VS})/1.8 \qquad \dots (20.4)$$

where, $$\%\text{VS} = 100 - \%\,\text{Ash} \qquad \dots (20.5)$$

Carbon content

Typical carbon content profiles is shown in the Fig. 20.16. According to which, during the period of active composting the carbon content of the waste decreased from an initial value of 44.4, 46.5 and 41.8 per cent to 40.4, 33.5 and 35.3 per cent within the period of two weeks. The Fig. 20.16 also shows that even after the continued composting for the third week, the decrease in carbon content was not significant. These carbon substrates are utilised by the micro-organisms for respiration and for their cell growth. As stated by Polprasert, 20 to 40 per cent of carbon substrate is eventually assimilated into new microbial cells in the composting and the remainder being converted to CO_2 in the energy process. The percentage of total carbon content was calculated from the Eq. 20.4 as proposed by Gottas. Though this equation does not provide us with the precise value, but for the purpose of pretreatment it was sufficient to check the initial and the final carbon content.

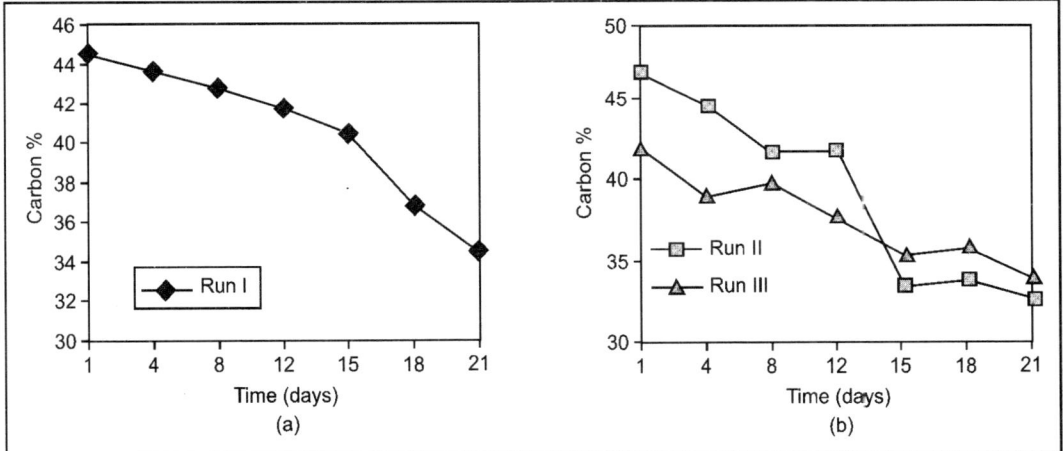

Fig. 20.16. Typical carbon profile during the raw waste composting.

The variation of carbon was greater during the thermophilic phase as compared to the variation obtained within the Mesophilic phase. Somehow within the experimental period the pile temperature remained within the thermophilic range. It was analysed that the carbon content of waste and other chemical parameters differs with each waste, this was due to the seasonal change and its effect on the generation of the waste. From this carbon reduction profile, we can reach to a simple conclusion that the degradation of readily available organic matter or carbon had occurred during the first two weeks and need not to continue further composting, as the organic load of the raw waste had been reduced.

Nitrogen content

Most municipal solid waste contains little nitrogen, and it has high carbon content to nitrogen. The variation of nitrogen content in the run 1 is shown in the Fig. 20.21. The initial nitrogen content in the compost piles were 1.05, 1.01 and 0.9 per cent for the pile 1, 2 and 3, respectively. During the first two weeks, the nitrogen contents in each pile slightly increased, this is due to the breakdown of the protein and some from the nitrogen fixation by the micro-organisms under favourable aerobic condition. Later the nitrogen content remained constant, by the end of the third week it started decreasing, this loss or the reduction of nitrogen maybe due to the increase in the pH of the pile. As reported by Snell denitrification of nitrate and nitrite occurs during the rise of pH in the system. According to the evolution of the nitrogen curve as shown in Fig. 20.17.

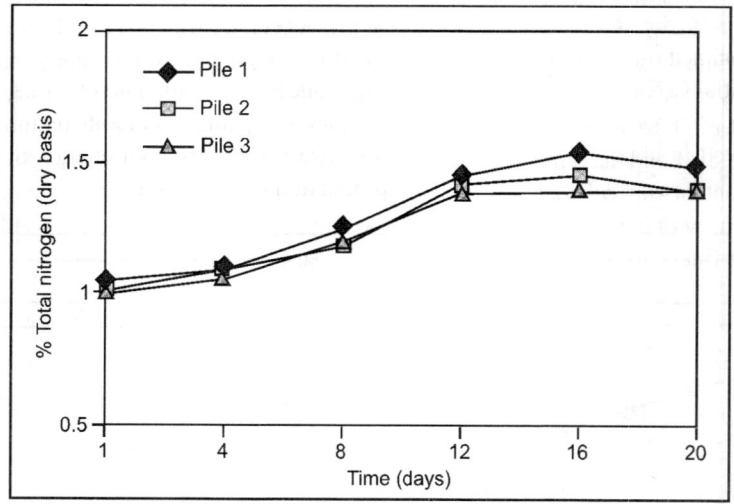

Fig. 20.17. Typical nitrogen profiles during composting (Run one).

The increase in the nitrogen content can be explained using a simple mass balance relationship of the biomass. Since the final fate of carbon in the composting heap is carbon dioxide and to assimilate in new microbial cells. The consumption of nitrogen and carbon by the composting microbes is not same. They need more carbon as to nitrogen, which might be the main reason behind the increase in the nitrogen percentage with respect to the remaining biomass. During the microbial growth, approximately 25 to 30 parts of carbon are needed for every unit of nitrogen. The concentration of carbon decreases in the form of as from the starting date of active composting due to bacterial respiration. The final fate of major part of nitrogen in a composting biomass is to get mineralised into nitrate and a little portion of it will be loss as ammonia gas during the rise of temperature.

Carbon to nitrogen ratio

The C/N ratio plays an important role in the nutrient balance in a composting heap, this ratio tells us the amount of carbon available with respect to nitrogen for the composting micro-organisms. The ideal C/N ratio for composting of Municipal solid waste generally falls in the range of 20:1 to 25:1 (Table 20.12).

Table 20.12. Comparison in the C/N ratio of different waste.

Waste type	Nitrogen %	C/N ratio
Rangsit	1.02	44.60
Food waste	1.52	34.8
Mixed paper	0.25	173
Vegetables	2.57	19.84
Tropic fruits	1.35	39.25
Yard waste	1.58	31.64

As stated by Gasser, most municipal solid waste are high in C/N ratio. The initial carbon to nitrogen ratio of the incoming waste was in the range of 40–47. This high ratio of the C/N ratio might be due to the presence of brown vegetation and plastics, which largely dominate the incoming waste stream. The other reason could be due to the absence of green leafy vegetable, which are high in nitrogen content. The C/N ratio obtained during composting run I, II and III are given in Table 20.13. From this table we know that the initial C/N ratio of the waste are being reduced due to the microbial respiration. Carbon (C), nitrogen (N), phosphorous (P), and potassium (K) are the primary nutrients required by the micro-organisms involved in composting. Micro-organisms use carbon for both energy and growth, while nitrogen is essential for protein production and reproduction (Fig. 20.18).

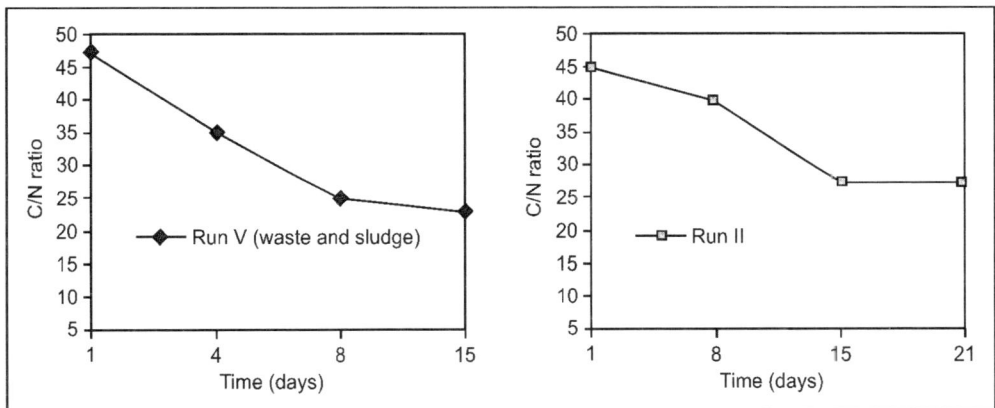

Fig. 20.18. C/N ratio profile obtained during the composting.

The reduction in the C/N ratio was due to the fast degradation of organic matter mainly the degradation of cellulose and other readily available carbon and consequent volatilisation of organic matter as the compost heats up. The nitrogen content in the composting material remained more or less same only it was transformed to inorganic forms, which might be the reason behind the drop in the C/N ratio.

C/N ratio decreases faster during the last run if compared with the one for the raw waste. This was due to the addition of nutrient rich sludge, which further aided in the degradation of organic matter and also a little contribution in the nitrogen pool. During the final run, highest reduction in the C/N ratio was obtained during the first week of its active composting. This was mainly due to the change in the proportion of carbon and nitrogen in the composting heap. As the composting process progresses the C/N ratio of the mass begins to drop, the main reason behind this fall was due to the active decomposition of carbon to carbon dioxide and new cells and transformation of mineralisation of nitrogen within the pile. This ultimately increases the nitrogen percentage with respect to carbon in the remaining biomass. The loss in the nitrogen was minimal in the form of ammonia and rests are retained in the composting heap in the form of inorganic nitrogen. So this decrease in the carbon content and the transformation of nitrogen is the main reason behind the fall in the C/N ratio of the composting material.

Carbon to nitrogen ratio was actually meant to check the nutrient balance in the composting heap. As reported by Basnayake, the C/N Ratio becomes a good indicator for the stability of the compost. He further added that, the initial C/N Ratio of the material determines the critical level for C/N ratio when it reaches stability. So here this parameter was used to see the maturity of the compost, which in other words also reflects the stabilisation of the mixed waste. It was analysed that the initial C/N Ratio of the raw waste was in the range of (40–44), which according to Polprasert, could need two weeks or more of active composting. The final C/N ratio of the precomposted waste was recorded 27 after two weeks, this reduction in the C/N ratio is due to the active degradation of readily available organic matter by the micro-organisms in the composting heap.

The reduction in the C/N ratio was from the Table 20.13. We can see that major reduction in the C/N has taken place during the second week except for the Run III, which might be due to the heterogeneous nature of the incoming waste. The data of C/N ratio value is given in Appendix B—Table 1, 2. The decrease in the C/N ratio especially in the third run was due to the increase in the nitrogen content of the waste. Which might be due to fixation of nitrogen by aerobic bacteria or might be due to the denaturing of protein from the dead microbes through biological succession, which over all increases the nitrogen content of the substrate. Since our objective lies in achieving the maximum degradation of the organic portion of the waste within the shortest time period, thus after evaluation it was seen that, should C/N ratio be used as a pretreatment indicator we can pretreat the waste in two weeks of time.

Table 20.13. Reduction in carbon to nitrogen ratio obtained during composting.

Composting	C/N initial	Reduction % 1 week	Reduction % 2 weeks	Reduction % 3 weeks
Run I	42.31	18.2	37.76	42.14
Run II	44.86	14.8	43.42	44.65
Run III	41.01	27.7	50.77	54

pH

Epstein, reported that the stabilisation efficiency decreases at lower or higher pH. This parameter somehow gives us an idea about the survival and the types of micro-organisms that might be dominating the compost pile, either acidophiles, alkalophiles or the neutrophiles. It was observed that the temperature of the pile has a direct effect on the pH variation in the pile. Higher the temperature, higher pH values were obtained.

The pH of raw wastes before composting lies in acidic range, which might have been due to the reduction of organic matter to mineral acids and thereby decreasing the acidity. From the curve shown in the Fig. 20.19 with a starting pH of 6, which was slightly acidic latter shifted towards an alkaline range within three weeks of composting.

As stated by Epstein the increase in the pH level was due to the reduction of the volatile acids and its further combination with the ammonia gas released from the denaturing of protein. In the other runs also the pH pattern shows a similar trend.

The pH drops to 5 or below in the first two to three days of composting and then begins to rise, this decrease in the initial stage is due to the formation of organic acids. After one week of composting the pH of the composting heap increases gradually and levels off at about 8.2 and remains there as long as aerobic condition is maintained. The initial pH of the raw waste before composting was recorded in the range of 5.5–6.

This which gradually increased as the waste gets stabilised. Epstein, stated that the low pH indicates the lack of maturity and as the material gets stabilised the pH increases. From our data it is shown that the pH of the raw waste initially being acidic, latter after ten days of active composting is shifted towards the alkaline phase as already indicated in the Fig. 20.17. Thus we can stop further composting after 10 days or preferably after two weeks.

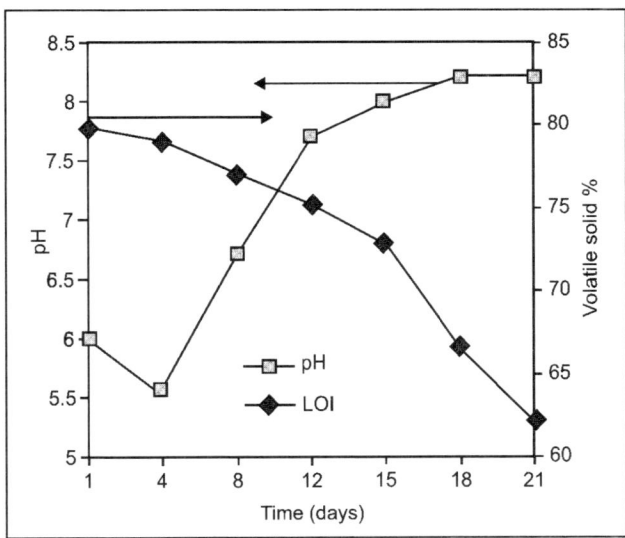

Fig. 20.19. pH profile and the volatile solid % during the windrow composting pile (Run one).

BOD/COD ratio

As per Sawyer, BOD or biochemical oxygen demand is usually defined as the amount of oxygen required by bacteria while stabilising the decomposable organic matter under aerobic condition. In other words it reflects that the BOD demand is directly related with the pollution strength, higher the demand more potential for contamination and vice versa. The BOD value here not actually represents the total BOD demand of the composting pile.

It was calculated from the compost filtrate using oxitop bottle to simulate the natural BOD load that might occur if the untreated waste was land filled or disposed off. Using this BOD value so obtained from the filtrate as a stabilisation indicator it was able to judge how effective our pretreatment operation was running. The BOD/COD ratio was also analysed to estimate the amount of degradable organic matter left as suggested by Katarina. Greater reduction BOD was recorded latter during the second week of the composting, this may be due to the presence of more lignin or other biologically resistance substance, which cannot be consumed by the compost micro-organisms (Fig. 20.20)

However from this graph (Fig. 20.20) we can interpret that the degradable organic portion of the waste has reduced in the course of time, with the maximum degradation achieved after the first week of active composting. Katarina reported that the ratio of BOD to COD of about 0.5 or above indicates the presence of readily degradable organic material, and a ratio of 0.5 or below indicates the presence of poorly degradable organic matter. Or in other words the ratio of BOD/COD will enable us to determine whether the sample still has organic matter to degrade further or not. But this indicator cannot stand alone to prove the substrate degradability due to the heterogeneity of the incoming waste. So the BOD/COD ratio of the waste was considered as an additional indicator supporting the pretreatment of MSW.

Temperature and pathogen inactivation

As per Polprasert, thermal death point of majority of the secondary pathogens occurred when the temperature of the pile reaches 60°C. It was also reported much earlier by Allan, that if the pile temperature remains above 55°C for 24 hours it will result in almost complete destruction of human and animal pathogens, including viruses, bacteria and parasites.

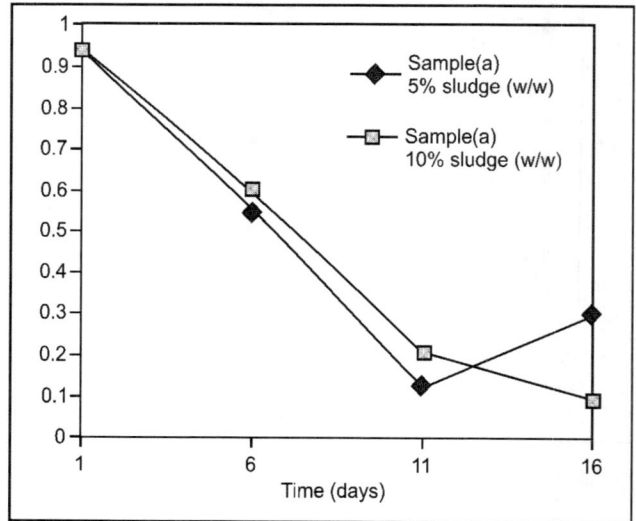

Fig. 20.20. Typical BOD/COD ratio obtained from the sorted waste (Run V).

Temperature gradient of the composting pile has a direct effect on the pathogen inactivation, which in other words, higher the temperature more will be the rate of pathogen die-off and more sooner the precomposted matter will be biomedically stabilised from any secondary pathogen. Since the microbial analysis of the pathogens are beyond the scope of this study, but still according to the temperature profile of the composting pile as shown in Fig. 20.21, we can rule out the possibility for the inactivation of the pathogen after the active composting.

It was recorded that the temperature profile during the Run IV was entirely different from the other runs. This was due to the composition of the waste in the fourth run, which was greatly dominated by large durian (an oval spiny tropical fruit containing a creamy pulp with a fetid smell), fruit peels. That non-uniform distribution of the unsorted raw waste further discourages the thermophilic micro-organisms to act on it. As stated by Epstein, microbial decomposition of the organic matter takes place on the surface of the particles and smaller the material greater is their surface area. It was also recorded that the temperature profile of each run was different from the other, this was mainly due to the unsorted waste stream and the seasonal occurrence of different fruits and vegetables.

Normally the pile temperature should start from the Mesophilic phase (25°–45°C) and then followed by the thermophilic phase (50°–65°C). Thermophilic temperatures are desirable because they destroy more pathogens, weed seeds and fly larvae in the composting material. But here the temperature of the pile started from the thermophilic phase since the Mesophilic phase took place at the generation source according to the period of collection and the time spent on transportation. A typical temperature profile recorded during the Run I is shown in Fig. 20.21. The peak temperature during the first run was recorded 60°C in the pile 2 and 3. Rise in the pile temperature within the first week of composting means that the micro-organisms are actively consuming the organic fractions of the litter and thermophilic organisms largely dominate the pile.

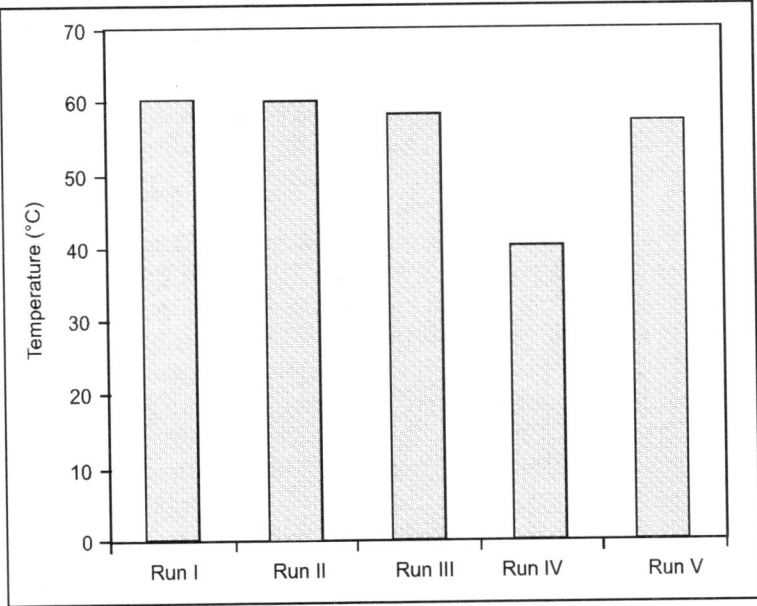

Fig. 20.21. Maximum temperature attained during each composting run.

As reported by Allan, the temperature of the pile rises due to the biological oxidation of carbon. Figure 20.22 shows the typical temperature profile recorded during the Run I of raw waste composting. From this temperature profile we can directly interpret that the maximum reduction in the organic matter have occurred during the first week of active composting. Since the heat generated within the system comes only from the bacterial respiration, where the organic part of the waste are being consumed.

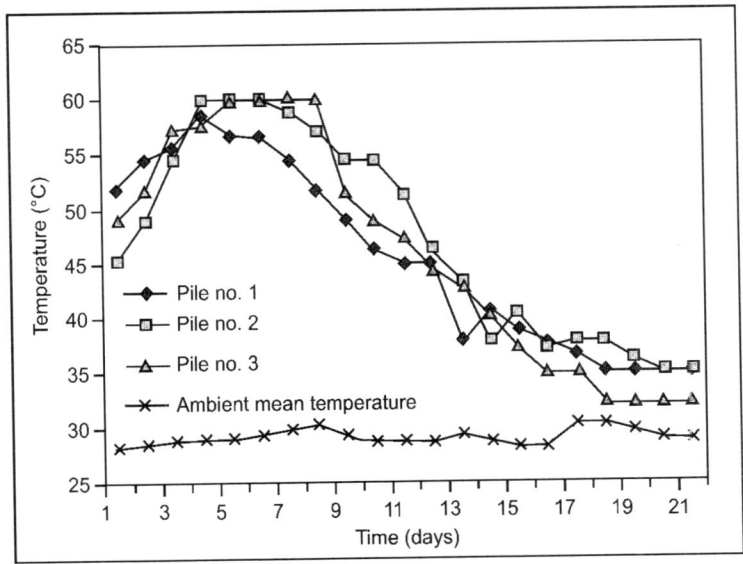

Fig. 20.22. Typical temperature profile during the composting Run I.

Bulk density and pile settlement study

The bulk density of the raw waste as received, ranged from (0.32–0.4) kg/l or (320 to 400) kg/m³. The bulk density of the waste was determined using a pen as described already. The seasonal variation and the shelf life of the vegetables directly effect the difference in the bulk density of the incoming waste. The bulk density of the waste was reduced from 320 kg/m³ to 150 kg/m³ within two weeks. Latter in the next run, when the waste was mixed with sludge, greater reduction was recorded in its bulk density. The bulk density of the waste (with sludge) was reduced from 420 kg/m³ to 168 kg/m³. In both cases, the reduction in the bulk density after two weeks was recorded as 53 per cent in the case of the raw waste and as high as 60 per cent for the waste mixed with sludge. These reductions in the bulk density may be due to the high initial moisture content of the raw waste. And that the degradation of the known amount of waste dumped in the cage might have taken place, like that of a conventional static pile. Since during the experimental period the cages were neither disturbed nor it was kept in a direct sunlight. Regarding the aeration, the structure of the cage promotes proper aeration system from all the sides. Such reduction in both the pile height and the bulk density clearly indicates the applicability of the method on a large scale to save the load on the landfill space and to minimise other related environmental issues.

The cubic cage or the pen employed in the determination of the bulk density, pile settlement was also studied with the raw mixed waste and then latter with the addition of sludge to the raw waste as another means of disposing the sludge. The results obtained through the pile settlement study are represented graphically in the Fig. 20.23. The initial height of the waste was set to 100 cm. The pile settlement rate was recorded highest during the first two days with an average of 12.5 cm/day for the waste which was mixed with 5 per cent (w/w) of sludge.

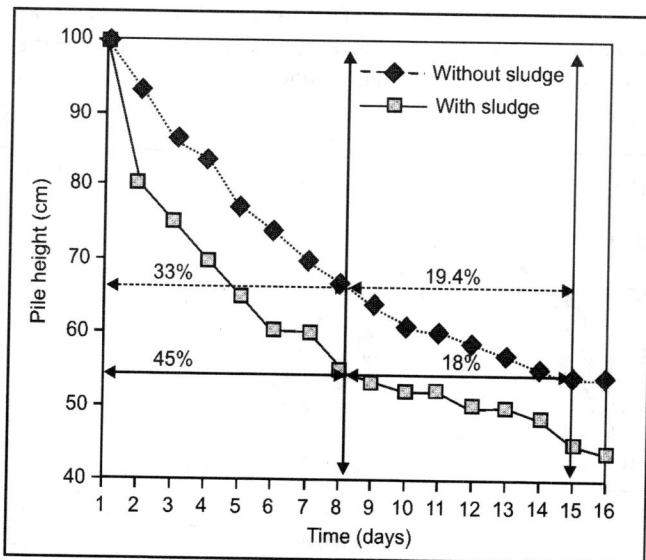

Fig. 20.23. Pile height reduction profile.

This greater reduction in the latter case is due to the presence of more nutrients for the composting bacteria, which further promoted the degradation of the organic matter present in the waste. This reduction in the pile height prophecies the situation that might have otherwise taken place in a static-pile composting where the pile is not disturbed throughout the experimental time frame. The design of the cage or the pen somehow enables an easy flow of the air as compared to the forced aeration and aerating poles inserted in the static pile. The picture of the cage used here is shown in Appendix (Fig. A-9). This reduction in the pile height and also in its bulk density indicates that this method will further save the landfill area by more than 50 per cent had the raw waste been pretreated prior to landfilling. Over all pile settlement percentage for waste mixture with and without the sludge were as follows, the one without sludge has reduced as much as 52 per cent. And the one with sludge has reduced to 63 per cent. This pile height data also tells us the volume reduction of the waste that might occur since the volume of the pen was 1000 l or 1 m³. From the pile settlement study it would be more advisable and more beneficial, had the raw waste be composted for one week of active composting only rather than two weeks.

Sorting results

As already discussed, the raw waste and the precomposted waste are sorted using the inclined mechanical vibrator to compare the sorting efficiency and its convenience. The sample weight of 10 kg and the operating time of three minutes were found to be the best set conditions needed to run the other precomposted samples through the sorter. It was recorded during Run II of the composting pile that the sorting efficiency of the raw waste increased as the days add up to its composting process as shown in Fig. 20.24. It also describes the increased efficiency of the sorter machine towards the precomposted waste with reference to the raw waste being sorted on the first day. It explains that the sorting efficiency rate was greater by 6 per cent during the first week of the composting. This may be due to the fact that during the first week of composting the maximum degradation of the waste had taken place due to its high average temperature.

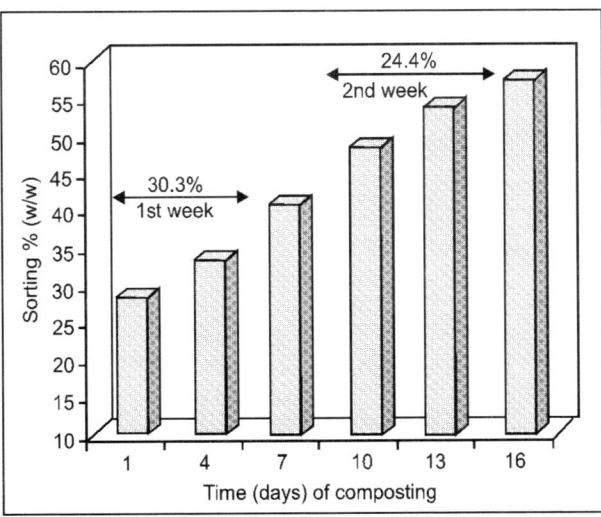

Fig. 20.24. Sorting results of precomposted waste, Run II.

From the sorted sample, on the topmost screening deck the particles retained are generally dominated by plastic matters and some large semi-degraded organic parts. The components of the particles on the topmost screen also varies according to the additional days of composting. Furthermore by the addition of the sludge during the Run V, the sorting efficiency increased much more, if compared to the ones without sludge. The sorting percentage of the vibrating screen was defined as the relative decrease in the amount of oversized waste retained on the topmost screen and increase in the smaller fraction from the total waste with respect to composting time. Where, lesser the waste retained on the topmost screen, the more it is being sorted into smaller fractions. The sorting efficiency of the mechanical shaker achieved better when the waste sample containing the sludge were passed through it. Where, one of the sample was mixed with 15 kg or 5 per cent (w/w) of sludge and other with 30 kg or 10 per cent (w/w) respectively.

It was clearly indicated from the above sorting result, that the addition proper amount of sludge not only promotes the biological degradation it also increases the sorting efficiency. Comparing Fig. 20.25 and Fig. 20.26 it was clear that, even by addition of 30 kg of sludge the sorting results were greater than the one which was operated without the sludge. Total sorting percentage of 28.5 per cent was achieved from the precomposted raw waste and a total of 32 per cent of sorting was recorded for the precomposted raw waste and sludge mixture.

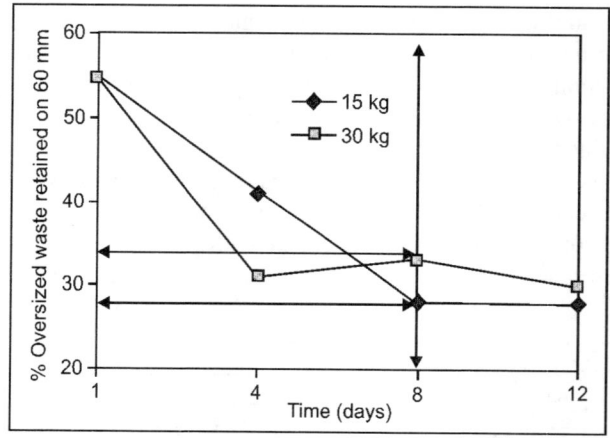

Fig. 20.25. Decrease in the oversized waste on the top screen with sludge.

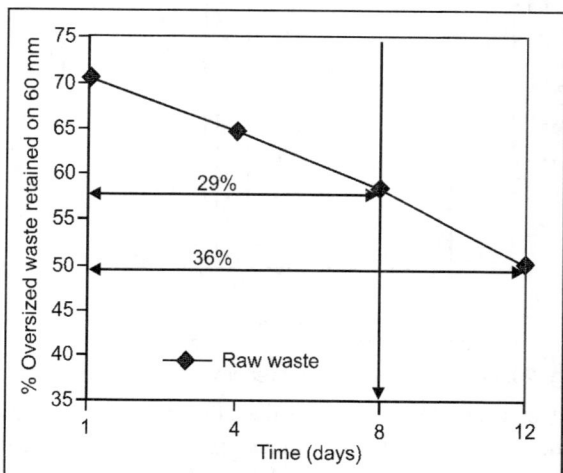

Fig. 20.26. Decrease in the oversized waste retained on the top screen without sludge.

Although the sorting efficiency was related with the microbial degradation of the organic matter, but it was observed during the fourth composting run that the moisture content and the particle size of the sample negatively affects the sorting efficiency of the mechanical shaker. Since back then, non-shredded durian peels have dominated the major part of the waste in the fourth run, which are not only hard to degrade but also out-bounds from the screen openings. In other words the sorting efficiency of the waste greatly depends on the particle size of the incoming raw waste, this was observed in the Run IV. The poor sorting efficiency was due to the presence of large unshredded durian peel which was not only hard to decompose but also hard to handle during the pile formation.

So from this sorting data we can assume that the raw waste can be best sorted out by active composting for one week and further more the sorting efficiency can be improved by adding correct amount of sludge. Since our objective lies in the favour of post sorting the MSW, therefore the optimum condition for the best sorting can be obtained by composting the raw waste for one week with sludge.

After testing all the mentioned pretreatment indicators it was experienced that the main driving force behind all the different testing method used here in aerobic composting was the temperature, it has direct effect on all the parameters governing the stability of the waste. It was recorded that all the optimum conditions for the pretreatment of raw waste took place during the thermophilic phase where the temperature of the pile was well above 45°C. Optimum moisture content of the initial raw waste for best pretreatment should fall in the range of 65–75 by wet basis. If the raw waste was too wet it may create anaerobic pockets and would further demote the degradation rate.

Using those pretreatment indicators we can reach to a promising conclusion that pretreatment of the incoming MSW from the source was achieved within a period of one week and there has been a substantial decrease in the organic load of the raw waste. The portion of post sorted pretreated waste retained on the third and last deck can also be further vermicomposted after its pH adjustment, for more safer and economical disposal, which will further reduce the organic load by reducing the disposable volume of the pretreated waste.

Vermicomposting Study

Vermicomposting study for the pretreatment of MSW was started by first breeding the worms in a suitable vermibins for an approximately six weeks. Then the worms were later shifted to an open bed of area 2 m^2 at the composting site.

Breeding of worms

Both the local worms and the redworms are breeded in the vermibins for a period of one and half months. It was found that the local worms could not breed that fast as compared to the redworms and were sensitive to variation in pH, temperature and the moisture of the bedding soil. The breeding potentials of the redworms greatly increased when they are shifted to an open vermibed. The pH, temperatures and the moisture are maintained and recorded thrice a week as shown in Table 20.14.

Table 20.14. Average measured range of parameters.

Parameters	Vermibin	Open bed	Unit	Method
Temperature	29–31	29–31	Celsius	Thermometer
Moisture	60–70	65–75	%	Standard method
pH	6.5–7.5	7–8	–	Glass electrode

Growth rate determination

It was not at all an easy task to count the initial and final worms on a 2 m² bed scale as thought previously, even if so the results so obtained would have lots of error. But it was seen that the reproductive rate of the red worms was much higher as compared to the local worms. However this same experiment was done on a smaller scale using two vermireactors, which will be discussed latter in the section. The result so obtained from the two reactors also showed that the worms grow better when they were fed with vegetable and fruit scraps as with the precomposted waste. Local worms were not used in the latter experiments due to their poor reproductive rate and their poor feeding habits. However local worms did showed an increase in their populations but very less or negligible as compared to the redworms.

Vermibed

Vermibed by the dimension of (1 × 2) metre was used to study the organic degradation of the waste as shown in Appendix (Fig. A-1), since the waste has to be spread uniformly on the vermibed. Therefore a total waste of 15–20 kg of raw waste and precomposted waste was fed to the worms to a maximum height of 3–4 cm per week. It was observed that the vermibed works better in reducing the organic parts of the mixed waste if compared to the aerobic composting. But during the sampling of waste, it gets mixed up with the bedding soil and on top of it the moisture content of the waste were never constant due to the watering of beds in order to maintain a moist environment for the worms. As a result a precise proximal analysis was not possible. When in the process of watering the beds it also increases the moisture content of the waste spreaded on the bed due to which it cannot be passed through the sorter, even if so the result cannot be compared with the precomposted waste sample due to its higher moisture content. So to have a comparative data regarding the applicability of pretreatment indicators developed, lab-scale reactors were used to compare the stabilisation rate amongst the two using red worms.

Vermireactor study

As explained above, the growth rate of the red worms was higher when the vegetable and fruit peels were fed as with the precomposted waste. The final number of living worms (including the adults and the young ones) were more in the reactor fed with vegetable and fruit peel but it was hard to compare the number of cocoons in both of the reactors. The growth rate of the redworms in the precomposted waste was 0.45 worms/day whereas the growth rate of the worms in the vegetable and fruit peel was 0.78 worms/day. The low growth rate of the worms in the first case may be due to the high alkalinity of the precomposted waste, which have a direct effect on the bedding soil and to the worms. Ronald reported that the worms perform best in neutral pH. And the red worms generally prefer the pH of their bedding soil slightly acidic or neutral then to alkaline. The raw data recorded from the two reactors are given in the Table 20.15.

Table 20.15. Growth rate of worms under two different feeds.

Vermireactor	Time (days)	Feed	Initial worm wt. (gm)	Final worm wt. (gm)	Growth rate worms/day
A	21	Waste sample	100	109.45	0.45
B	21	Vegetable scraps	100	116.38	0.78

In the next run, using the same reactor other parameters namely, volatile solid, nitrogen content and carbon content were analysed; the total mass reduction was also studied.

It was found that the reduction in volatile solid and the C/N ratio of the vermicomposted waste were not that high as compared to the ones analysed from the aerobic composting. Total of 19.1 per cent of reduction was obtained in the case of volatile solid and carbon. Whereas increase in 6.87 per cent of nitrogen that resulted in the total decrease of C/N ratio to 24.6 per cent. Table 20.16 explains the data obtained from the vermireactor operated. The mass reduction in the waste was also studied using the same reactor, total mass loss of 0.635 kg was recorded within a period of 12 days and this mass reduction value was recorded daily.

The moisture content of the reactor was also maintained and the water added to that system was being taken into consideration when adding up the total mass loss. Total of 555 ml of water was added to maintain the proper moisture in the reactor. Loading rate of the precomposted waste was 1750 mg VS/gm of worm/week.

Table 20.16. Vermireactor analysis of the precomposted waste.

Dated	Day	VS%	Carbon%	N%	C/N
14/4/2002	1	68	37.77	1.76	21.46
17	4	67	37.22	1.76	21.14
20	7	63	35	–	–
23	10	64	35.55	1.8	19.75
29/4/2002	16	55	30.55	1.89	16.16
Total reduction %		19.11	19.11	–6.87	24.68

It was known from the vermireactor study that only two indicators namely volatile solid and C/N profile can be studied effectively that too on a reactor scale. The drawback in using vermicomposting as a method to pretreatment lies in the less throughput of the pretreated waste if compared to aerobic composting.

Comparison

The results obtained from the aerobic composting and vermicomposting can be compared on the basis of their suitability on a pilot scale, the results may differ if applied on a large scale. Regarding the stabilisation of the waste it was found that volatile solid reduction during the vermicomposting proved better than the aerobic composting, the difference of 7.3 per cent was recorded higher in the former case. Other chemical parameters so analysed between the two are given in Table 20.17.

Table 20.17. Comparative study between vermicomposting and composting.

Day	Volatile solid reduction		Nitrogen		C/N ratio	
	Composted	Vermicomposted	Composted	Vermicomposted	Composted	Vermicomposted
1	68	68	1.76	1.76	21.46	21.46
4	63.6	67	1.75	1.76	20.19	21.14
7	64	63	–	–	–	–
10	60	64	1.77	1.8	18.83	19.75
16	60	55	1.77	1.89	18.83	16.16
Reduction %	11.76	19.11	–0.56	–7.38	12.25	24.69

Note: The data above are analysed from the composting and the vermireactor, Run II.

Greater increase in the nitrogen content of the vermicomposted waste is the main reason behind the production of vermicompost, the nitrogen content in the waste can only be used to study the decrease in the carbon to nitrogen ratio (Table 20.18).

Table 20.18. Comparative evaluation of composting and vermicomposting.

Pretreatment indicators	Applicability		Remarks
	Aerobic composting	Vermicomposting	
Volatile solid	Applicable	Applicable	Greater reduction was obtained during vermicomposting but less through put
C/N ratio	Applicable	Applicable	Greater reduction was obtained during vermicomposting but less through put
pH	Applicable	Not applicable	Vermibeds have to be maintained at favourable pH
Volume reduction	Applicable	Applicable	Not comparable, since waste input in the vermibed has to be precomposed.
Temperature	Applicable	Not applicable	Vermibed beds have to be maintained at favourable temperature

Limitations in the sampling and chemical analysis

1. Due to the heterogeneity of the non-pulverised waste, it was rather cumbersome to get the representative sample for the various analyses, though a best effort was made.
2. The determination of total carbon content in the waste was calculated using Eq. 20.4. The total carbon content of the waste however, does not necessarily represent the total available carbon, as the lignin content of the waste was not determined.
3. Volatile solid was analysed using procedures given in standard method, by oven heating the sample for one hour at 550°C. The biodegradable volatile solid (BVS) of the sample was not determined, as it was beyond the scope of the study.
4. The post sorting to erroneous results, when the waste out bounds from the topmost screening deck. Such results were discarded and not considered.

CONCLUSIONS

1. From the results obtained from this study it can be concluded that:
    (a) The most suitable method for the pretreatment of non-pulverised and unsorted MSW is to run an active windrow composting of 300–400 kg of waste for one week. Whereas, for vermicomposting the most applicable method was to spread the waste over the bed to a thickness of 8–10 cm and simultaneously maintain the pH, temperature and the moisture of the bedding soil within the favourable range of the worms.
    (b) Maximum degradation in the organic portion of the raw waste took place during the first week of active windrow composting. When the temperature of the pile was within the thermophilic range and the moisture of the raw waste drops down to 65–67 per cent of the total weight.
    (c) Most conducive method in sorting out the inorganic and organic parts from the raw waste was to run an active windrow composting of the raw waste for one week with 5 per cent (w/w) of nutrient rich sludge and latter passing it through the mechanical sorter.

(d) Optimum operating condition for the mechanical shaker towards the commingled MSW was reached, when it was operated for 3 minutes with 10 kg of the precomposted waste.

(e) Optimum sorting efficiency of the mechanical shaker in separating the pretreated waste was reached when the raw waste was put to active composting with 5 per cent (w/w) sludge for one week.

2. From the study that was conducted, following advantages of pretreating the raw waste can be concluded which are:

(a) Reduction in the organic load of the raw waste within a period of one week should it be landfilled or disposed off.

(b) Easier separation of the plastics and other refractory particles from the raw waste stream.

(c) Less emission of landfill gas and leachate due to reduction in the organic parts from the waste stream should the waste be pretreated.

(d) Volume reduction of the raw waste to 50 per cent by pretreating it for one week, which will save the overall landfill space required.

3. By conducting the above study we can develop a strategic plan to a sustainable solid waste management (Fig. 20.27).

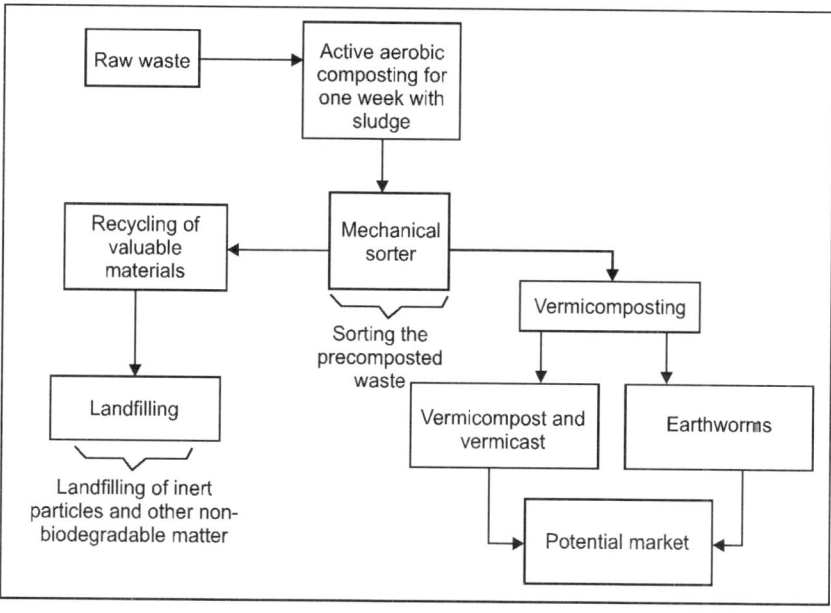

Fig. 20.27. Strategic plan for solid waste management.

RECOMMENDATIONS FOR FUTURE WORK

The following aspects are recommended for future investigations and study:

1. To study the parameters governing the stabilisation and the optimum conditions needed for the shredded and nonshredded special waste using vermicomposting and aerobic composting.

2. To investigate the possibilities of composting and pretreating the raw market waste mixed with industrial or demolition waste for the purpose of stabilising the latter one.

3. To run a kinetic study on the leachate quality generated from a precomposted and vermicomposted waste using pilot scale lysimeter.
4. To study the biological succession in a raw waste composting heap, especially to see its effect on the BOD reduction and overall stabilisation of the waste.
5. To investigate the possibilities and optimum conditions needed, in the reclamation of chemically polluted soil by vermiculturing.
6. To investigate the effect and role of different bulking agents in the pretreatment of municipal solid waste.
7. To study the nitrogen transformations during the pretreatment of municipal solid waste by windrow composting.

APPENDIX A

Fig. A-1. Vermibed at the composting site.

Fig. A-2. Raw market waste piled-up for composting.

Fig. A-3. Four different particle size sorted from the precomposted waste.

Fig. A-4. Inclined mechanical vibrator used for the sorting.

Fig. A-5. Sorted waste retained on the topmost screen.

Fig. A-6. Sorted waste retained on the second screening deck.

Fig. A-7. Sorted waste retained on the last screening deck.

Fig. A-8. Sorted waste retained inside the pan.

Fig. A-9. Pen used during the pile settlement study.

Fig. A-10. The pen (cage) used to study the bulk density and pile settlement.

Fig. A-11. Composting site, AIT research station, Thailand.

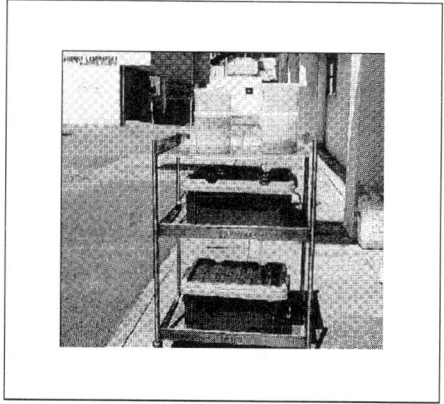

Fig. A-12. Vermireactors used in the study.

Fig. A-13. Vermireactor, a closer view.

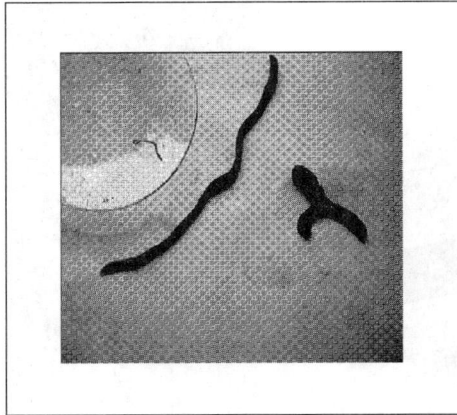

Fig. A-14. An adult (clitellate) and a juvenile red worm.

Fig. A-15. Vermibins at the site used for the breeding of worms.

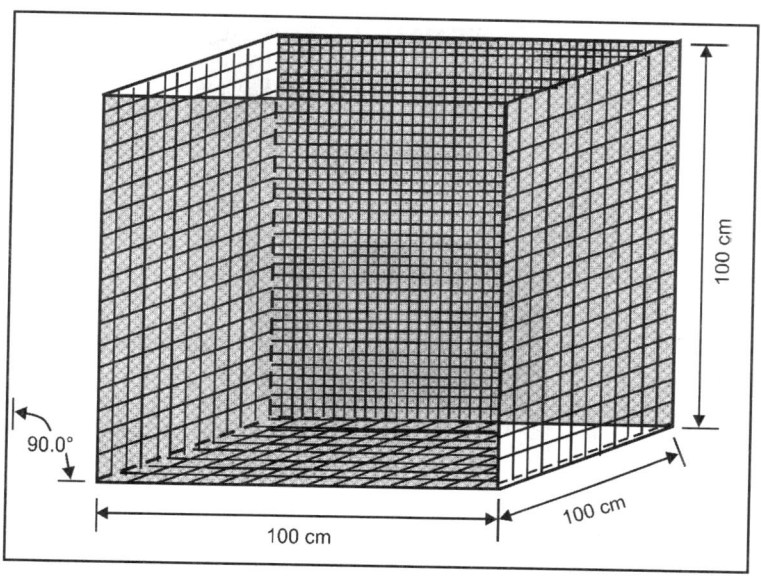

Fig. A-16. Dimensions of the pen (cage) used.

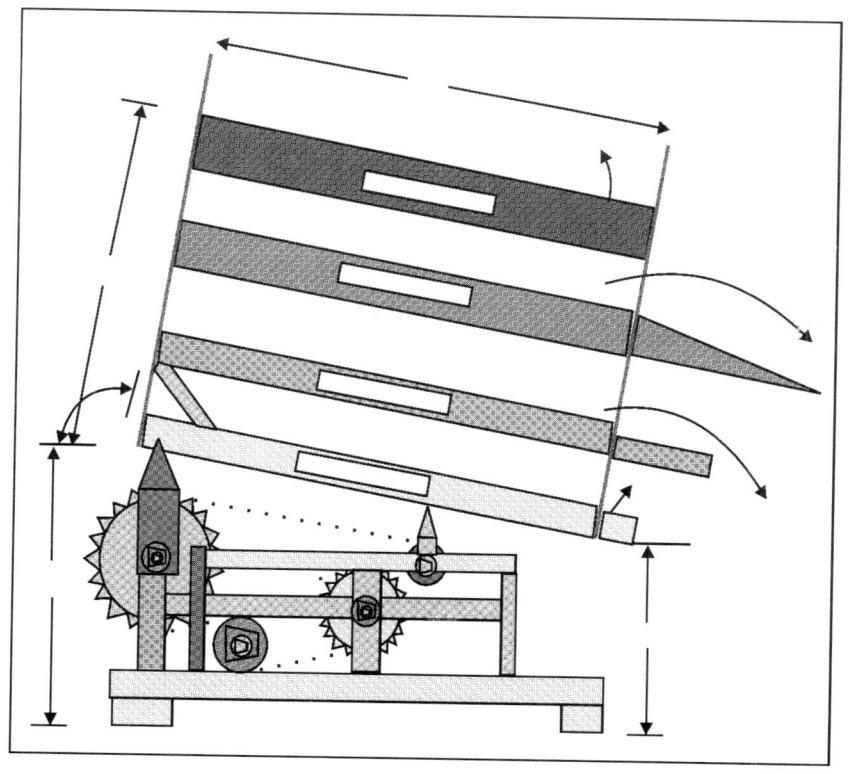

Fig. A-17. Side view of the sorter machine and its general dimensions.

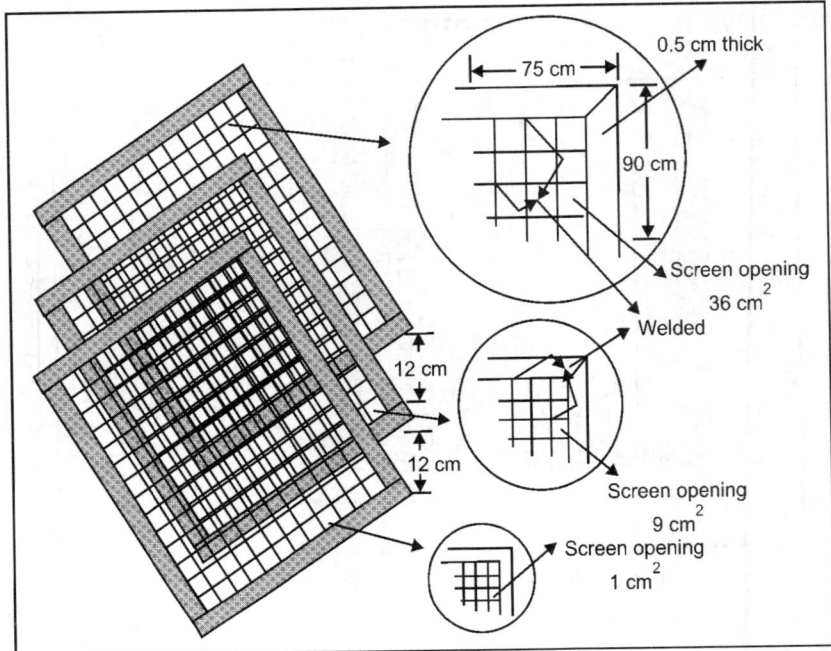

Fig. A-18.

APPENDIX B

Compost Analysis

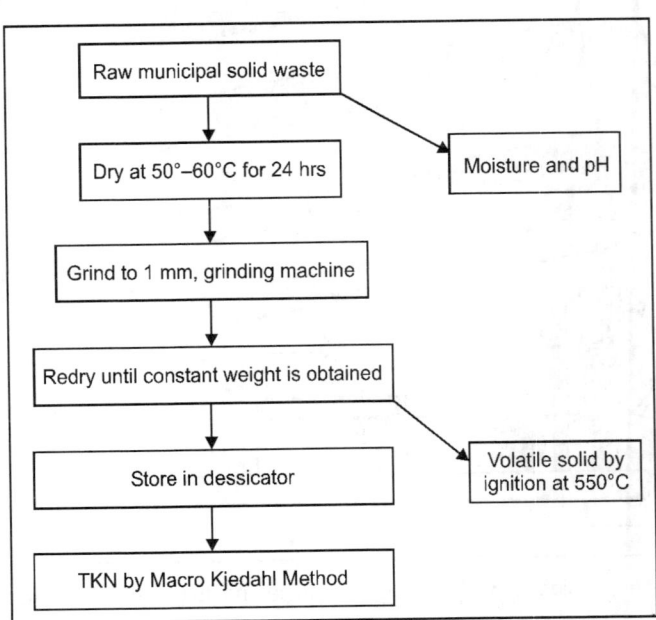

Sampling of Compost for BOD and COD Analysis

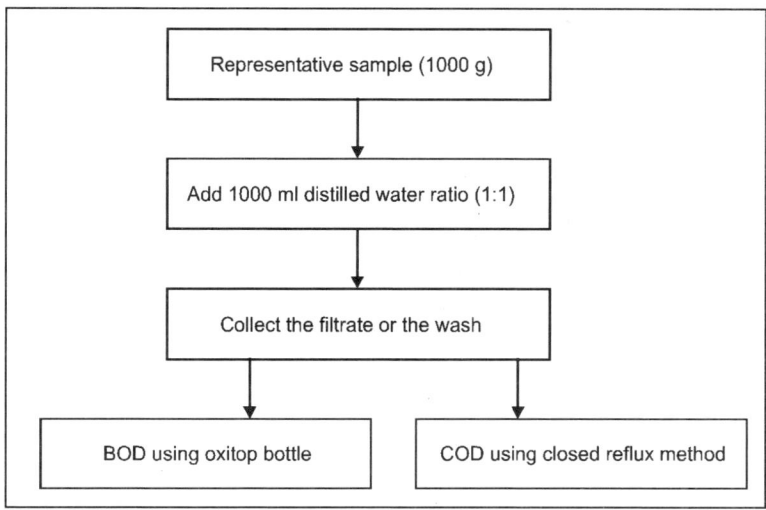

Nitrogen Per cent Determination (Macro Kjedahl Method, E-105 ASTM, 1996)

The procedure measures free ammonia or ammonia formed from the conversion of organic nitrogenous compounds such as amino acids and proteins.

1. Digestion of the sample:
   (a) Take 1 g of the sample and carefully transfer it to 500 or 800 ml Kjedahl flask containing 7 to 10 g of K_2SO_4 and 0.6 to 0.8 g of mercury.
   (b) Add 30 ml of sulphuric acid to the mixture by pouring it down the neck of the flask while rotating the flask to wash any samples adhering to the walls.
   (c) Continue the digestion until all the sample particles are oxidised, as evidenced by a nearly colourless solution. The total time for digestion will require 3 to 6 hours.
   (d) When the digestion is completed and the solution has cooled, a few crystals of $KMnO_4$ may be added to ensure the complete oxidation.
2. Distillation of the digestate (acid titration):
   (a) Dilute the cooled digestion mixture to about 300 ml with water and remove any heat of the dilution by cooling the flask under running water or by allowing it to stand until cool.
   (b) Add 1 to 2 g of granular zinc to the digestion mixture and slowly add 30 ml to 100 ml of alkali solution according to the dilution and distillate the mixture.
   (c) Accurately pipette 20 ml of 0.2 N H_3BO_4 into a 250 ml Erlenmeyer flask as a receiving solution.
   (d) Add 6 drops of mixed indicator solution in it.
   (e) Set up the distillation unit.
   (f) Collect 100 ml of distillate.
   (g) Titrate the ammonia collected in the Erlenmeyer flask containing boric acid to the mixed indicator end point using 0.2 N H_2SO_4 as the titrant.

(h) Run a blank determination in the same manner using approximately 1 g of sucrose as the sample material.

3. Calculation:

(a) Calculate the per cent of nitrogen in the analysis sample as follows:

$$\text{Nitrogen \%} = \{(A-B) \times N \times 0.014/C\} \times 100$$

where,

A	= millilitres of H_2SO_4 required for the titration of the sample.
B	= millilitres of H_2SO_4 required for the titration of the blank.
N	= normality of the H_2SO_4.
C	= grams of sample used.
0.014	= milli-equivalent weight of nitrogen.

Chapter 21

Composting Toilet and Aerated Static Pile Composting

INTRODUCTION

A composting toilet is an aerobic processing system that treats excreta, typically with no water or small volumes of flush water, via composting or managed aerobic decomposition. This is usually a faster process than the anaerobic decomposition at work in most waste-water systems, such as septic systems.

Composting toilets are often used as an alternative to central waste-water treatment plants (sewers) or septic systems. Typically they are chosen: (i) to alleviate the need for water to flush toilets, (ii) to avoid discharging nutrients and/or potential pathogens into environmentally sensitive areas, and (iii) to capture nutrients in human excreta. Several manufactured composting toilet models are on the market, and construct-it-yourself systems are also popular.

These should not be confused with pit latrines (see latrine, pit latrine, and arborloo or tree bog), all of which are forms of less controlled decomposition, and may not protect groundwater from nutrient or pathogen contamination or provide optimal nutrient recycling.

TYPES OF COMPOSTING TOILET

Manufactured Composting Toilet Systems

'Self-contained' composting toilets complete or begin the composting in a container within the receiving fixture (Fig. 21.1). 'Remote', 'central' or 'underfloor' units collect excreta via a toilet stool, either waterless or micro-flush, from which it drains to a composter. 'Vacuum-flush systems' can flush horizontally or upward with a small amount of water to the composter. 'Micro-flush toilets' use a small amount of water usually 0.5 litre (0.88 imp pt) per use.

'Self-contained' composting toilets are slightly larger than a flush toilet, but use roughly the same floor space. Some units use fans for aeration, and optionally, heating elements to maintain optimum temperatures to hasten the composting process and to evaporate urine and other moisture. Operators of composting toilets commonly add a small amount of absorbent carbon material (such as untreated sawdust, coconut coir, peat moss) after each use to create air pockets for better aerobic processing, to absorb liquid, and to create an odour barrier. This additive is sometimes referred to as 'bulking agent'. Some owner-operators use microbial 'starter' cultures to ensure composting bacteria are in the process, although this is not critical.

'Remote', 'central', and 'underfloor' models each feature a chamber below the toilet stool (such as in a basement or outside) where composting takes place. These are typically used for high-volume and year-round applications as well as to serve multiple toilet stools. Several systems are available as well as many build-it-yourself options. In contrast, 'desiccating toilets' dry the excreta to destroy pathogens, though one study suggested that drying can result in rehydration of pathogens when in contact with moisture later.

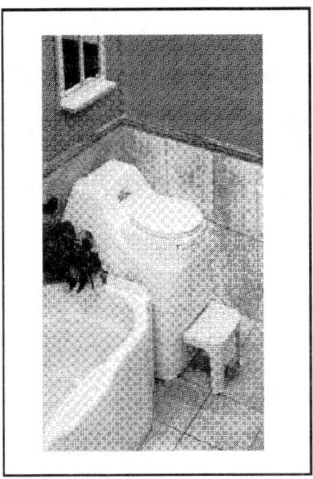

Fig. 21.1. Self-contained composting toilet system.

The performance testing standard for composting toilets in the United States is American National Standard/NSF International Standard ANSI/NSF 41-1998: Non-Liquid Saturated Treatment Systems. An updated version of ANSI/NSF Standard 41 was published in 2005. Systems might also be listed with CSA, cETL-US, and other standards programs.

Build-it-Yourself, Site-built and Owner-built Design

Site-built indoor composting toilet designs vary, ranging from rollaway containers fitted with aerators to large concrete sloped-bottom tanks.

These are not to be confused with 'direct outdoor composting', which typically uses a collector bucket, where each deposit is covered with sawdust or other dry organic material, with the collector periodically being hand transported to an outdoor composting bin, where it may be added to yard waste or other organic material being composted (Fig. 21.2).

Public use

Increasingly, composting toilet systems are commonly used in water closets in public facilities. One example is the three-storey CK Choi Building at the University of British Columbia (Canada), which features five composting toilet systems with 12 toilet stools that serve 300 employees. They may also be found in various places around Europe, like many of the roadside facilities in Sweden (Fig. 21.3).

Composting toilets greatly reduce the volume of excreta on site through psychrophilic, thermophilic or mesophilic composting and yield a soil amendment that can be used in horticultural or agricultural applications as local regulations allow. In combination with a constructed wetland these even require only the half area.

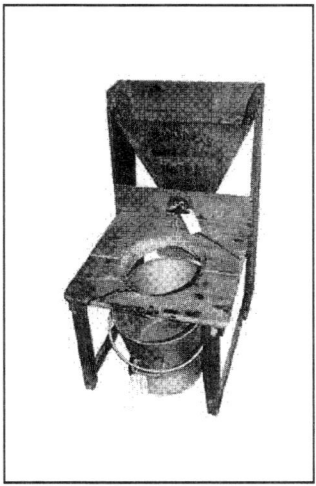

Fig. 21.2. Henry Moule's earth closet, patented in 1873. Example from around 1875. Rear chamber for dispensing cover material.

Fig. 21.3. Public composting toilet facility on E6 highway in Sweden

Operating Process

Although there are many designs, the process factors at work are the same. Rapid aerobic composting will be thermophilic decomposition in which bacteria that thrive at high temperatures (40°–60°C/104°–140°F) oxidises (breaks down) the waste into its components, some of which are consumed in the process, reducing volume, and eliminating potential pathogens.

Drainage of excess liquid or leachate via a separate drain at the bottom of the composter is featured in some manufactured units, as the aerobic composting process requires moisture levels to be controlled (ideally 50 per cent ±/10): too dry, and the mass decomposes slowly or not at all; too wet and anaerobic organisms thrive, creating undesirable odours. This separated liquid may be diverted to a graywater system or collected for other uses.

An approach that is becoming more common is the 'dry' toilet or urine-separating (also: urine-diverting) toilet. Where solar heat is used, this might be called a 'solar' toilet. These systems depend on desiccation to achieve sanitation safety goals features systems that make use of the separated liquid fraction for immediate area fertilisation.

Urine can contain up to 90 per cent of the N (nitrogen), up to 50 per cent of the P (phosphorus) and up to 70 per cent of the K (potassium) present in human excreta. In healthy individuals it is usually pathogen free, although undiluted it may contain levels of inorganic salts and organic compounds at levels toxic to plants. The other requirement critical for microbial action (as well as drying) is oxygen. Commercial systems provide methods of ventilation that move air from the room, through the waste container, and out a vertical pipe, venting above the enclosure roof. This air movement (via convection or fan forced) will vent carbon dioxide and odours.

Most units require manual methods for periodic aeration of the solid mass such as rotating a drum inside the unit or working an 'aerator rake' through the mass. Composting toilet brands have different provisions for emptying the 'finished product', and supply a range of capacities based on volume of use. Frequency of emptying will depend on the speed of the decomposition process and capacity, from a few months (active hot composting) to years (passive, cold composting). With a properly sized and managed unit, a very small volume (about 10 per cent of inputs) of a humus-like material results, which can be suitable as soil amendment for agriculture, depending on local public health regulations.

AERATED STATIC PILE COMPOSTING

Aerated static pile (ASP) composting, refers to any of a number of systems used to biodegrade organic material without physical manipulation during primary composting. The blended admixture is usually placed on perforated piping, providing air circulation for controlled aeration. It may be in windrows, open or covered, or in closed containers. With regard to complexity and cost, aerated systems are most commonly used by larger, professionally managed composting facilities, although the technique may range from very small, simple systems to very large, capital intensive, industrial installations (Fig. 21.4).

Fig. 21.4. Channelled concrete floor of a composting pad for perforated piping that delivers oxygen to the composting mass.

Aerated static piles offer process control for rapid biodegradation, and work well for facilities processing wet materials and large volumes of feedstocks. ASP facilities can be under roof or outdoor windrow composting operations or totally enclosed in-vessel composting, sometimes referred to tunnel composting.

Aeration

The aeration system uses fans to push and/or pull air through the composting mass. Rigid or flexible perforated piping, connected to fans, delivers the air. The pipes can be installed in channels, on top of a floor, or included throughout the pile during build-up (Fig. 21.5).

Fig. 21.5. Aeration system for a closed chamber composting facility.

In large-scale systems, forced aeration is accompanied with a computerised monitoring system responsible for controlling the rate and schedule of air delivery to the composting mass, although meters and manual monitoring techniques may also be used in smaller scale operations.

Advantages of this composting method include the ability to maintain the proper moisture and oxygen levels for the microbial populations to operate at peak efficiency to reduce pathogens, while preventing excess heat, which can crash the system. Aerated systems also facilitate the use of biofilters to treat process air to remove particulates and mitigate odours prior to venting. However, aerated systems can dry out quickly and must be monitored closely to maintain desired moisture levels. In Thailand this system has been used by 470 farmer groups. The process required 30 days to finish without turning, with 10 metric tons of compost (10 piles) obtained each time. A 15-inch squirrel-cage blower was used to force the air through 10 static piles of compost, one at a time, for 15 minutes period twice a day. The raw materials consisted of agricultural wastes and animal manure in the ratio of 3:1 by volume.

IN-VESSEL COMPOSTING

In-vessel composting is an industrial form of composting biodegradable waste that occurs in enclosed reactors. These generally consist of metal tanks or concrete bunkers in which air flow and temperature can be controlled, using the principles of a 'bioreactor'. Generally the air circulation is metered in via buried tubes that allow fresh air to be injected under pressure, with the exhaust being extracted through a biofilter, with temperature and moisture conditions monitored using probes in the mass to allow maintenance of optimum aerobic decomposition conditions. Figure 21.6 shows typical in-vessel composting flow chart.

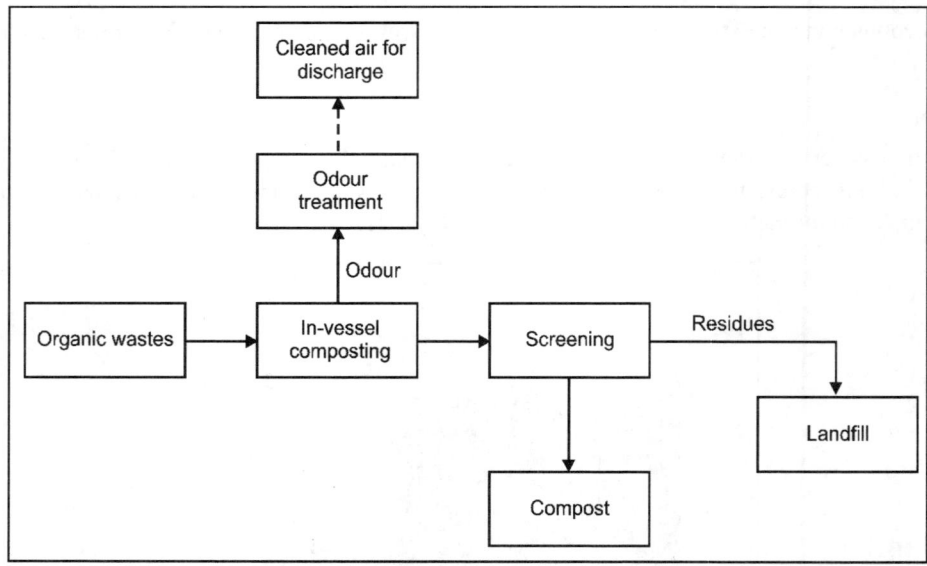

Fig. 21.6. Typical in-vessel composting flow chart.

This technique is generally used for municipal scale organic waste processing, including final treatment of sewage biosolids, to a safe stable state for reclamation as a soil amendment. In-vessel composting can also refer to aerated static pile composting with the addition of removable covers that enclose the piles, as with the system in extensive use by farmer groups in Thailand, supported by the National Science and Technology Development Agency there.

Offensive odours are caused by putrefaction (anaerobic decomposition) of nitrogenous animal and vegetable matter gassing off as ammonia. This is controlled with a higher carbon to nitrogen ratio, or increased aeration by ventilation, and use of a coarser grade of carbon material to allow better air circulation. Prevention and capture of any gases naturally occurring (volatile organic compounds) during the hot aerobic composting involved is the objective of the biofilter, and as the filtering material saturates over time, it can be used in the composting process and replaced with fresh material.

SECTION VII

Waste Landfill

Chapter 22

Sanitary Landfill: A Review

INTRODUCTION

The term 'sanitary landfill' was first used in the 1930s to refer to the compacting of solid waste materials. Initially adopted by New York City and Fresno, California, the sanitary landfill used heavy earth-moving equipment to compress waste materials and then cover them with soil. The practice of covering solid waste was evident in Greek civilisation over 2000 years ago, but the Greeks did it without compacting.

Today, the sanitary landfill is the major method of disposing waste materials in North America and other developed countries, even though considerable efforts are being made to find alternative methods, such as recycling, incineration, and composting. Among the reasons that landfills remain a popular alternative are their simplicity and versatility. For example, they are not sensitive to the shape, size, or weight of a particular waste material. Since they are constructed of soil, they are rarely affected by the chemical composition of a particular waste component or by any collective incompatibility of co-mingled wastes. By comparison, composting and incineration require uniformity in the form and chemical properties of the waste for efficient operation. About 67 per cent of the solid waste generated in the United States is still dumped in landfills. This corresponds to several tons of waste per landfill daily, considering 4.5 lb (2 kg) of solid waste is generated each day per person in this country. Americans have created approximately 220 million tons of solid waste in the year 2000. The many tons of solid waste dumped in a landfill today will not decompose until 30 years from now. In order to create environmentally friendly landfills, new sites are being engineered to recover the methane gas that is generated during decomposition, and some older landfills are being mined for useful products.

About 70 per cent of materials that are routinely disposed of in landfills could be recycled instead. More than 30 per cent of bulk municipal garbage collections consist of paper that could be remanufactured into other paper products. Other materials like plastic, metal and glass can also be reused in manufacturing, which can greatly reduce the amount of waste materials disposed in landfills, as well as preserving sources of nonrenewable raw materials.

LANDFILL—SANITARY LANDFILL

Sanitary landfills involve well-designed engineering methods to protect the environment from contamination by solid or liquid wastes. A necessary condition in designing a sanitary landfill is the availability of vacant land that is accessible to the community being served and has the capacity to handle several years of waste material. In addition, cover soil must be available. Of course, the location must also be acceptable to the local community. Historically, landfills were placed in a particular location more for convenience of access than for any environmental or geological reason. Now more care is taken in determining the location of new landfills. For example, sites located on faulted or highly permeable rock are passed over in favour of sites with a less-permeable foundation. Rivers, lakes, floodplains, and groundwater recharge zones are also avoided. It is believed that the care taken in the initial location of a landfill will reduce the necessity for future clean-up and site rehabilitation. Locations near airports are avoided because the landfill usually attracts birds that can interfere with aircraft. Due to these and other factors, it is becoming increasingly difficult to find suitable locations for new landfills. Easily accessible open space is becoming scarce and many communities are unwilling to accept the building of a landfill within their boundaries. Since 1978, over 14,000 landfills have been filled up and shut down. Many major cities have already exhausted their landfill capacity and must export their trash, at significant expense, to other communities or even to other states and countries.

The three basic procedures that are carried out in sanitary landfills are: spreading the solid waste materials in layers; compacting the wastes as much as possible; and covering the material with dirt at the end of each day. This method reduces the breeding of rats and insects at the landfill, reduces the threat of spontaneous fires, prevents uncontrolled settling of the materials, and uses the available land efficiently. Although this method does help control some of the pollution generated by the landfill, the fill dirt also occupies up to 20 per cent of the landfill space, reducing its waste-holding capacity. Another important consideration for landfill design is the use of the site after it is filled. Some sites have become parks, housing projects or sites for agriculture. Under pressure from the government, environmentalists, and the public, and with diminishing natural and financial resources available to them, municipalities are now planning their landfills carefully to avoid some of the later costs of clean-up or containment.

LANDFILL—METHOD TYPES

Trench and area methods, along with combinations of both, are used in the operation of landfills. Both methods operate on the principle of a 'cell', which in landfills comprises the compacted waste and soil covering for each day. The trench method is good in areas where there is relatively little waste, low groundwater, and the soil is over 6 ft (1.8 m) deep. The area method is usually used to dispose of large amounts of solid waste.

In the trench method, a channel with a typical depth of 15 ft (4.6 m) is dug, and the excavated soil is later used as a cover over the waste. Grading in the trench method must accommodate the drain-off of rainwater. Another consideration is the type of subsurface soil that exists under the topsoil. Clay is a good source of soil because it is nonporous. Weather and the amount of time the landfill will be in use are additional considerations.

In the area method, the solid wastes and cover materials are compacted on top of the ground. This method can be used on flat ground, in abandoned strip mines, gullies, ravines, valleys or any other suitable land. This method is useful when it is not possible to create a landfill below ground.

A combination method is called the progressive slope or ramp method, where the depositing, covering, and compacting are performed on a slope. The covering soil is excavated in front of the daily cell. Where there is no cover material at the site, it is then brought in from outside sources.

LANDFILL—DECOMPOSITION

A landfill has three stages of decomposition. The first one is an aerobic phase. The solid wastes that are biodegradable react with the oxygen in the landfill and begin to form carbon dioxide and water. Temperature during this stage of decomposition in the landfill rises about 30°F (16.7°C) higher than the surrounding air. A weak acid forms within the water and some of the minerals are then dissolved. The next stage is anaerobic, in which micro-organisms that do not need oxygen break down the wastes into hydrogen, ammonia, carbon dioxide, and inorganic acids.

In the third stage of decomposition in a landfill, methane gas is produced. Sufficient amounts of water and warm temperatures have to be present in the landfill for the micro-organisms to form the gas. About half of the gas produced during this stage will be carbon dioxide, but the other half will be methane. Systems of controlling the production of methane gas are either passive or active. In a passive system the gas is vented into the atmosphere naturally, and may include venting trenches, cutoff walls, or gas vents to direct the gas. An active system employs a mechanical method to remove the methane gas and can include recovery wells, gas collection lines, a gas burner or a burner stack. Both active and passive systems have monitoring devices to prevent explosions or fires.

LANDFILL—OPERATING PRINCIPLES

While landfills may outwardly appear simple, they need to operate carefully and follow specific guidelines that include where to start filling, wind direction, the type of equipment used, method of filling, roadways to and within the landfill, the angle of slope of each daily cell, controlling contact of the waste with groundwater, and the handling of equipment at the landfill site.

Considerations have to be made regarding the soil that is used as a daily cover, which is usually 6 in (15.2 cm) thick, an intermediate cover of 1 ft (30.5 cm), and a final cover of 2 ft (61 cm). The compacting of the solid waste and soil has to be considered as well, so that the biological processes of decomposition can take place properly.

Shredding of solid wastes is one method of saving space at landfills. Another method is baling of wastes. The advantages to shredding are twofold. The material can be compacted to a greater density, thereby extending the life of the landfill, and it can be compacted more quickly as well. Less cover is required and there is also less danger of spontaneous fire. Landfills using shredded materials produce more organic decomposition than those disposing of unshredded solid wastes. The advantages of baling are an increase in landfill life because of an increase in waste density. Hauling times are reduced, as are litter, dust, odour, fires, traffic, noise, earth moving, and land settling. Less heavy equipment is needed for the cover operation and the amount of time it takes for the land to stabilise is reduced. Using biodegradable materials also helps save space in landfills because micro-organisms can break down these materials more quickly. Trash bags made of biodegradable materials are of particular use because micro-organisms cause holes to form in the bags, allowing the material inside to break down more quickly as well.

When the secure landfill reaches capacity, it is capped by a cover of clay, plastic, and soil, much like the bottom layers. Vegetation is planted to stabilise the surface and make the site more attractive. Sump pumps collect any fluids that filter through the landfill either from rainwater or from waste leakage. This liquid is purified before it is released. Monitoring wells around the site ensure that the groundwater does not become contaminated. In some areas where the water table is particularly high, above-ground storage may be constructed using similar techniques. Although such facilities are more conspicuous, they have the advantage of being easier to monitor for leakage.

The uses to which closed landfills have been put are varied. Efforts to limit what goes into the landfill reflect particular concerns of different communities across the country. They include industrial parks, airport runways, recreational parks, ski slopes, ball fields, golf courses, playgrounds, and many others. When it has been determined that the bearing capacity of the landfill surface is adequate, buildings can also be erected. The antiquated view of landfills as 'garbage dumps' has given way to a science to engineer the establishment, maintenance, closure, and reuse of the area for the community.

LANDFILL—ALTERNATIVES TO LANDFILLS

The United States Environmental Protection Agency (EPA) requires all new landfills to include a leachate collection system. Recirculation of leachate accelerates the decomposition of solid waste. Another alternative use of landfills is to capture the methane gas produced during decomposition to generate electricity. For example, in Yolo County, California, a landfill releases 1.4 million cubic feet of gas a day used to generate electricity.

Landfill mining is another process that is used to reclaim the materials of the landfill for other purposes. More than 65 per cent of the product from a landfill is usable soil. Small percentages of other materials, such as rock, metal, wood, aluminium, glass, plastic, polystyrene, and other items, can also be extracted from a landfill that is ready to be closed. The soil can be used as daily cover at other landfills and for grading roads and other construction projects. This process can only take place in landfills that are free of toxic wastes. Other landfill mining projects use the material to turn waste into energy. Another alternative to landfill disposal for many areas has been the incineration of solid wastes. This method is often criticised because it has the potential of polluting the air, and the residual ash still has to be buried in a secure landfill. Dumping in the ocean has also come under attack by environmentalists who cite pollution of marine ecosystems and destruction of recreational beaches as reasons against ocean dumping.

LANDFILL—RECYCLING

As a method of reducing the costs of solid waste disposal in landfills and of solving the problem of finding suitable landfill sites, many communities have initiated recycling programs. Some programs are carried out by segregating and collecting the recyclables separately from the materials destined for the landfill. There are also many drop-off programs for specific items such as bottles, plastics, cans, and newspapers.

Some communities require individual households to separate glass, plastic, and paper, while other programs have installed systems to separate the items at a plant and then sell them to manufacturers. The special collection of hazardous chemical wastes has also been initiated in communities that either recycle them or dispose of them more safely than in a landfill. Several things, besides saving space in landfills, are then accomplished with recycling programs. One is a cost benefit to the municipality and another is a decrease in the exploitation of natural resources, such as trees, metals, and petroleum.

LANDFILL—COMPOSTING

The composting of organic materials for reuse in gardening and in agriculture can help alleviate the problem of using land to dispose of waste material. Plant and food substances are biodegradable, which means they are capable of decomposing through the agency of bacteria, fungi, and other living organisms. Temperature and sunlight play a role in the decomposition of biodegradable substances as well. When substances are not biodegradable, they may remain in the environment and may be capable of polluting the soil and water of an area if they are toxic. Some biodegradable pollutants may also be capable of causing harm to the environment.

Substances that in the past were freely disposed of by dumping are now being considered by many municipalities for recycling as compost, such as weeds, leaves, and cut grass. Many communities throughout the country encourage people to compost plant material and use it as humus in their gardens. Since plant material is biodegradable this is a significant way to reduce solid waste problems for towns and cities. Other significant efforts involve the use of composted sewage sludge for soil application on farms, yards, and golf courses.

LANDFILL LEACHATE: PERSISTENT THREATS TO AQUATIC ENVIRONMENT

Landfill is one of the most widely employed methods for the disposal of municipal solid waste (MSW) around the world. After being landfilled, the refuse decomposes through a series of combined physico-chemical and biological processes, which may take a period of more than 50 years. During the degradation process, one ton of landfilled solid waste generates about 0.2 m^3 of highly contaminated waste-water, called 'leachate', depending on the type of waste and seasonal climate. This waste-water primarily results from the degradation of the organic portion of the waste in combination with percolating rainwater and moisture that leaches out organic and inorganic constituents through the waste layer in the landfill. Depending on the rainfall conditions, the colour of leachate varies from black to brown.

The common features of raw leachate from a local landfill are its high concentrations of ammoniacal nitrogen (NH_3-N) (2000–5000 mg/l) and moderately high strength of recalcitrant compounds (as reflected by its COD value) (5000–20,000 mg/l), as well as a low ratio of BOD_5/COD of less than 0.1. Of the toxic pollutants that are present in landfill leachate, NH_3-N, resulting from the decomposition process of organic nitrogen, has been identified not only as a major long-term pollutant, but also as the primary cause of acute toxicity. Because NH_3-N is stable under anaerobic conditions, it typically accumulates in the leachate. With a concentration of higher than 100 mg/l, untreated NH_3-N is highly toxic to aquatic organisms, as confirmed by toxicity tests using zebrafish.

If allowed to migrate, the contaminant released from a landfill would also pose potentially serious threats to the surrounding soil and the underlying groundwater. Since groundwater is the major source of drinking water worldwide, in recent years, the risk of groundwater pollution has become one of the most important environmental concerns, particularly in developing countries, where most of the landfills have been built without any sound engineering design such as engineered liners and leachate interception and collection system. Unless properly treated, leachate that seeps from a landfill can infiltrate and contaminate the underlying groundwater. Once the leachate escapes to the groundwater, it is difficult and expensive to have it controlled and cleaned up, thus posing potentially serious hazards not only to living organisms, but also to public health in the long-term. In most cases, it is extremely difficult to restore the polluted groundwater to its former state. For this reason, in recent years, the risk of groundwater pollution due to leachate seepage has become a major environmental concern worldwide.

MUNICIPAL WASTE AND LANDFILLS

Municipal waste, commonly known as trash or garbage, is a combination of all of a city's solid and semisolid waste. It includes mainly household or domestic waste, but it can also contain commercial and industrial waste with the exception of industrial hazardous waste (waste from industrial practices that causes a threat to human or environmental health). Industrial hazardous waste is excluded from municipal waste because it is typically dealt with separately based on environmental regulations.

Five Categories of Municipal Waste

The types of trash that are included in municipal waste are grouped into five different categories. The first of these is waste that is biodegradable. This includes things like food and kitchen waste such as meat trimmings or vegetable peelings, yard or green waste and paper. The second category of municipal waste is recyclable materials. Paper is also included in this category but non-biodegradable items like glass, plastic bottles, other plastics, metals and aluminium cans fall into this section as well.

Inert waste is the third category of municipal waste. For reference, when discussed with municipal waste, inert materials are those that are not necessarily toxic to all species but can be harmful or toxic to humans. Therefore, construction and demolition waste is often categorised as inert waste.

Composite waste is the fourth category of municipal waste and includes items that are composed of more than one material. For example, clothing and plastics such as children's toys are composite waste.

Household hazardous waste is the final category of municipal waste. This includes medicines, paint, batteries, light bulbs, fertiliser and pesticide containers and e-waste like old computers, printers, and cellular phones. Household hazardous waste cannot be recycled or disposed of with other waste categories so many cities offer residents other options for hazardous waste disposal.

Municipal Waste Disposal and Landfills

In addition to the different categories of municipal waste, there are a number of different ways in which cities dispose of their waste. The first and most well known however, are dumps. These are open holes in the ground where trash is disposed of and has little environmental regulations. More commonly used today to protect the environment however, are landfills. These are areas that are specially created so waste can be put into the ground with little or no harm to the natural environment through pollution.

Today, landfills are engineered to protect the environment and prevent pollutants from entering the soil and possibly polluting groundwater in one of two ways. The first of these is with the use of a clay liner to block pollutants from leaving the landfill. These are called sanitary landfills while the second type is called a municipal solid waste landfill. These types of landfills use synthetic liners like plastic to separate the landfill's trash from the land below it.

Once trash is put into these landfills, it is compacted until the areas is full, at which time the trash is buried. This is done to prevent the trash from contacting the environment but also to keep it dry and out of contact with air so it will not quickly decompose. About 55 per cent of the waste generated in the United States goes to landfills while around 90 per cent of waste created in the United Kingdom is disposed in this manner.

In addition to landfills, waste can also be disposed using waste combustors. This involves the burning of municipal waste at extremely high temperatures to reduce waste volume, control bacteria, and sometimes generate electricity. Air pollution from the combustion is sometimes a concern with this type of waste disposal but governments have regulations to reduce pollution. Scrubbers (devices that spray liquids on smoke to reduce pollution) and filters (screens to remove ash and pollutant particles) are commonly used today.

Finally, transfer stations are the third type of municipal waste disposal currently in use. These are facilities that where municipal waste is unloaded and sorted to remove recyclables and hazardous materials. The remaining waste is then reloaded onto trucks and taken to landfills while the waste that can be recycled for example, is sent to recycling centres.

Municipal Waste Reduction

On top of the proper disposal of municipal waste, some cities promote programs to reduce overall waste. The first and most widely used program is recycling through the collection and sorting of materials that can be remanufactured as new products. Transfer stations aid in sorting recyclable materials but city recycling programs sometimes work to ensure that its residents separate their own recyclable materials from the rest of their trash.

Composting is another way cities can promote municipal waste reduction. This type of waste is comprised solely of biodegradable organic waste like food scraps and yard trimmings. Composting is generally done on the individual level and involves the combination of organic waste with micro-organisms like bacteria and fungi that break down the waste and create compost. This can then be recycled and used as a natural and chemical free fertiliser for personal plants.

Along with recycling programs and composting, municipal waste can be reduced via source reduction. This involves the reduction of waste through the alteration of manufacturing practices to reduce the creation excess materials which get turned into waste.

Future of Municipal Waste

To further reduce waste, some cities are currently promoting policies of zero waste. Zero waste itself means reduced waste generation and the 100 per cent diversion of the remainder of waste from landfills to productive uses via materials reuse, recycling, repair and composting. Zero waste products should also have minimal negative environmental impacts over their life cycles.

Chapter 23

Landfilling Methods and Operations

INTRODUCTION

A landfill, also known as a dump, rubbish dump or both, rubbish landfill dump (and historically as a midden), is a site for the disposal of waste materials by burial and is the oldest form of waste treatment. Historically, landfills have been the most common methods of organised waste disposal and remain so in many places around the world. Landfills may include internal waste disposal sites (where a producer of waste carries out their own waste disposal at the place of production) as well as sites used by many producers. Many landfills are also used for other waste management purposes, such as the temporary storage, consolidation and transfer or processing of waste material (sorting, treatment or recycling).

A landfill also may refer to ground that has been filled in with soil and rocks instead of waste materials, so that it can be used for a specific purpose, such as for building houses. Unless they are stabilised, these areas may experience severe shaking or liquefaction of the ground in a large earthquake. Figure 23.1 shows the images of of landfilling.

OPERATIONS

Typically, in nonhazardous waste landfills, in order to meet predefined specifications, techniques are applied by which the wastes are:

1. Confined to as small an area as possible.
2. Compacted to reduce their volume.
3. Covered (usually daily) with layers of soil.

During landfill operations the waste collection vehicles are weighed at a weighbridge on arrival and their load is inspected for wastes that do not accord with the landfill's waste acceptance criteria. Afterward, the waste collection vehicles use the existing road network on their way to the tipping face or working front where they unload their load. After loads are deposited, compactors or dozers are used to spread and compact the waste on the working face. Before leaving the landfill boundaries, the waste collection vehicles pass through the wheel cleaning facility. If necessary, they return to the weighbridge in order to be weighed without their load. Through the weighing process, the daily incoming waste tonnage can be calculated and listed in databases. In addition to trucks, some landfills may be equipped to handle railroad containers. The use of 'rail-haul' permits landfills to be located at more remote sites, without the problems associated with many truck trips.

Typically, in the working face, the compacted waste is covered with soil daily. Alternative waste-cover materials are several sprayed-on foam products and temporary blankets. Blankets can be lifted into place with tracked excavators and then removed the following day prior to waste placement. Chipped wood and chemically 'fixed' biosolids may also be used as an alternate daily cover. The space that is occupied daily by the compacted waste and the cover material is called a daily cell. Waste compaction is critical to extending the life of the landfill. Factors such as waste compressibility, waste layer thickness and the number of passes of the compactor over the waste affect the waste densities.

Fig. 23.1. Landfilling.

IMPACTS

A large number of adverse impacts may occur from landfill operations. These impacts can vary: fatal accidents (e.g. scavengers buried under waste piles); infrastructure damage (e.g. damage to access roads by heavy vehicles); pollution of the local environment (such as contamination of groundwater and/or aquifers by leakage and residual soil contamination during landfill usage, as well as after landfill closure); offgassing of methane generated by decaying organic wastes (methane is a greenhouse gas many times more potent than carbon dioxide, and can itself be a danger to inhabitants of an area); harbouring of disease vectors such as rats and flies, particularly from improperly operated landfills, which are common in Third-world countries; injuries to wildlife; and simple nuisance problems (e.g. dust, odour, vermin or noise pollution).

Environmental noise and dust are generated from vehicles accessing a landfill as well as from working face operations. These impacts are best to intercept at the planning stage where access routes and landfill geometrics can be used to mitigate such issues. Vector control is also important, but can be managed reasonably well with the daily cover protocols. Most modern landfills in industrialised countries are operated with controls to attempt to manage problems such as these. Analyses of common landfill operational problems are available. Some local authorities have found it difficult to locate new landfills. Communities may charge a fee or levy in order to discourage waste and/or recover the costs of site operations. Many landfills are publicly funded, but some are commercial businesses, operated for profit.

Trash Dump Communities

In many developing countries around the world, communities exist in and around landfills. Residents of these communities, such as La Chureca in Nicaragua, often live in conditions of extreme poverty and use the landfills as a source of food and income. Scavengers work in the garbage in search of recyclables and other valuables.

LANDFILLING METHODS AND OPERATIONS

To use the available area at a landfill site effectively, a plan of operation for the placement of solid wastes must be prepared. Various operational methods have been developed primarily on the basis of field experience. The principal methods used for landfilling dry areas may be classified as (i) area, (ii) trench, and (iii) depression.

1. Area method: The area method is used when the terrain is unsuitable for the excavation of trenches in which to place the solid wastes. The filling operation is usually started by building an earthen levee against which wastes are placed in thin layers and compacted (Fig. 23.2). Each layer is compacted as the filling progresses until the thickness of the compacted wastes reaches a height varying from 2 to 3 m (6 to 10 ft). At that time and at the end of each day's operation, a 150 to 300 mm (6 to 12 in) layer of cover material is placed over the completed fill. The cover material must be hauled in by truck or earth moving equipment from adjacent land or from borrow pit areas. In some newer landfill operations, the daily cover material is omitted. A completed lift, including the cover material, is called a 'cell' (Fig. 23.3). Successive lifts are placed on top of one another until the final grade called for in the ultimate development plan is reached. A final layer of cover material is used when the fill reaches the final design height.

2. Trench method: The trench method (Fig. 23.4) of landfilling is ideally suited to areas where an adequate depth of cover material is available at the site and where the water table is well below the surface. To start the process, a portion of the trench is dug and the dirt is stockpiled to form an embankment behind the first trench. Wastes are then placed in the trench, spread into thin layers and compacted. The operation continues until the desired height is reached. Cover material is obtained by excavating an adjacent trench or continuing the trench that is being filled.

3. Depression method: At locations where natural or artificial depressions exist, it is often possible to use them effectively for landfilling operations. Canyons, ravines, dry burrow pits, and quarries have all been used for this purpose. The techniques to place and compact solid wastes in depression landfills vary with the geometry of the site, the characteristics of the cover material, the hydrology and geology of the site, and access to the site.

   In a canyon, site-filling starts at the head end of the canyon (Fig 23.5) and ends at the mouth. This practice prevents the accumulation of water behind the landfill. Wastes are usually deposited on the canyon floor and from there are pushed up against the canyon face at a slope of about 2 to 1. In this way, a high degree of compaction can be achieved.

4. Landfills in wet areas: Because of the problems associated with contamination of local groundwater, the development of odours and structural stability, landfills are seldom used in wet areas. If wet areas such as swamps and marshes, tidal areas, and ponds, pits or quarries must be used as landfill sites, special provisions must be made to contain or eliminate the movement of leachate and gases from completed cells. Usually, this is accomplished by first draining the site and then lining the bottom with a clay liner or other appropriate sealants. If a clay liner is used, it is important to continue operation of the drainage facility until the site is filled to avoid the creation of uplift pressures that can cause the liner to rupture from heaving.

Occurrence of Gases and Leachate in Landfills

The following biological, physical and chemical events occur when solid wastes are placed in a sanitary landfill: (i) biological decay of organic materials, either aerobically or anaerobically, with the evolution of gases and liquids, (ii) chemical oxidation of waste materials, (iii) escape of gases from the fill, (iv) movement of liquids caused by differential heads, (v) dissolving and leaching of organic and inorganic materials by water and leachate moving through the fill, (vi) movement of dissolved material by concentration gradients and osmosis, and (vii) uneven settlement caused by consolidation of material into voids.

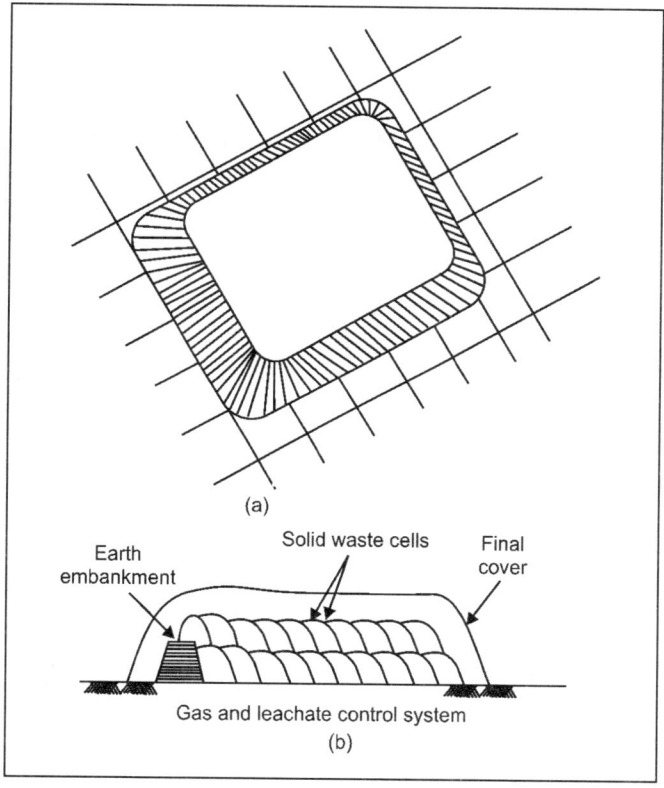

(a)

Earth embankment   Solid waste cells   Final cover

Gas and leachate control system

(b)

Fig. 23.2. Area method for landfilling, solid-wastes (a) pictorial view of completed landfill, (b) section through landfill.

With respect to biological decay, bacterial decomposition initially occurs under aerobic conditions because a certain amount of air is trapped within the landfill. However, the oxygen in the trapped air is exhausted within days, and long-term decomposition occurs under anaerobic conditions.

1. Gases in landfills: Gases found in landfills include air, ammonia, carbon dioxide, carbon monoxide, hydrogen, hydrogen sulphide, methane, nitrogen and oxygen. Carbon dioxide and methane are the principal gases produced from the anaerobic decomposition of the organic solid waste components.

The anaerobic conversion of organic compounds is thought to occur in three steps: the first involves the enzyme mediated transformation (hydrolysis) of higher weight molecular compounds into compounds suitable for use as a source of energy and cell carbon; the second is associated with the bacterial conversion of the compounds resulting from the first step into identifiable lower molecular weight intermediate compounds; and the third step involves the bacterial conversion of the intermediate compounds into simpler end products, such as carbon dioxide (CO_2) and methane (CH_4). The rate of decomposition in unmanaged landfills, as measured by gas production, reaches a peak within the first two years and then slowly tapers off, continuing in many cases for periods up to 25 years or more.

2. Leachate in landfills: Leachate may be defined as the liquid that has percolated through solid waste and has extracted dissolved or suspended materials from it. In most landfills, the liquid portion of the leachate is composed of the liquid produced from the decomposition of the wastes and liquid that has entered the landfill from external sources, such as surface drainage, rainfall, groundwater and water from underground springs.

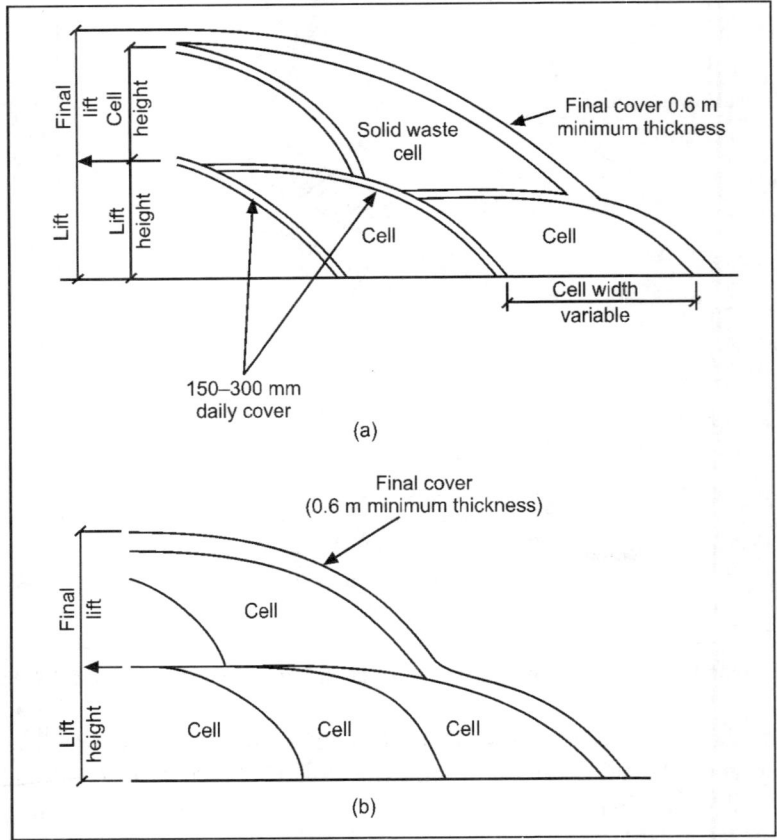

Fig. 23.3. Typical section through a landfill (a) with daily or intermediate cover, (b) without daily or intermediate cover.

Gas and leachate movement and control

Under ideal conditions, the gases generated from a landfill should be either vented into the atmosphere or, in larger landfills, collected for the production of energy. The leachate should either be contained within the landfill or removed for treatment.

1. Gas movement: In most cases, over 90 per cent of the gas volume produced from the decomposition of solid wastes consists of methane and carbon dioxide. Although most of the methane escapes into the atmosphere, both methane and carbon dioxide have been found in

concentrations of up to 40 per cent at lateral distances of up to 120 m (400 ft) from the edges of landfills. If vented into the atmosphere in an uncontrolled manner, methane can accumulate (because its specific gravity is less than that of air) below buildings or in other enclosed spaces on or close to a sanitary landfill. With proper-venting, methane should not pose a problem. Because carbon dioxide is about 1.5 times as dense as air and 2.8 times as dense as methane, it tends to move toward the bottom of the landfill. As a result, the concentration of carbon dioxide in the lower portions of landfill may be high for years. Ultimately, because of its density, carbon dioxide will also move downward through the underlying formation until it reaches the groundwater. Because carbon dioxide is readily soluble in water, it usually lowers the pH, which, in turn, can increase the hardness and mineral content of the groundwater through the solubilisation of calcium and magnesium carbonates.

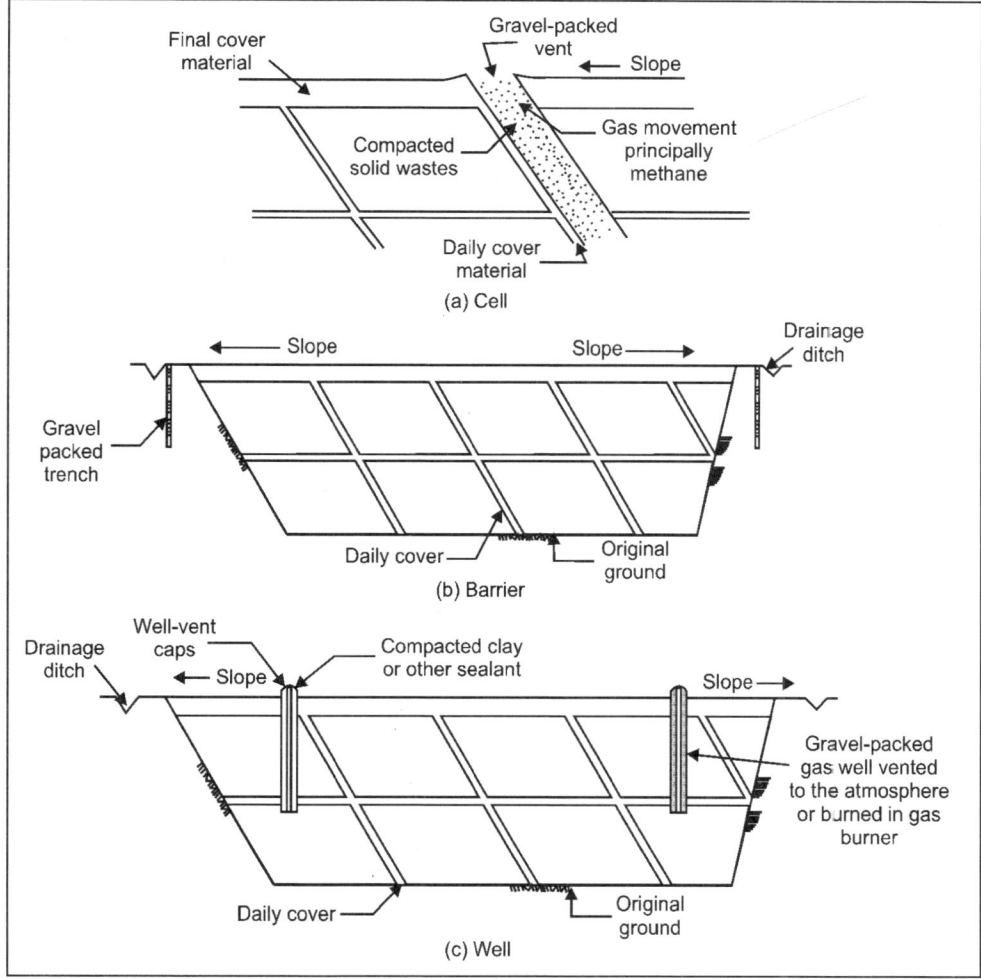

Fig. 23.4. Vents used to control the lateral movement of gases in landfills. (a) Cell, (b) Barrier, (c) Well.

2. Control of gas movement: The lateral movement of gases produced in a landfill can be controlled by installing vents made of materials that are more permeable than the surrounding soil. Typically, as shown in Fig. 23.4(a), gas vents are constructed of gravel. The spacing of cell vents depends on the width of the waste cell but usually varies from 18 to 60 m (60 to 200 ft). The thickness of the gravel layer should be such that it will remain continuous even though there may be differential settling; 300 to 450 mm (12 to 18 in) is recommended. Barrier or well vents [Fig 23.4(b)] can also be used to control the lateral movement of gases. Well vents are often used in conjunction with lateral surface vents buried below grade in a gravel trench [Fig 23.4(c)]. Control of the downward movement of gases can be accomplished by installing perforated pipes in the gravel layer at the bottom of the landfill. If the gases cannot be vented laterally, it may be necessary to install gas wells and vent the pumped gas into the atmosphere.

   The movement of landfill gases through adjacent soil formations can be controlled by constructing barriers of material that are more impermeable than the soil. Some of the landfill sealants that are available for this are identified in Table 23.1. Of these, the use of compacted clay is the most common. The thickness will vary depending on the type of clay and the degree of control required; thicknesses ranging from 0.15 to 1.25 m (6 to 48 in) have been used.

3. Control of gas movement by recovery: The movement of gas in landfills can also be controlled by installing gas recovery wells in completed landfills [Fig. 23.4(b)]. Clay and other liners are used when landfill gas is to recovered. In some gas recovery systems, leachate is collected and recycled to the top of the landfills and reinjected through perforated lines located in drainage trenches. Typically, the rate of gas production is greater in leachate recirculation systems. Although gas recovery systems have been installed in some large municipal landfills, the economics of such operations are, at present, not well defined. The cost of the gas clean-up and processing equipment may limit the recovery of landfill gases, especially from small landfills.

4. Leachate movement: Under normal conditions, leachate is found at the bottom of landfills. From there, it moves through the underlying strata, although some lateral movement may also occur, depending on the characteristics of the surrounding material. The rate of seepage of leachate from the bottom of a landfill can be estimated by *Darcy's law* by assuming that the material below the landfill to the top of the water table is saturated and that a small layer of leachate exists at the bottom of the fill. Under these conditions, the leachate discharge rate per unit area is equal to the value of the coefficient of permeability K, expressed in metres per day. The computed value represents the maximum amount of seepage that would be expected, and this value should be used for design purposes. Under normal conditions, the actual rate will be less than this value, because the soil column below the landfill will not be saturated.

5. Control of leachate movement: As leachate percolates through the underlying strata, many of the chemical and biological constituents originally contained in it will be removed by the filtering and adsorptive action of the material composing the strata. In general, the extent of this action depends on the characteristics of the soil, especially the clay content. Because of the potential risk involved in allowing leachate to percolate to the groundwater, best practice calls for its elimination or containment. Ultimately, it may be necessary to collect and treat the leachate.

   The use of clay has been the favoured method of reducing or eliminating the percolation of leachate (Table 23.1). Membrane liners have also been used, but they are expensive and require care so that they are not be damaged during filling operations. Equally important in controlling the movement of leachate is the elimination of surface water infiltration, which is the major contributor to the total volume of leachate. With the use of an impermeable clay layer, an appropriate surface slope (1 to 2 per cent) and adequate drainage, surface infiltration can be controlled effectively.

6. Settlement and structural characteristics of landfills: The settlement of landfills depends on the initial compaction, characteristics of wastes, degree of decomposition and effects of consolidation when the leachate and gases are formed. The height of the completed fill will also influence the initial compaction and degree of consolidation. The degree of consolidation can be modelled with a first order equation.

Table 23.1. Landfill sealant for the control of gas and leachate movement.

Sealant		Remarks
Classification	Representative types	
Compacted soil		Should contain some clay or fine silt
Compacted clay	Bentonites, illites, kaolinites	Most commonly used sealant for landfills; layer thickness varies from 0.15 to 1.2 m (6 to 48 in); layer must be continuous and not be allowed to dry out and crack
Inorganic chemicals	Sodium carbonate, silicate, or pyrophosphate	Use depends on local soil characteristics
Synthetic chemicals	Polymers, rubber latex	Experimental, use not well established
Synthetic membrane liners	Polyvinyl chloride, butyl rubber Hypalon, polyethylene, nylon reinforced liners	Expensive, may be justified where gas is to be recovered
Asphalt	Modified asphalt, asphalt covered polypropylene fabric, asphalt concrete	Layer must be thick enough to maintain continuity under differential settling conditions
Others	Gunite concrete, soil cement, plastic soil cement	

Design and operation of landfills

Important design considerations in the design and operation of landfills include (i) land requirements, (ii) types of wastes that must be handled, (iii) evaluation of seepage potential, (iv) design of drainage and seepage control facilities, (v) development of a general operation plan, (vi) design of solid waste filling plan, and (vii) determination of equipment requirements. The more important individual factors that must be considered in the design of a landfill are reported in Table 23.2. The last three items are considered further in the following discussion:

1. Landfill operation plan: The layout of the site and the development of a workable operating schedule are the main features of a landfill operation plan. In planning the layout of a landfill site, the location of the following must be determined: (i) access roads, (ii) equipment shelters, (iii) scales, if used, (iv) storage sites for special wastes, (v) topsoil stockpile sites, (vi) landfill areas, and (vii) plantings.

2. Solid waste filling plan: The specific method of filling will depend on the characteristics of the site, such as the amount of available cover material, the topography and local hydrology and geology. To assess future development plans, it will be necessary to prepare a detailed plan for the layout of the individual solid waste cells. On the basis of the characteristics of the site or the method of operation (e.g. gas recovery), it may be necessary to incorporate special features for the control of the movement of gases and leachate from the landfill.

3. Equipment requirements: The types of equipment that have been used at sanitary landfills include both crawler and rubber-tyred tractors, scrapers, compactors draglines, and motor graders. The size and amount of equipment required will depend primarily on local site conditions, the size of the landfill operation and the method of operation.

Table 23.2. Important factors that must be considered in design and operation of solid waste landfills.

Factor	Remarks
Design	
Access	Paved allweather access roads to landfill site; temporary roads to unloading areas
Cell design and construction	Will vary depending on terrain, landfilling method, and whether gas is to be recovered
Cover material	Maximise use of on site earth materials; approximately 1 m^3 of cover material will be required for every 4 to 6 m^3 of solid wastes; mix with sealants to control surface infiltration. In some designs, intermediate cover is not used
Drainage	Install drainage ditches to divert surface water runoff; maintain 1 to 2 per cent grade on finished fill to prevent ponding
Equipment requirements	Vary with size of landfills
Fire prevention	Water on site; if nonpotable, outlets must be marked clearly; proper cell separation prevents continuous burn through if combustion occurs
Groundwater protection	Divert any underground springs; if required, install sealants for leachate control; install wells for gas and groundwater monitoring
Land area	Area should be large enough to hold all wastes for a minimum of one year, but preferably for 5 to 10 years
Landfilling method	Selection of method will vary with terrain and available cover
Litter control	Use movable fences at unloading areas; crews should pick up the litter at least once per month or as required
Operation plan	With or without the codisposal of treatment plant sludges and the recovery of gas
Spread and compaction	Spread and compact waste in 0.6 m (2 ft) layers
Unloading area	Keep small, generally under 30 m (100 ft)
Operation	
Communications	Telephone for emergencies
Days and hours of operation	Usual practice is 5 to 6 days/week and 8 to 10 h/day
Employee facilities	Rest rooms and drinking water should be provided
Equipment maintenance	A covered shed should be provided for field maintenance of equipment
Operational records	Tonnage, transactions, and billing if a disposal fee is charged
Salvage	No scavenging; salvage should occur away from the unloading area; no salvage storage on site
Scales	Essential for record keeping

Landfilling of Hazardous Wastes

In most states, the only disposal option available for most hazardous wastes is landfilling. In general, disposal sites for hazardous wastes should be separate from sites for municipal solid wastes. If separate sites are not possible, great care must be taken to ensure that separate disposal operations are maintained.

Requirements

From a design standpoint, two of the most important requirements are (i) complete leachate containment, and (ii) control of the surface water on and around the site.

Site selection

Factors that must be considered in evaluating potential sites for the disposal of hazardous waste are currently in a state of flux. To qualify as a class I site, it must be shown that:

1. Geological conditions are naturally capable of preventing vertical hydraulic continuity between liquids and gases emanating from the waste in the site and usable surface or groundwater.
2. Geological conditions are naturally capable of preventing lateral hydraulic continuity between liquids and gases emanating from wastes in the site and usable surface or groundwater, or the disposal area has been modified to achieve such capability.
3. Underlying geological formations which contain rock fractures or fissures of questionable permeability must be permanently sealed to provide a competent barrier to the movement of liquids or gases from the disposal site to usable water.
4. Inundation of disposal areas shall not occur until the site is closed in accordance with requirements of the regional board.
5. Disposal areas shall not be subject to washout.
6. Leachate and subsurface flow into the disposal areas shall be continued within the site unless other disposition is made in accordance with requirements of the regional board.
7. Sites shall not be located over zones of active faulting or where other forms of geological change will impair the competence of natural features or artificial barriers which prevent continuity with usable waters.
8. Sites made suitable for use by man-made physical barriers shall not be located where improper operations or maintenance of such structures could permit the waste, leachate or gases to contact usable groundwater or surface water.
9. Sites which comply with the above noted clauses and are subject to inundation by tides or floods of greater than 100 year frequency may be considered by the regional board as limited Class I disposal sites.

Landfilling Methods and Operation

Operation of a landfill for hazardous wastes is quite different from that of a conventional landfill. The specific details will vary depending on whether the wastes are containerised, liquid or solid. When containerised hazardous wastes are to be disposed of, they are unloaded and placed in position individually to avoid rupturing the containers. To avoid the codisposal of incompatible wastes, separate storage areas within the total landfill site should be designated for various classes. Liquid wastes are usually placed in containment areas and allowed to dry by evaporation. In many sites, liquid wastes are processed before disposal. In some cases, liquid wastes are injected into the soil. For dry wastes, conventional landfilling methods are used.

Design of Hazardous Waste Landfills

Currently, many of the regulations governing the design of hazardous waste landfills are unresolved. Although specific requirements will vary, the factors identified in Table 23.2 can be used as a design guide. Some special precautions that can be taken to prevent contamination of underlying strata are shown in Fig. 23.5.

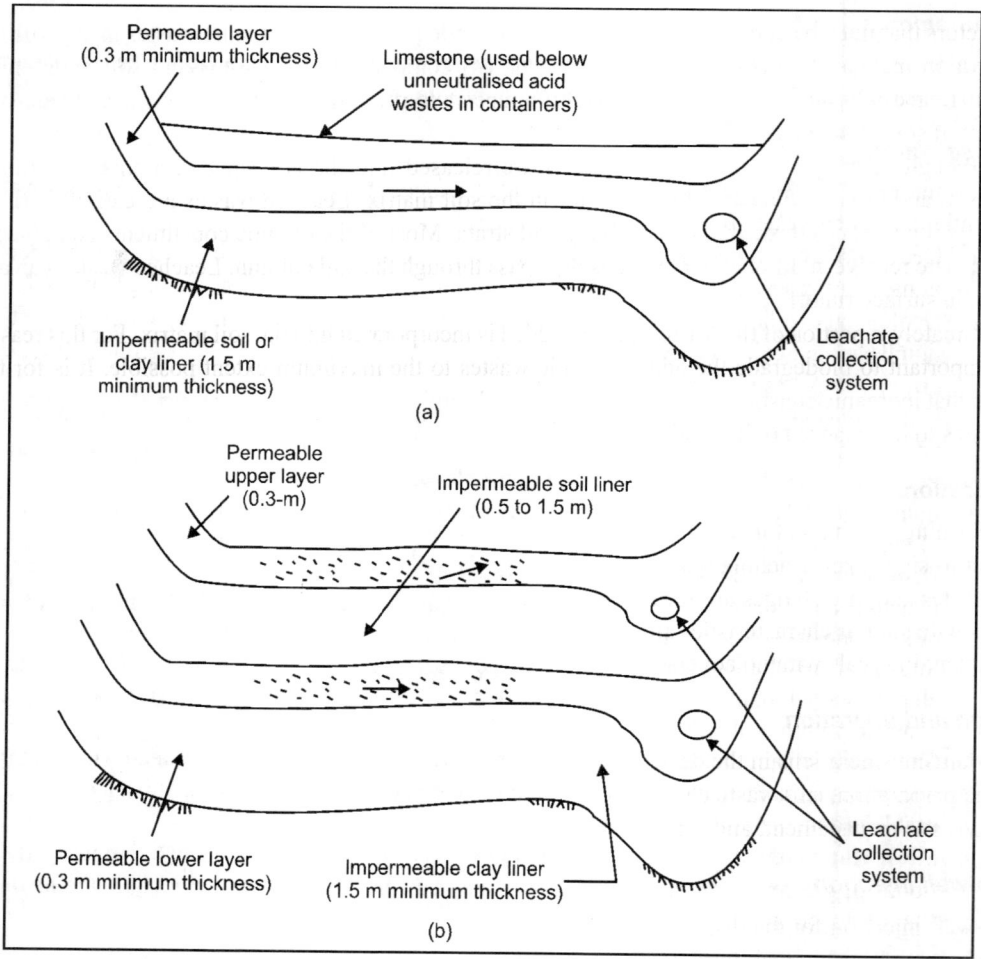

Fig. 23.5. Special design features for hazardous waste landfills to prevent the contamination of the underlying strata. (a) Single liner system, (b) double liner system.

Landfarming

Landfarming is a waste disposal method in which the biological, chemical and physical processes that occur in the surface of the soil are used to treat biodegradable industrial wastes. Wastes to be treated are either applied on top of the land which has been prepared to receive the wastes or injected below the surface of the soil.

Process description

When organic wastes are added to the soil, they are subjected simultaneously to the following processes: (i) bacterial and chemical decomposition, (ii) leaching of water soluble components in the original wastes and from the decomposition products, and (iii) volatilisation of selected components in the original wastes and from the products of decomposition.

Factors that must be considered in evaluating the biodegradability of organic wastes in a landfilling application include: (i) composition of the waste, (ii) compatibility of wastes and soil micro flora, (iii) environmental requirements, including oxygen, temperature, pH, and inorganic nutrients, and (iv) moisture content of soil waste mixture.

Although most of the volatile components are released into the atmosphere, a small fraction is dissolved and/or carried away with the water in the soil matrix. Leached wastes are carried with the water, as it percolates through the underlying soil strata. Most of the organic constituents contained in the leachate receive additional treatment as they pass through the soil column. Leached wastes can also be lost in surface runoff.

Ultimately, a portion of the wastes that are added is incorporated into the soil matrix. For this reason, it is important to biodegrade the added organic wastes to the maximum extent possible. It is for this reason that inorganic constituents such as cadmium, chromium, copper and lead must also be controlled in wastes to be disposed of by landfarming.

Applications

Landfarming is suitable for wastes that contain organic constituents that are biodegradable and are not subject to significant leaching while the bioconversion process is occurring. For example, petroleum oily wastes and oily sludges are ideally suited for disposal by landfarming. A variety of other organic wastes with similar characteristics are also suitable. Properly managed landfarming sites can be reused at frequent intervals with no adverse effects.

Design and operation

Important considerations in the design and operation of landfarming systems include (i) site selection, (ii) site preparation, (iii) waste characteristics, (iv) method of waste application, (v) waste application rate, (vi) site management, and (vii) monitoring.

Deep-well injection

Deep-well injection for the disposal of liquid solid wastes involves injecting the wastes deep into the ground into permeable rock formations (typically limestone or dolomite) or underground caverns.

Process description

The installation of deep wells for the injection of wastes closely follows the practices used for the drilling and completion of oil and gas wells. To isolate and protect potential water supply aquifers, the surface casing must be set well below such aquifers and cemented at the surface of the well. The drilling fluid should not be allowed to penetrate the formation that is to be used for waste disposal. To prevent clogging of the formation, the drilling fluid is replaced with a compatible solution. Also, in some cases, it may be necessary to acid treat the formation before the injection of wastes is initiated.

Applications

Deep-well injection has been used principally for liquid wastes that are difficult to treat and dispose of by more conventional methods and for hazardous wastes. Chemical, petrochemical and pharmaceutical wastes are those most commonly disposed of with this method. The waste may be liquid, gaseous or solid. Gases and solids are either dissolved in the liquid or are carried along with the liquid.

Design and operation

Important design and operation considerations for deep-well injection are related to (i) well-site selection, (ii) pretreatment, (iii) installation of an injection well, and (iv) monitoring. These are shown in Table 23.3. As noted in the table, wastes are usually treated prior to injection to prevent clogging of the formation and damage to equipment. Particles greater than about 1 to 5 μm must be removed. Typically, treated wastes must be filtered prior to injection. Wastes must also be compatible with the characteristics of the aquifer. This may require pH adjustment and the use of compatible buffers. Design operation considerations for deep-well used for waste injection are shown in Table 23.3.

Table 23.3. Important design and operation considerations for deep-wells used for wastes injection.

Items	Remarks
Well-site selection	Criteria for assessing the feasibility of a deep-well-injection site include (i) uniformity, (ii) large extent, (iii) substantial thickness, (iv) high porosity and permeability, (v) low pressure, (vi) saline aquifer, (vii) separation from potable-water horizons, (viii) adequate overlying and underlying aquicludes, (ix) no poorly plugged wells nearby, and (x) compatibility between the mineralogy and fluids of the reservoir and the injected wastes.
Waste pretreatment	Suspended solid less than 10 to 15 mg/L; particle sizes equal to or less than 1 to 5 μm (depends on injection formation). Adjustment of pH and buffering of the waste may be necessary.
Deep-well installation	Well depths vary from 550 to 3660 m (1800 to 1200 ft); well-injection rates vary from 4 to 60 l/s; rates in the range from 15 to 20 l/s are typical. Operation pressures up to 27,600 kPa (4000 psig) are used.
Monitoring	Continuous monitoring facilities should be installed when wells are put into operation. Irregularities in the pressure may require changes in operating procedures.

LANDFILL GAS

Gases are produced in landfills due to the anaerobic digestion by microbes on any organic matter. This gas can be collected and flared off or used to generate electricity in a gas fired power plant. Landfill gas monitoring can be carried out to alert for the presence of a build-up of gases to a harmful level.

Regional Practice

United Kingdom

Landfilling practices in the UK have had to change in recent years to meet the challenges of the European Landfill Directive. The UK now imposes landfill tax upon biodegradable waste which is put into landfills. In addition to this the Landfill Allowance Trading Scheme has been established for local authorities to trade landfill quotas in England. A different system operates in Wales where authorities are not able to 'trade' between themselves, but have allowances known as the Landfill Allowance Scheme.

United States

In the US, landfills are regulated by the state's environmental agency that establishes minimum guidelines; however, none of these standards may fall below those set by the United States Environmental Protection Agency (EPA); such as was the case with the Fresh Kills Landfill in Staten Island, which is claimed by many to not only be the world's largest landfill, but the world's largest human structure. The landfill has since been closed and is being transformed into a park.

Reclaiming Materials

Landfills can be regarded as a viable and abundant source of source materials and energy. In the developing world, this is widely understood and one may thus often find waste pickers scavenging for still usable materials. In a commercial context, landfills sites have also been discovered by companies and many have begun harvesting materials and energy. Well known examples are gas recovery facilities. Other commercial facilities include fossil fuel power plants and waste incinerators which have built-in material recovery. This material recovery is possible through the use of filters (electro filter, active carbon and potassium filter, quench, HCL-washer, SO_2-washer, bottom ash-grating, etc.). An example of these is the AEB waste fired power plant.

The AEB waste incinerator is hereby able to recover a large part of the burned waste in source materials. According to Marcel van Berlo (who helped build the plant), the processed waste contained higher percentages of source materials than any mine in the world. He also added that when the plant was compared to a Chilean copper mine, the waste fired plant could recover more copper. However, because of the high concentration of gases and the unpredictability of the landfill contents, which often include sharp objects, landfill excavation is generally considered dangerous. Furthermore, the quality of materials residing within landfills tends to degrade and such materials are thought to be not worth the risks required to recover them.

Alternatives

The alternatives to landfills are waste reduction and recycling strategies. Secondary to not creating waste, there are various alternatives to landfills. In the late 20th century, alternative methods of waste disposal to landfill and incineration have begun to gain acceptance. Anaerobic digestion, composting, mechanical biological treatment, pyrolysis and plasma-arc gasification have all began to establish themselves in the market.

In recent years, some countries, such as Germany, Austria, Belgium, the Netherlands, and Switzerland, have banned the disposal of untreated waste in landfills. In these countries, only the ashes from incineration or the stabilised output of mechanical biological treatment plants may still be deposited.

LANDFILLS: IMPACTS ON GROUNDWATER

Solid waste landfills are a necessity in modern-day society, because the collection and disposal of waste materials into centralised locations helps minimise risks to public health and safety. Solid waste landfills, which are regulated differently than hazardous waste landfills, may accept a variety of solid, semisolid, and small quantities of liquid wastes. Landfills generally remain open for decades before undergoing closure and postclosure phases, during which steps are taken to minimise the risk of environmental contamination.

Municipal solid waste (MSW) landfills accept nonhazardous wastes from a variety of sources, such as households, businesses, restaurants, medical facilities, and schools. Many MSW landfills also can accept contaminated soil from gasoline spills, conditionally exempted hazardous waste from businesses, small quantities of hazardous waste from households, and other toxic wastes. Industrial facilities may utilise their own captive landfill (i.e. a solid waste landfill for their exclusive use) to dispose of nonhazardous waste from their processes, such as sludge from paper mills and wood waste from wood processing facilities.

Concern Over Landfill Impacts

Although landfills are an indispensable part of everyday living, they may present long-term threats to groundwater and also surface waters that are hydrologically connected. In the United States, federal standards to protect groundwater quality were implemented in 1991 and required some landfills to use plastic liners and collect and treat leachate. However, many disposal sites were either exempted from these rules or grandfathered (excused from the rules owing to previous usage).

Although the federal rules marked a significant improvement in the management of solid waste, some think that these rules do not go far enough. There is an increasing belief among solid waste experts that unless further steps are taken to detoxify landfilled materials, today's society will be placing a burden on upcoming generations to address future landfill impacts. Much of the concern revolves around leachate, the watery solution that results after water passes through a landfill.

Leachate generation and composition

The precipitation that falls into a landfill, coupled with any disposed liquid waste, results in the extraction of the water-soluble compounds and particulate matter of the waste, and the subsequent formation of leachate. The creation of leachate, sometimes deemed 'garbage soup', presents a major threat to the current and future quality of groundwater. (Other major threats include underground storage tanks, abandoned hazardous waste sites, agricultural activities, and septic tanks.) Table 23.4 given typical leachate quality of municipal waste.

Table 23.4. Typical leachate quality of municipal waste (excludes volatile and semi-volatile organic compounds).

Parameter	Typical range (milligrams per litre, unless otherwise noted)	Upper limit (milligrams per litre, unless otherwise noted)
Total Alkalinity (as $CaCO_3$)	730–15,050	20,850
Calcium	240–2330	4080
Chloride	47–2400	11,375
Magnesium	4–780	1400
Sodium	85–3800	7700
Sulphate	20–730	1826
Specific conductance	2000–8000 μmhos/cm	9000 μmhos/cm
Total Dissolved Solids	1000–20,000	55,000
Chemical Oxygen Demand	100–51,000	99,000
Biological Oxygen Demand	1000–30,300	1,95,000
Iron	0.1–1700	5500
Total Nitrogen	2.6–945	1416
Potassium	28–1700	3770
Chromium	0.5–1.0	5.6
Manganese	Not detected – 400	1400
Copper	0.1–9.0	9.9
Lead	Not detected – 1.0	14.2
Nickel	0.1–1.0	7.5

Leachate composition varies relative to the amount of precipitation and the quantity and type of wastes disposed. In addition to numerous hazardous constituents, leachate generally contains nonhazardous parameters that are also found in most groundwater systems. These constituents include dissolved metals (e.g. iron and manganese), salts (e.g. sodium and chloride), and an abundance of common anions and cations (e.g. bicarbonate and sulphate). However, these constituents in leachate typically are found at concentrations that may be an order of magnitude (or more) greater than concentrations present in natural groundwater systems.

Leachate from MSW landfills typically has high values for total dissolved solids and chemical oxygen demand, and a slightly low to moderately low pH. MSW leachate contains hazardous constituents, such as volatile organic compounds and heavy metals. Wood-waste leachates typically are high in iron, manganese, and tannins and lignins. Leachate from ash landfills is likely to have elevated pH and to contain more salts and metals than other leachates.

Leachate release and migration

A release of leachate to the groundwater may present several risks to human health and the environment. The release of hazardous and nonhazardous components of leachate may render an aquifer unusable for drinking-water purposes and other uses. Leachate impacts to groundwater may also present a danger to the environment and to aquatic species if the leachate-contaminated groundwater plume discharges to wetlands or streams.

Once leachate is formed and is released to the groundwater environment, it will migrate downward through the unsaturated zone until it eventually reaches the saturated zone. Leachate then will follow the hydraulic gradient of the groundwater system.

Monitoring wells at landfills allow scientists to determine whether contaminants in leachate are escaping into the local groundwater system. The wells are placed downgradient of the landfill at appropriate depths and at various intervals to intercept any contaminants and monitor their movement.

Monitoring wells at landfills allow scientists to determine whether contaminants in leachate are escaping into the local groundwater system. The wells are placed downgradient of the landfill at appropriate depths and at various intervals to intercept any contaminants and monitor their movement.

A number of forces may act on or react with the migrating leachate, resulting in changes of chemistry and a general reduction of strength from the original release. These forces are physical (filtration, sorption, advection, and dispersion), chemical (oxidation-reduction, precipitation-dissolution, adsorption-desorption, hydrolysis, and ion exchange), and biological (microbial degradation). The extent of these reactions depends on the materials underlying the landfill, the hydraulics of the groundwater system, and the chemistry of the leachate.

Although many of these reactions have the capability to reduce the potential impact to groundwater, some (such as microbial degradation) can actually increase the toxicity by producing by-products that are more hazardous than the original contaminant. This can be seen, for example, in the creation of vinyl chloride from the degradation of trichloroethene.

Old and new viewpoints

Today's landfills are constructed with liners that contain leachate, and leachate collection systems that collect it. But historically, many landfills were constructed without liners or leachate collection systems.

Two philosophies previously existed regarding the placement of unlined landfills. One viewpoint was to place most of these facilities in moderately permeable materials and as close as possible to rivers or streams (i.e. surface water). This type of siting would allow the landfills to slowly (and deliberately) leak leachate into the groundwater, minimising the length and size of the leachate plume, which would ultimately discharge into surface water.

The other theory was the exact opposite: to place landfills as far away from surface water as possible. This may have spared impacts to the surface water, but locating unlined facilities in this fashion usually resulted in creating significant leachate-contaminated groundwater plumes that followed the direction of groundwater flow.

Design standards

The US Environmental Protection Agency issued standards for MSW landfill design and operation in 1991. These new rules were adopted to protect groundwater from the release of leachate by MSW landfills. Landfill owners who could not meet these new design standards requiring plastic liners and leachate collection were required to close and to conduct groundwater monitoring for 30 years. Groundwater protection standards were developed for all MSW landfills, setting a national precedent for solid waste management that may eventually become the standard for all types of landfills.

As the twenty-first century opened, there was a controversy whether the new rules for MSW landfills offer adequate environmental protection, especially as these facilities age. Landfill designs that utilise plastic liners below the waste and then are covered with plastic when the landfill stops placing waste into the active cell (area of waste input) are referred to as 'dry tombs'. These engineered systems are designed to minimise leachate generation by restricting the introduction of moisture, primarily precipitation.

It is widely recognised that even the best-installed plastic liner will succumb to deterioration and eventually will allow leachate to be created and released. However, this may not happen within the required 30 years of postclosure groundwater monitoring. Moreover, it may not be detected during the time the landfill operators are actively involved and financially obligated.

Opponents of dry-tomb landfills advocate for recycling the collected leachate through the waste, which will enhance the rate of chemical reactions inside the landfill and eventually stabilise the waste material prior to covering the landfill. This could reduce the toxic nature of the waste materials in the closed landfill and minimise the future threat posed to groundwater from these facilities.

BIOREACTOR LANDFILL

Landfills are the primary method of waste disposal in many parts of the world, including United States and Canada. Bioreactor landfills are expected to reduce the amount of and costs associated with management of leachate, to increase the rate of production of methane (natural gas) for commercial purposes and reduce the amount of land required for landfills. Bioreactor landfills are monitored and manipulate oxygen and moisture levels to increase the rate of decomposition by microbial activity.

Traditional Landfills and Associated Problems

Landfills are the oldest known method of waste disposal. Waste is buried in large dug out pits [unless naturally occurring locations are available) and covered. Bacteria decompose the waste over several decades producing several by-products of importance, including methane gas (natural gas), leachate fluid and volatile organic compounds [such as hydrogen sulphide (H_2S), N_2O_2, etc.].

Methane gas, a strong greenhouse gas, can build up inside the landfill leading to an explosion unless released from the pit. Leachate are fluid metabolic products from decomposition and contain various types of toxins and dissolved metallic ions. If leachate escapes into the groundwater it can cause health problems in both animals and plants. The volatile organic compounds (VOCs) are associated with causing smog and acid rain. With the increasing amount of waste produced, appropriate places to safely store it have become difficult to find.

Working of a Bioreactor Landfill

There are three types of bioreactors: aerobic, anaerobic and a hybrid (using both aerobic and anaerobic method). All three mechanisms involve the reintroduction of collected leachate supplemented with water to maintain moisture levels in the landfill. The micro-organisms responsible for decomposition are thus stimulated to decompose at an increased rate with an attempt to minimise harmful emissions.

In aerobic bioreactors air is pumped into the landfill using either vertical or horizontal system of pipes. The aerobic environment decomposition is accelerated and amount of VOCs, toxicity of leachate and methane are minimised. In anaerobic bioreactors with leachate being circulated the landfill produces methane at a rate much faster and earlier than traditional landfills. The high concentration and quantity of methane allows it to be used more efficiently for commercial purposes while reducing the time that the landfill needs to be monitored for methane production. Hybrid bioreactors subject the upper portions of the landfill through aerobic-anaerobic cycles to increase decomposition rate while methane is produced by the lower portions of the landfill. Bioreactor landfills produce lower quantities of VOCs than traditional landfills, except H_2S. Bioreactor landfills produce higher quantities of H_2S. The exact biochemical pathway responsible for this increase is not well studied.

Advantages of Bioreactor Landfills

Bioreactor landfills accelerate the process of decomposition. As decomposition progresses, the mass of the landfill declines, creating more space for dumping garbage. Bioreactor landfills are expected to increase this rate of decomposition and save up to 30 per cent of space needed for landfills. With increasing amounts of solid waste produced every year and scarcity of landfill spaces, bioreactor landfill can thus provide a significant way of maximising landfill space. This is not just cost effective, but since less land is needed for the landfills, this is also better for the environment.

Furthermore, most landfills are monitored for at least 3 to 4 decades to ensure that no leachate or landfill gases escape into the community surrounding the landfill site. In contrast, bioreactor landfills are expected to decompose to level that does not require monitoring in less than a decade. Hence, the landfill land can be used for other purposes such as reforestation or parks, depending on the location at an earlier date. In addition, reusing leachate to moisturise the landfill filters it. Thus, less time and energy is required to process the leachate, making the process more efficient.

Disadvantages of Bioreactor Landfills

Bioreactor landfills are a relatively new technology. For the newly developed bioreactor landfills initial monitoring costs are higher to ensure that everything important is discovered and properly controlled. This includes gases, odours and seepage of leachate into the ground surface. The increased moisture content of bioreactor landfill reduces the structural stability of the landfill. The landfill can become too soft too quickly and end up collapsing in on itself due to its weight.

Another consequence of rapid decomposition is the rapid accumulation of landfill gases, primarily methane. Traditional landfills have exhaust pipes dug into them to release methane as it is produced. Bioreactor landfills may produce enough landfill gases at a fast enough rate that pipes are not be able vent them, causing an explosion.

In addition, the types of gases bioreactor landfills produce in excess compared to traditional landfills, such as H_2S, have excessively putrid smell (H_2S smells like rotten eggs). Hence, there is a chance that bioreactor landfill land may not be used for other projects due to the presence of these odorous gases. Since the target of bioreactor landfills is to maintain a high moisture content, gas collection systems can be effected by the increased moisture content of the waste.

Implementation of Bioreactor Landfills

Bioreactor landfills being a novel technology are still in the development phase. Pilot projects for bioreactor landfills are showing promise and more are being experiment with in different parts of the world. Despite the potential benefits of bioreactor landfills there are no standardised and approved designs with guidelines and operational procedures.

LANDFILL MINING

Landfill mining and reclamation (LFMR) is a process whereby solid wastes which have previously been landfilled are excavated and processed. The function of landfill mining is to reduce the amount of landfill mass encapsulated within the closed landfill and/or temporarily remove hazardous material to allow protective measures to be taken before the landfill mass is replaced. In the process, mining recovers valuable recyclable materials, a combustible fraction, soil, and landfill space. The aeration of the landfill soil is a secondary benefit regarding the landfill's future use. The combustible fraction is useful for the generation of power. The overall appearance of the landfill mining procedure is a sequence of processing machines laid out in a functional conveyor system. The operating principle is to excavate, sieve and sort the landfill material.

The concept of landfill mining was introduced as early as 1953 at the Hiriya landfill operated by the Dan Region Authority next to the city of Tel Aviv, Israel. Waste contains many resources with high value, the most notable of which are nonferrous metals such as aluminium cans and scrap metal. The concentration of aluminium in many landfills is higher than the concentration of aluminium in bauxite from which the metal is derived.

Practical Applications

Landfill mining is also possible in countries where land is not available for new landfill sites. In this instance landfill space can be reclaimed by the extraction of biodegradable waste and other substances then refilled with wastes requiring disposal.

Mining construction landfill sites is the simplest form of landfill mining. Construction landfills contain three basic components, wood, scrap metal and gypsum or drywall, along with a minimal amount of other construction materials. The wood collected can be used as fuel in coal burning power plants and the scrap metal reprocessed.

Mining of municipal landfills is more complicated and has to be based on the expected content of the landfill. Older landfills, in the United States before 1994, were often capped and closed, essentially entombing the waste. This can be beneficial for waste recovery. It can also create a higher risk for toxic waste and leachate exposure as the landfill has not fully processed the stewing wastes. Mining of bioreactor landfills and properly stabilised modern sanitary landfills provides its own benefits. The biodegradable wastes are more easily sieved out, leaving the non biodegradable materials readily accessible. The quality of these materials for recycling and reprocessing purposes is not as high as initially recycled materials, however materials such as aluminium and steel are usually excluded from this.

Landfill mining is most useful as a method to remediate hazardous landfills. Landfills that were established before landfill liner technology was well established often leak their unprocessed leachate into underlying aquifers. This is both an environmental hazard and also a legal liability. In the US, Environmental Protection Agency fines can tax the local economy up to 30 years after the site has closed. Mining the landfill simply to lay a safe liner is a last, but sometimes necessary resort.

Tools and Machinery

The parts of the mining process are the different mining machines. Depending on the complexity of the process more or fewer machines can be used. Machinery is easily transported on trucks from site to site, mounted on trailers. The following machines are added in order in increase of mining complexity:

1. Excavators.
2. Moving floor and elevator conveyor belts.
3. A coarse rotating trommel screen.
4. A fine rotating trommel screen.
5. A magnet.
6. Front end loader.
7. Odour control sprayer.

Mechanics of Mining

An excavator or front end loader uncovers the landfilled materials and places them on a moving floor conveyor belt to be taken to the sorting machinery. A trommel is used to separate materials by size. First, a large trommel separates materials like appliances and fabrics. A smaller trommel then allows the biodegraded soil fraction to pass through leaving non-biodegradable, recyclable materials on the screen to be collected. An electromagnet is used to remove the ferrous material from the waste mass as it passes along the conveyor belt. A front end loader is used to move sorted materials to trucks for further processing. Odour control sprayers are wheeled tractors with a cab and movable spray arm mounted on a rotating platform. A large reservoir tank mounted behind the cab holds neutralising agents, usually in liquid form, to reduce the smell of exposed wastes.

Operational Flow

Excavators dig up waste mass and transport it, with the help of front end loaders, onto elevator and moving floor conveyor belts. The conveyor belts empty into a coarse, rotating trommel. The large holes in the screen allow most wastes to pass through, leaving behind the oversized, non-processable materials. The oversized wastes are removed from inside the screen. The coarse trommel empties into the fine rotating trommel. The fine rotating trommel allows the soil fraction to pass through, leaving mid-sized, non-biodegradable, mostly recyclable materials. The materials are removed from the screen. These materials are put on a second conveyor belt where an electromagnet removes any ferromagnetic debris. Depending on the level of resource recovery, material can be put through an air classifier which separates light organic material from heavy organic material. The separate streams are then loaded, by front end loaders, onto trucks either for further processing or for sale. Further manual processing can be done on site if processing facilities are too far away to justify the transportation costs.

LAND RECLAMATION

Land reclamation, usually known as reclamation, is the process to create new land from sea or riverbeds. The land reclaimed is known as reclamation ground or landfill.

Habitation

The creation of new land was for the need of human activities. Notable examples in the West include large parts of the Netherlands, parts of New Orleans (which is partially built on land that was once swamp); much of San Francisco's waterfront has been reclaimed from the San Francisco Bay; Mexico City (which is situated at the former site of Lake Texcoco); Helsinki (of which the major part of the city center is built on reclaimed land); the Cape Town foreshore; the Chicago shoreline; the Manila Bay shoreline; Back Bay, Boston, Massachusetts; Battery Park City, Manhattan; Liberty State Park, Jersey City; the port of Zeebrugge in Belgium; the southwestern residential area in Brest, Belarus, the polders of the Netherlands; and the Toronto Islands, Leslie Street Spit, and the waterfront in Toronto. In the Far East, Hong Kong, Macau, Japan, the southern Chinese cities of Shenzhen, the Philippine capital Manila, and the city-state of Singapore, where land is in short supply, are also famous for their efforts on land reclamation. One of the earliest and famous project was the Praya Reclamation Scheme, which added 50 to 60 acres (2,40,000 m^2) of land in 1890 during the second phase of construction. It was one of the most ambitious projects ever taken during the Colonial Hong Kong era. Some 20 per cent of land in the Tokyo Bay area has been reclaimed. Monaco and the British territory of Gibraltar are also expanding due to land reclamation. The city of Rio de Janeiro was largely built on reclaimed land, as was Wellington, New Zealand.

Artificial islands are an example of land reclamation. Creating an artificial island is an expensive and risky undertaking. It is often considered in places with high population density and a scarcity of flat land. Kansai International Airport (in Osaka) and Hong Kong International Airport are examples where this process was deemed necessary. The Palm Islands, The World and hotel Burj al-Arab off Dubai in the United Arab Emirates are other examples of artificial islands.

Agriculture

Agriculture was a drive for land reclamation before industrialisation. In South China, farmers reclaimed paddy fields by enclosing an area with a stone wall on the seashore near river mouth or river delta. The species of rice that grow on these grounds are more salt tolerant. Another use of such enclosed land is creation of fish ponds. It is commonly seen on the Pearl River Delta and Hong Kong. These reclamation also attracts species of migrating birds.

A related practice is the draining of swampy or seasonally submerged wetlands to convert them to farmland. While this does not create new land exactly, it allows commercially productive use of land that would otherwise be restricted to wildlife habitat. It is also an important method of mosquito control.

Beach Restoration

Beach rebuilding is the process of repairing beaches using materials such as sand or mud from inland. This can be used to build up beaches suffering from beach starvation or erosion from longshore drift. It stops the movement of the original beach material through longshore drift and retains a natural look to the beach. Although it is not a long-lasting solution, it is cheap compared to other types of coastal defences.

Landfill

As human overcrowding of developed areas intensified during the 20th century, it has become important to develop land reuse strategies for completed landfills. Some of the most common usages are for parks, golf courses and other sports fields. Increasingly, however, office buildings and industrial uses are made on a completed landfill. In these latter uses, methane capture is customarily carried out to minimise explosive hazard within the building. An example of a Class A office building constructed over a landfill is the Dakin Building at Sierra Point, Brisbane, California. The underlying fill was deposited from 1965 to 1985, mostly consisting of construction debris from San Francisco and some municipal wastes. Aerial photographs prior to 1965 show this area to be tidelands of the San Francisco Bay. A clay cap was constructed over the debris prior to building approval. A notable example is Sydney Olympic Park, the primary venue for the 2000 Summer Olympic Games, which was built atop an industrial wasteland that included landfills. Another strategy for landfill is the incineration of landfill trash at high temperature via the plasma-arc gasification process, which is currently used at two facilities in Japan, and will be used at a planned facility in St. Lucie County, Florida.

Environmental Impact

Draining wetlands for ploughing, for example, is a form of habitat destruction. In some parts of the world, new reclamation projects are restricted or no longer allowed, due to environmental protection laws.

Environmental legislation

Hong Kong legislators passed the Protection of the Harbour Ordinance in 1996 in an effort to safeguard the increasingly threatened Victoria Harbour against encroaching land development.

LANDFILL DIVERSION

Waste diversion or landfill diversion is the process of diverting waste from landfill. The success of landfill diversion can be measured by comparison of the size of the landfill from one year to the next. If the landfill grows minimally or remains the same, then policies covering landfill diversion are successful. For example, currently in the United States there are 3000 landfills. A measure of the success of landfill diversion would be if that number remains the same or is reduced.

Landfill diversion can occur through recycling. Recycling refers to taking used materials and creating new products in order to prevent the disposal of these products in landfills. Recycling material can include glass, paper, metal, plastic, textiles, and electronics.

In addition to reusing materials waste products can go through biological treatment. There are two types of biological treatments anaerobic digestion or composting. Simply stated, biological treatment is the breaking down of material through the action of micro-organisms. Materials are broken down to carbon dioxide, water and biomass. Some materials easily break down, others do not. The environment in which the material is placed determines the speed of breakdown.

Another method of landfill diversion is thermal treatment (such as Incineration). Approximately sixteen per cent (16 per cent) of waste is incinerated yearly in the United States. One-fifth (1/5) of municipal solid waste is recycled into usable fuel. Incineration, however, can lead to other environmental issues that may have positive or negative results.

European waste legislation focuses upon the diversion of biodegradable waste from landfill, due to its potential to add to the effects of climate change.

MATERIALS RECOVERY FACILITY

A materials recovery facility or materials reclamation facility or materials recycling facility (MRF— pronounced 'murf') is a specialised plant that receives, separates and prepares recyclable materials for marketing to end-user manufacturers. Generally, there are two different types—clean and dirty MRFs.

Clean MRF

A clean MRF accepts recyclable commingled materials that have already been separated at the source from municipal solid waste generated by either residential or commercial sources. There are a variety of clean MRFs. The most common are single stream where all recyclable material is mixed or dual stream MRFs, where source-separated recyclables are delivered in a mixed container stream (typically glass, ferrous metal, aluminium and other nonferrous metals, PET [No. 1] and HDPE [No. 2] plastics) and a mixed paper stream, (including OCC, ONP, OMG, Office packs, junk mail, etc.). Material is sorted to specifications, then baled, shredded, crushed or otherwise prepared for shipment to market.

Dirty MRF

A dirty MRF accepts a mixed solid waste stream and then proceeds to separate out designated recyclable materials through a combination of manual and mechanical sorting. The sorted recyclable materials may undergo further processing required to meet technical specifications established by end-markets while the balance of the mixed waste stream is sent to a disposal facility such as a landfill.

The percentage of residuals (unrecoverable recyclable or non-program materials) from a properly operated clean MRF supported by an effective public outreach and education program should not exceed 10 per cent by weight of the total delivered stream and in many cases it can be significantly below 5 per cent. A dirty MRF recovers between 5 and 45 per cent of the incoming material as recyclables, then the remainder is landfilled or otherwise disposed. A dirty MRF can be capable of higher recovery rates than a clean MRF, since it ensures that 100 per cent of the waste stream is subjected to the sorting process, and can target a greater number of materials for recovery than can usually be accommodated by sorting at the source. However, the dirty MRF process is necessarily labour-intensive, and a facility that accepts mixed solid waste is usually more challenging and more expensive to site.

Wet MRF

New mechanical biological treatment technologies are now beginning to utilise wet MRFs. This combines a dirty MRF with water which acts to density separate and clean the output streams. It also hydrocrushes and dissolves biodegradable organics in solution to make them suitable for anaerobic digestion.

MECHANICAL BIOLOGICAL TREATMENT

A mechanical biological treatment system is a form of waste processing facility that combines a sorting facility with a form of biological treatment such as composting or anaerobic digestion. MBT plants are designed to process mixed household waste as well as commercial and industrial wastes.

Process

The terms 'mechanical biological treatment' or 'mechanical biological pretreatment' relate to a group of solid waste treatment systems. These systems enable the recovery of materials contained within and the stabilisation of the biodegradable component of the material (Fig. 23.6).

The sorting component of the plants resemble a materials recovery facility. This component is either configured to recover the individual elements of the waste or produce a refuse-derived fuel that can be used for the generation of power. The components of the mixed waste stream that can be recovered include: (i) metals, (ii) plastics, and (iii) glass. MBT is also sometimes termed BMT—biological mechanical treatment, however, this simply refers to the order of processing, i.e. the biological phase of the system precedes the mechanical sorting. MBT should not be confused with MHT—mechanical heat treatment, which does not include any form of biological degradation or stabilisation.

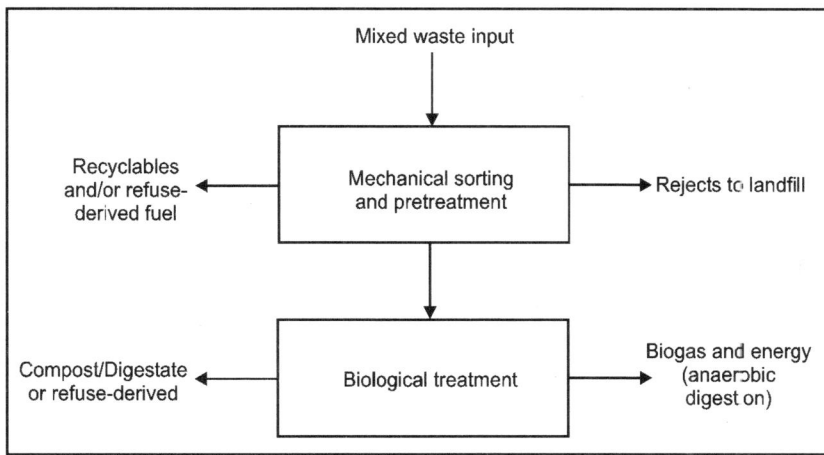

Fig. 23.6. Process flow chart.

Mechanical Sorting

The 'mechanical' element is usually an automated mechanical sorting stage. This either removes recyclable elements from a mixed waste stream (such as metals, plastics, glass and paper) or processes them. It typically involves factory style conveyors, industrial magnets, eddy current separators, trommels, shredders and other tailor made systems or the sorting is made by hand. The mechanical element has a number of similarities to a materials recovery facility (MRF). Some systems integrate a wet MRF to recover and wash the recyclable elements of the waste in a form that can be sent for recycling. MBT can alternatively process the waste to produce a high calorific fuel given the term refuse derived fuel (RDF). RDF can be used in cement kilns or power plants and is generally made up from plastics and biodegradable organic waste. Systems which are configured to produce RDF include the Herhof and Ecodeco Processes. It is a common misconception that all MBT processes produce RDF. This is not the case and depends strictly on system configuration and suitable local markets for MBT outputs.

Biological Processing

The 'biological' element refers to either:
1. Anaerobic digestion.
2. Composting.
3. Biodrying.

Anaerobic digestion breaks down the biodegradable component of the waste to produce biogas and soil improver. The biogas can be used to generate electricity and heat. Biological can also refer to a composting stage. Here the organic component is treated with aerobic micro-organisms. They break down the waste into carbon dioxide and compost. There is no green energy produced by systems employing only composting treatment for the biodegradable waste.

In the case of biodrying, the waste material undergoes a period of rapid heating through the action of aerobic microbes. During this partial composting stage the heat generated by the microbes result in rapid drying of the waste. These systems are often configured to produce a refuse-derived fuel where a dry, light material is advantageous for later transport combustion.

Some systems incorporate both anaerobic digestion and composting. This may either take the form of a full anaerobic digestion phase, followed by the maturation (composting) of the digestate. Alternatively a partial anaerobic digestion phase can be induced on water that is percolated through the raw waste, dissolving the readily available sugars, with the remaining material being sent to a windrow composting facility. By processing the biodegradable waste either by anaerobic digestion or by composting MBT technologies help to reduce the contribution of greenhouse gases to global warming.

Usable wastes for this system:

1. Municipal solid waste.
2. Sewage sludge.

Products of this system:

1. Recycable materials such as metals, paper, plastics, glass, etc.
2. Organic fertiliser (separate collection of organic waste).
3. Unusable materials prepared for their unharmful final deposit (compaction > 1.3 T/m^3).
4. Carbon credits—additional revenues.
5. High calorific fraction (refuse derived fuel—RDF)—additional revenues.

Further advantages:

1. The finally deposited waste is inert.
2. Reduction of the waste volume to be deposited to at least a half (density > 1.3 T/m^3), thus the lifetime of the landfill is at least twice as long as usually.
3. Utilisation of the leachate in the process.
4. No unbidden guests such as birds, dogs, vermin, rats on site.
5. No additional facilities for the collection and combustion of biogas as there is no biogas.
6. Daily covering not necessary.
7. Aftercare 3 to 5 years.

Consideration of applications

MBT systems can form an integral part of a region's waste treatment infrastructure. These systems are typically integrated with curbside collection schemes. In the event that a refuse-derived fuel is produced as a by-product then a combustion facility would be required.

Alternatively MBT solutions can diminish the need for home separation and curbside collection of recyclable elements of waste. This gives the ability of local authorities and councils to reduce the use of waste vehicles on the roads and keep recycling rates high.

Position of environmental groups

Friends of the earth suggests that the best environmental route for residual waste is to firstly maximise removal of remaining recyclable materials from the waste stream (such as metals, plastics and paper). The amount of waste remaining should be composted or anaerobically digested and disposed of to landfill, unless sufficiently clean to be used as compost.

A report by Eunomia undertook a detailed analysis of the climate impacts of different residual waste technologies. It found that an MBT process that extracts both the metals and plastics prior to landfilling is one of the best options for dealing with our residual waste, and has a lower impact than either MBT processes producing RDF for incineration or incineration of waste without MBT. Friends of the Earth does not support MBT plants that produce refuse derived fuel (RDF), and believes MBT processes should occur in small, localised treatment plants.

Chapter 24

Landfill Gas Emission to the Atmosphere

INTRODUCTION

The storage of municipal solid waste (MSW) in landfills contributes to the GHG effect. Methane (CH_4) gas is one of the most important GHGs because its global worming potential more than 20 times carbon dioxide (CO_2). Atmospheric CH_4 gas has more than doubled in concentration over the last 150 years. Landfill gas (LFG) is known to be produced both in managed 'landfill' and 'open dump' sites. It is forming during the decomposition process of waste organic content under anaerobic conditions. It consists on 50–60 vol. per cent CH_4, 30–40 vol. per cent CO_2 and others trace amount. Therefore, CH_4 and CO_2 gases are considered the main end products of solid waste biodegradation under anaerobic conditions.

The evaluation of the areal emission rate of gaseous pollutants from landfills is a very difficult to be control, owing to the high number of factors affecting the emission process. These factors included the gas production rate, the gas migration properties through the waste layers and through the top layer of the landfill, the gas collection efficiency, and the factors affecting transfer of gas to the atmosphere (meteorological factors).

The CH_4 oxidation activity in landfills cover soil as well has a significant affect on CH_4 emission. The gas production rate is influenced by local environmental factors and landfill operation procedures. The gas migration through the waste layers and through soil layer depends on the properties of the gas, the gas collection layout and the characteristics of buried waste and soil cover. The collection efficiency of LFG depends on the spacing between gas collection wells and the maintenance of the landfill cover, which affect the surface emission rate. The meteorological factors control the transfer of LFG to the atmosphere and lead in some cases to enhance lateral migration of LFG that causes gas explosion accident. This chapter reviews the generation process of landfill gas (LFG), the methods of measurement and calculation of LFG emission rate, and the factors affecting LFG emission rate.

GAS GENERATION PROCESS IN LANDFILLS

The gas generation from landfills is resulted from the process of waste decomposition and related to the waste landfilled and landfill technologies used. The waste receiving in landfills included amount of organic waste, the microbial conversion of biodegradable organic carbon to CH_4, CO_2, and tracer gases such as: hydrogen sulphide (H_2S), volatile organic compounds (VOC). The degradation processes of biodegradable waste were divided into five stages. Figure 24.1 shows the decomposition pathways of the major organic and inorganic components of biodegradable wastes.

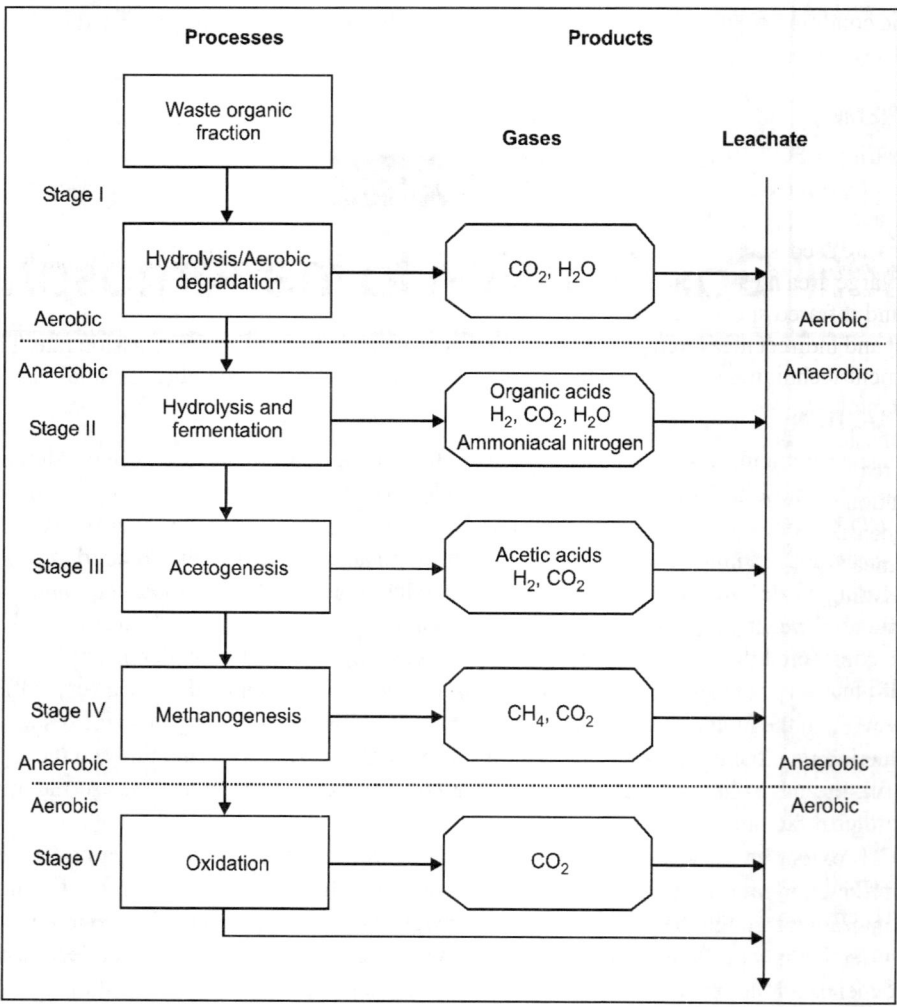

Processes   Products

Fig. 24.1. Major stages of waste degradation in landfills.

In the first stage, aerobic bacteria are responsible for degradation of organic matter and produces CO_2, water (H_2O) and heat. CO_2 may be releases as a gas or adsorbs in the H_2O to form carbonic acid (H_2CO_3), which gives acidity for the leachate generation. A facultative bacterium grows during the second stage, which can survive in aerobic and anaerobic conditions. Carbohydrates, proteins and lipids hydrolysed to sugars, which decomposed to CO_2, hydrogen (H_2), ammonia (NH_3) and organic acids. Organic acids from the second stage convert to acetic acid (CH_3COOH), H_2 and CO_2 by acetogen micro-organisms available in the third stage under anaerobic conditions, as well H_2S may produced by the reduction of sulphate (SO_4^{-2}) compounds in the waste by SO_4^{-2} reduction bacteria. The fourth stage, considers the main stage for LFG production and the longest time stage. Methanogenic micro-organisms under anaerobic conditions degrade the organic acids produced from the third stage to produce CH_4 and CO_2, while another micro-organism directly converted H_2 and CO_2 to CH_4 and H_2O. In the final stage, an aerobic condition occurred with aerobic micro-organisms convert the CH_4 generated in the previous stage to CO_2 and H_2O; as well H_2S gas may forms in waste with high concentration of SO_4^{-2}.

MEASUREMENT METHODS OF LFG EMISSION

For measuring LFG emission rate, there is no precise method exist. Few methods used to measure emission rate, some of them are used to quantify the emission rate for small areas, while others used for large surface area (e.g. for the entire landfill). For measuring the emission from small area, some techniques are used, such as chamber method, method of subsurface vertical gradient of the concentration, while for large area measurements, micrometeorological methods, the isotope ratio technique, the trace method and infrared spectroscopy.

Direct and indirect measurements techniques can be used for quantify LFG emission rate. The direct measurement techniques involve passive sampling methods, and flux chamber methods. The passive sampling methods involve the utilisation of sorbent probes in order to trap gaseous that diffuse upwards through the landfill, while flux chamber methods have been utilised to measure emission rate from typical areal sources. The indirect measurement techniques involve measurement of ambient air concentrations of pollutants around the source, these techniques are depend mainly on the accurate measurements of wind speed and direction during the sampling. A comparison between different methods used for measuring CH_4 emission rate from landfill sites were reported, however each technique has a unique advantages and disadvantages, and the choice will be depended on economic constraints and measurement objectives.

LFG EMISSION BY CALCULATION

Default methodology (DM) method can be applied to whole regions or countries for estimation of LFG emission rate. This methodology depend on estimating the degradable organic carbon (DOC) content of the solid waste, and using this estimate to calculate the amount of CH_4 that can be generated from the waste. Furthermore, this methodology assumes that all potential of CH_4 is released from waste in the year that the waste is disposed of. The DM methodology is the most widely accessible, easy to apply methodology for calculating country-specific emission of CH_4 from solid waste disposal sites. The annual CH_4 emission estimation for each region or country can be calculated from Eq. 24.1.

$$\text{Emission} = (MSW_T \times MSW_F \times MCF \times DOC \times DOC_F \times F \times 16/12 - R) \times (1 - OX) \quad ... (24.1)$$

In Eq. 24.1 MSW_T is the total MSW generated (Gg/yr; 'Gg' equal 10 g), MSW_F is the fraction of MSW disposed to solid waste disposal sites. MCF is the CH_4 correction factor (fraction) which based on the categories of solid waste disposal sites (SWDS) as shown in Table 24.1, while DOC is the degradable organic carbon (fraction), which based on the waste composition. DOC_F is the dissimilated organic fraction, which means the portion of DOC that is converted to LFG. F is the fraction of CH_4 in LFG, with 0.5 default value. R is the value of CH_4 recovered (Gg/yr) and OX is the oxidation factor fraction, with default value equal to 0.

Table 24.1. SWDS classification and MCF.

Type of site	MCF default values
Managed	1.0
Unmanaged-deep (≥ 5 m waste)	0.8
Unmanaged-shallow (≤ 5 m waste)	0.4
Default value-uncategorised SWDSs	0.6

Another method used to estimate LFG emission rate was reported, which based on gas production rate, refuse density; gas collection efficiency and depth of landfill see Eq. 24.2 below:

$$E = G \cdot \gamma_R \cdot L \cdot (1 - \eta) \qquad \qquad \ldots (24.2)$$

Here, E is the emission rate ($m^3 m^{-2} yr^{-1}$), G is the mean specific gas production rate ($m^3 t^{-1} yr^{-1}$), γ_R is the refuse density ($t\ m^{-3}$), η is the gas collection efficiency of the extraction system, and L (m) is the depth of landfill.

FACTORS INFLUENCING LFG EMISSION RATE

Several factors affecting quantity of the LFG dispersed to the atmosphere through the top cover. These factors included the gas production rate, the gas migration properties through the waste layers and through the top layer of the landfill, the gas collection efficiency, and the factors affecting transfer of gas to the atmosphere. Moreover, the CH_4 oxidation activity in landfills cover soil has a significant control role on CH_4 emission rate. The following paragraphs briefly reviewing the influence of each factor on LFG emission rate.

LFG Production Rate

LFG production rate influenced by local environment as characterised by the abiotic factors, as well as landfill operation procedures. The abiotic factors, such as: oxygen (O_2), H_2, pH, SO_4^{2-}, nutrients, inhibitors, temperature, and water content, while the landfill operation procedures are waste composition, sewage sludge addition, shredding, compaction, soil cover, recirculation of leachate, and pre-composting. The influence of these factors will be briefly described in the following:

1. Oxygen: O_2 can enter the landfill by diffusion from the atmosphere during the soil cover in the upper part of the landfill or can enter it by advection if substantial vacuum is created in the landfill by extensive gas extraction. The O_2 is consider as inhibitory for CH_4 formation, the micro-organisms are required very low redox (reduction-oxidation reaction) potential roughly (below -330 mV).

2. Hydrogen: Fermentative and the acetogenic bacteria produce H_2. At low H_2 pressure the fermentative bacteria are produce hydrogen, CO_2 and CH_3COOH, while at high H_2 pressure produce H_2, CO_2 and ethanol (C_2H_5OH), butyric acid ($CH_3C_2H_4COOH$), and propionic acid (CH_3CH_2COOH). In case H_2 pressure is low, the last three organic compounds (C_2H_5OH, $CH_3C_2H_4COOH$, and CH_3CH_2COOH) can be converted by acetogenic bacteria, while if the H_2 pressure is high the organic compound will not be converted which causes the accumulation of volatile organic acids, reduction in pH, and as a result will cause the inhibition of CH_4 formation.

3. pH: The methanogenic bacteria operate only within 6–8 pH. The accumulation of H_2 and CH_3COOH decrease the pH value, which inhibits the methanogenic bacteria to format CH_4.

4. Sulphate: SO_4^{2-} originated from the gypsum board ($CaSO_4 \cdot 2H_2O$) presents in the building construction and demolition material fraction of MSW, incinerator slag and fly ashes. SO_4^{2-} reduction bacteria dominated by desulphovibrio and desulphotomaculum resemble the methanogenic bacteria in many ways. A high activity of SO_4^{2-} reduction bacteria reduces the amount of organics available for CH_4 production, due to the ability to converts of H_2, CH_3COOH and fatty acids, and the defects of organic carbon caused by oxidation.

5. Nutrients: Anaerobic ecosystem is requires much less nitrogen (N_2) and phosphorous (P) than the aerobic system. The optimal ratio for organic matter (express as chemical oxygen demand;

COD), N_2 and P content, is 100:0.44:0.08. Nutrients and metal supplementation have a positive effect on biogas production rate.

6. Inhibitors: Beside of the inhibition caused by O_2, H_2 and SO_4^{2-}, there are other inhibitors appear in landfill sites. CO_2 found to be inhibited the conversion of CH_3COOH at CO_2 partial pressure as low as 0.2; typical CO_2 partial pressure is 0.5 in the methanogenic phase. Macroions such as sodium (Na), potassium (K), calcium (Ca), magnesium (Mg) and ammonium (NH_4) are having inhibitory effects on CH_4 formation if they are exceed 2000 mg/l each.

7. Temperature: The active temperature for methanogenic micro-organisms in the range 30°–50°C. The temperature for mesophilic bacteria in the range 30°–35°C, while 45°–65°C for the thermophilic bacteria. The optimum temperature range of gas generation between 30°–45°C during the main landfill gas generation phase. The change of temperature will have an impact on the growth of biomas and the activity of the micro-organisms.

8. Water content: Moss, reported that the range of moisture content in a typical landfill is 15 to 40 per cent with a typical average 30 per cent. Some studies have indicated that refuse samples containing greater than 55 per cent (wt/wt) moisture content produced increased amounts of CH_4 while those that contained less than 33 per cent moisture content did not produce any CH_4. The rate of gas generation increases with the increment of moisture in landfill site. The water content in landfill sites assists to exchange of substrate, nutrients, buffer, and dilution of inhibitors and spreading of micro-organisms.

9. Waste composition: The amount of biodegradation in waste is depending on the availability of the biodegradable portion. The composition of organic components (cellulose, proteins and lipids) effects the degradation of waste and as a result affects gas generation process. Cellulose-to-lignin ratio (CLR) has a negative effect on CH_4 production due to the presence of easily degradable carbon sources, and has a negative relation with age of solid waste samples which indicate that the older samples are methanogenically active.

10. Sewage sludge addition: Sewage sludge has a positive effect which increases CH_4 formation if it added with neutral pH.

11. Buffer addition: The buffer addition increases pH and assists the CH_4 formation process. It has a positive effect on CH_4 formation.

12. Shredding: Shredding the waste prior to landfilling has a positive and negative effect on CH_4 formation process. It affects waste stabilisation by increasing the homogeneity of solid waste through size reduction and mixing, increasing the surface area of waste, removing water barriers such as plastic bags and foil, and improving the water content and distribution in landfilled waste. It motivates the acid phase of landfills stabilisation, which prevents or postpones the start of CH_4 formation.

13. Compaction: In relatively wet waste, lack of compaction may affect negatively to acid phase, while improve the start of CH_4 formation. In dry waste, compaction improves CH_4 formation by improvement the water content, and reduces diffusion of O_2.

14. Soil cover: The amount of CH_4 that vent from landfills is dependent on physical, chemical and biological components of the soil cover. Thickness and permeability of soil cover have effects on LFG emission, through elongate the retention time of transported LFG and thereby increase the oxidation probability. The type and amount of bacteria and nutrients present in the soil cover play a major role in CH_4 oxidation capacity. A positive effect of soil daily cover if provides important buffer capacity to landfill, to avoid pH value which inhibited CH_4 formation process.

15. Leachate recirculation: Leachate recirculation has a positive effect on CH_4 formation. Attention must pay to leachate composition and pH value which affect CH_4 formation. Leachate recirculation is enhancing CH_4 formation by increasing the water content, supply and distribution of nutrients and biomass, and diluted high concentrations of inhibitors. Leachate recirculation and subsurface irrigation enhance LFG generation and boosts its emission from cover soil.

16. Precomposting: By allowing partial stabilisation of the waste by aerobic process, the methanogenic phase establish faster and acid phase become less vigorous, such as in German, the concept of *in situ* pre-composting of the bottom layer of the landfill to prevent too vigorous of the acid phase was used.

LFG Migration Properties through Waste Layers and through the Top Layer of Landfill

The movement of LFG through the waste layers depends on the properties of the gas itself (diffusivity and viscosity), the physical and chemical characteristics of buried waste (permeability, moisture content and temperature) and the layout and the efficiency of the gas collection system, while its movement through the top layer of landfill (soil cover) depends on the previous properties of the gas and properties of soil cover (permeability, moisture content, thickness).

Diffusion flux is caused by variation in gas concentration, and depends on the gas filled porosity, diffusion coefficient and the concentration gradient. The diffusion coefficient depends on diffusion coefficient in air, which is a function of pressure and temperature. The gas viscosity is a function of temperature and gas composition. The permeability of waste is related to the compaction degree, which restrict the movement of the gas. On another hand, the variability of permeability in soil cover is highly and depends on the grain size distribution, organic matter content. Soil moisture influences both permeability and the diffusivity of soil, by affects the available pore space for gaseous transport and diffusion. Thickness and permeability of soil cover elongate the retention time of transported LFG and increase the probability of methane oxidation.

LFG Collection System

Many previous studies indicated that LFG collection system has a strong effect on surface LFG emission rate; LFG collection system reduces the gas emission rate to the atmosphere. The CH_4 emission rate observed in two different locations (5 m, and 20 m) from gas collection pipes, and the results show that the average of CH_4 emission rate from the nearest locations (5 m) almost doubled that at far location (20 m). LFG emission amount from the landfill flatter surface is about 30 per cent of the LFG generation amount without extraction process, while if the LFG was extracted using the blower, the surface emission rate of LFG decreased by more than 80 per cent.

Oxidation Activity

Microbial oxidation of CH_4 gas from landfill sites has been reported to effectively minimise landfill CH_4 emission. CH_4 oxidation is the primary factor that control CH_4 emission from landfills, where gas collection system is not available. The efficiency of oxidation activity is related to the environmental soil conditions. Microbial CH_4 oxidation in landfills soil cover is controlled by several soil properties such as: soil moisture, ammonium and nitrite content, pH, temperature and nitrogen turnover.

Encouraging of CH_4 oxidation in the soil cover of landfills is a much cheaper and more effective option for reducing emission in smaller and older landfills with lower amounts of CH_4 generation, compared with gas extracting, which becomes inefficient at low CH_4 contents. A famous method to quantify CH_4 oxidation is by measuring isotope fractionation because it is a noninvasive technique.

Meteorological Conditions

Several meteorological factors influence the LFG migration to the atmosphere. These factors are reviewed as follows:

1. Atmospheric pressure: The influence of LFG emission by atmospheric pressure resulted from the pressure gradient between landfill and atmosphere, thus any variation on atmospheric pressure cause difference on LFG emission rate. Inverse linear relationship was observed between LFG emission rate and atmospheric pressure. The barometric pressure considers one of the external controlling factors that control methane emission from landfills. In some cases the barometric pressure raises, thus surface LFG emission decreases and even become negative, which cause atmospheric air intrusions to the top layer of landfill.

2. Precipitation: Soil water content influences both permeability and the diffusivity of soil. Therefore, the amount of precipitation has an effect on LFG migration process. It has a significant controlling role for surface methane emission. By increasing soil moisture content the available pore space for gaseous transport and diffusion is reduced, as well as reduces the diffusion of oxygen from above, while low moisture content reduced the biological activity in soil cover. The combination of drying due to low precipitation and the heat generated by the oxidation in soil cover are likely to reduce the pore water content of soil. This may facilitate LFG transport, if the soils cover shallow, and reduce the oxidation capacity due to the inhibition of the microbiological activity that requires optimum moisture content. The desirable moisture content for high methane oxidation activity ranges between 11–25 per cent by volume. There is no significant relationship between soil moisture content and surface methane emission in temporal variation was reported.

3. Air temperature: Surface emission rate has close relationship to the surface temperature; the surface emission rate was peak when the surface temperature during the day period was at peaks. Air temperature has direct effects on soil temperature, which has significant effects on oxidation activity in soil cover. Methane emission rate showed good temporal correlation with soil temperature. Two sites (A and B) with different operation in Tianziling MSW landfill in China were tested for diurnal variations of surface methane emission rate. There were no significant correlations between either soil water content, soil temperature or air temperature and methane fluxes in site A ($p > 0.05$), although peaks methane flux were largely observed in the daytimes when higher air and soil temperatures were observed. Site B, showed there were significantly negatively correlated between methane flux with soil temperature ($r = -0.598$, $p < 0.01$) and air temperature ($r = -0.491$, $p < 0.05$), while no correlation was observed with soil water content ($r = 0.316$, $p = 0.174$).

4. Other meteorological parameters: Snow during winter season creates a sealing layer of ice, thus inhibit oxidising organisms, due to low temperature and drought. A landfill gas explosion accident at a landfill in eastern France was reported, where the soil was frozen to a depth 50 cm, which led to enhanced lateral migration of gas.

On the other hand, the change in wind speed has an important effect on the exchange of the soil gas with the atmosphere for the top few centimetres of the soil layer.

To sum up, methane and carbon dioxide gases are the main end products of the decomposition of solid waste under anaerobic process, and they have global warming effects. There are many factors effects on surface LFG emission rate, such as gas production rate, LFG migration properties throw the waste layers and through the top layer of landfill, LFG collection system, oxidation activity in the landfill cover and metrological conditions. These factors have significant controlling role for surface methane emission rate. LFG production rate depend on many factors controlling the quantity of gas production, while LFG collection system using blower can decreased 80 per cent of surface LFG emission rate. Microbial oxidation activity in soil cover of landfills is a much cheaper and more effective option for reducing emission in smaller and older landfills with lower amounts of CH_4 generation, compared with gas extracting, which becomes inefficient at low CH_4 contents. Some metrological factors have important effects on the surface LFG emission rate, due to prevent of upward LFG migration which led to enhancement of lateral migration of LFG that cause gas explosion accident. Therefore, development of predictable models for surface LFG emission rate should be taken on to their account those factors and their effects on LFG emission rate.

Chapter 25

Leachate

INTRODUCTION

Leachate is any liquid that, in passing through matter, extracts solutes, suspended solids or any other component of the material through which it has passed. Rain falling on the top of the landfill is the main contributor to the generation of leachate, and is by far the largest contributor for modern sanitary landfills which do not accept liquid waste. In old unlined and unengineered landfills, some leachate is produced from groundwater entering the waste. Some, additional leachate volume is produced during waste decomposition, and some additional surface water will sometimes run onto waste from its surroundings.

The decomposition of carbonaceous material produces some additional water, and a wide range of other materials including methane, carbon dioxide and a complex mixture of organic acids, aldehydes, alcohols and simple sugars, which dissolve in the leachate cocktail. The precipitation percolates through the waste and takes in dissolved and suspended components from the biodegrading waste, through physical and chemical reactions.

Most landfills are designed to minimise the amount of leachate they create during their lifetimes. However, there are good scientific reasons to suggest that it would be better to flush all landfills out and to do this, would produce more leachate, faster. Landfills where the latter philosophy is adopted, are called, 'bioreactor' landfills. In Europe, bioreactor landfills are effectively prohibited by EU directives, leading them to be called 'dry tombs' by some, due to their rapid capping, and minimised leachate production.

The environmental risks of leachate generation arise from it escaping into the environment around landfills, particularly to watercourses and groundwater. These risks can be mitigated by properly designed and engineered landfill sites. Such sites are those that are constructed on geologically impermeable materials or sites that use impermeable liners made of geotextiles or engineered clay. The use of linings is now mandatory within both the United States and the European Union, except where the waste closely controlled and genuinely inert.

Most toxic and difficult materials are now specifically excluded from landfill. However, despite much stricter statutory controls the leachates from modern sites are currently stronger than ever. They also contain a huge range of contaminants. In fact, anything soluble in the waste disposed will enter the leachate. Within the lists of substances present in leachate are very low concentrations of 'trace contaminants' which can have quite strongly contaminating effects. These are nowadays most often derived from materials in household and domestic retail products which enter the waste stream perfectly legally. Unfortunately, the leachate draining from most landfills will continue to reflect the contaminants of past years, when regulatory controls were less.

These substances include extremely low concentrations of heavy metals (for example from batteries), herbicides and pesticides (as used in gardens), etc. However, leachate is becoming less contaminated with difficult substances as time goes forward, and public awareness, recycling and increased statutory control over these substances, throughout the industrialised world is making leachate less harmful in this respect.

LEACHATE HAS A VERY HIGH AMMONIACAL NITROGEN CONCENTRATION

The concern about environmental damage from waste leachate, largely arises from its high organic contaminant concentrations and much higher ammoniacal nitrogen than commonly found in any other organic effluent. Pathogenic micro-organisms and toxic substances that might be present in it have in the past been described as the most important. However, pathogenic organism counts reduce rapidly with time in the landfill, so this only applies to the youngest leachate and leachate is seldom removed from the landfill in this condition.

LANDFILL LEACHATE

Leachate from a landfill varies widely in composition depending on the age of the landfill and the type of waste that it contains. It can usually contain both dissolved and suspended material. The generation of leachate is caused principally by precipitation percolating through waste deposited in a landfill. Once in contact with decomposing solid waste, the percolating water becomes contaminated and if it then flows out of the waste material it is termed leachate. Additional leachate volume is produced during this decomposition of carbonaceous material producing a wide range of other materials including methane, carbon dioxide and a complex mixture of organic acids, aldehydes, alcohols and simple sugars.

The risks of leachate generation can be mitigated by properly designed and engineered landfill sites, such as sites that are constructed on geologically impermeable materials or sites that use impermeable liners made of geomembranes or engineered clay. The use of linings is now mandatory within both the United States and the European Union except where the waste is deemed inert. In addition, most toxic and difficult materials are now specifically excluded from landfilling. However despite much stricter statutory controls leachates from modern sites are found to contain a range of contaminants that may either be associated with some level of illegal activity or may reflect the ubiquitous use of a range of difficult materials in household and domestic products which enter the waste stream legally.

Composition of Landfill Leachate

When water percolates through the waste, it promotes and assists process of decomposition by bacteria and fungi. These processes in turn release by-products of decomposition and rapidly use up any available oxygen creating an anoxic environment. In actively decomposing waste the temperature rises and the pH falls rapidly and many metal ions which are relatively insoluble at neutral pH can become dissolved in the developing leachate. The decomposition processes themselves release further water which adds to the volume of leachate. Leachate also reacts with materials that are not themselves prone to decomposition such as fire ash, cement based building materials and gypsum based materials changing the chemical composition. In sites with large volumes of building waste, especially those containing gypsum plaster, the reaction of leachate with the gypsum can generate large volumes of hydrogen sulphide which may be released in the leachate and may also form a large component of the landfill gas.

In a landfill that receives a mixture of municipal, commercial, and mixed industrial waste, but excludes significant amounts of concentrated specific chemical waste, landfill leachate may be characterised as a water-based solution of four groups of contaminants; dissolved organic matter (alcohols, acids, aldehydes, short chain sugars, etc.), inorganic macro components (common cations and anions including sulphate, chloride, iron, aluminium, zinc and ammonia), heavy metals (Pb, Ni, Cu, Hg), and xenobiotic organic compounds such as halogenated organics (PCBs, dioxins, etc.).

The physical appearance of leachate when it emerges from a typical landfill site is a strongly odoured black, yellow or orange coloured cloudy liquid. The smell is acidic and offensive and may be very pervasive because of hydrogen, nitrogen and sulphur rich organic species such as mercaptans.

Leachate Management

In older landfills and those with no membrane between the waste and the underlying geology, leachate is free to egress the waste directly into the groundwater. In such cases high concentrations of leachate are often found in nearby springs and flushes. As leachate first emerges it can be black in colour, anoxic and may be effervescent with dissolved and entrained gases. As it becomes oxygenated it tends to turn brown or yellow because of the presence of iron salts in solution and in suspension. It also quickly develops a bacterial flora often comprising substantial growths of *Sphaerotilus*.

Goals of Leachate Collection Systems

The primary criterion for design of the leachate system is that all leachate be collected and removed from the landfill at a rate sufficient to prevent an unacceptable hydraulic head occurring at any point over the lining system.

Components of Leachate Collection Systems

There are many components to a collection system including pumps, manholes, discharge lines and liquid level monitors. However, there are four main components which govern the overall efficiency of the system. These four elements are liners, filters, pumps and sumps.

Liners

Natural and synthetic liners may be utilised as both a collection device, and as a means for isolating leachate within the fill to protect the soil and groundwater below. The chief concern is a liners ability to maintain integrity and impermeability over the life of the landfill. Subsurface water monitoring, leachate collection, and clay liners are commonly included in the design and construction of a waste landfill. To effectively serve the purpose of containing leachate in a landfill, a liner system must possess a number of physical properties. The liner must have high tensile strength, flexibility, and elongation without failure. It is also important that the liner resists abrasion, puncture, and chemical degradation by leachate. Lastly the liner must withstand temperature variation, be black (to resist UV light), easily installed, and economical. There are several types of liners used in leachate control and collection. These types include geomembranes, geosynthetic clay liners, geotextiles, geogrids, geonets, and geocomposites. Each style of liner has specific uses and abilities. Geomembranes are used to provide a barrier between mobile polluting substances released from wastes, and the groundwater. In the closing of landfills, geomembranes are used to provide a low-permeability cover barrier to prevent the intrusion of rainwater. Geosynthetic clay liners (GCLs) are fabricated by distributing sodium bentonite in a uniform thickness between woven and non-woven geotextiles. Sodium bentonite has a low permeability which makes GCLs a suitable alternative to clay liners in a composite liner system. Geotextiles are used as separation between two different types of soils to prevent contamination of the lower layer by the upper layer. Geotextiles also act as a cushion to protect synthetic layers against puncture from underlying and overlaying rocks. Geogrids are structural synthetic materials used in slope veneer stability to create stability for cover soils over synthetic liners or as soil reinforcement in steep slopes. Geonets are synthetic drainage materials which are often used in lieu of sand and gravel. Geonets can replace 12 inches of drainage sand, thus increasing the landfill space for waste. Geocomposites are a combination of synthetic materials ordinarily used singly. A common type of geocomposite is a geonet heat bonded to two layers of geotextile, one on each side. The geocomposite serves as a filter and drainage medium. Geosynthetic clay liners are a type of combination liner. One advantage to using a geosynthetic clay liner (GCL) is the ability to order exact amounts of the liner. Ordering precise amounts from the manufacturer prevents surplus and overspending. Another advantage to GCL's is the liner can serve appropriately in areas without an adequate clay source. Conversely, GCL's are heavy, cumbersome, and installation is very labour intensive. In addition to be arduous and difficult under normal conditions, installation can be cancelled during damp conditions because the bentonite absorbs the water making it even more burdensome and tedious.

Leachate Drainage System

The leachate drainage system is responsible for the collection and transport of the leachate collected inside the liner. The pipe dimensions, type, and layout must all be planned with the weight and pressure of waste, and transport vehicles in mind. The pipes are located on the floor of the cell. Above the network, lies an enormous amount of weight and pressure. To support this, the pipes can either be flexible or rigid. However, the joints to connect the pipes yield better results if the connections are flexible. An alternative to placing the collection system underneath the waste is to position the conduits in trenches or above grade. The collection pipe network of a leachate collection system drains, collects, and transports leachate through the drainage layer to a collection sump where it is removed for treatment or disposal. The pipes also serve as drains within the drainage layer to minimise the mounding of leachate in the layer.

Filters

The filter layer is used above the drainage layer in leachate collection. There are two types of filters typically used in engineering practices: granular and geotextile. Granular filters consist of one or more soil layer or multiple layers having a coarser gradation in the direction of the seepage than the soil to be protected.

Sumps

As liquid enters the landfill cell, it moves down the filter, passes through the pipe network, and rests in the sump. As collection systems are planned, the number, location, and size of the sumps are vital to an efficient operation. When designing sumps, the amount of leachate and liquid expected is the foremost concern. Areas in which rainfall is higher than average typically have larger sumps. A further criterion for sump planning is accounting for the pump capacity. The relationship of pump capacity and sump size is inversed. If the pump capacity is low, the volume of the sump should be larger than average. It is critical for the volume of the sump to be able to store the expected leachate between pumping cycles. This relationship helps maintain a healthy operation. Sump pumps can function with preset phase times. If the flow is not predictable, a predetermined leachate height level can automatically switch the system on. Other conditions for sump planning are maintenance and pump drawdown. Collection pipes typically convey the leachate by gravity to one or more sumps, depending upon the size of the area drained.

Leachate collected in the sump is removed by pumping to a vehicle, to a holding facility for subsequent vehicle pickup or to an on-site treatment facility.

Sump dimensions are governed by the amount of leachate to be stored, pump capacity, and minimum pump drawdown. The volume of the sump must be sufficient to hold the maximum amount of leachate anticipated between pump cycles, plus an additional volume equal to the minimum pump drawdown volume. Sump size should also consider dimensional requirements for conducting maintenance and inspection activities. Sump pumps may operate with preset cycling times or, if leachate flow is less predictable, the pump may be automatically switched on when the leachate reaches a predetermined level.

Membrane and Collection for Treatment

More modern landfills in the developed world have some form of membrane separating the waste from the surrounding ground and in such sites there is often a leachate collection series of pipes laid on the membrane to convey the leachate to a collection or treatment location. All membranes are porous to some limited extent so that over time low volumes of leachate will cross the membrane. The design of landfill membranes is at such low volumes that they should never have a measurable adverse impact on the quality of the receiving groundwater. A more significant risk may be the failure or abandonment of the leachate collection system. Such systems are prone to internal failure as landfills suffer large internal movements as waste decomposes unevenly and thus buckles and distorts pipes. If a leachate collection system fails, leachate levels will slowly build in a site and may even over-top the containing membrane and flow out into the environment. Rising leachate levels can also wet waste masses that have previously been dry triggering further active decomposition and leachate generation. Thus what appears to be a stabilised and inactive site can become reactivated and restart significant gas production and exhibit significant changes in finished ground levels.

Reinjection into Landfill

One method of leachate management that was more common in uncontained sites was leachate recirculation in which leachate was collected and re-injected into the waste mass. This process greatly accelerated decomposition and therefore gas production had the impact of converting some leachate volume into landfill gas and reducing the overall volume of leachate for disposal. However it also tended to increase substantially the concentrations of contaminant materials making it a more difficult waste to treat.

Treatment

The most common method of handling collected leachate is on-site treatment. When treating leachate on-site, the leachate is pumped from the sump into the treatment tanks. The leachate may then be mixed with chemical reagents to modify the pH and to coagulate and settle solids and to reduce the concentration of hazardous matter. Further treatment is typically a modified form of activated sludge to substantially reduce the dissolved organic content. Nutrient imbalance can cause difficulties in maintaining an effective biological treatment stage. The treated liquor is rarely of sufficient quality to be released to the environment and may be tankered or piped to a local sewage treatment facility.

Removal to Sewer System

In some older landfills, leachate was directed to the sewers, but this can cause a number of problems. Toxic metals from leachate passing through the sewage treatment plant concentrate in the sewage sludge, making it difficult or dangerous to dispose of the sludge without incurring a risk to the environment. In

Europe, regulations and controls have improved in recent decades and toxic wastes are now no longer permitted to be disposed of to the Municipal Solid Waste landfills, and in most developed countries the metals problem has diminished. Paradoxically, however, as sewage treatment works discharges are being improved throughout Europe and many other countries, the sewage treatment works operators are finding that leachates are difficult waste streams to treat. This is because leachates contain very high ammoniacal nitrogen concentrations, they are usually very acidic, they are often anoxic and, if received in large volumes relative to the incoming sewage flow, they lack the phosphorus needed to prevent nutrient starvation for the biological communities that perform the sewage treatment processes. The result is that leachates are a difficult-to-treat waste stream. However, within ageing municipal solid waste landfills, this may not be a problem as the pH returns close to neutral after the initial stage of acidogenic leachate decomposition. Many sewer undertakers limit maximum ammoniacal nitrogen concentration in their sewers to 250 mg/l to protect sewer maintenance workers, as the WHO's maximum occupational safety limit would be exceeded at above pH 9 to 10, which is often the highest permitted pH of permitted sewer discharges.

Many older leachate streams also contained a variety of synthetic organic species and their decomposition products, some of which had the potential to be acutely damaging to the environment.

ENVIRONMENTAL IMPACT

The risks from waste leachate are due to its high organic contaminant concentrations and high concentration of ammonia. Pathogenic micro-organisms that might be present in it are often cited as the most important, but pathogenic organism counts reduce rapidly with time in the landfill, so this only applies to the most fresh leachate. Toxic substances may however be present in variable concentration and their presence is related to the nature of waste deposited.

Most landfills containing organic material will produce methane, some of which dissolves in the leachate. This could in theory be released in weakly ventilated areas in the treatment plant. All plants in Europe must now be assessed under the EU ATEX Directive and zoned where explosion risks are identified to prevent future accidents. The most important requirement is the prevention of discharge of dissolved methane from untreated leachate when it is discharged into public sewers, and most sewage treatment authorities limit the permissible discharge concentration of dissolved methane to 0.14 mg/l, or 1/10th of the lower explosive limit. This entails methane stripping from the leachate.

The greatest environmental risks occur in the discharges from older sites constructed before modern engineering standards became mandatory and also from sites in the developing world where modern standards have not been applied. There are also substantial risks from illegal sites and adhoc sites used by criminal gangs to dispose of waste materials. Leachate streams running directly into the aquatic environment have both an acute and chronic impact on the environment which may be very severe and can severely diminish biodiversity and greatly reduce populations of sensitive species. Where toxic metals and organics are present this can lead to chronic toxin accumulation in both local and far distant populations. Rivers impacted by leachate are often yellow in appearance and often support severe overgrowths of sewage fungus.

PROBLEMS AND FAILURES

Leachate collection systems can experience myriad of major and minor problems. However, some troubles are more common than others. One familiar hindrance is clogging from mud and silt. The clogging can be attributed, in some cases, to the growth of micro-organisms in the conduit. The damp, dark, and waste-laden conditions of pipe are ideal living spaces for micro-organisms to spread and grow. As leachate travels through the pipeline, residue left on the walls of the duct is a perfect food source for the bacteria inhabiting the space. Another potential reason for clogging is chemical reactions. Often leachate contains a concoction of chemicals and as the leachate travels it mixes. Sometimes these mixtures can leave a product behind. Without adequate cleaning and maintenance, the problem will grow as new blends leave behind more products. Also, the chemicals are capable of weakening the pipe walls. This can cause the pipes to fail structurally.

OTHER TYPES OF LEACHATE

Leachate can also be produced from land that was contaminated by chemicals or toxic materials used in industrial activities such as factories, mines or storage sites. Composting sites in high rainfall also produce leachate. Leachate is also associated with stockpiled coal and with waste materials from metal ore mining and other rock extraction processes, especially those in which sulphide containing materials are exposed to air and thus to oxygen generating acidic, sulphur rich liquors, often with elevated metal concentrations.

In the context of civil engineering (more specifically reinforced concrete design), leachate refers to the effluent of pavement wash-off (that may include melting snow and ice with salt) that permeates through the cement paste onto the surface of the steel reinforcement, thereby catalysing its oxidation and degradation. Leachates can be genotoxic in nature.

Chapter 26

Electrochemical Oxidation of Landfill Leachate Treatment

INTRODUCTION

This chapter aims at providing an overview of electrochemical oxidation processes used for treatment of landfill leachate. The typical characteristics of landfill leachate are briefly reviewed, and the reactor designs used for electro-oxidation of leachate are summarised. Electrochemical oxidation can significantly reduce concentrations of organic contaminants, ammonia, and colour in leachate. Pretreatment methods, anode materials, pH, current density, chloride concentration, and other additional electrolytes can considerably influence performance. Although high energy consumption and potential chlorinated organics formation may limit its application, electrochemical oxidation is a promising and powerful technology for treatment of landfill leachate.

Laboratory studies to determine the effectiveness of various biological, physical and chemical treatment processes on sanitary landfill leachates have been investigated since the early 1970s. Biological treatment processes, including anaerobic and aerobic processes, are quite effective for leachate generated in the early stage with a high BOD_5/COD. However, they generally fail to treat a leachate with a rather low BOD_5/COD or high concentrations of toxic metals. Hence, physical-chemical processes are mostly used for pretreatment or full treatment for this sort of landfill leachate. The physical-chemical processes used for leachate treatment mainly include flocculation/precipitation, activated adsorption, membrane technologies, and chemical oxidation.

Since the late 1970s, electrochemical oxidation has been successfully applied to the treatment of textile waste-water, tannery waste-water, coke-plant waste-water, coffee curing waste-water, and other waste-waters containing cyanides and phenol. This process has been used for landfill leachate treatment over the past 10 years. This chapter reviews the current state-of-the-art understanding of electrochemical oxidation of landfill leachate in both laboratory and pilot scales. Emphasis is placed on removal of organic constituents, ammonia nitrogen, and colour, rather than removal of heavy metals.

CHARACTERISTICS OF LANDFILL LEACHATE

An understanding of the characteristics of landfill leachate is needed to interpret the variable performance found when treating leachates with electrochemical oxidation. Typical characteristics of landfill leachate are listed in Table 26.1. Organic compounds and ammonia nitrogen in landfill leachate are two principal chemical characteristics of environmental concern. Organic contaminants in leachate are described mainly using global parameters such as chemical oxygen demand (COD), 5-day biochemical oxygen demand (BOD_5), and total organic carbon (TOC).

Table 26.1. Typical data on characteristics of municipal landfill leachate.

Constituents	Landfill leachate	
	From young landfills (<1–2 yr old)	*From old landfills (>5–10 yr old)*
pH	6	6.6–7.5
Chemical oxygen demand (COD)(mg/l)	3000–60,000	100–500
BOD$_5$ (5-day biochemical oxygen demand)/COD	0.6–1.0	0–0.3
Total organic carbon (TOC) (mg/l)	1500–20,000	80–160
Total suspended solids (TSS) (mg/l)	200–2000	100–400
Ammonia nitrogen (NH$_3$–N) (mg/l)	10–800	20–40
Organic nitrogen (mg/l)	10–800	80–120
Calcium (Ca^{2+}) (mg/l)	200–3000	100–400
Magnesium (Mg^{2+}) (mg/l)	50–15,000	50–200
Sulphate (SO$_4^{2-}$) (mg/l)	50–1000	20–50
Chloride (Cl$^-$) (mg/l)	200–3000	100–400

As shown in Table 26.1, in young landfills (typically <1–2 yr old), leachate is characterised by a high COD (typical 18,000 mg/l) and a high BOD$_5$/COD (typical >0.6). In contrast, leachate in old landfills (typically >5–10 yr old) is characterised by a relatively low COD (typical 100–500 mg/l), and a low BOD$_5$/COD (typical <0.3). Determination of the molecular weight distribution of leachate organics, as well as the functional groups, further enhances understanding of the behaviour and reactivity of leachate organic substances. Calace reported that old leachate contained a wide range of molecular weights with a large high molecular weight fraction (77 per cent >10,000 Da), but that young leachate had a narrow range with a large lower molecular weight fraction (70 per cent <500 Da).

Ammonia, released from wastes mainly by decomposition of protein, has been reported to be in leachate at concentrations from 500 to 2000 mg NH$_4$–N/L. It may disrupt biological units for leachate treatment due to its toxicity. Moreover, ammonia does not have an obvious decreasing trend in concentration with time except due to leaching. Therefore, ammonia has been identified as the most significant component in leachate in the long-term.

ELECTROCHEMICAL OXIDATION OF LANDFILL LEACHATE

A basic conceptual diagram of electrolysis is shown in Fig. 26.1, including a direct current (DC) power supply, a cathode, an anode, and the electrolyte (a medium that provides the ion transport mechanism between the anode and the cathode necessary to sustain the electrochemical process). At the cathode, an electrode at which reduction occurs and from which electrons are repelled, metal-cations (mostly heavy metals) can be removed; and at the anode, a electrode at which oxidation occurs and to which electrons travel, some pollutants (e.g. organic compounds) can be directly oxidised. Additionally, an oxidation reaction may occur in bulk solution by an oxidant generated by the electrodes. Electrochemical oxidation has been widely investigated as an efficient means of controlling pollution in water and waste-water treatment. An important advantage of electrochemical oxidisation is to oxidise organic pollutants into CO$_2$ and water to avoid a problem of contaminants shifting from one phase to another. Also, the operation at room temperature and atmospheric pressure prevents volatilisation and discharge of unreacted wastes, and the reaction can be simply terminated in seconds by cutting off the power.

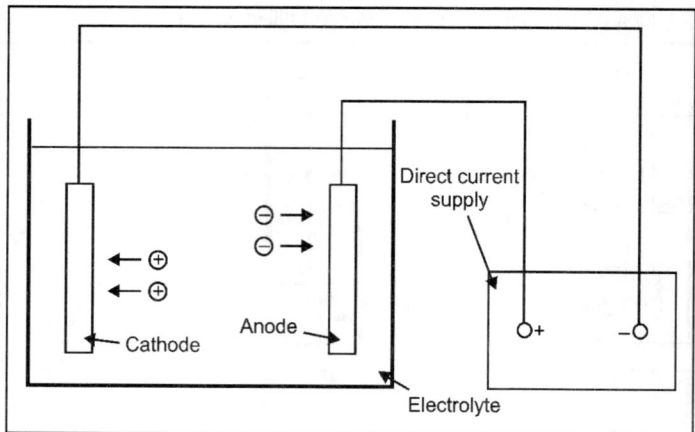

Fig. 26.1. Conceptual diagram of electrolysis.

Reactor Designs

Experiments on electrochemical oxidation of landfill leachate have been conducted at the laboratory or pilot plant scale. Bipolar cell and cylindrical electrode cell are two major reactor designs used in current studies of electrochemical oxidation of leachate as shown in Figs 26.2 and 26.3, although more designs have been employed in other waste-water electro-oxidation. Electrochemical oxidation of landfill leachate is generally operated in a galvanostatic state (a constant current is maintained in electrolyte) for both batch reactors and continuously stirred tank reactors. Recirculation may be used in batch mode through a peristaltic pump. The required electrical power is provided through a DC power supply with current–voltage monitoring.

The useful volume of the electrolytic cell in lab scale experiments is commonly below 2 litre, and in two pilot scale experiments around 3.8 and 6 litre are used. A magnetic stirrer keeps the electrolyte well mixed through a constant and rapid mixing; pH and temperature controllers may adjust pH and cell temperature, respectively. Cathode material is usually stainless steel, except copper used by Smith and titanium used by Moraes and Bertazzoli, while anode (working electrode) materials vary extensively. A reference electrode can be connected to the working electrode through a Luggin probe in order to measure its potential.

The Luggin probe is a small tube filled with electrolyte, terminating close to the metal surface under study, and used to provide an ionically conducting path without diffusion between an electrode under study and a reference electrode. In bipolar cell reactors, anodes and cathodes are two sheets, vertical and parallel to each other. Their geometric surface areas vary from 12 cm^2 to 160 cm^2, and their gap is kept constant.

For example, Chiang set the electrode gap at 1.5 cm during batch electro-oxidation mode and 0.5 cm during continuous electro-oxidation mode.

In cylindrical electrode cell reactors, anodes are located inside a stainless steel cylinder serving as the cathode. More recently, a single compartment tubular flow reactor for leachate treatment has been used for electro-oxidation of leachate. The anode was a 0.07 m diameter, 1.0 m long titanium tube coated with TiO$_2$ and RuO$_2$, and a 0.055 m diameter titanium tube as the cathode was concentric to it. Their gap was just 3 mm.

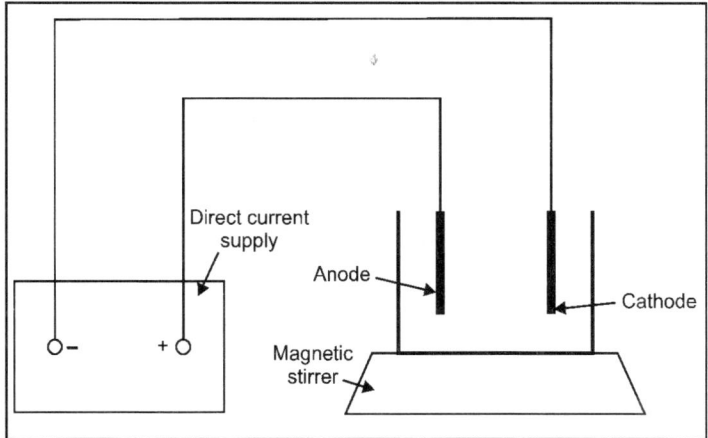

Fig. 26.2. Schematic diagram of a bipolar cell.

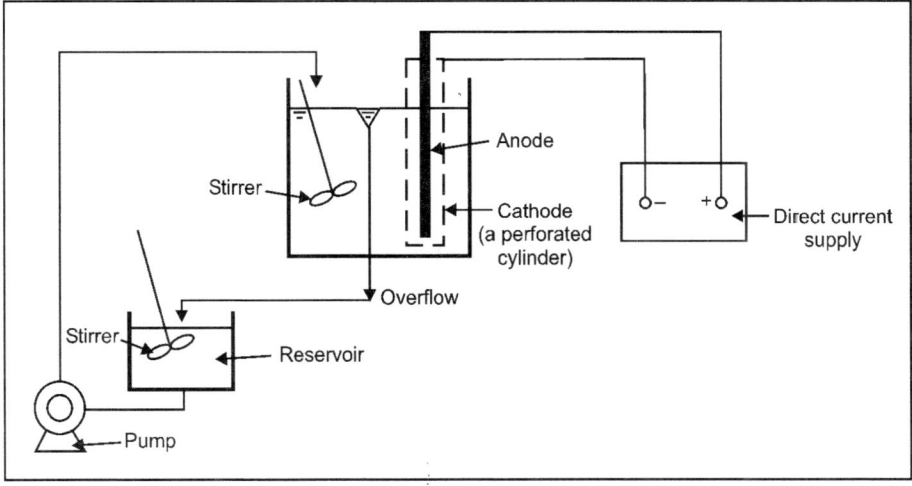

Fig. 26.3. Schematic diagram of a cylindrical electrode cell.

The choice of electrode material is a key issue in reactor design, focusing on high activation energies to avoid undesired side-reactions. Cathode materials should have high over-voltages for hydrogen-evolution, while anode materials should have over-voltages for oxygen-evolution. Moreover, the Space–Time–Gain (∂) can be used to evaluate the economic efficiency of an electrochemical reactor:

$$\partial = \frac{M}{nF} \bullet a \bullet j \bullet A_v \qquad\qquad ...(26.1)$$

where ∂, Space–Time–Gain (kg/l h); M, molar mass (kg/mol); n, number of electrons; F, Faraday constant (coulomb/mol); a, gain factor; j, current density (A/m^2); A_v, specific electrode surface (cm^{-1}).

The Space–Time–Gain (∂) is described as the product gain per cell-volume and operating hour. And a high Space–Time–Gain is related to a high current density (j) and/or a high specific electrode surface (A_v).

Indirect Oxidation and Direct Anodic Oxidation

Electro-oxidation of pollutants in waste-water is fulfilled through two different approaches, as shown in Fig. 26.4: indirect oxidation, where a mediator is electrochemically generated to carry out the oxidation, and direct anodic oxidation, where pollutants are destroyed on the anode surface. During indirect oxidation, the agents generated anodically, which are responsible for oxidation of inorganic and organic pollutants, may be chlorine and hypochlorite, hydrogen peroxide, ozone, and metal mediators such as Ag^{2+}. Furthermore, hydroxyl radicals can also be generated to enhance oxidation through electro-Fenton reactions where added ferrous ion reacts with electrochemically generated hydrogen peroxide.

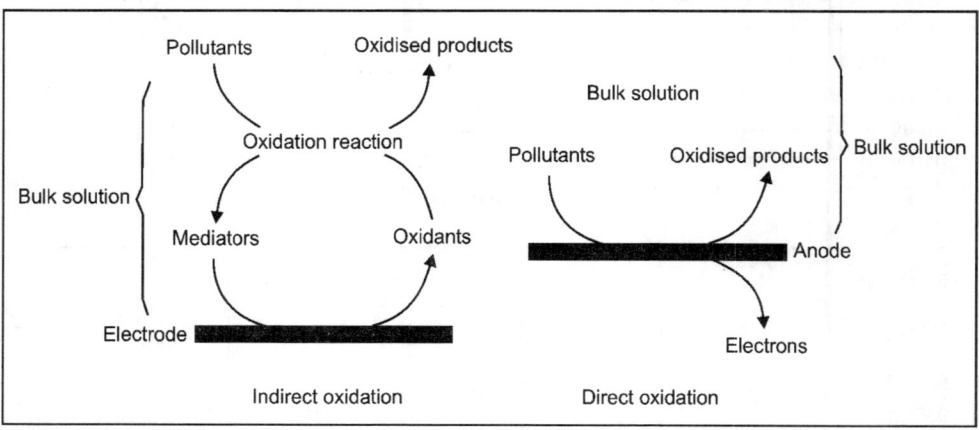

Fig. 26.4. Pollutant removal pathways in electrochemical oxidation (indirect and direct oxidation).

Direct anodic oxidation is achieved through two different pathways: electrochemical conversion and electrochemical combustion. During electrolysis, two species of active oxygen can be electrochemically generated on oxide anodes (MO_x). One is the chemisorbed 'active oxygen' (oxygen in the oxide lattice, MO_{x+1}), responsible for electrochemical conversion through Eq. 26.2, while the other is the physisorbed 'active oxygen' (adsorbed hydroxyl radicals, $\cdot OH$), responsible for electrochemical combustion through Eq. 26.3.

$$R + MO_{x+1} \rightarrow RO + MO_x \qquad \qquad \dots (26.2)$$
$$R + MO_x\,(\cdot OH)_z \rightarrow CO_2 + zH^+ + ze + MO_x \qquad \qquad \dots 26.3)$$

where, R, organic compounds; z, number of absorbed $\cdot OH$ on anode.

During electrochemical conversion, organic compounds are only partially oxidised so that a subsequent biological treatment may be required. In contrast, electrochemical combustion yields CO_2 and water to complete full purification. During the electrochemical oxidation of landfill leachate, pollutant removal may be primarily due to indirect oxidation, utilising chlorine/hypochlorite formed by anodic oxidation of chlorine originally existing or added in the leachate, although direct anodic oxidation may to some extent destroy pollutants adsorbed on the anode surface. A series of reactions involving indirect oxidation during electro-oxidation are shown in Eqs 26.4–26.10.

Anodic reactions:

$$2Cl^- \rightarrow Cl_2 + 2e^- \qquad \qquad \dots (26.4)$$
$$6HOCl + 3H_2O \rightarrow 2ClO_3^- + 4Cl^- + 12H^+ + 1.5O_2 + 6e^- \qquad \qquad \dots (26.5)$$
$$2H_2O \rightarrow O_2 + 4H^+ + 4e^- \qquad \qquad \dots (26.6)$$

Bulk reactions:

$$Cl_2 + H_2O \rightarrow HOCl + H^+ + Cl^- \qquad \text{... (26.7)}$$
$$HOCl \rightarrow H^+ + OCl^- \qquad \text{... (26.8)}$$

Cathodic reactions:

$$2H_2O + 2e^- \rightarrow 2OH^- + H_2 \qquad \text{... (26.9)}$$
$$OCl^- + H_2O + 2e^- \rightarrow Cl^- + 2OH^- \qquad \text{... (26.10)}$$

Hypochlorite (OCl^-) generated in Eqs 26.7 and 26.8 is a strong oxidant that can oxidise aqueous organic compounds. Chiang found that anode material, current density, and chloride concentration had similar effects on chloride/hypochlorite production efficiency for electrolysis of both saline water and landfill leachate. They also found that COD and NH_3–N removal efficiencies in electro-oxidation of leachate increased with increases in current density. However, in a direct electrochemical oxidation, pollutant removal efficiency at the same charge loading (in coulomb per litre) is independent of current density. Hence, Chiang suggested that indirect oxidation was the main process during electrochemical oxidation of leachate. Moreover, they proposed that NH_3–N removal could be due to a series of reactions between hypochlorite and ammonia, similar to 'breakpoint reactions' described by White, instead of air stripping or direct anodic oxidation.

Removal Efficiencies

During electro-oxidation of leachate, COD reduction efficiency ranges from 70 per cent up to above 90 per cent, and NH_3–N removal efficiency almost reaches 100 per cent under appropriate conditions. The result for removal of organic compounds in electro-oxidation of leachate is superior to those reported in coagulation/flocculation, light-enhanced oxidation, combination of UV and O_3/H_2O_2, Fenton process, ultrasound, and other physical/chemical processes. Kinetic data and competition of removals for organics and ammonia have been investigated in past research efforts. Different results on kinetic tests have been reported. Chiang found that COD and NH_3–N removal had pseudo-first-order and zero-order kinetic rate constants, respectively.

However, Moraes and Bertazzoli reported that decaying profiles of COD and TOC both followed pseudo-second-order kinetics, and that a competition between COD and NH_3–N removal existed during the electrochemical oxidation process. Li reported that when indirect oxidation predominated, most of NH_3–N and only 30 per cent of COD were removed in the first hour, and subsequently the remaining 70 per cent of COD began to decline during electro-oxidation of a leachate subsequent to a sequencing batch reactor (SBR).

Chiang also found that removal of NH_3–N was obviously dominant when in competition with removal of COD under indirect oxidation during electro-oxidation of an old leachate. Cossu reported that the removal rate of NH_3–N was lower than that of COD at the initial stage of electro-oxidation of an old leachate when direct oxidation predominated, and then NH_3–N was substantially removed in the subsequent electro-oxidation stage when indirect oxidation became prevalent. Marinci and Leitz also found that a direct anodic oxidation of ammonia was a fairly slow procedure. Based on the above reports, the rule of competition between removal of COD and NH_3–N seems to be that the removal of NH_3–N is greater than that of COD when indirect oxidation is dominant, while removal rate of COD takes priority under direct anodic oxidation.

Removal of colour during electro-oxidation of leachate also has been investigated by Moraes and Bertazzoli. The researchers reported that 86 per cent of colour was removed within 180 min. of electrochemical oxidation, and colour removal had a pseudo-second-order kinetic constant.

Influence of Operating Factors

The effects of a number of operating factors on electro-oxidation of leachate have been investigated, including pretreatment, anode materials, pH, current density, Cl⁻ concentration, as well as electrolytes added. These factors variously influence pollutant removal efficiency, current efficiency and energy consumption.

Pretreatments

Pretreatments for electrochemical oxidation of landfill leachate used in previous research have included SBR, upflow anaerobic sludge blanket (UASB), coagulation, carbon adsorption, and electrocoagulation combined with magnetic separation. The goal of both biological processes, SBR and UASB, is to remove most of biodegradable organics and NH_3–N in order to lessen the electro-oxidation loading and energy consumption when treating a young leachate with a high BOD_5/COD. Wang, in experiments of electrochemical oxidation of leachate pretreated by a UASB, reported that energy consumption was 55 kW hr/kg COD, greatly below a value of 80 kW hr/kg COD during direct electrochemical oxidation of the leachate.

Coagulation and activated carbon adsorption as pretreatments for electro-oxidation of an old leachate were compared by Chiang. Electrochemical oxidation gave a total COD removal and energy consumption of 88.4 per cent and 99 kW hr/m³ for leachate pretreated by coagulation, and 90.3 per cent and 101 kW hr/m³ for leachate pretreated by carbon adsorption. These are considerably superior to 58.5 per cent COD removal and 159 kW hr/m³ energy consumption with only electrochemical oxidation of their untreated leachate. It was further noted that coagulation preferentially removed high molecular weight organics of the landfill leachate, and, in contrast, carbon adsorption mostly removed low molecular weight organics. Since electrochemical oxidation seemed to favour the destruction of high molecular weight organics, electro-oxidation combined with adsorption was superior to that when combined with coagulation. Electrocoagulation combined with magnetic separation was used by Ihara as a pretreatment for electro-oxidation of leachate. This pretreatment removed 99.6 per cent of total phosphorus, although it only removed around 10 per cent of the COD. Additionally, air stripping was also recommended as a pretreatment to remove NH_3–N and avoid the competition with COD removal.

Anode materials

The behaviour of anode materials is a major experimental concern during electro-oxidation of wastewater. During electro-oxidation of leachate, the effect of cathode materials has not been extensively investigated, although they may have a considerable influence on electro-oxidation of organic compounds. However, various anode materials have been investigated for electrochemical oxidation of leachate, including ternary Sn–Pd–Ru oxide-coated titanium (SPR), binary oxide-coated titanium Ru–Ti oxide (DSA), PbO_2-coated titanium (PbO_2/Ti), graphite, SnO_2-coated titanium (SnO_2/Ti), iron (Fe) and aluminium (Al). Chiang found that COD removal efficiencies when using the anode materials followed the order of SPR > DSA > PbO_2/Ti > graphite. In their Taguchi array experiments (a kind of fractional factorial experimental design to evaluate the importance and optimal level of every operating factor) of electrolysing saline water, they also found that chlorine/hypochlorite production efficiencies of these four anode materials followed the same order, indicating that indirect oxidation might be dominant in electrochemical oxidation of leachate.

Li also reported a similar order of anode materials not only for COD removal efficiency, but also for NH_3–N removal efficiency. Among these anode materials, the high removal efficiency of the SPR anode may be attributed to a high current efficiency owing to its high electrocatalytic activity and high anodic oxygen evolution potential. Cossu reported that SnO_2/Ti and PbO_2/Ti anodes did not have substantial differences on COD and NH_3–N removal, mostly because both materials have close oxygen evolution potentials of around 1.9 V.

In addition, Tsai employed Fe and Al anodes to simultaneously utilise electrocoagulation, which is responsible for removal of high molecular weight organics, and oxidation during treatment of a raw leachate. Between the two mechanisms, oxidation was the main contributor to COD removal, and Fe provided better COD removal at low applied voltages when compared with Al.

pH

A review of previous experiments does not allow a conclusion on whether increasing or decreasing pH favours COD removal in electrochemical oxidation of landfill leachate. These discrepancies probably derive from complex compositions of leachates and different pH ranges used. Some researchers found that pH variation did not considerably alter COD removal in electro-oxidation of leachate. Chiang reported that the pH effect on chlorine/hypochlorite production efficiency was insignificant within the pH range of 4–10 in saline water electrolysis experiments, conducted to help understand the mechanism involved in electrolysing leachate.

However, they did not report results in leachate electro-oxidation experiments. Cossu found that the COD pseudo-first-order rate constant only had a slight increase at pH 3, compared with at pH 8.3. Also, Wang reported that treatment at pH 8.9 and 10 just achieved approximately 4 per cent higher COD removal than at a pH of 7.5.

In contrast, other researchers found that pH has a significant effect on COD reduction. Li reported that pH 4.0 achieved at least 20 per cent higher COD removal than pH 8.0 after 4-hr electro-oxidation. And Vlyssides even found that pH was the most significant controlling parameter in electro-oxidation of leachate, compared with Cl^- concentration, temperature, applied voltage, SO_4^{2-} concentration, and leachate input rate. They reported that a low pH favoured COD removal and energy consumption within pH 5.5–7.5. Further research, probably involving study of the oxidation mechanisms, is required to better understand the effect of pH on COD removal for leachate electro-oxidation.

Theoretically, acidic and alkaline conditions during electrochemical oxidation could help organics removal. Acidic conditions significantly decrease the concentration of CO_3^{2-} and HCO_3^-, both well-known scavengers of ·OH generated on anodes, and so enhances oxidation, while alkaline condition boosts the $Cl^- \rightarrow Cl_2 \rightarrow ClO^- \rightarrow Cl^-$ redox circulation to enhance indirect oxidation.

Current density

Current density, the current per unit area of electrode, may be the most frequently referred term in an electrochemical process because it controls the reaction rate. Current densities used in the publications reviewed range from 5 up to 540 mA/cm^2 during electro-oxidation of landfill leachate. Generally, 5 mA/cm^2 is the minimum current density required to achieve an effective oxidation of organics. Otherwise, the leachate solution may become darker, and brown precipitates may form at the anode surface under weak oxidative conditions.

An increase in current density improves COD and NH_3–N treatment efficiencies under the same charge loading (in coulomb per metre). Chiang reported that during electro-oxidation of leachate with 25 mA/cm^2, COD removal was approximately 50 per cent higher than that by electro-oxidation with 6.25 mA/cm^2 under the same charge loading of 1.178×10^5 coulombs/l. A possible explanation is that an increased current density during electro-oxidation enhanced chlorine generation, which was probably responsible for subsequent removal of pollutants. Li further reported that when between approximately 30 and 120 mA/cm^2, the effect of current density on removal of pollutants in leachate was not obvious under a low Cl$^-$ concentration (1650 mg/l), but was noticeable under a high Cl$^-$ concentration (5000 mg/l). This phenomenon again proves the key role of indirect oxidation during electro-oxidation of leachate.

Additionally, Moraes and Bertazzoli reported that colour removal from leachate also strongly depended on current density. Removal efficiency of colour at 116 mA/cm^2 was five times that at 13 mA/cm^2 with a 180 min. electrochemical process.

Chloride ion concentration

Adding extra Cl$^-$ generally improves electro-oxidation of landfill leachate due to enhanced indirect oxidation through higher chlorine/hypochlorite production efficiency, but the addition of Cl$^-$ may produce hazards of chlorinated organics formation in effluent. Typically, Cl$^-$ concentrations range between 200 and 3000 mg/l in young leachates, and range between 100 and 400 mg/l in old leachates as shown in Table 26.1. However, these Cl$^-$ concentration ranges are generally not enough for effective indirect oxidation. Waste-water indirect oxidation using chlorine/hypochlorite requires a high chloride concentration, typically larger than 3000 mg/l. Hence, additional Cl$^-$ is frequently necessary in electro-oxidation of waste-waters including landfill leachate.

Generally, a high Cl$^-$ concentration causes a high removal of pollutants. However, an overall Cl$^-$ concentration most beneficial to removing COD from leachate by electro-oxidation seems to exist, above which COD removal rate does not significantly increase. Li found that COD removal rose between 2500 and 5000 mg/l Cl$^-$, but slowed down between 5000 and 10,000 mg/l Cl$^-$. Wang also found that COD removal efficiencies of leachate obviously increased within 2010–4010 mg/l Cl$^-$, but higher Cl$^-$ concentration did not cause further removal. Different results are reported by Cossu and Vlyssides. Cossu found that there was no significant variation on COD removal with 1600 and 3600 mg/l Cl$^-$. The dissimilarity may be due to different current densities used. The result of Cossu was attained at a current density of 5 mA/cm^2, greatly lower than the 150 mA/cm^2 used by Chiang, 100 mA/cm^2 used by Li, and 32.2 mA/cm^2 used by Wang.

Vlyssides also found that Cl$^-$ concentrations from 20,000 to 40,000 mg/l almost did not influence COD removal and power consumption. The possible reason is that organic removal and power consumption are insensitive at such a high concentration range. Although addition of extra Cl$^-$ can enhance oxidation efficiency, the potential formation of chlorinated organic intermediate and final compounds, probably more hazardous organic components, may hinder wide application of electrochemical oxidation. Naumczyk found formation of many chloroorganics in high concentration during electro-oxidation of textile waste-water containing high concentration Cl$^-$. Chiang reported that many chlorinated by-products (as TOX) were formed at the beginning during electrolysing a coke-plant waste-water.

Up to now, investigation on formation of chlorinated organic compounds during electro-oxidation of leachate is rare.

Additional electrolytes

Various electrolytes have been added to enhance removal of pollutants from landfill leachate, including sulphate (SO_4^{2-}), hydrogen peroxide (H_2O_2) and ferrous ion (Fe^{2+}). Sulphate itself does not take part in oxidation reactions on the electrodes, but may improve electrochemical oxidation as a supporting electrolyte. Different effects of organics removal are reported after addition of SO_4^{2-} during electro-oxidation of leachate. Wang found that the addition of 5000 mg/l SO_4^{2-} increased COD removal efficiency from 36 to 47 per cent but did not have a significant improvement on NH_3–N reduction. In contrast, Chiang found that COD and NH_3–N removal efficiencies both dropped below 30 per cent with 5000 mg/l SO_4^{2-}, probably because SO_4^{2-} improved the anodic oxygen evolution to suppress chlorine/hypochlorite production.

During electro-oxidation, H_2O_2 may decompose into hydroxyl radical to enhance pollutant removal. Wang reported that addition of 200 mg/l H_2O_2 increased COD removal from 36 per cent up to 47 per cent, and NH_3–N removal from 47 per cent up to 53 per cent. The addition of H_2O_2 also may initiate Fenton oxidation during electrolysis when the anode is zero-valent iron. In the so-called electro-Fenton oxidation, additional H_2O_2 and Fe^{2+} released from the anode serve as Fenton reagents to generate hydroxyl radicals ($\cdot$OH) shown as Eq. 26.11:

$$Fe^{2+} + H_2O_2 \rightarrow Fe^{3+} + \cdot OH + OH^- \qquad \qquad \dots (26.11)$$

Lin and Chang investigated the electro-Fenton method for leachate oxidation after chemical coagulation. They found that the electro-Fenton process achieved 67.3 per cent COD removal, while an electrochemical method with identical conditions except no addition of H_2O_2 just attained 26.7 per cent COD removal. Also, they found that the optimal pH was around 4.0 for electro-Fenton, slightly higher than reported optimal pH of 2–3, pH 3.0, and pH 3.5 using conventional Fenton process (directly add Fe^{2+} salt and H_2O_2 solution).

The summary of influence of operating factors mentioned above is listed in Table 26.2. To evaluate the synthetic effect of all operating factors, linear models can be established by implementing a factorial experiment in terms of the COD and BOD_5 reduction, as well as in terms of energy consumption. During this modelling, the importance of every operating factor can be assessed, and the optimal operating conditions for maximum pollutant removal can be found. Vlyssides employed such a method and reported that pH and leachate input rate were the prevailing factors in electro-oxidation of a leachate, when compared with the applied voltage, the amount of NaCl added, reaction temperature, and the concentration of $FeSO_4$ added. However, it seems to be unnecessary to develop a linear model for ammonia reduction, because a near 100 per cent ammonia removal can be readily achieved.

Table 26.2. Influence of operating factors in electro-oxidation of leachate.

Operating factor	Influence
Pretreatment	Pretreatments lessen pollutant loads and save energy consumption for electro-oxidation.
Anode materials	Anode materials with high electrocatalytic activity and high anodic oxygen evolution potential cause a high COD and NH_3–N removal efficiency; usage of metal anode such as Fe and Al causes simultaneous electro-oxidation and electrocoagulation.
pH	The influence of pH is unclear. Reported results are inconsistent.
Current density	Increase in current density causes increase in removal efficiencies of COD and colour.
Chloride ion concentration	Increase in Cl^- concentration improves removal of pollutants, but increase the hazard of formation of chlorinated organics.
Additional electrolytes	Effects of additional electrolytes depend on their species and properties.

Thus, electrochemical oxidation of landfill leachate under appropriate conditions can remove most COD and almost all ammonia, and also significantly remove colour. During electro-oxidation, reduction of pollutants appears to be primarily due to indirect oxidation. Pretreatment methods, anode materials, pH, current density, chloride concentration and electrolytes added all influence removal efficiencies of pollutants and energy consumption. Two drawbacks of electro-oxidation may limit its wide application for landfill leachate treatment: high energy consumption, and potential for formation of chlorinated organics. Especially because of its expensive operating costs compared with other available technologies (for example, biological processes), electro-oxidation will be favoured as a finishing step in a combined process or an auxiliary unit in emergency situations, instead of a full treatment for landfill leachate. However, electrochemical oxidation is still a promising and powerful technology, especially for low BOD_5/COD or high toxic landfill leachate where biological processes suffer. Future work is expected to clarify the influence of pH, and the risk of formation of chlorinated organics during electro-oxidation of landfill leachate. Additionally, development of material science for more economical and effective electrodes is expected to improve the application of electro-oxidation of landfill leachate.

Chapter 27

Pretreatment of Industrial Landfill Leachate by Fenton's Oxidation

INTRODUCTION

The main feature of current EU regulations on solid waste management (91/156 CE and EU 91/689 CE) is essentially based on 'recovery, recycling and reuse' criteria. Nevertheless landfilling is still the most important technology for solid waste management. In the average, 60 per cent of municipal solid wastes (MSW) produced, and even a larger fraction of industrial solid wastes (ISW), are still disposed-off in controlled landfilling.

Leachate formation, i.e. the liquid waste formed after weathering (rain, snowfalls) on the solid wastes, beyond the fraction released by intrinsic humidity and fermentation of solid wastes, is the main environmental impact during landfilling operation. The composition of such effluents is variable depending on type, origin and composition of the wastes, the structure, management and the 'age' of the landfill. Above considerations are strictly applicable to MSW leachate, the problem is amplified in reference to ISW leachate being the origin of the landfilled solids diversified depending on the productive activity of origin.

Treatment of such highly concentrated waste-water is based on physico-chemical methods, i.e. chemical co-precipitation of heavy metals coupled to sorption of bulk COD on cheap coagulants [$Ca(OH)_2$, polyelectrolytes] or, in more advanced configurations, by the use of biosorbents such as lignocellulosic residues from the agri-food industry (olive-oil milling, winery, etc.). In this latter case biosorbents are used 'once through' and the exhausted materials, containing massive quantities of contaminants, are disposed-off in the same landfill.

The use of more expensive sorbents such as activated carbon, adsorbent resins is possible with related economic implications. Alternatively, landfill leachate may be evaporated, incinerated or, in arid-semiarid areas, it may be re-circulated on the waste thus taking advantage of the evaporation favoured by the dry climatic conditions. Odour emissions is however the main limit of the operation.

Application of conventional biological treatment to reference liquors is limited by the presence of toxics (e.g. heavy metals) and/or recalcitrant organics (pharmaceuticals, polyphenols, endocrine disrupters). More than the generalised biotoxicity, the technological problem is related to biorefractory nature of the organic matter due to the presence of high molecular weight substrates. Possibly, the autothrophic biomasses mediating hydrolysis of organic macromolecules are more sensitive to the mentioned toxicity and generalised nonviable conditions for biomass proliferation in the biological reactors. Growing interest is lately focussed on the advanced oxidation process (AOP) of landfills leachate with a multipurpose goal associated with: (i) abatement of refractory COD load, with related

411

enhancement of biodegradation after raising of the BOD_5/COD ratio, (ii) simultaneous removal of toxic contaminants by sorption-coprecipitation.

AOP are based on the formation of hydroxyl radicals (OH ·), an extremely strong oxidant resulting from several reactions such as the synergistic action of two oxidants: $O_3 + H_2O_2$; a catalyst and an oxidant: $Fe^{2+} + H_2O_2$ (the Fenton's reagent); a photocatalyst and an oxidant: $TiO_2 + H_2O_2$; irradiation plus oxidation: $UV + O_3/H_2O_2$), etc.

Generally speaking, full scale application of the Fenton's reagent does not requires supplementary operative costs beyond chemicals strictly needed for the oxidation reaction. In acidic media hydroxyl radicals are very efficiently formed for technical purposes. Chemicals are cheap, process layouts simple.

For proper design and control of full scale installations a basic knowledge of process principles and operative condition is needed. In the present chapter, through laboratory experiments carried-out on the real leachate from Grottaglie industrial landfill, is evaluated the influence of process operative parameters (pH, initial chemicals concentration and Fe^{2+}/H_2O_2 ratio) on BOD_5/COD ratio for enhancement of conventional biological treatment of the resulting liquids.

GROTTAGLIE INDUSTRIAL LANDFILL

Grottaglie landfill site is located in the Apulia Region, 4 km south of the town. It is classified as a former 2nd Category type B installation, for 'special-non-hazardous' industrial wastes according to the Italian legislation.

Figure 27.1 shows a schematic view of the installation including two lots, with capacities exceeding 3,30,000 m³ (lot no. 1) and 15,00,000 m³ (lot no.2) and surface 15,000 m² (lot no. 1), 28,000 (lot no. 2, basin A) and 23,000 m² (lot no.2, basin B). A 3rd lot is underway and will be soon in operation.

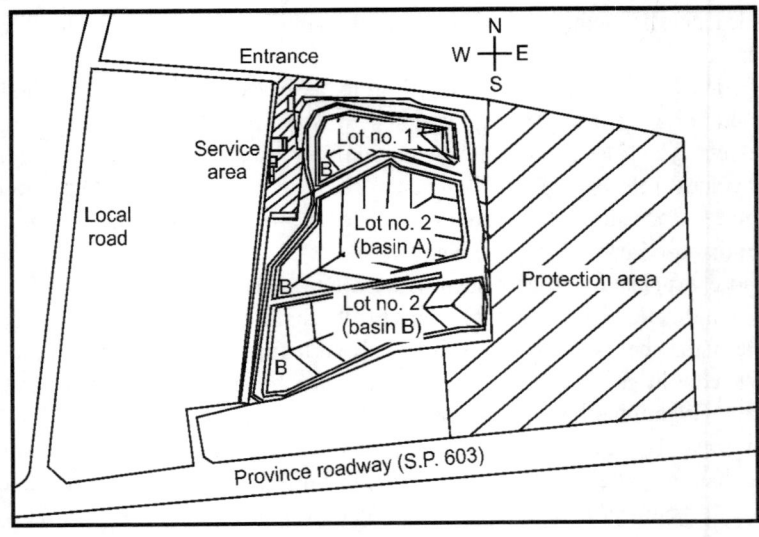

Fig. 27.1. Layout of the Grottaglie industrial landfill.

The annual leachate production for lot no. 1 exceeded 7600 m³. The average composition is shown in Table 27.1.

Table 27.1. Average composition of the leachate produced at Grottaglie landfill.

Parameter	Concentration (mg l^{-1})	Parameter	Concentration (mg l^{-1})
COD	13,500	Al	0.5
BOD$_5$	3000	Cr$_{tot}$	3
BTEX	<5	Cr(VI)	<0.5
Organic solvents	<3	Ni	0.5
N-NH$_4$	5500	Pb	0.5
Oil-grease	<10	Hg	<0.5
Dry residue (105°, 600°C)	545, 285	Cu	0.5
Suspended solids	700	Sn	<0.5
Chlorinated organics	<5	Cd	<0.5
pH	8,5	Te	<0.5
BOD$_5$/COD	0.22	Fe	5
Conductivity	60,000 μS cm^{-1}	Zn	<0.5

EXPERIMENTAL PROCEDURE

Test leachate was characterised by the following relevant parameters: pH 8.7; COD = 10,915 mg l^{-1}; BOD$_5$ = 2400 mg l^{-1}; N-NH$_4$ = 2880 mg l^{-1}; Cr(III) = 9.8 mg l^{-1}; conductivity = 58,700 μS/cm; and a BOD$_5$/COD ratio as low as 0.22.

The leachate sampled at the landfill was preliminarily neutralised at pH 3.2 by the addition of H$_2$SO$_4$ (98 per cent w/v). The amount of the added acid was predetermined by separate titration. To optimise operative H$_2$O$_2$/Fe^{2+} ratio, in a first set of experiments the acidic leachate (pH 3.2) was contacted with arbitrary amounts of H$_2$O$_2$ (i.e. 9900 and 5280 mg l^{-1} respectively), from stock solution (30 per cent w/v), followed by increasing amounts of Fe^{2+} from a stock solution (FeSO$_4$·7H$_2$O 25 per cent w/v). After 2 hr contact time, stirring was stopped and the solution allowed to settle (1 hr). The supernatant solution was analysed for residual COD, BOD$_5$ and Cr(III).

Alternatively, at the end of the oxidation reaction and settling the mixture was added with Ca(OH)$_2$ slurry or 5 M NaOH till neutralisation; a cation polyelectrolyte (Dryfloc 652 from Nymco Waters, Milan, Italy) was also added (3 mg l^{-1}) to favour clarification and metals coprecipitation.

In a second set of experiments, constant the H$_2$O$_2$/Fe^{2+} ratio previously optimised, were carried-out tests at increasingly higher concentrations to verify the influence of potential limitations on the oxidation reaction operated by the chemicals addition to the liquid phase. In this case also, the oxidation reaction was completed by the neutralisation step using the mentioned alkalising agents and polyelectrolyte.

A final experiment was carried-out in a gas tight system for collection and quali-quantitative characterisation of different outcoming products, i.e. gas, liquid, solid, resulting from the oxidation reaction.

RESULTS AND DISCUSSION

Figure 27.2 shows COD abatement (residual fraction in the liquid-phase) after optimisation of the H$_2$O$_2$/Fe^{2+} ratio, for tests carried-out at 9900 and 5280 mg H$_2$O$_2$ l^{-1} respectively. In spite of different initial amounts of hydrogen peroxide dosed, the curves converge to steady values of residual COD after dosage of about 400 and 700 mg Fe^{2+} l^{-1}, corresponding to a constant ratio in the range 14 w/w.

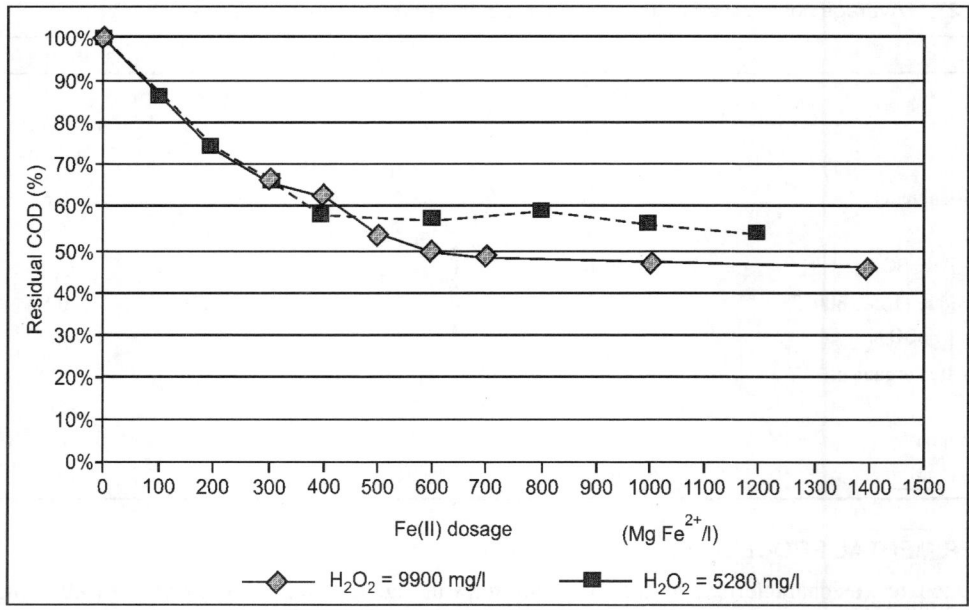

Fig. 27.2. Fractional abatement of COD vs. Fe^{2+} dosage for Grottaglie leachate.

By assuming the following reactions presiding the Fenton's oxidation process:

$$Fe^{2+} + H_2O_2 + H^+ \rightarrow Fe^{3+} + OH\bullet + H_2O \quad \text{(radicals formation)}$$
$$Fe^{2+} + OH\bullet \rightarrow Fe^{3+} + OH^- \quad \text{(side reaction)}$$
$$H_2O_2 + OH\bullet \rightarrow H_2O + HO_2\bullet \quad \text{(side reaction)}$$
$$HO\bullet + SO \rightarrow \text{oxidation products} \quad \text{(oxidation reaction)}$$
$$(SO = \text{organic matter})$$

any excess dosage of chemicals should theoretically lead to a sensible reduction of free hydroxyl radicals (see side reactions above) thus inducing a minimum in the residual COD vs. Fe^{2+} curve at the optimal ratio, followed by immediate raising due to depletion of OH• with corresponding sensitive reduction of the COD abatement. This is actually not verified in Fig. 27.7 and, accordingly, some other oxidation reaction mechanism(s) should be postulated for the reference Fenton's process, based on a more accurate chemical characterisation of the leachate.

Fenton's oxidation of the Grottaglie leachate leads to ~50 per cent abatement of the initial COD and 63 per cent enhancement of the effluent biodegradability (BOD_5/COD ratio from 0.22 to 0.36) not too far from the indicated minimum figure (0.4) for efficient implementation of the aerobic biological treatment.

The addition of alkalising agents (lime slurry or caustic soda) and coagulation adjuvant (cationic polyelectrolyte) improves sensibly the quality of the oxidised liquid, in terms of reduced toxicity, after abatement of 40–70 per cent trivalent chromium by using caustic or lime respectively, thus inducing sensible advantages to biological treatment to follow.

Based on the above results the following process scheme (Fig. 27.3) may be proposed for potential upscaling of the Fenton's oxidation of the Grottaglie leachate.

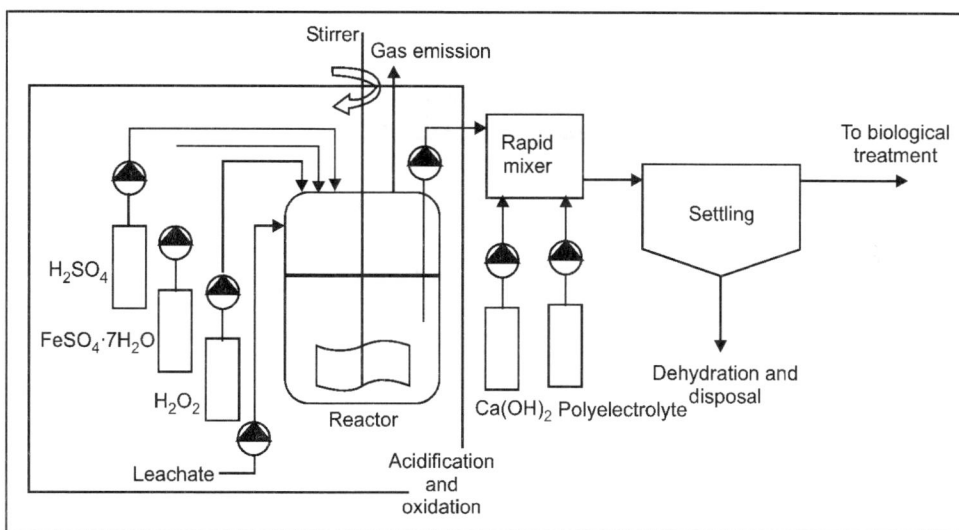

Fig. 27.3. Schematic flow-sheet of the proposed Fenton's oxidation of Grottaglie leachate.

Preliminary economic evaluation of the process indicates a tentative figure in the range of 9 € m^{-3} for the overall running costs of the process, including chemicals, power and maintenance. The figure appears definitely competitive as an 'off-site' treatment of the landfill leachates based on conventional technologies, whose operative costs may exceed 20–30 €/m^3 depending on the final COD concentration.

CONCLUSION

A laboratory investigation was carried-out to verify potentialities of AOP (Fenton) pretreatment of an industrial landfill leachate, with the aim of improving its biodegradability in view of conventional biological treatment to follow. To this aim the real leachate from Grottaglie landfill was tested under different operative conditions (i.e. initial pH, H_2O_2/Fe^{2+} ratio, addition of coagulants, final neutralisation). Possible layouts for process upscale are now open for development at a larger scale which is underway.

As a consequence of the present work the following main conclusions may be drawn:

1. COD abatement ranging 50 per cent of the initial figure is obtained, by the adoption of a H_2O_2/Fe^{2+} ratio exceeding 14 w/w.
2. Enhancement of leachate biodegradation exceeding 60 per cent (BOD_5/COD from 0.22 to 0.36) close to the threshold value set at 0.4.
3. Final clarification of the oxidised effluent by the use of cheap coagulants/precipitating agents [i.e. $Ca(OH)_2$, NaOH] and adjuvants (cationic polyelectrolytes) reduces final toxicity of the effluent after sensible reduction (70 per cent) of the residual Cr(III) concentration.
4. Based on a preliminary technical/economic evaluation, O&M costs of the proposed process layout is in the range 9–10 € m^{-3}, a figure highly competitive with conventional treatments costs (e.g. inertisation) in the same landfill.

Chapter 28

Combined Landfill Gas and Leachate Extraction Systems

INTRODUCTION

One of the main impetuses for the installation of landfill gas collection and disposal systems is the already considerable weight of environmental legislation, driven by an increasingly aware public opinion. For leachate, however, although much has been written on the subject, legislation has, as yet, failed to produce any concrete guidelines for treatment. There remains confusion concerning environmentally acceptable leachate strengths and its concomitant polluting potential, and consequently the cost of treatment remains unrealistically biased towards off-site methods. The situation is unlikely to remain static; change will come, and the well prepared operator will be looking for efficient and cost effective collection and disposal systems which will allow him to fulfil his/her environmental obligations into the twenty-first century. This chapter briefly reviews methods of evaluating production of both landfill gas and leachate and focuses on available technology for a method of co-collection and disposal of both landfill gas and leachate.

ASSESSMENT OF GAS AND LEACHATE PRODUCTION

Important in the decision pathway as to which method of collection and disposal is most suitable for any particular site, is the evaluation of whether the proposed design will meet its objectives in terms of capacity. It is unfortunately too often the case that specifiers leave this crucial evaluation exercise until the competitive tendering stage, relying on the integrity of contractors to judge the necessity or otherwise of active site investigations. This can potentially result in costly errors of judgement if the winning contractor deems it unnecessary to conduct such trials to achieve the prime commercial objective of winning the work.

There are basically two methods for determining the production of toxic emissions from a landfill site. Both of these can be subject to considerable margins of error relying essentially on the experience of the investigator. However, they do provide an empirical basis for specification which reduces the likelihood of error.

The first method is by using mathematical models, ideally done as part of a desk study of the site; while the second is field based pumping trials which are designed to determine as much as possible about the behaviour of the aqueous and gaseous phases within the landfill. In the case of landfill gas, pumping trials are especially important when utilisation is being considered. Data gathered from the trials can be entered into a cumulative spreadsheet and the real-time production rate compared with predicted values. This also allows for extrapolation of gas production rates into the future.

Preinstallation Studies

It is critical at this stage to learn as much as possible about the site. In many cases this work will have been undertaken by the client. The objective is to ensure a clear understanding as to the nature of the problem. Studies centre around collating as much information about a site as possible.

In order to construct a realistic model of gas and leachate production from the site, it is necessary to gather the following information:

1. Period, method and rate of landfilling.
2. Mass of waste in place.
3. Site dimensions
4. Waste types infilled, i.e. domestic, industrial, commercial, inert and hazardous.
5. Waste configuration, i.e. baled, shredded, compacted.
6. Packing density and moisture content of waste in place.
7. Internal temperature and pH of the waste.
8. Site geology including any borehole logs available.
9. Type of capping and potential gas recovery effectiveness.
10. Gas and leachate monitoring results.
11. Available assays from leachate samples.
12. Meteorological conditions and rainfall data.
13. Site hydrogeology and hydrological data, including:
    (a) Surface water run-off.
    (b) Rainfall and evapotranspiration.
    (c) Local water utilisation.

Modelling landfill gas production

In modelling landfill gas production the amount of gas that can potentially be generated by a unit of waste is an important parameter, yet difficult to define. Previous studies have arrived at an average figure for gas yield of about 150 m^3 for every wet ton, although this figure varies depending on source. Of this it is stated that about 70–80 per cent would be recoverable. A rule of thumb estimate is that between 6 to 10 m^3 of landfill gas will be produced per ton per year for ten to fifteen years from placement.

Inputting the raw data concerning waste fractions, amounts deposited and periods of deposition into a computer program results in exponential decay curves. These curves can also model varying concentrations of methane within the gas stream and an input of maximum potential collection efficiency will allow an initial assessment as to the size of system required.

Landfill Gas Pumping Trials

A landfill gas pumping trial is generally conducted after all the known data concerning the site has been gathered and a theoretical exercise such as described above has been carried out. One of the main objectives of the trial is to ensure that the design under consideration is of the correct order of magnitude in terms of capacity, and provides a test for the calculations carried out previously. If conducted correctly this relatively inexpensive measure will result in avoidance of costly remedial actions. It may allow savings to be made at the beginning of the project by reducing the capacity specification. It may also offset many of the direct expenses, such as the drilling of boreholes which can be used again, from the final cost.

During the pumping trial data is gathered concerning the amounts and quality of gas being produced. The objectives of conducting a pumping trial on a landfill site can be summarised as follows:

1. Investigation the radius of the zone (or cone) of influence being effected by individual extraction wells. This is obtained using an array of piezometers around individual wells; from this the optimum spacing of wells can be deduced. Alternatively, existing monitoring boreholes can be used, with the 3-D resolution being increased if multi-level piezometers are used.

2. Defining wellhead gas flow characteristics. These describe the behaviour of individual wells under passive and active conditions.

3. Quantifying probable landfill gas production rates. These are obtained by active abstraction over a period of time, increasing flow rate until air is introduced into the system, and the quality of the recovered gas falls. This test can, however, be dependent on the quality of the capping. It is also critical to allow for seasonal variation as gas production may be higher in the summer months.

4. Defining required abstraction pressures, which in turn enables correct sizing of the permanent abstraction equipment. The final sizing of the equipment will be a function of gas production rate versus flow rate, which should be extrapolated to include the whole site at the post-trial stage.

5. Investigating the effect of active extraction on the incidence of far-field migration from the boundaries of a landfill site. This is of particular significance where the landfill site is located within or close to a sensitive area.

Tests may be designed to gather data from local or restricted parts of the site or from individual gas wells, and can be expanded to cover the entire site by connecting to networks of pipe and wells.

The most common method is to test a small part of the site and then to extrapolate the results to incorporate the whole site.

Strategy for the trials

Trials consist of installing extraction wells within a representative section of the site, and connecting these to an extraction rig. Ideally, at least two wells should be drilled in each worked phase of the site in order to allow for possible blinding of extraction wells and in order to gather representative data from the whole site.

Extraction wells will ideally be drilled to the base of the site and be lined with a perforated liner and suitable filter medium. 110 mm liner is generally sufficient for short-term trials, although the diameter will depend on the use to which the well will be put after the pumping trial is finished. If the test is to be run in tandem with a pumping trial for leachate then it will be necessary to install larger bore liner in order to allow enough room for the installation of the pumping mechanism as detailed below.

Once conditions within the site have stabilised following well drilling, and before connecting the wells to an active abstraction system, it is advisable to conduct static tests on each of the wells to establish the following:

1. Static pressure in millibars to monitor whether internal pressures exist.

2. Average gas temperature at the top, middle and bottom of the well.

3. Percentage by volume of CH_4, CO_2, O_2 (at this stage some samples should be taken for gas chromatographic analysis).

4. Atmospheric pressure, again in millibars. (Over time this often shows a mirror-type relationship with gas pressures, although a definitive relationship has yet to be shown.)

After connecting to the extraction rig, the dynamic phase may commence. Tests during this stage are designed to investigate the physical characteristics of the wells and of the surrounding waste when under active pressure. During the tests gas quality should be frequently monitored to note the relative change as the test proceeds.

Dynamic tests may include

1. Flow against suction pressure: The test has a duration of about five minutes and the results are plotted as a graph of gas flow (m^3/hr) versus applied suction. The test is designed to show how easily gas can be extracted from the vicinity of the well. In general it is expected that wastes with low permeabilities will require greater rates of suction than those with high permeability. Flow rate is usually measured either at the extraction wells or at the flare stack, although it is important to ensure that there is laminar flow within the gas stream at the point of measurement.
2. Pumping test: This can be subdivided as follows: single and multiple well testing. Single well testing is designed to calculate the zone of influence of an individual extraction well. This is obtained by measuring pressures within the sample borehole and comparing these with readings taken from a series of observation wells located at preset distances from the active well. The sphere of influence is defined when relative pressure within the observation well is measured to be the same as the static pressure established earlier. The zone of influence can be further defined by installing an array of piezometers radially around the well. These may also be installed vertically to obtain a 3-D representation of the zone of influence within the waste. This test defines the extent of influence within a particular body of waste and may have a direct bearing on the final number of extraction wells which will be installed for the permanent system.

Multiple-well testing is carried out by connecting a network of wells and is used to determine the production rate within the site. In this test the emphasis is on determining the maximum abstraction pressure which the site can sustain before significant air is pulled into the system. The test is carried out over a number of weeks (six weeks is usual) and the total gas flow rate is recorded. As the methane concentration falls away the flow rate is stabilised to fulfil the equation:

$$\text{Flow rate} = \text{Generation rate}$$

This will be evidenced by constant methane concentrations in the region of 50 per cent by volume methane.

Leachate

One of the objectives of investigatory trials may be to define leachate production rates and flow characteristics within the landfill. To this end a leachate pumping trial will form part of an hydrogeological study of the site with the objective of determining a method of removing leachate from the lower part of the hydraulic system and directing it to the point of treatment and/or point of disposal.

During the desk study all available hydrogeological data should be collected, as listed above. It may be possible to arrive at a first appraisal of basic hydrogeological parameters such as groundwater contours, leachate production rates, water balance and the categorisation of hydrogeological regime, i.e. is the site saturated below the local water table, damp water retained within the body of the waste, or dry—especially in the case of contained site?

Model predictions of leachate production rates generally centre around a water balance analysis, which is given by the equation:

$$Q = I - E - aW$$

where,

Q = Free leachate generated (m³/year).

I = Total liquid input including liquid waste (m³/year).

E = Actual evaporation losses (m/year).

a = Absorptive capacity of the waste (m³/ton).

W = Weight of waste deposited (tons/year)

Essentially, this balances liquid inputs against liquid outputs. Refinements to the model introduce considerations of the effects of preferential pathways through the waste. Once this type of exercise is conducted, leachate production curves can be produced. This phase will result in a report with recommendations for further action which will ideally include a pumping trial.

Field investigations

Typically, this phase will encompass all the operations needed to complete a hydrogeological model. In particular this may include:

1. Pumping trials: Drilling one or more test wells within a representative area of the site, together with a pattern of observation boreholes (ideally two or more to each well). The latter would be used to observe the range over which the effect of pumping is noticeable. The objective of this is to define the transmissivity and retentive capacity of the waste, transmissivity of the country rock if relevant, specific yields and recharge rates of test wells.

   Specific tests that would be carried out on each well would normally include a step-draw-down test (one working day duration), a constant discharge test (minimum 24 hours) and a draw-down and recovery test. It should be noted that obtaining a reliable overall value for permeability will be difficult for a landfill situation, although the value of the trial will be to study the rate of recovery of the piezometric level within the waste.

   Leachate pumping trials would normally be carried out at the same time as gas extraction trials. The pumping equipment can be installed in the same boreholes and the tests can also define the behaviour of the leachate piezometric level when under active extraction conditions.

2. Collection of samples for chemical analysis and instigation of a medium to long-term monitoring program.

3. Detailed analysis of dispersion patterns by the use of tracers such as pyronine or fluorescene.

Once this phase is complete it will be possible to put forward an analysis of the problem and from this to recommend a design for the collection and treatment of leachate.

LANDFILL GAS AND LEACHATE CO-COLLECTION AND TREATMENT

Components of the System

In a combined leachate/landfill gas system it is important to ensure that the collection method can be easily serviced in the event of failure or modification. Extraction systems rely on a series of boreholes drilled to the base of the site. These are suitably lined and wellheads, adapted for both gas and leachate extraction, are installed.

Borehole construction

Drilling boreholes within landfilled refuse is now common practice. Experienced contractors, while ever conscious of the cost implications of various methods of drilling, are moving away from conventional techniques such as shell and auger, to more reliable methods of achieving the required diameter without incurring expensive stoppages. These include rotary air flush drilling techniques and barrel auger drilling. A novel way of installing gas and leachate extraction wells uses techniques borrowed from the piling industry. This method uses a string of concrete liners, inserted by a large percussion drilling rig. This technique ensures that a pre-lined well is installed and obviates the need either to insert a gravel pack or to remove spoil resulting from drilling operations. The system is highly attractive, because, although mobilisation costs are high, the rate of drilling is rapid; a typical drilling schedule would include 30 boreholes to ten metres depth in about five days.

Whatever technique is used it is recommended that a minimum borehole diameter of 300 mm is achieved. This ensures that there is plenty of space for insertion of pumps or eductors, which will require liner of at least 160 mm.

Liners

With conventional drilling techniques it is usual to install HDPE liners slotted at between 10 and 20 per cent of the surface area. Some operators have opted for the insertion of steel liners, however, unless the expensive option of stainless steel is chosen, the lifetime of such liners is limited due to continued exposure to highly aggressive chemicals.

In a standard water well, designed to service public consumption, stainless steel liners are commonly used. These are screened to prevent blinding of the liner by fine material. In a saturated landfill site, however, there is potentially such a high concentration of fine material that the use of screens may be counteractive; blinding may occur more rapidly. Experience shows that a conventional well construction with a slotted liner and suitable annulus fill is sufficient to ensure a long-term operational lifetime. Should it later be observed that the efficiency of the well is being reduced by blinding it may be possible to reinstate the well with high pressure water jets, either mounted on a rotary drilling rig or inserted on the end of a hose.

The third form of liner system is by the use of a column of concrete shells. These are installed as described above, and the concrete can be mixed to ensure that the grade will withstand the aggressive nature of the environment into which it is being introduced.

As the operation of each extraction well is of paramount importance to the system efficiency, it is recommended that wells are tested periodically by short duration pumping trials, both step-draw-down and constant discharge tests and, if necessary, routine maintenance performed.

Leachate Collection

Leachate pumps

There are many types of leachate pump available on the market today, from narrow bore pumps designed for down-the-hole operation, to pumps that can control a series of extraction wells. It is worth noting the various criteria that such a unit must meet when used as part of a gas and leachate extraction scheme:

1. It must be easy to install and remove.
2. It must be of robust manufacture, body stainless steel, hoses MDPE.
3. It must have minimum moving parts.
4. It should be capable of a variable flow rate, as conditions will vary widely throughout the seasonal cycle. A range of from about 60 litres an hour to greater than 1 m^3/hr would be realistic for a pump located within a borehole.
5. The pump should ideally be capable of running dry with no harm being done to its operation.

6. The pump should be designed for low maintenance.
7. It should be capable of handling varying quantities of fine material and sludge that often accompanies leachate production, perhaps with additional protection being afforded by filtration.
8. It must have sufficient head.

Eductors

An alternative to leachate pumps is a unit known as an eductor or ejector. This unit is positioned within and close to the base of the borehole. It is connected to a surface mounted pump which in turn controls up to ten individual eductors. The advantages of units such as these are: firstly, they are lightweight and streamlined for use within relatively narrow boreholes, secondly, there are no moving parts and thirdly they are relatively inexpensive.

Eductors operate by circulation of water within a semi-closed system through a venturi located within the body of the eductor. This creates a pressure differential within the bottom of the borehole and causes leachate to be sucked through the eductor to the surface. There are basically two types available: these are concentric eductors, where flow and return water is directed within a concentric hose, and non-concentric eductors, fabricated as a single cast iron body in which flow and return water are contained within separate hoses. An additional criterion that may be considered when deciding on which type of eductor to use may be that the parts within the eductor can be replaced should fouling or failure occur.

Pneumatic pumps

Pneumatic pumps have become a favourite method of leachate collection for many companies extracting leachate from landfill sites. They have many advantages over the electric motor and the eductor systems. These include the following:

1. The pumps are built from materials that ensure long-term trouble free operation.
2. The pumps will only operate when there is liquid in the location of the pump. Pneumatic pumps operate with a level sensor that shuts them down when there is no liquid around them.
3. The motive power is pneumatic, resulting in no electrical cables around a site and fewer pressurises liquid pipes.

The disadvantages are:

1. The individual pumps are considerably more expensive than electric pumps or eductors.
2. A system with only one or two pneumatic pumps is very expensive, and difficult to justify as a compressor is required to deliver the air supply.

Combined Gas and Leachate Wellheads

Within the system described above, wellheads are adapted to accept both gas control valves and pump or eductor hoses. These require relatively straightforward attachments which for the leachate comprise flow and return hoses, control valves assemblies and air release valves for use during surging (which may happen if the pumps stop for any reason) within the pipelines, and for priming the system. For gas extraction, wellheads are equipped with control valves, either ball, gate, butterfly or sleeve type valves are acceptable, although should a fine control of the gas flow be required, it is recommended that ball or sleeve valves are used as these allow fine adjustments to be made. All valves should be made of non-corrosive material. On newly completed sites wellheads should be constructed to allow for settlement of the landfill site.

The basic unit is relatively simple although optional attachments include devices for measuring flow rates both of gas and leachate. Ports to measure gas concentration, pressure and to dip the leachate level can also be included in the construction.

Pipeline and Collection Networks

The design of the combined gas extraction system will not be discussed here as there is a large body of literature dealing with the physical attributes of both gas and leachate collection systems. Essential in the design stage of the collection pipeline are hydraulic calculations, surge analysis and, in the case of gas carrier mains, pressure drop calculations. The object of these investigations is firstly, within the leachate collection lines, to calculate head losses contributed by the various components within the system. These are calculated depending on the number of bends, branches and junctions within the system, which are totalled to arrive at a suitable head requirement for the main pump. Secondly, to ascertain whether or not appropriate pressure surge control devices are required in order to counteract the formation of sub-atmospheric pressure should pumping stop for any reason.

Within the gas extraction system pressure drop calculations are designed to ensure that the design diameter of the pipeline is adequate and that sufficient pressure will be exerted over the whole of the system to ensure that all extraction wells are effective in gas removal.

DISPOSAL OF GAS AND LEACHATE

Within the system described above the method of collection and transport of gas and leachate to the point of disposal is effected via common collection points and pipes laid within common trenches. At the point of disposal, however, the method of disposal obviously differs markedly and installations have to be constructed to ensure safe, effective disposal. The actual method of disposal will have been decided upon at the desk study and field investigation stages.

Landfill gas is basically disposed of by two methods: either by incineration in a flare stack or by use in a utilisation plant. The suitability of the gas for utilisation will depend on the following factors:

1. Quality of the gas stream in terms of methane concentration and other gases.
2. Size and potential gas reservoir of the site.
3. Final use of the electricity produced, either local use or sale to the national grid.
4. Viability of the scheme in terms of capital expenditure versus royalty returns.

The technology involved in landfill gas disposal is well understood and guidelines are in place to aid in the construction of schemes that will meet both current and future legislation.

The situation is not so clear cut for leachate disposal. There remains confusion as to what is and what is not acceptable in terms of leachate strength and environmental polluting potential, and the short-term effects are not so apparent as are those for landfill gas. Consequently, many operators do not treat leachate on-site, preferring to allow the water companies to dispose of their effluent. There are, however, many methods of treatment and disposal of leachate. These are briefly listed below, with the qualifying criteria added. It is emphasised that in order to properly design a leachate treatment and disposal system, the specifier must be aware of all the site specific facts including the variability of leachate composition with time.

1. Recirculation: (i) Reduces leachate volume by pervading unused absorptive capacity, (ii) promotes rapid breakdown of wastes, and (iii) uses the landfill as storage.
2. Spray irridation: Combines evapotranspiration with physical, chemical and biological mechanisms. Not known what the long-term effects on the soil horizon are.

3. Chemical treatment:
   (a) Precipitants and flocculants: Mainly to remove suspended solids that may block pipeline or result in scale formation.
   (b) Chemical oxidants: Used to destroy odour forming compounds ~ hydrogen peroxide to remove sulphides.
4. Aerobic biological treatment:
   (a) Effective, economical means of reducing BOD and ammonia.
   (b) Aerated lagoons have been used successfully to remove BOD and nitrify ammonia. Can be adversely affected by low values of BOD.
5. Rotating biological contactor: Bacteria grow on a plastic surface and are rotated through the incoming leachate. Useful for treating low BOD and for nitrification of ammonia. Short retention time. Can be adversely affected by variations in leachate quality.
6. Anaerobic biological treatment: Used extensively in sewage sludge treatment. Particularly effective for acetogenic leachates, although may become inoperative once methanogenesis commences. Ammonia removal not always achieved.
7. Reed beds: This method has the potential to remove COD, ammonia, nitrate and suspended solids, although research has yet to establish optimum conditions of operation.
8. Reverse osmosis: Used to remove non-degradable organic compounds, although often used after some prior form of treatment to 'polish' effluent. The technique is, however, expensive and can produce a high-strength concentrate which must in turn be disposed of.
9. Activated carbon adsorbtion: This can potentially: (i) remove toxic organic constituents, (ii) treat trace levels of adsorbable halogencompounds (AOXs), and (iii) treat colour and low levels of COD and TOC after leachate treatment. Exhausted carbon must be disposed of or thermally regenerated.

To sum up, the installation of a combined landfill gas and leachate extraction system can have the following advantages: (i) there is only one set of contractors working on the site, ensuring that any disturbance due to trenching operations and drilling of boreholes is kept to a minimum, (ii) all restoration work is completed as part of a single combined project, (iii) the system can be easily adapted and added to if required, ensuring minimum additional disturbance to the site, and (iv) the system uses the same set of extraction boreholes and trench lines. In short, the headaches of two tender proposals and submissions, two requests for financial approval, two sets of mobilisation and attendant difficulties are distilled into one.

Any combined landfill gas and leachate system has to be thoroughly investigated prior to installation. This type of investigation includes desk studies, pumping trials and design recommendations. Once a system has been decided upon it is essential that adequate mathematical investigation as to the physical characteristics of the design is undertaken. This exercise will ensure that the pumps, both for landfill gas and for leachate, are correctly sized. Too often, it is the case that these stages are dispensed with in order to keep the cost of the installation down. It is stressed that this is false economy. Comprehensive investigation as to the sizing of the intended works should be included as a major part of the works in any project and should not be optional.

SECTION VIII

Hospital and Biomedical Wastes

Chapter 29

Hospital Waste and Its Management

INTRODUCTION

Hospital is a place of almighty, a place to serve the patient. Since beginning, the hospitals are known for the treatment of sick persons but we are unaware about the adverse effects of the garbage and filth generated by them on human body and environment. Now it is a well established fact that there are many adverse and harmful effects to the environment including human beings which are caused by the 'hospital waste' generated during the patient care. Hospital waste is a potential health hazard to the health care workers, public and flora and fauna of the area. Hospital acquired infection, transfusion transmitted diseases, rising incidence of Hepatitis B, and HIV, increasing land and water pollution lead to increasing possibility of catching many diseases. Air pollution due to emission of hazardous gases by incinerator such as Furan, Dioxin, Hydrochloric acid, etc. have compelled the authorities to think seriously about hospital waste and the diseases transmitted through improper disposal of hospital waste. This problem has now become a serious threat for the public health and, ultimately, the Central Government had to intervene for enforcing proper handling and disposal of hospital waste and an act was passed in July 1996 and a biomedical waste (handling and management) rule was introduced in 1998.

A modern hospital is a complex, multidisciplinary system which consumes thousands of items for delivery of medical care and is a part of physical environment. All these products consumed in the hospital leave some unusable leftovers, i.e. hospital waste. The last century witnessed the rapid mushrooming of hospital in the public and private sector, dictated by the needs of expanding population. The advent and acceptance of 'disposable' has made the generation of hospital waste a significant factor in current scenario. Hospital waste refers to all waste generated, discarded and not intended for further use in the hospital.

CLASSIFICATION OF HOSPITAL WASTE

1. General waste: Largely composed of domestic or household type waste. It is non-hazardous to human beings, e.g. kitchen waste, packaging material, paper, wrappers, plastics.
2. Pathological waste: Consists of tissue, organ, body part, human foetuses, blood and body fluid. It is hazardous waste.
3. Infectious waste: The wastes which contain pathogens in sufficient concentration or quantity that could cause diseases. It is hazardous, e.g. culture and stocks of infectious agents from laboratories, waste from surgery, waste originating from infectious patients.

4. Sharps: Waste materials which could cause the person handling it, a cut or puncture of skin, e.g. needles, broken glass, saws, nail, blades, scalpels.
5. Pharmaceutical waste: This includes pharmaceutical products, drugs, and chemicals that have been returned from wards, have been spilled, are outdated or contaminated.
6. Chemical waste: This comprises discarded solid, liquid and gaseous chemicals, e.g. cleaning, housekeeping, and disinfecting product.
7. Radioactive waste: It includes solid, liquid, and gaseous waste that is contaminated with radionucleides generated from *in vitro* analysis of body tissues and fluid, *in vivo* body organ imaging and tumour localisation and therapeutic procedures.

Amount and Composition of Hospital Waste Generated

Amount

Country	Quantity (kg/bed/day)
UK	2.5
USA	4.5
France	2.5
Spain	3.0
India	1.5

Hazardous/non-hazardous

Hazardous	15%
Hazardous but non-infective	5%
Hazardous and infective	10%
Non-hazardous	85%

Composition

By weight

	Plastic	14%
Combustible		
	Dry cellublostic solid	45%
	Wet cellublostic solid	18%
Noncombustible		20%

Biomedical Waste

Any solid, fluid and liquid or liquid waste, including its container and any intermediate product, which is generated during the diagnosis, treatment or immunisation of human being or animals, in research pertaining thereto or in the production or testing of biological and the animal waste from slaughter houses or any other similar establishment. All biomedical waste are hazardous. In hospital it comprises of 15 per cent of total hospital waste.

Rationale of Hospital Waste Management

Hospital waste management is a part of hospital hygiene and maintenance activities. In fact only 15 per cent of hospital waste, i.e. 'biomedical waste' is hazardous, not the complete. But when hazardous waste is not segregated at the source of generation and mixed with nonhazardous waste, then 100 per cent waste becomes hazardous. The question then arises that what is the need or rationale for spending so much resources in terms of money, manpower, material and machine for management of hospital waste ? The reasons are:

1. Injuries from sharps leading to infection to all categories of hospital personnel and waste handler.
2. Nosocomial infections in patients from poor infection control practices and poor waste management.
3. Risk of infection outside hospital for waste handlers and scavengers and at time general public living in the vicinity of hospitals.
4. Risk associated with hazardous chemicals, drugs to persons handling wastes at all levels.
5. 'Disposable' being repacked and sold by unscrupulous elements without even being washed.
6. Drugs which have been disposed of, being repacked and sold off to unsuspecting buyers.
7. Risk of air, water and soil pollution directly due to waste or due to defective incineration emissions and ash.

Approach for Hospital Waste Management

Based on Biomedical Waste Rules 1998, notified under the Environment Protection Act by the Ministry of Environment and Forest (Government of India).

Segregation of waste

Segregation is the essence of waste management and should be done at the source of generation of biomedical waste, e.g. all patient care activity areas, diagnostic services areas, operation theatres, labour rooms, treatment rooms, etc. The responsibility of segregation should be with the generator of biomedical waste, i.e. doctors, nurses, technicians, etc. (medical and paramedical personnel). The biomedical waste should be segregated as per categories mentioned in the rules.

Collection of biomedical waste

Collection of biomedical waste should be done as per biomedical waste (Management and Handling) Rules. At ordinary room temperature the collected waste should not be stored for more than 24 hours.

Transportation

Within hospital, waste routes must be designated to avoid the passage of waste through patient care areas. Separate time should be earmarked for transportation of biomedical waste to reduce chances of its mixing with general waste. Desiccated wheeled containers, trolleys or carts should be used to transport the waste/plastic bags to the site of storage/treatment.

Trolleys or carts should be thoroughly cleaned and disinfected in the event of any spillage. The wheeled containers should be so designed that the waste can be easily loaded, remains secured during transportation, does not have any sharp edges and is easy to clean and disinfect. Hazardous biomedical waste needing transport to a long distance should be kept in containers and should have proper labels. The transport is done through desiccated vehicles specially constructed for the purpose having fully enclosed body, lined internally with stainless steel or aluminium to provide smooth and impervious surface which can be cleaned. The drivers compartment should be separated from the load compartment with a bulkhead. The load compartment should be provided with roof vents for ventilation.

Treatment of hospital waste

Treatment of waste is required:

1. To disinfect the waste so that it is no longer the source of infection.
2. To reduce the volume of the waste.

3. Make waste unrecognisable for aesthetic reasons.
4. Make recycled items unusable.

General waste

The 85 per cent of the waste generated in the hospital belongs to this category. The, safe disposal of this waste is the responsibility of the local authority.

Biomedical waste: 15 per cent of hospital waste

1. Deep burial: The waste under category 1 and 2 only can be accorded deep burial and only in cities having less than 5 lakh population.
2. Autoclave and microwave treatment: Standards for the autoclaving and microwaving are also mentioned in the biomedical waste Rules 1998. All equipment installed/shared should meet these specifications. The waste under category 3, 4, 6, 7 can be treated by these techniques. Standards for the autoclaving are also laid down.
3. Shredding: The plastic, sharps (needles, blades, glass, etc.) should be shredded but only after chemical treatment/microwaving/autoclaving. Needle destroyers can be used for disposal of needles directly without chemical treatment.
4. Secured landfill: The incinerator ash, discarded medicines, cytotoxic substances and solid chemical waste should be treated by this option.
5. Incineration: The incinerator should be installed and made operational as per specification under the BMW Rules 1998 and a certificate may be taken from CPCB/State Pollution Control Board and emission levels etc should be defined. In case of small hospitals, facilities can be shared. The waste under category 1, 2, 3, 5, 6 can be incinerated depending upon the local policies of the hospital and feasibility. The polythene bags made of chlorinated plastics should not be incinerated.
6. It may be noted that there are options available for disposal of certain category of waste. The individual hospital can choose the best option depending upon the facilities available and its financial resources. However, it may be noted that depending upon the option chosen, correct colour of the bag needs to be used.

Safety measures

All the generators of biomedical waste should adopt universal precautions and appropriate safety measures while doing therapeutic and diagnostic activities and also while handling the biomedical waste.

It should be ensured that:

1. Drivers, collectors and other handlers are aware of the nature and risk of the waste.
2. Written instructions, provided regarding the procedures to be adopted in the event of spillage/accidents.
3. Protective gears provided and instructions regarding their use are given.
4. Workers are protected by vaccination against tetanus and hepatitis B.

Training

1. Each and every hospital must have well planned awareness and training program for all category of personnel including administrators (medical, paramedical and administrative).
2. All the medical professionals must be made aware of biomedical waste (Management and Handling) Rules 1998.
3. To institute awards for safe hospital waste management and universal precaution practices.
4. Training should be conducted to all categories of staff in appropriate language/medium and in an acceptable manner.

Management and administration

Heads of each hospital will have to take authorisation for generation of waste from appropriate authorities as notified by the concerned State/Central government, well in time and to get it renewed as per time schedule laid down in the rules. Each hospital should constitute a hospital waste management committee, chaired by the head of the Institute and having wide representation from all major departments. This committee should be responsible for making hospital specific action plan for hospital waste management and its supervision, monitoring and implementation. The annual reports, accident reports, as required under BMW rules should be submitted to the concerned authorities as per BMW rules format.

Measures for waste minimisation

As far as possible, purchase of reusable items made of glass and metal should be encouraged. Select non-PVC plastic items. Adopt procedures and policies for proper management of waste generated, the mainstay of which is segregation to reduce the quantity of waste to be treated. Establish effective and sound recycling policy for plastic recycling and get in touch with authorised manufactures.

Coordination between, hospital and outside agencies

1. Municipal authority: As quite a large percentage of waste (in India upto 85 per cent), generated in Indian hospitals, belong to general category (nontoxic and nonhazardous), hospital should have constant interaction with municipal authorities so that this category of waste is regularly taken out of the hospital premises for landfill or other treatment.
2. Co-ordination with pollution control boards: Search for better methods technology, provision of facilities for testing, approval of certain models for hospital use in conformity with standards laid down.
3. To search for cost effective and environmental friendly technology for treatment of biomedical and hazardous waste. Also, to search for suitable materials to be used as containers for biomedical waste requiring incineration/autoclaving/microwaving.
4. Development of non-PVC plastics as a substitute for plastic which is used in the manufacture of disposable items.

TREATMENT OF MEDICAL WASTE

Medical waste treatment technologies used to decontaminate waste can be classified under four basic processes — thermal process, chemical processes, irradiative processes, biological processes, mechanical processes, and incineration. Thermal processes rely on heat to destroy pathogens (disease-causing micro-organisms). Chemical processes employ disinfectants to destroy pathogens or chemicals to react with the waste. Irradiation involves ionising radiation to destroy micro-organisms while biological processes use enzymes to decompose organic matter. Mechanical processes, such as shredders, mixing arms or compactors, are added as supplementary processes to render the waste unrecognisable, improve heat or mass transfer or reduce the volume of treated waste. Incineration involves high temperature combustion of waste to convert the waste into ash. Note that medial waste incineration is not recommended as the world health organisation (WHO) considers it a significant source of highly toxic dioxin, a known carcinogen that has been linked to birth defects, immune system disorders and other harmful health effects. The US environmental protection agency (EPA) identified medical waste incineration as the single largest source of dioxin air emissions in the United States. A brief description of medical waste treatment processes are provided below.

Thermal Process

Thermal processes are those that rely on heat (thermal energy) to destroy pathogens in the waste. This category is further subdivided into low-heat, medium-heat, and high-heat thermal processes. This further sub-classification is necessary because physical and chemical mechanisms that take place in thermal processes change markedly at medium and high temperatures.

Low-heat thermal process: Low-heat thermal processes are those that use thermal energy to decontaminate the waste at temperatures insufficient to cause chemical breakdown or to support combustion or pyrolysis. In general, low-heat thermal technologies operate between 200°F to about 350°F (93°–177°C). The two basic categories of low-heat thermal processes are: (i) wet heat (steam), and (ii) dry heat (hot air) disinfection.

Wet heat treatment involves the use of steam to disinfect waste and is commonly done in an autoclave. Microwave treatment is essentially a steam disinfection process since water is added to the waste and disinfection occurs through the action of moist heat and steam generated by microwave energy.

In dry heat process, no water or steam is added. Instead, the waste is heated by conduction, natural or forced convection, and/or thermal radiation using infrared heaters.

Medium-heat thermal process: Medium-heat thermal processes take place at temperatures between 350° to 700°F (177°–370°C) and involve the chemical breakdown of organic material. These processes are the basis for relatively new technologies. They include reverse polymerisation using high-intensity microwave energy and thermal depolymerisation using heat and high pressure.

High-heat thermal process: High-heat thermal processes generally operate at temperatures ranging from around 1000° to 15,000°F (540°–8300°C) or higher. Electrical resistance, induction, natural gas, and/or plasma energy provide the intense heat. High-heat processes involve chemical and physical changes to both organic and inorganic material resulting in total destruction of the waste. A significant change in the mass and volume of the waste also occurs. For example, low-heat thermal technologies that rely on shredders or grinders to reduce size decrease waste volume by about 60 to 70 per cent, compared to 90 or 95 per cent with high-heat thermal processes.

Incineration

According to the EPA, 90 per cent of medical waste is incinerated. Incineration is the controlled burning of the medical waste in a dedicated medical waste incinerator. The waste generally passes through the incinerator on a belt, and because most medical waste can be incinerated, the waste is not sorted or separated prior to treatment. Incineration has the benefit of reducing the volume of the waste, sterilising the waste, and eliminating the need for pre-processing the waste before treatment. The resulting incinerated waste can be disposed of in traditional methods, such as brought to a landfill. The downside of incineration is potential pollution from emissions generated during incineration. The EPA has stringent requirements on emissions from medical incinerators. The incineration process can be applied to almost all medical waste types, including pathological waste, and the process reduces the volume of the waste by up to 90 per cent.

Modern incinerators can provide a secondary benefit by harnessing the heat created by the incineration process to power boilers in the facility. The flames in the primary chamber can ignite fossil fuels in a secondary chamber and power facility boilers.

The largest concern associated with incineration is pollution. The EPA has reported that at least 20 per cent of medical waste is plastic. The biggest concern is the incineration of chemicals that are released from combusting plastics. While incineration provides the advantage of reducing the volume of waste into ash and the ability to dispose recognisable waste and sharps, the incinerator may contain toxic gases. Dioxins and furans can be produced when these plastics burn. The majority of older medical waste incinerators contain no pollution control equipment. As new federal and state emission regulations are instituted that have more stringent requirements, medical incinerators are often not being replaced at the end of their service life. Over time, the amount of waste being incinerated will be reduced as other technologies replace on-site incinerators. Another concern is related to the contents of incinerator ash. As incinerators are designed or retrofit with pollution prevention equipment, more of the potentially toxic chemicals that previously ended up in emissions now remain in the ash. Incinerator ash is generally disposed of in landfills, and little data is available on the effects of ash on the environment.

As additional requirements are added to the emissions for medical waste incinerators, the cost of incinerating medical waste increases, and alternative treatments have increased their market share.

Autoclaves

Autoclaves are closed chambers that apply both heat and pressure, and sometimes steam, over a period of time to sterilise medical equipment. Autoclaves have been used for nearly a century to sterilise medical instruments for reuse. Autoclaves are used to destroy all micro-organisms that may be present in medical waste before disposal in a traditional landfill. The autoclave lowers the pressure within the chamber, which shortens the amount of time required to generate steam.

Medical waste that is subjected to an autoclave is often also subjected to a compaction process, such as shredding, after treatment so that it is no longer recognisable and cannot be reused for other purposes. The compaction process reduces the volume of the treated waste significantly. After treatment and compaction, the treated waste can be combined with general waste and disposed of in traditional manners. Waste that is treated using an autoclave is still recognisable after treatment, and therefore must be shredded after treatment to allow for disposal with general waste. Autoclaves are not recommended for the treatment of pathological waste, due to the recognisability factor after treatment, and that pathological waste may contain low levels of radioactive material or cytotoxic compounds. The autoclave process can aerosolise chemicals present in the waste and depending on the design of the autoclave, these chemicals can be released into the air when the autoclave is opened.

Autoclaves can be used to process up to 90 per cent of medical waste, and are easily scaled to meet the needs of any medical organisation. Small counter-top autoclaves are often used for sterilising reusable medical instruments. Large autoclaves are used to treat large volumes of medical waste at once.

Steam sterilisation provides generators a way to treat waste in a cost-efficient manner. The destruction of the micro-organisms is highly effective, but the problem comes when transportation is required. Many landfills and general incineration facilities are reluctant to accept the waste, fearing the waste is infectious. Recent work in Japan has found a method of chemically stabilising heavy metals in fly-ash from medical waste incinerators. Much development goes on in Japan, including recent work on a dual torch plasma arc furnace.

Mechanical/Chemical Disinfection

Chemical disinfection, primarily through the use of chlorine products, is another method to treat medical waste. The use of chlorine bleach for cleaning and disinfecting is well-known and this method has been in use for many years. The mechanical/chemical disinfection process provides control and consistency to the disinfection process. The EPA identifies chemical disinfection as the most appropriate method to treat liquid medical waste. Chemical disinfection processes are often combined with a mechanical process, such as shredding or maceration, to ensure sufficient exposure of the chemicals to all portions of the waste. The disinfectant is usually combined with a large amount of water to assist with the disinfection process and to cool the mechanical equipment in the shredding process. Liquid waste treated with a mechanical/chemical disinfection process can usually be discharged into the sewer system, as long as the organisation has obtained the proper sewer discharge permits from their city. Mechanical/chemical disinfection treatment devices are primarily on-site installations, rather than mobile treatment units, though these devices are available in different sizes based on the amount of waste to be treated.

Microwave

The use of microwaves to disinfect medical waste has only recently been introduced in the United States. Microwave treatment units can be either on-site installations or mobile treatment vehicles. In this type of disinfection process, the waste is first shredded. The shredded waste is then mixed with water and subjected to microwaves. The microwaves internally heat the waste, rather than applying heat externally, as in an autoclave. The heat generated in this method provides even heating over all portions of the waste, and the high-temperature steam that is generated effectively neutralises all biologicals. The shredding operation reduces the volume of the waste by up to 80 per cent, and the treated waste can be disposed of in a landfill. The entire process takes place within a single vessel, and the system can be operated by unskilled workers. Treatment of medical waste through exposure to microwaves is less expensive than incineration. This method is not recommended by the EPA for the treatment of pathological waste.

Irradiation

Another method used to sterilise medical equipment or waste is irradiation, generally through exposure of the waste to a cobalt source. The gamma radiation generated by the cobalt source inactivates all microbes that may be present in the waste. Dedicated sites are required for this form of treatment, as opposed to the mobile versions available for other non-incineration methods. One private company that specialises in this form of treatment shreds the treated waste after irradiation, then ships the waste to a cement kiln, where it is burned as fuel. The cost of developing a dedicated facility for this method is quite high, and therefore this method is not as widely used as other treatment methods at this time. The risk of radiation exposure by workers operating the facility, while low, is also a factor. Also, pathological waste cannot be treated using irradiation. Questions have been raised about the effectiveness of irradiation to provide consistent treatment across a batch of waste.

Biological Process

Biological processes employ enzymes to destroy organic matter. Only a few non-incineration technologies have been based on biological processes.

BIOMEDICAL WASTE (MANAGEMENT AND HANDLING) RULES, 1998

Background: With a view to control the indiscriminate disposal of hospital waste/biomedical waste, the Ministry of Environment and Forest, Govt. of India has issued a notification on Biomedical Waste Management under the Environment (Protection) Act. Government of NCT of Delhi in its notification dated 6th July, 1999 has authorised Delhi Pollution Control Committee (DPCC) for the purpose of granting authorisation for collection, reception, storage, treatment and disposal of biomedical waste to implement the Biomedical Waste Management Rules, 1998. Government of NCT of Delhi has also constituted advisory committee, appellate authority in exercise of powers conferred under Biomedical rules. Some of the salient features of these rules are:

Rules are applicable to: These Rules will apply to hospitals, nursing homes, veterinary hospitals, animal houses, pathological labs and blood banks, generating hospital wastes [except such occupier of clinics, dispensaries, pathological labs, blood banks providing treatment/service to less than 1000 (one thousand) patients per month].

Duty: It shall be the duty of the every occupier of an institution generating biomedical waste which includes a hospital, nursing home, clinic, dispensary, veterinary institution animal house, pathological laboratory, blood bank by whatever name called to take all steps to ensure that such waste is handled without any adverse effect to the human health and the environment.

Management of biomedical waste: Every occupier generating the biomedical waste need to install an appropriate facility in the premises or set up a common facility to ensure requisite treatment of waste by 30-6-2000 in accordance with Schedule-I and in compliance with standards prescribed with Schedule-V.

The biomedical waste need to be segregated into container/bags at the point of generation in accordance with Schedule-II, prior to its storage, transportation, treatment and disposal. The container shall be labelled according to Schedule-III.

Mandatory/legal requirement: Every occupier of an institution, generating, collecting, receiving, storing, transporting, treating, disposing and/or handling biomedical waste in any other manner, shall make an application in Form-I along with the following fee structure to the Delhi Pollution Control Committee for grant of authorisation. The Form-I can be obtained after paying an amount of Rs. 100/- in the form of Draft in favour of DPCC. It can also be downloaded from this website but an additional draft for Rs. 100/- in favour of DPCC may also be attached with the application at the time of submission of application.

Category	*Structure* Fee (in Rupees)
1. Clinics, pathological laboratories and blood banks	1000/- per annum
2. Veterinary institutions, dispensaries and animal houses	1000/- per annum
3. Hospitals, nursing homes and health care establishments	1000/- per annum up to 4 beds and additional Rs. 100 per bed per annum from fifth bed onwards
4. Operator of the facility of biomedical waste	10,000/- per annum (excluding transportation)
5. Transporter of biomedical waste	7500/- per annum

An operator of biomedical waste facility may also engage in transportation of biomedical waste on payment of additional fees prescribed for a transporter of biomedical waste. An application in Form-I appended to the aforesaid rule shall be made to the prescribed authority, i.e. the Chairman, Delhi Pollution Control Committee, for grant of authorisation along with the checking of documents as given in check list, wherever applicable.

An authorisation shall be granted for a period of 3 years, including an initial trial period of one year for which a provisional authorisation will be granted. All authorisation shall be for a period of three years. Fee shall be payable for three years at time. The above fee structure is subject to revision from time to time.

The Government's notification No. F.23(522)/95-Env/99 dated the 6th July 1999, issued in pursuance of rule 8(3) *ibid* shall stand superseded with immediate effect. An operator of a facility shall make an application form in Form-I with the fee as applicable for grant of authorisation.

In addition, they shall also submit an annual report to DPCC in Form-II by 31st January every year to include information about the categories and quantities of biomedical wastes handled during the proceding year and also maintain records related to the generation, collection, reception, storage, transportation, treatment, disposal, and/or any form of handling of biomedical waste in accordance with rules and guidelines issued. All records shall be subject to inspection and verification by the DPCC at any time. The transporter, operator of a facility shall label the biomedical strictly in accordance with the procedure given in Schedule-IV.

Penalty: The defaulting hospitals/nursing homes, etc. are liable to be penalised as per the provisions of Environment (Protection) Act, 1986 and other pollution control Acts.

Appeal: Any person aggrieved by an order made by the DPCC under these rules may within thirty days from date on which the order is communicated to him, prefer an appeal to the Financial Commissioner, Govt. of NCT of Delhi who is appointed as Appellate Authority under the rules.

With the objective to provide the common facility as envisaged under the rules, Delhi Pollution Control Committee has authorised the operator of a facility who collects, transports, treats and disposes the waste in accordance with the provisions of the rules.

Chapter 30

Biomedical Waste Management

INTRODUCTION

This chapter focuses upon the importance and the purpose of biomedical waste management, definition of biomedical waste, risks associated and dangers of improper management of biomedical waste.

Hospitals and other healthcare establishments have a 'duty of care' for the environment, public health and have particular responsibilities in relation to the waste they produce (i.e. biomedical waste). Negligence in terms of biomedical waste management significantly contributes to polluting the environment and affects the health of human beings. The waste generated by any hospital/health care facilities consists of general waste like packaging material, eatables, paper, wrapper, etc. hazardous and infectious waste like outdated medicines, cytotoxic drugs, soiled dressing, swabs, cotton with blood and body fluid, dissected body organs and tissues, disposable syringes, intravenous fluid bottles, catheters, gloves, injection vials, needles, blades, scalpels, etc. Quantity wise around 70–80 per cent is general waste and 20–30 per cent is hazardous and infectious waste which poses risk to human health and environment. These two basic category of wastes (hazardous and infectious) should be segregated otherwise the whole waste, the entire volume of waste will become infectious.

As per biomedical waste rules, 1998 and amendments, any waste, which is generated during the diagnosis, treatment or immunisation of human beings or animals or in research activities pertaining thereto or in the production of testing of biological and including categories mentioned in schedule 1 of the Rule, is the biomedical waste.

As per WHO norms the healthcare waste includes all the waste generated by healthcare establishments, research facilities, and laboratories. In addition, it includes the waste originating from minor or scattered sources such as that produced in the course of healthcare undertaken in the home (dialysis, insulin injections, etc.).

RISKS TO PERSONNEL DUE TO BIOMEDICAL WASTE

Poor biomedical waste management exposes hospital and other healthcare facility workers, waste handlers and community to infection, toxic effects and injuries. Doctors, nurses, paramedical staff, sanitary staff, hospital maintenance personnel, patients receiving treatment, visitors to the hospital, support service personnel, workers in waste disposal facilities, scavengers, general public and more specifically the children playing with the items they can find in the waste outside the hospital when it is directly accessible to them are potentially at risk of being injured or infected when they are exposed to biomedical waste.

Risk to all those who generate, collect, segregate, handle, package, store, transport, treat and dispose waste (an occupational hazard). Occupational exposure to blood can result from percutaneous injury (needle stick or other sharps injury), mucocutaneous injury (splash of blood or other body fluids into the eyes, nose or mouth) or blood contact with non-intact skin. Over 20 blood born diseases can be transmitted but particular concern is the threat of spread of infectious and communicable diseases like AIDS, Hepatitis B & C, Cholera, Tuberculosis, Diphtheria, etc. Waste chemicals, radioactive waste and heavy metals also finds its way in waste stream which are also hazardous to health.

Dangers of Improper Management of Biomedical Waste

There is public health hazard due to poor management of biomedical waste which can cause a number of disease. Serious situations are very likely to happen when biomedical waste is dumped on uncontrolled sites where it can be easily accessed by public. Children and rag pickers are particularly at risk to come in contact with infectious waste. Inappropriate treatment and disposal contributes to environmental pollution (uncontrolled incineration causes air pollution, dumping in drains, tanks and along the river bed causes water pollution and unscientific landfilling causes soil pollution).

In many parts of the country biomedical waste is neither segregated nor disinfected. It is being indiscriminately dumped into municipal bins, along the roadsides, into water bodies or is being burnt in the open air. All this is leading to rapid proliferation and spreading of infectious, dangerous and fatal communicable diseases. The improper handling and mismanagement of biomedical waste is posing serious problems, few of the problems due to improper disposal are as follows:

1. The infectious waste which is only 20–25 per cent of the entire waste from hospitals is not segregated and is mixed with general waste by doing so the whole of waste may turn up to infectious waste. If the same is dumped into the municipal bin then there are fair chances of the waste in municipal bin to become infectious.
2. The disposal of sharps will lead to needle stick injuries, cuts, and infections among hospital staff, municipal workers, rag pickers and the general public. This will lead to transmission of diseases like Hepatitis B, C, E and HIV, etc.
3. The needles and syringes which are not mutilated or destroyed are being circulated back through traders who employ the poor and the destitute to collect such waste for repackaging and selling in the market.
4. One of the reasons for spreading of infection is reuse of disposable items like syringes, needles, catheters, and dialysis sets, etc.
5. The dumping of untreated biomedical waste in municipal bins may increase the possibility of survival, proliferation and mutation of pathogenic microbial population in the municipal waste. This leads to epidemics and increased incidence and prevalence of communicable diseases in the community.
6. Chances of vectors are high, like cats, rats, mosquitoes, flies and stray dogs getting infected or becoming carriers which also spread diseases among the public.

REGULATIONS ON BIOMEDICAL WASTE MANAGEMENT

This section discusses various legal provisions governed on waste management. The salient features of biomedical waste (Management and Handling) Rules, 1998 and amendments has been provided.

Establishment of a sustainable biomedical waste management system gets benefit from a national legal framework that regulates and organises the different elements of a waste management system.

Legislation usually places obligations and controls on what is permitted and prescribes sanctions on those that deviate from accepted practice. In reality, a law will remain ineffective if sources (finance, material and knowledge) are not available in the hospitals or healthcare sectors to implement it and/or if enforcement is weak.

The five guiding principles governing in waste-related laws are the 'polluter pays' principle, this requires any waste producer to be made legally and financially responsible for the safe and environmentally sound disposal of their waste. The responsibility to ensure that the disposal of waste causes no environmental damage is placed upon each waste generator, the 'precautionary' principle, the rationale of the principle is that if the outcome of a potential risk is suspected to be serious, but may not be accurately known, it should be assumed that this risk is high. This has the effect of obliging health care waste generators to operate a good standard of waste collection and disposal, as well as provide health and safety training, protective equipment and clothing for their staff, the 'duty of care' principle, this recognises that any person managing or handling healthcare waste, or waste-related equipment, is morally responsible to take good care of the waste while it is under their responsibility, the 'proximity' principle, the philosophy behind this principle is that treatment and disposal of hazardous waste (including healthcare waste) should take place at the nearest convenient location to its place of generation, in order to minimise the risks to the general population. This does not necessarily mean treatment or disposal has to take place at each healthcare establishment; instead it could be done at a facility shared locally or at a regional or national location. An extension to this principle is the expectation that every country should make arrangements to dispose of all wastes in an acceptable manner inside its own national borders and prior informed consent principle/also known as 'cradle to grave' control, this principle introduces the concept that all parties involved in the generation, storage, transport, treatment and disposal of hazardous wastes (including healthcare waste) should be licensed or registered to receive and handle named categories of waste. In addition, only licensed organisations and sites are allowed to receive and handle these wastes. No hazardous wastes (including healthcare waste) should leave a place of waste generation until the subsequent parties (e.g. transport, treatment and disposal operators and regulators) are informed that a waste consignment is ready to be moved.

National Legislations Governing Waste Management

National legislation is the basis for biomedical waste management practices in the country. It establishes control and permits for the disposal. The regulatory framework which governs the management of waste is as follows:

1. The water (prevention and control of pollution) Act, 1974 (for liquid waste)
2. The air (prevention and control of pollution) Act, 1981 (for air quality).
3. The environment (protection) Act, 1986.
4. Hazardous wastes (Management, Handling and Transboundary Movement) Rules, 2008 (for hazardous waste).
5. The biomedical wastes (management and handling) Rules, 1998 (for hospital waste).
6. The municipal solid wastes (management and handling) Rules, 2000 (for domestic municipal waste).
7. Battery (management and handling) Rules, 2001 (for used batteries waste).

Excerpts from Biomedical Waste (Management and Handling) Rules, 1998 and as Amended

The biomedical waste management and handling rules regulate biomedical waste management at local, regional and national level. The rules provides a general foundation for improving biomedical waste management systems by indicating in broad terms what is regarded as good and acceptable practice in the hospitals or healthcare institutions. The main benefit of a national law covering hospital waste is that it can give a uniform basis for a country to develop good practices by providing the definition of waste, its categories, defined legal obligations of waste producers, requirements for record-keeping and reporting to regulatory agencies, authority for an inspection system, establishment of procedures to permit or prohibit some waste handling, treatment and disposal practices and the courts with powers to settle disputes and impose penalties on offenders.

This rule has 14 sections, 6 schedules and 5 forms and is applied to all persons who generate, collect, receive, store, transport, treat, dispose, or handle biomedical waste in any form. As per the rule 'Occupier' means in relation to any institution generating biomedical waste, which includes a hospital, nursing home, clinic dispensary, veterinary institution, animal house, pathological laboratory, blood bank by whatever name called, means a person who has control over that institution and/or its premises. The duty of every occupier of an institution generating biomedical waste is to take all steps to ensure that such waste is handled without any adverse effect to human health and the environment.

No untreated biomedical waste shall be kept stored beyond a period of 48 hours, provided that if for any reason it becomes necessary to store the waste beyond such period, the authorised person must take permission from the prescribed authority and take measures to ensure that the waste does not adversely affect human health and the environment. 'Authorised person' means an occupier or operator authorised by the prescribed authority to generate, collect, receive, store, transport, treat, dispose and/or handle biomedical waste in accordance with these rules and any guidelines issued by the Central Government. The 'Prescribed Authority' for the enforcement of provisions of these rules shall be the State Pollution Control Boards in respect of states and the Pollution Control Committees in respect of the Union territories. The 'Prescribed Authority' for the healthcare establishments of Armed Forces under the Ministry of Defence shall be the Director General, Armed Forces Medical Services. Table 30.1 describes categories of biomedical wastes.

Every occupier of an institution generating, collecting, receiving, storing, transporting, treating, disposing and/or handling biomedical waste in any other manner, shall make an application in form 1 to the prescribed authority for grant of authorisation. Occupier of clinics, dispensaries, pathological labs, blood banks providing treatment/services to less than 1000 patients per month are exempted for taking authorisation. Every authorised person shall maintain records related to the generation, collection, reception, storage, transportation, treatment, disposal and/or any form of handling of biomedical waste in accordance with these rules and any guidelines issued. All these records can be subjected to inspection and verification by the prescribed authority at any time. When any accident occurs at any institution or facility or any other site where biomedical waste is handled or during transportation of such waste, the authorised person shall report the accident to the prescribed authority. The segregation, packaging, transportation and storage is as follows:

1. Biomedical waste shall not be mixed with other wastes.
2. Biomedical waste shall be segregated into containers/bags at the point of generation in accordance with Table 30.2 prior to its storage, transportation, treatment and disposal.
3. The containers shall be duly labelled as per Table 30.3.
4. If a container is transported from the premises where biomedical waste is generated to any waste treatment facility outside the premises, the container shall, apart from the label prescribed also carry information prescribed in Table 30.4 and it describes the type of waste where it is generated and to where it is being transferred.

The treatment and disposal of biomedical waste shall be in accordance with Table 30.1 and in compliance with Table 30.4. Table 30.4 presents the standards for incinerators, autoclave, liquid waste, microwave and deep burial.

Table 30.1. Categories of biomedical waste.

Waste category	Category waste	Category type treatment and disposal option
1	Human anatomical waste (body parts, organs, human tissues, etc.)	Incineration/deep burial*
2	Animal waste (animal tissues, organs, body parts carcasses, bleeding parts, fluid, blood and experimental animals used in research, waste generated by veterinary hospitals, colleges, discharge from hospitals, animal houses)	Incineration/deep burial*
3	Microbiology and biotechnology waste (wastes from laboratory cultures, stocks or micro-organisms live or vaccines, human and animal cell culture used in research and infectious agents from research and industrial laboratories, wastes from production of biologicals, toxins, dishes and devices used for transfer of cultures)	Local autoclaving/micro waving/ incineration
4	Waste sharps (needles, syringes, scalpels, blade, glass, etc. that may cause puncture and cuts. This includes both used and unused sharps)	Disinfection (chemical treatment/ autoclaving/micro waving and mutilation/shredding
5	Discarded medicines and cytotoxic drugs (waste comprising of outdated, contaminated and discarded medicines)	Incineration/destruction and drugs disposal in secured landfills
6	Soiled waste (items contaminated with blood, and body fluids including cotton, dressings, soiled plaster casts, lines, bedding, other material contaminated with blood)	Local autoclaving/micro waving/ incineration
7	Solid waste (waste generated from disposal items other than the sharps such a tubings, catheters, intravenous sets, etc.)	Disinfection by chemical treatment of autoclaving/micro waving and mutilation/shredding
8	Liquid waste (waste generated from laboratory and washing, cleaning, housekeeping and disinfecting activities)	Disinfection by chemical treatment and discharge into drains
9	Incineration ash (ash from incineration of any biomedical waste)	Disposal in municipal landfill
10	Chemical waste (chemicals used in production of biologicals, chemicals used in disinfection, as insecticides, etc.).	Disinfection by chemical treatment and discharge into drains for liquids and secured landfill for solids

*Deep burial shall be an option available only in towns with population less than five lakhs and in rural areas.

Table 30.2. Colour coding and type of container for disposal of biomedical waste.

Colour coding	Type of container	Waste category	Treatment options as per Table 30.1
Yellow	Plastic bag	Cat. 1, Cat. 2, and Cat. 3, Cat. 6	Incineration/deep burial
Red	Disinfected container/ plastic bag	Cat. 3, Cat. 6, Cat. 7	Autoclaving/micro waving/chemical treatment

(Contd...)

Colour coding	Type of container	Waste category	Treatment options as per Table 30.1
Blue/white translucent	Plastic bag/puncture proof container	Cat. 4, Cat. 7	Autoclaving/micro waving/chemical treatment and destruction/shredding
Black	Plastic bag	Cat. 5 and Cat. 9 and Cat. 10 (solid)	Disposal in secured landfill

Notes: 1. Colour coding of waste categories with multiple treatment options as defined in Table 30.1, shall be selected depending on treatment option chosen, which shall be as specified in Table 30.1.

2. Waste collection bags for waste types needing incineration shall not be made of chlorinated plastics.

3. Categories 8 and 10 (liquid) do not require containers/bags.

4. Category 3 if disinfected locally need not be put in containers/bags.

Table 30.3. Label for transport of biomedical waste containers/bags.

Day .... Month ....

Year ....

Date of generation ....

Waste category no. ....

Waste class ....

Waste description ....

Sender's name and address   Receiver's name and address

Phone no. ....   Phone no. ....

E-mail ....   E-mail ....

Fax no. ....   Fax no. ....

Contact person ....   Contact person ....

In case of emergency please contact

Name and address: ....

Phone no............

Note: Label shall be non-washable and prominently visible.

Table 30.4. Standards for treatment and disposal of biomedical wasts standards for incinerators.

All incinerators shall meet the following operating and emission standards

A. Operating standards

1. Combustion efficiency (CE) shall be at least 99.00%.

2. The Combustion efficiency is computed as follows:

$$C.E. = \frac{\%CO_2}{\%CO_2 + \%CO} \times 100$$

3. The temperature of the primary chamber shall be $800° \pm 50°C$.

4. The secondary chamber gas residence time shall be at least 1 (one) second at $1050 \pm 50°C$, with minimum 3% oxygen in the stack gas.

B. Emission standards

Parameters	Concentration mg/Nm^3 at (12% CO_2 correction)
(1) Particulate matter	150

(Contd...)

(2) Nitrogen Oxides	450
(3) HCl	50
(4) Minimum stack height shall be 30 metres above ground	
(5) Volatile organic compounds in ash shall not be more than 0.01%	

Note:

1. Suitably designed pollution control devices should be installed/retrofitted with the incinerator to achieve the above emission limits, if necessary.

2. Wastes to be incinerated shall not be chemically treated with any chlorinated disinfectants. Chlorinated plastics shall not be incinerated.

3. Toxic metals in incineration ash shall be limited within the regulatory quantities as defined under the Hazardous Waste (Management and Handling) Rules, 1989.

4. Only low sulphur fuel shall be used as fuel in the incinerator.

ROLE OF DOCTORS, MEDICAL SUPERINTENDENT AND ADMINISTRATORS IN BIOMEDICAL WASTE MANAGEMENT

This section deals with the role of doctors, medical superintendents and administrators of hospitals in planning and designing of Biomedical waste management. Unit-wise generation of waste, its audit and minimisation techniques, items and equipments required to manage the waste and their placement has been mentioned. Financial management as per methodology adopted for disposal is explained.

Dealing biomedical waste in safe manner is the responsibility of all medical staff. Every person including doctors and medical superintendents producing waste items are responsible for ensuring its safe segregation at the point of generation itself. Biomedical waste is poorly managed in many hospitals not only in India, worldwide. Identifying the causes and then supporting improvements in the system are key skills that doctors, medical superintendents and administrators of hospitals need to develop. Assess the waste handling and treatment system of biomedical waste and its mandatory compliance with regulatory notifications. Estimate the amount of non-infectious and infectious waste, preferably categorywise, generated in different wards or sections. Analyse the biomedical waste management system, including policy, practice, storage, collection, transportation, treatment, disposal and compliance with the standards prescribed under the regulatory framework. In order to develop a model biomedical waste management system in the hospital, create awareness among all the stakeholders about the importance of biomedical waste management, related regulations and how to dispose off the waste. The doctors, medical superintendents and the administrators should have 'will' to improve waste management from a poor standard of performance to a better one. To improve the performance, develop policy, plan, look for waste minimisation options, provide the required materials for waste management and implement sustainable waste management system. Arrange for regular training for all the staff and organise refresher courses, monitor and oversee biomedical waste management system regularly. While monitoring care should be taken that it is necessary to ensure that each type of segregated waste are kept intact in separate specific containers and disposed off in separate specific ways, otherwise medical staff will loose confidence in the benefit of waste segregation if all wastes are remixed in subsequent handling and disposal. Doctors, superintendents and administrators are responsible, have to play a vital role in planning, designing and implementing biomedical waste management system, reducing risk of disease transmission by taking appropriate measures, response to accidental spillage and financial management.

Planning and Designing of Biomedical Waste Management

All the medical staff should realise that it is part of their duty to tackle biomedical waste management problems. To plan and design biomedical waste management one should know how much and what type of waste is generated and from which unit. Is waste minimisation possible if so in which unit and for what type of waste? What all items and equipments and their quantities are required for managing the waste? What type of disposal methodology is to be adopted to suit to their facility? Ascertain whether common biomedical waste facility is available in the area or not. Forming a waste management committee will enhance the waste management practice. For planning and designing of biomedical waste management, unitwise generation of waste, its audit and minimisation, items and equipments required for managing the waste and its appropriate placement, defining route of movement of waste and finance management needs to be taken into consideration.

Unitwise generation of biomedical waste

Depending on the services offered by hospitals or healthcare establishments, there exist the type of facilities or units in the hospitals. Activities in each unit and number of units should be identified along with the generation of type of biomedical waste expected. A preliminary study should be taken up before attempting to waste audit and its minimisation. In general various types of units available in any hospitals are outdoor patient, injection room, general ward, labour room, operation theatre, intensive care unit, casualty or emergency, laboratory and pharmacy, etc. Observe waste generation in each unit, segregate the waste as per rule at its generation place, weigh it daily for one week, aggregate it and then predict for monthly waste generation. If the segregation is not good then take the total weight of waste unitwise and 10 to 25 per cent will be the infectious waste. Depending on the activities performed in each unit, different types of waste is generated. The expected type of waste generated unitwise is as follows:

Unit	Waste generation
Outdoor patient	Soiled waste (gauze, bandages, etc.), solid waste (plastic) and sharps
Injection room	Soiled waste, sharps and solid waste
General ward	Sharps waste, solid waste and soiled waste
Labour room	Body part (placenta, etc.), sharps waste, solid waste and soiled waste
Operation theatre	Body parts, sharps waste, solid waste and soiled waste
Intensive care unit	Sharps waste, solid waste, soiled waste
Casualty/emergency	Sharps waste, solid waste and soiled waste
Laboratory	Sharps waste, solid waste, soiled waste, biologicals (culture/media)
Pharmacy	Discarded medicines

Waste audit and waste minimisation

After knowing the waste generation in all units in a hospital, perform waste audit and then minimise the generation of waste. This is one of the main step in planning and designing of biomedical waste management. The audit will give the clear picture of what type of waste, how much and from where it is generated. This information will be helpful to opt for waste minimisation, items and equipments required for segregation and treatment of waste and their placement in different units. To know how much and what type of waste is generated in each medical area, segregate the waste at the point of generation categorywise in specific colour codes as per Biomedical Waste (Management and Handling) Rules. The following steps will help in finding the waste generated quantitywise/categorywise and unitwise.

1. Ascertain how many medical areas produce biomedical waste. List all the departments and study on its activities, production of waste and quantity.
2. Find the composition of the waste in each place. Segregate waste categorywise, weigh it daily at least for one week and then average to monthly. The waste generated is not same in all the areas producing waste.
3. Keenly look for waste minimisation options in all the departments.
4. Along with the solid waste generation assessment, liquid waste assessment is also necessary.

Waste minimisation benefits the waste producers. The costs for the purchase of goods, waste treatment and disposal are reduced and the liabilities associated with the disposal of waste are lessened. By implementing policies and practices such as purchasing restrictions to ensure the selection of methods or supplies that are less wasteful or generate less hazardous waste can lead to source reduction. Use such materials which can be recycled either on-site or off-site. Careful segregation (separation) of waste into the ten categories (solid and liquid) as per rule helps to minimise the quantities of hazardous/harmful waste. Careful management of stores will prevent the accumulation of large quantities of outdated chemicals or pharmaceuticals and limit the waste to the packaging (boxes, bottles, etc.) plus residues of the products remaining in the containers. These small amounts of chemical or pharmaceutical waste can be disposed of easily and relatively cheaply, whereas disposing of larger amounts requires costly and specialised treatment, which underlines the importance of waste minimisation. Suppliers of chemicals and pharmaceuticals can also become responsible partners in waste minimisation. The health service can encourage this by ordering only from suppliers who provide rapid delivery of small orders, who accept the return of unopened stock, and who offer off-site waste management facilities for hazardous wastes.

Medical and other equipment used in a hospital may be reused provided that it is designed for the purpose and will withstand the sterilisation process. Reusable items may include certain sharps, such as scalpels and hypodermic needles, syringes, glass bottles and containers, etc. After use, these should be collected separately from nonreusable items, carefully washed (particularly in the case of hypodermic needles, in which infectious droplets could be trapped), and may then be sterilised. Although reuse of hypodermic needles is not recommended, it may be necessary in establishments that cannot afford disposable syringes and needles. Plastic syringes and catheters should not be thermally or chemically sterilised, they should be discarded for recycling industries. Long-term radionuclides conditioned as pins, needles or seeds and used for radiotherapy may be reused after sterilisation. Special measures must be applied in the case of potential or proven contamination with the causative agents of transmissible diseases.

Care should be taken while opting for recycle or reuse of materials, medical and other equipments. Ensure that effective sterilisation is attained. Sterilisation can be achieved by thermal sterilisation and chemical sterilisation. Dry sterilisation is an exposure to 160°C for 120 minutes or 170°C for 60 minutes in an oven. Wet sterilisation is an exposure to saturated steam at 121°C for 30 minutes in an autoclave. Sterilisation by ethylene oxide is done by exposing to an atmosphere saturated with it for 3–8 hours, at 50°– 60°C, in a reactor tank 'gas-steriliser', the tank should be dry before injection of the ethylene oxide. Ethylene oxide is a very hazardous chemical, this process should therefore be undertaken only by highly trained and adequately protected technical personnel. Exposure to a glutaraldehyde solution for 30 minutes will sterilise the material and this process is safer for the operators than the use of ethylene oxide but is microbiologically less efficient. The effectiveness of thermal sterilisation may be checked by the *Bacillus stearothermophilus* test and for chemical sterilisation by the *Bacillus subtilis* test.

Items and equipments required for biomedical waste management

The items and facilities required for managing the biomedical waste are as follows:

1. Protective aids like gloves, boots, over garment/apron, etc. (for self-protection against infection/injury).
2. Coloured bins and bags (yellow, red/blue and white puncture proof translucent, black and green). The biohazard label should be on all bins and bags except on black and green. The cytotoxic label should be on black bin and bag. The green colour bin should be used for general waste which is like domestic waste (for segregation of waste).
3. Big blue or red container (for storing mutilated and disinfected plastic waste).
4. Temporary central storage room (to keep all categories of waste after segregation before disposal).
5. Trolley (to carry the waste to temporary central storage place).
6. Needle cutter or needle burner (for destroying injection needle).
7. Scissors or knife (for destroying plastic waste).
8. Incinerator where mommon biomedical waste treatment facility is not available (for incinerating waste, but having individual incinerator is discouraged).
9. Deep burial pit where population is less than 5,00,000 and in rural areas where common biomedical waste treatment facility is not available (for burial of waste category 1 and 2).
10. Sharp pit where common biomedical waste treatment facility is not available (for encapsulating disinfected mutilated sharps).
11. Autoclave/microwave (for disinfection).
12. Sodium hypochlorite solution (for disinfection).
13. Soap (to wash hands).
14. Secured landfill.
15. Waste water treatment plant [for chemical (liquid) and liquid (lab and washing, etc.) waste].

Placement of required items

Waste will be generated depending on the activity of each individual unit. After ensuring the category and quantity of waste generation, required items to manage the biomedical waste should be placed appropriately. In general the requirement of biomedical waste management items and its placement in each unit is as follows:

Unit	Requirement and placement in units
Outdoor patient	Yellow/red, blue and white puncture proof translucent bag and bin/container, needle cutter, scissor and disinfection chemical
Injection room	Yellow/red, blue and white puncture proof translucent bag and bin/container, needle cutter, scissor and disinfection chemical
General ward	Yellow/red, blue and white puncture proof translucent bag and bin/container, needle cutter, scissor and disinfection chemical
Labour room	Yellow/red, blue and white puncture proof translucent bag and bin/container, needle cutter, scissor and disinfection chemical
Operation theatre	Yellow/red, blue and white puncture proof translucent bag and bin/container, needle cutter, scissor and disinfection chemical
Intensive care unit	Yellow/red, blue and white puncture proof translucent bag and bin/container, needle cutter, scissor and disinfection chemical

(Contd...)

Unit	Requirement and placement in units
Casualty/emergency	Yellow/red, blue and white puncture proof translucent bag and bin/container, needle cutter, scissor and disinfection chemical
Laboratory	Yellow/red, blue and white puncture proof translucent bag and bin/container, needle cutter, scissor and disinfection chemical
Pharmacy	Black bin or bag

Designing the movement of biomedical waste

The movement of waste should be such that after segregating the waste in specific colour coded bags from individual units, it should be placed in dedicated trolleys to transport the waste to temporary storage place for onward transmission to final disposal place. The route should be predefined, that is it should neither be through inter units nor from crowded places. Care should be taken that there should not be any spillages from bins/bags/trolleys while movement or transporting the biomedical waste.

Formation of committee for biomedical waste management

A committee to be constituted with representative members drawn from all the departments of hospital (doctor/specialist doctor, nurse, paramedical staff, etc.) representative from each cadre and one from common biomedical waste treatment facility if available. The committee should meet once in a week to discuss on continual improvement of biomedical waste management and its minimisation. The coordinator will be on turnwise basis for a period of one month from each department who will be in charge for biomedical waste management and allocates resources to support the system and ensures arrangements are in place to deal with emergencies and investigates any waste-related accidents. Heads of medical departments ensure that all their staff are aware of the waste segregation and local storage procedures, encourage good practices and enforce compliance. Matron or head nurse will be responsible for a continual training and also to new nurses and new recruits on good biomedical waste handling practices. They should oversee the handling of biomedical waste by class IV employees, like there should not be any spillage along the way, should carry the waste through predefined routes, etc. and ensures that supplies of consumable items are available (e.g. waste bags, etc.).

The committee should ensure technically feasible, environmentally sound, economically viable and socially acceptable system for management of biomedical waste. The committee members should guide the staff in assessing the waste generation in hospital with frequent intervals of time, details of assessment should include minimum weight of biomedical waste in each unit of hospital and composition of which to be determined by segregating the waste at the point of generation itself. A person to be designated to assess the level of scavenging if any or recycling taking place inside the hospital, along transportation routes and at final disposal sites, also determine social issues in relation to scavenging taking place. The committee to meet once in fortnight and review and analyse existing biomedical waste generation, storage, collection and its frequency and disposal system with due regards to level of segregation. Review existing awareness on biomedical waste management among all cadres of staff and prepare training need analysis (TNA) and organise programs. Committee should also oversee whether the consent of operation has been obtained or not and other regulatory parameters.

Reducing Risk of Disease Transmission and Response to Accidents

Diseases can be transmitted from doctors and nurses to patient (due to unwashed hands, contaminated sharps or improperly cleaned reusable equipment). Patient to health worker (due to being accidentally needle stick or sharps that have been used on patients. Also due to blood or body fluids accidentally splashing onto or coming in contact with broken skin). Health worker to family and community (health workers with unclean hands or contaminated clothing or shoes can carry infection home to family members). Health facility to community (improper disposal of biomedical waste can lead to transmission of disease to community members due to needle stick injury or needle reuse, droplet infection, respiratory route, skin contacts, etc.). The risk can be reduced by following the guidelines mentioned below:

1. Handle all sharps with care to minimise needle stick injury.
2. Instruct the staff that while handling waste they should wear appropriate protective clothing, including a water resistant apron, thick gloves, boots or closed-toe shoes, and eye protection.
3. Do not allow to sort waste or open waste containers to sort waste.
4. Educate the staff to wash hands after working with waste or infected material.
5. Before and after examining patient or in between two patients wash hands.
6. Be aware of procedures for treatment of injuries, cleaning of contaminated areas and reporting sharps injuries or accidents.
7. Report sharps injuries to the appropriate personnel.
8. Injuries should be followed up by post-exposure prevention treatment.
9. Head nurse should maintain a log of all accidents.
10. A full course of hepatitis B and tetanus vaccination will protect from the hepatitis -B virus and tetanus.

Health workers are at risk of accidental needle stick or other injuries from sharps. World health organisation (WHO) recommends the following steps after a needle stick injury:

1. Wounds and skin sites exposed to blood or body fluids should be washed with soap and water, and mucous membranes flushed with water.
2. If blood or body fluids have gotten into eyes, splash eyes with clean water.
3. Immediately report the incident to a designated person or head nurse.
4. Retain, if possible, the item involved in the incident, get details of its source for identification of possible infection.
5. Seek additional medical attention in an emergency health department as soon as infection identified (based on body substance and severity of exposure).
6. Get blood tests or other tests and counselling, if indicated.
7. Record the incident.
8. Investigate the incident and identify and implement remedial action to prevent similar incidents in the future.

Health workers need to protect themselves by establishing a barrier between themselves and the infective agent. The type of protection needed depends on the worker's activities. Protective clothing must be worn at all times when handling biomedical waste. It must be properly maintained and kept clean. The clothing should not be taken home, must remain at the health facility to avoid possible contamination of the community. Protective clothing includes:

1. Gloves: Always wear gloves when contaminated items are handled. Puncture-resistant gloves should be used when handling sharps containers or bags with unknown contents.
2. Boots or closed-toe shoes: Rubber boots or leather shoes provide extra protection to the feet from injury by sharps or heavy items that may accidentally fall. They must be kept clean. When possible, avoid wearing sandals or shoes made of soft materials.

3. Aprons: Rubber or plastic aprons provide a protective, waterproof barrier to the body.
4. Goggles: Plastic goggles can protect the eyes from accidental splashes.
5. Hand washing: Wash with soap and antiseptic detergent.

The measures that could/should be taken in case of accidental spillages in hospitals is as follows:

1. Evacuate the contaminated area.
2. Decontaminate the eyes and skin of exposed personnel immediately.
3. Inform the designated person who should coordinate the necessary actions.
4. Determine the nature of the spill.
5. Evacuate all the people not involved in cleaning up.
6. Provide first aid and medical care to injured individuals.
7. Secure the area to prevent exposure of additional individuals.
8. Provide adequate protective clothing to personnel involved in cleaning-up.
9. Limit the spread of the spill.
10. Neutralise or disinfect the spilled or contaminated material if indicated.
11. Collect all spilled and contaminated material. Sharps should never be picked up by hand, brushes and pans or other suitable tools should be used. Spilled material and disposable contaminated items used for cleaning should be placed in the appropriate waste bags or containers.
12. Decontaminate or disinfect the area, wiping up with absorbent cloth. The decontamination should be carried out by working from the least to the most contaminated part, with a change of cloth at each stage. Dry cloths should be used in the case of liquid spillage, for spillages of solids, cloth impregnated with water (acidic, basic or neutral as appropriate) should be used.
13. Rinse the area, and wipe dry with absorbent cloths.
14. Decontaminate or disinfect any tools that were used.
15. Remove protective clothing and decontaminate or disinfect it if necessary.
16. Seek medical attention if exposure to hazardous material has occurred during the operation.

If the spillage of mercury occurs then collection of mercury spill and storage aspect is as follows:

1. Remove everyone from the area that has been contaminated with mercury. Keep the heat below 20°C and ventilate the area if possible.
2. Put on face mask in order to prevent breathing of mercury vapour.
3. Remove all jewellery from hands and wrists so that the mercury cannot combine (amalgamate) with the precious metals.
4. Appropriate personal protective equipment (rubber gloves, goggles/face shields and clothing) should be used while handling mercury.
5. Locate all mercury beads carefully. Cardboard sheets should be used to push the spilled beads of mercury together. Mercury should be placed carefully in a container with some water.
6. Never use a broom or a vacuum cleaner.
7. It should not be swept down the drain and wherever possible, it should be disposed off at a hazardous waste facility or given to a mercury-based equipment manufacture.

Financial Management

According to the 'polluter pays principle', all organisations are financially liable for the safe management of any waste it generates. The costs of separate collection, appropriate packaging. and on-site handling are internal to the establishment and paid as labour and supplies costs. The costs of off-site transport, treatment, and final disposal are external and paid to the contractors who provide the service (common biomedical waste treatment facilitator). Where common biomedical waste treatment facility is not available, the costs of construction, operation, and maintenance of systems for managing the waste can represent a significant part of the overall budget of a hospital. They should be covered by a specific allotment from the hospital budget. Certain basic principles should always be respected in order to minimise these costs. Waste minimisation, segregation, and recycling are recommended which can greatly reduce disposal costs. The benefits of producing less waste are evident, and segregation prevents the unnecessary treatment of general waste by the costly methods necessary for waste management.

For government-owned hospitals, the government may use general revenues to pay the cost of the waste management system. For private organisations, they need to implement waste management system from their own resources. Since few years privatisation of waste management system (common biomedical waste treatment facility) is gaining importance and it should be encouraged to reduce environmental pollution in the vicinity of hospitals. For cost estimation all hospitals need to establish accounting procedures to document the costs they incur in managing waste. Accurate record keeping and cost analysis must be undertaken which helps to reduce management cost. All the activities of biomedical waste management should be observed without compromise in its cost involved.

Costs of waste management system where common biomedical waste treatment facility is not available

An initial capital investment is necessary for management of biomedical waste. Cost on the following items has to be taken into account. Plant and equipment (steriliser, shredder, incinerator/deep burial where population is less than 5000 i.e. population in rural areas), utility requirements (fuel, electricity, water, etc.), operation and maintenance, consumables, incinerator building, waste storage room, offices, waste collection trucks, bins/containers/bags for transporting waste from hospitals to incinerator site, trolleys for collecting waste bags from wards, bag holders to be located at all sources of waste in hospitals, weighing machines for weighing waste bags, protective clothing, disinfecting solution, soap to wash hands and mutilating agents. The indirect operating costs involves training, replacement of parts, consumables, vehicle maintenance, uniforms and safety equipment, ash disposal, compliance monitoring of flue-gas emissions, project management and administrative costs for the organisation responsible for the execution and long-term operation of the project.

Costs of waste management system where common biomedical waste treatment facility is available

When the common biomedical waste treatment facility is available, the cost of coloured bags/bins/containers, trolley for transporting the waste to temporary storage place, mutilating agents, protective clothing, disinfecting solution, soap to wash hands needs to be considered. The treatment like autoclave, shredding, etc. will be taken care by common biomedical waste treatment facility along with final disposal of waste.

IMPLEMENTATION OF BIOMEDICAL WASTE MANAGEMENT PLAN

This section explains the implementation of biomedical waste management plan in hospitals where common biomedical waste treatment facility is not available and/or available, in primary health centers and in small hospitals situated in rural areas.

The biomedical waste management is a crucial one which starts from point of generation and ends at point of disposal. Policy on biomedical waste management needs to be evolved on the feasibility option and optimal sustainable treatment technologies. There are various options available for managing biomedical waste and the selection of treatment and disposal depends on the availability and nonavailability of common biomedical waste treatment facility, nature of hospital (large scale or small scale) place where it is situated, etc. The implementation plan for biomedical waste management with various options is as follows:

Where Common Biomedical Waste Treatment Facility is Not Available

The biomedical waste management starts from the point of generation. Waste minimisation options should be considered and adopted. After the waste is generated the immediate step is segregation followed by collection, storage, transportation, treatment and disposal. The path between the two points (cradle to grave) can be segmented schematically as:

1. Identification of areas of waste generation.
2. Categorisation, quantification of waste and minimisation.
3. Segregation, handling and storage.
4. Treatment, destruction and disposal.

The detailed implementation of biomedical waste management plan where common biomedical waste treatment facility is not available is as follows.

Identification of areas of waste generation

To identify areas of waste generation, list out units available in the hospital and a survey of all the units will help to identify waste generation. In almost all the units (out patient, wards, operation theatre, labour room, laboratories, intensive care units, etc.), waste is generated, only difference will be in quantity and category.

Categorisation, quantification of waste and minimisation

Categorise the waste according to biomedical waste (management and handling) rules. The quantification will help in placing the bins/bags of appropriate size, quantity and at appropriate places as close to the source of waste generation. Waste minimisation helps in reducing the burden of waste management in special way.

Waste minimisation practice should be adopted at source of generation (reuse, recycle and reduction). Reuse of chemicals, medical equipments, etc. translates into cost saving. Recycling of specific materials like disinfected and shredded plastic helps a secondary industry. Reduction in waste generation decreases waste disposal costs. The ten categories of biomedical waste mentioned above are as follows:

1. Human anatomical waste (body parts, organs, human tissues, etc.).
2. Animal waste (animal tissues, organs, body parts, carcasses, bleeding parts, fluid, blood and experimental animals used in research, waste generated by veterinary hospitals, colleges, discharge from hospitals, animal houses).
3. Microbiology and biotechnology waste (wastes from laboratory cultures, stocks or micro-organisms live or attenuated vaccines, human and animal cell culture used in research and infectious agents from research and industrial laboratories, wastes from production of biologicals, toxins, dishes and devices used for transfer of cultures).
4. Waste sharps (needles, syringes, scalpels, blade, glass, etc. that may cause puncture and cuts. This includes both used and unused sharps).
5. Discarded medicines and cytotoxic drugs (waste comprising of outdated, contaminated and discarded medicines).

6. Soiled waste (items contaminated with blood, and body fluids including cotton, dressings, soiled plaster casts, lines, beddings, other material contaminated with blood).
7. Solid waste (waste generated from disposable items other than the waste sharps such as tubings, catheters, intravenous sets, etc.).
8. Liquid waste (waste generated from laboratory and washing, cleaning, housekeeping and disinfecting activities).
9. Incineration ash (ash from incineration of any biomedical waste).
10. Chemical waste (chemicals used in production of biologicals, chemicals used in disinfection, as insecticides, etc.).

Segregation, handling and storage

Segregation

Segregation is a very important factor in waste management system. Depending upon the treatment and disposal option for various categories of wastes, specific coloured containers are required to segregate and store it at temporary central storage place till it is disposed off. The disposal should be within 48 hrs. The waste which goes for incinerator or deep burial, should be collected in yellow plastic bag or bin. The waste which is planned for autoclaving or microwaving or chemical treatment and finally to find its way in secured landfill or for recycling, should be collected in red or blue bin or bag. The waste sharps such as needles, blades, etc. which is for disinfection, destruction or shredding should be collected in white puncture proof translucent container, which will be encapsulated or can go for recycling as final disposal. The chemical waste (solid), outdated medicines and cytotoxic drugs which goes for disposal in secured landfill should be collected in black bin or bag with cytotoxic label. All the bins and bags should have biohazard label except on black coloured bin or bag on which cytotoxic label to be inserted. The details of segregation of waste into specific colour coded bags or bins, as per treatment and disposal option planned is presented below. Maximising segregation is very effective in reducing waste management costs, environmental impacts and also complexity of management.

Handling

Handling of waste needs attention. As soon as the waste is generated it should be segregated into specific colour coded containers or bags. When these are 3/4th filled then it should be picked up from the neck and placed so that bags can be picked up by the neck again for further handling. While handling care should be taken to reduce the risk of needle prick injury and infection. No other forms of waste should be mixed with biomedical waste. The waste should not be overloaded while transporting. The movement of waste in the wheeled trolleys, containers or carts should be through predefined route within the hospital till it reaches central temporary storage place. These trolleys should not be used for any other purpose and need to be cleaned daily.

Storage

Storage location for hospitals/healthcare waste should be designated inside its premises. The waste in the bags or containers should be stored in central storage place in an area or room of a size appropriate to the quantities of waste produced and the frequency of collection. Recommendation for storage facilities within the hospitals is as follows:

1. The storage area should have an impermeable, hard-standing floor with good drainage; it should be easy to clean and disinfect.

2. There should be a water supply for cleaning purposes.
3. The storage area should afford easy access for staff in charge of handling the waste.
4. It should be possible to lock the store to prevent access by unauthorised persons.
5. Easy access for waste-collection vehicles is essential.
6. There should be protection from the sun.
7. The storage area should be inaccessible for animals, insects, and birds.
8. There should be good lighting and at least passive ventilation.
9. The storage area should not be situated in the proximity of fresh food stores or food preparation areas.
10. A supply of cleaning equipment, protective clothing, and waste bags or containers should be located conveniently close to the storage area.

Cytotoxic waste should be stored separately from other healthcare waste in a designated secure location.

Treatment, destruction and disposal

The various treatment, destruction and disposal methods for each category of waste as per biomedical waste management and handling rules are mentioned below:

1. Human anatomical waste (human tissues, organs, body parts): As soon as it is segregated in yellow coloured bin or bag, before 48 hours it should be incinerated or deep burial. The deep burial option is for towns where population is less than five lakh and in rural areas.
2. Animal waste (animal tissues, organs, body parts, bleeding parts, etc.): As soon as it is segregated in yellow coloured bin or bag, before 48 hours it should be incinerated or deep burial. The deep burial option is for towns where population is less than five lakh and in rural areas.
3. Microbiology and biotechnology waste (waste from Lab, cultures, stocks or specimens human and animal cells, etc.): As soon as it is segregated before 48 hours it should be incinerated or deep burial. The deep burial option is for towns where population is less than five lakh and in rural areas. Other option is disinfect and put it in secured landfill.
4. Waste sharps (needles, syringes, scalpels, blades, glass, etc. that may cause puncture and cuts. This includes both used and unused sharps): After the injection is administered the needles should be cut from the hub by a needle cutter, both the needle and the syringe become useless and can't be reused. The cut needle gets segregated in the pot which is fixed to the needle cutter. The cut syringe goes in the plastic bucket with sieve, which has 1 per cent sodium hypochlorite solution or any other equivalent chemical agent. Metal needle from the pot can be stored in the puncture proof translucent container having 1 per cent sodium hypochlorite solution or any other equivalent chemical agent. It must be ensured that chemical treatment ensures disinfection. The disinfected needle can be encapsulated for disposal into municipal secured landfill or can be given to authorised metal recycler. If auto disabled syringes are provided it prevents the reuse of nonsterile syringes as itself locks after single use. The waste syringes will follow the same route of management of sharps waste.
5. Discarded medicines and cytotoxic drugs (waste comprising of outdated, contaminated and discarded medicines): Either directly incinerate or after destruction put it in secured landfill.
6. Soiled waste (items contaminated with blood, and body fluids including cotton, dressings, soiled plaster casts, lines, beddings, other material contaminated with blood): Either incinerate or disinfect by autoclaving/microwaving and put it in secured landfill.

7. Solid waste (waste generated from disposable items other than waste sharps such as tubings, catheters, intravenous sets, etc.): Destroy the plastic waste to ensure prevention of reuse and disinfect by keeping in 1 per cent sodium hypochlorite solution or any other equivalent chemical agent. It must be ensured that chemical treatment ensures disinfection. If recycling of plastic waste is planned, care should be taken to give to authorised recycler only after disinfection and shredding.

8. Liquid waste (waste generated from laboratory and washing, cleaning, housekeeping and disinfection activities): The liquid waste generated from labs and washing, cleaning and house keeping need to be treated to the standards prescribed and flush in the drains.

9. Standards for liquid waste: The effluent generated from the hospital should conform to the following limits: pH 6.3–9.0, suspended solids 100 mg/l, oil and grease 10 mg/l, BOD 30 mg/l, COD 250 mg/l, bioassay test 90 per cent survival of fish after 96 hrs in 100 per cent effluent. These limits are applicable to those hospitals which are either connected with sewers without terminal sewage treatment plant or not connected to public sewers. For discharge into public sewers with terminal facilities, the general standards as notified under the Environment (Protection) Act, 1986 shall be applicable.

10. Chemical waste (chemical used in production of biological, chemicals used in disinfection, as insecticides, etc.): Chemical waste that is chemical used in production of biological, chemicals used in disinfection, as insecticides, etc. should be treated by using 1 per cent sodium hypochlorite solution or any other equivalent chemical agent. It must be ensured that chemical treatment ensures disinfection. After treatment discharge into drains for liquids and secured landfill for solid.

As per the guidelines issued by central pollution control board disposal of biomedical waste by individual hospitals is discouraged and common biomedical waste treatment facilities are encouraged. Pictorial representation of detail implementation plan of action with various technological options categorywise is presented below.

Provision of common biomedical waste treatment facility (CBMWTF) if in course of time comes up has also been considered and provided in the implementation plan. The details of categorywise treatment and disposal methods are presented in Table 30.5.

Table 30.5. Categorywise treatment and disposal.

Category	Treatment and disposal
1. Human anatomical waste	No treatment required
2. Animal waste	No treatment required
3. Microbiology and biotechnology waste	No treatment required
	Autoclaving/microwaving, municipal secured landfill
4. Waste sharps	Mutilating/shredding/disinfection and encapsulation municipal secured landfill
	Mutilating/shredding/disinfection and nonencapsulation, possibility of recycling shall be explored
5. Discarded medicines and cytotoxic	No treatment required
	Destruction, municipal secured landfill
6. Soiled waste (cotton dressings, etc.)	No treatment required
	Autoclaving/microwaving, municipal secured landfill

(Contd...)

Category	Treatment and disposal
7. Solid waste (tubing, catheters, etc.)	Disinfection/autoclaving/microwaving/mutilating/shredding recycling or municipal secured landfill
8. Liquid waste	Disinfection by chemical treatment, and discharge into drain
9. Incineration ash	No treatment required, disposal in municipal landfill/secured landfill
10. Chemical waste (chemicals used in production of biological, chemicals used in disinfection, etc.)	Chemical treatment, and discharge into drains for liquids and secured landfill for solids

Hospital wastes are unique in many ways. The waste volume is small relative to that of industrial facilities, but there are a large variety of wastes. Hospitals use toxic chemicals and hazardous materials for diagnostic and treatment purposes. The wastes generated by general and surgical hospitals include:

1. Chemotherapy and antineoplastic chemicals
2. Formaldehyde
3. Photographic chemicals
4. Radionuclides
5. Solvents
6. Mercury
7. Anaesthetic gases
8. Infectious wastes
9. Other toxic, corrosive and miscellaneous chemicals.

Chemotherapy wastes, including concentrated antineoplastic chemicals mixed with other inert materials, represent the highest volume of hazardous waste in a typical hospital.

MEDICAL AND HOSPITAL WASTE HAZARDS

It is ironical that hospitals which provide succour to the ailing, can also create health hazards. Indiscriminate disposal of hospital wastes is indeed one of the major sources for spread of pollution and infection. Biomedical wastes from hospitals, nursing homes and clinics include a variety of wastes, such as hypodermic needles, scalpel blades, surgical gloves, cotton, bandages, clothes, medicines, blood and body fluid, human tissues and organs, body parts, radioactive substances and chemicals. Some of these contain harmful organisms and disease causing agents. For instance, reuse of discarded syringes/ needles without disinfection can transmit lethal diseases, like AIDS and hepatitis. Similarly, indiscriminate recycling of used cotton, clothes and medicines can pose a host of health hazards. Also, improper incineration of wastes, particularly chlorinated organic compounds, can result in noxious emissions, including dreaded dioxin.

Hence, there is no option but to ensure that the wastes of different categories are properly segregated and rendered harmless through physical separation, disinfection, are disposed of in secure landfill and incinerated, depending on the nature of the wastes. However, biomedical wastes are not handled with the clinical care as needed to avoid the problems otherwise caused. More often than not, mixed biomedical wastes of different kinds are discarded together for disposal along with municipal wastes. A good amount of reusable materials, such as needles, syringes, plastics and bottles, are picked up by rag-pickers, and recycled back into the market, without any treatment for disinfection. The consequences of this can be anybody's guess.

Biomedical waste may be defined as any solid, fluid or liquid waste, including its container and any intermediate product, which is generated during the diagnosis, treatment or immunisation of human beings or animals, in research pertaining thereto or in the production or testing of biologicals, and the animal waste from slaughter houses or any other like establishments. It is of utmost importance that medical waste is managed in an environmentally-sound manner. Environmentally-sound management of medical waste requires proper understanding of risks associated with the disposal of such wastes, and methods for proper segregation, storage, handling, treatment and disposal.

The medical wastes can be categories into—general waste, sharps, bulk human blood and blood products, pathological wastes, isolation wastes, radioactive waste, chemical waste.

Segregation of wastes is the most important prerequisite in the process of wastes management. For the purpose of segregation, biomedical wastes can be broadly classified into the following categories: (i) general or nonhazardous/noninfectious medical waste; (ii) infectious/hazardous medical waste.

Proper segregation of infectious/hazardous wastes from the general waste is important for efficient and economic treatment and disposal. Infectious/hazardous waste is characterised by: (i) the potential of the waste to transmit infection, and (ii) properties of toxicity and/or low level radioactivity.

Medical wastes, except the general and nonhazardous wastes, should never be transported with general municipal wastes, and these should be kept separate at all stages. Special vehicles must be used so as to prevent access to, and direct contact with the waste by the transportation operators, the scavengers and the public. The transport containers should be properly enclosed.

There are five broad categories of medical waste treatment technologies: (i) mechanical; (ii) thermal; (iii) chemical, (iv) irradiation, and (v) biological.

Emissions from medical waste incinerators are generated from either waste constituents, components of combustion air or by-products of the combustion process itself. Pollutants that provide cause for concern, include particulate matter, toxic metals, toxic organics, carbon monoxide and acid gases (hydrogen chloride, sulphur dioxide, and nitrous oxide). The infectious wastes which cannot be effectively treated and safely disposed of through other methods should be subjected to incineration.

RESEARCH AND EDUCATIONAL INSTITUTIONS (REI)

Hazardous waste generation at research and educational institutions is different from that of most industrial generators. These institutions use small quantities of a broad spectrum of chemicals. This difference in waste characteristics requires research and educational institutions to employ unique pollution prevention techniques. Pollution prevention options are mostly limited to improved housekeeping and operating practices, rather than process modifications or recovery and recycling. Wastes are generated from:

1. Chemistry and biology laboratories
2. Physics, geology, psychology and engineering laboratories
3. Art programs
4. Plant and maintenance operations
5. Photography and printing operations (e.g. school newspaper, fine arts program, etc.)
6. Agricultural programs.

Waste Description

Most chemical wastes in REI are generated by laboratories. They include hundreds of thousands of different chemicals in quantities of 1 gram to 55-gallon drums. Chemistry departments generate the largest quantity and the broadest spectrum of chemicals.

Typical wastes generated in the chemistry laboratory are listed below:
1. Weak aqueous acid solutions (<10 wt %).
2. Weak aqueous basic solutions (<10 wt %).
3. Concentrated aqueous acid and basic solutions, and related compounds.
4. Flammable, non-halogenated organic solvents.
5. Flammable, halogenated organic solvents.
6. Nonflammable halogenated and non-halogenated organic solvents.
7. Organic acids.
8. Organic bases.
9. Inorganic oxidisers, peroxides.
10. Organic oxidisers, peroxides.
11. Toxic heavy metals.
12. Toxic herbicides and pesticides.
13. Aqueous solutions of reducing agents and related compounds.
14. Pyrophobic and hydrophobic chemicals.
15. Cyanide, sulphide and ammonia containing wastes.
16. Explosive materials.
17. Radioactive materials.
18. Empty containers.
19. Trash, inert chemicals (nontoxic, non-reactive, non-ignitable, non-corrosive solids).

The biology department is the second highest waste-generating department. Chemicals used in a biology department are less diverse and less in quantity (typically less than 1 gallon per waste type). Other departments with laboratories are geology, physics, psychology and engineering.

Art, printing, photography and maintenance departments also generate hazardous wastes. Wastes from art departments include paints, thinners, various solvents and heavy metals.

Printing operations generate waste inks, solvents and paper. Photographic processing produces waste silver, bath chemicals and cleaning solutions. Maintenance operation generates waste oils, solvents, pesticides and waste treatment chemicals.

Pollution Prevention Options

Because of the nature of the operation and the nature of the wastes generated in educational and research institutions, the pollution prevention options focus on improved housekeeping and operating practices. Methods of pollution prevention through better management practices include:
1. Identifying common users of particular chemicals. This will increase the sharing of chemicals between departments and instructors.
2. Identifying high volume chemical users.
3. Notifying users of chemicals with limited shelf life to use up old stock before ordering or using new stock.
4. Implementing efficient purchasing policies, such as arranging for staggered deliveries and the purchase of chemicals in several small bottles rather than single large bottles.
5. Encouraging responsible participation by chemical suppliers by ordering chemicals from suppliers who can provide quick delivery of small orders, who will accept returned unopened stock and who offer offsite waste management outlets or cooperatives for laboratory wastes.

6. Monitoring chemical flows from the time they arrive to the time they leave as waste.
7. Developing up-to-date listings of unused reagent chemicals that can be recycled for use by other laboratories.
8. Providing education on opportunities for pollution prevention.

Methods of pollution prevention through better laboratory operating practices include:

1. Reducing or eliminating the use of benzene, carbon tetrachloride, methylene chloride, mercury, lead, formaldehyde and other highly toxic chemicals in experiments.
2. Scaling down the volumes of chemicals used in experiments.
3. Increasing the degree of instrumentation in the laboratory.
4. Treating or destroying hazardous waste products as the last step in experiments. This may not require a permit if the step is part of the educational or research program.
5. Replacing oil-based paints with water-based paints in art instruction and maintenance operations.
6. Keeping individual waste streams segregated. Keep hazardous waste segregated from nonhazardous waste, recyclable waste segregated from non-recyclable waste. Minimise the dilution of hazardous wastes.

Most large research institutions have extensive waste management programs. The pollution prevention program should include the following guidelines:

1. Identify all waste streams; review and make recommendations for procedural modifications; provide incentive mechanisms for new ideas, procedures and information exchange.
2. Establish effective planning and procurement practices.
3. Set goals to meet quantitative reduction levels.
4. Establish a baseline with which to compare progress in pollution prevention.
5. Prioritise possible waste treatment options according to cost and environmental problems.
6. Develop a waste reduction plan for each waste stream.
7. Assess economic, technical and regulatory feasibility of plans.
8. Implement those plans that meet cost/benefit goals.

To reduce radioactive wastes, a nuclide with a shorter half-life, low energy, a stable and nontoxic decay product, and minimal amount of extraneous radiation should be selected. Extraneous radiation refers to the production of a type of radiation that is not required in the testing procedure. For example, if a beta emitter is required for a certain test, a nuclide that produces minimal gamma radiation should be chosen. Gamma radiation is hazardous to the patient and is more difficult to contain during handling. Properties of common nuclides used in hospitals are summarised in Table 30.6.

Table 30.6. Properties of radio-nuclides used in hospitals.

Nuclide	Type of radiation	Energies (MeV)	Physical half-Life	Effective half-Life	Daughters
Carbon-14	beta⁻, no gamma	0.156 max	5,730 yrs.	12 days	Nitrogen-14 (stable)
Phosphorus	beta⁻, no gamma	1.7 max	14 days	14 days	Sulphur-32 (stable)
Chromium-51	gamma	0.32	28 days	27 days	Vanadium-51 (stable)
Gallium-67	gamma	0.093 (40%)	78 hrs.		Zinc-67 (stable)
Technetium-99	gamma	0.14	6 hrs.	5 hrs.	Technetium-99 (radioactive)
					Ruthenium-99 (stable)

(Contd ...)

Nuclide	Type of radiation	Energies (MeV)	Physical half-Life	Effective half-Life	Daughters
Indium-111	gamma	0.173	2.8 days		Cadmium-111 (stable)
Iodine-125	gamma	0.035	60 days	42 days	Tellurium-125 (stable)
Tritium	beta⁻, no gamma	0.0186 max	12.3 yrs.	12 days	Helium-3 (stable)
Iodine-131	beta⁻	0.606 max	8 days	8 days	Xenon-131 (stable)
	gamma	0.365			
Cesium-137	beta⁻	1.176 max (7%)	30 yrs.	70 days	Barium-137 (stable)
		0.514 max			
	gamma	0.662			
Barium-137m	gamma	0.662	2.5 min		Barium-137 (stable)
Iridium-192	beta⁻	0.666 max	74 days		Platinum-192 (stable)
	gamma	0.317, 0.468			
Radium-226	alpha	4.78	1,600 yrs.	44 yrs.	Radon-222 (radioactive)
	gamma	0.186			
Cobalt-60	beta⁻	0.318 max	5.27 yrs.	10 days	Nickel-60 (stable)
	gamma	1.17, 1.33			

The effluents generated from the hospital shall conform to the following limits:

Parameter	Permissible limit
pH	6.5 – 9.0
Suspended solids	100 mg/l
Oil and grease	10 mg/l
BOD (3 days at 27°C)	30 mg/l
COD	250 mg/l
Bio-assay test	90 per cent survival of fish after 96 hours in 100% effluent

Note. These limits are applicable to those hospitals which are either connected with sewers without terminal sewage treatment plant or not connected to public sewers at all. For discharge into public sewers having terminal treatment facility, the general standards as notified under the Environment (Protection) Act, 1986, shall be applicable.

SECTION IX

Special Topics

Chapter 31

Environmental Audit of Municipal Solid Waste Management

INTRODUCTION

The management of municipal solid waste has become an acute problem due to enhanced economic activities and rapid urbanisation. Increased attention has been given by the government in recent years to handle this problem in a safe and hygienic manner. In this regard, municipal solid waste management (MSWM) environmental audit has been carried out for Bangalore city through the collection of secondary data from government agencies, and interviews with stakeholders and field surveys. Field surveys were carried out in seven wards (representative samples of the city) to understand the practice and identify the lacunae. The MSWM audit that was carried out functional-element-wise in selected wards to understand the efficacy and shortfalls, if any, is discussed in this chapter.

Solid waste generation is a continually growing problem at global, regional and local levels. Solid wastes are those organic and inorganic waste materials produced by various activities of the society, which have lost their value to the first user. Improper disposal of solid wastes pollutes all the vital components of the living environment (i.e. air, land and water) at local and global levels. Urban society rejects and generates solid material regularly due to rapid increase in production and consumption. The problem is more acute in developing nations than in developed nations, as their economic growth as well as urbanisation is more rapid. This necessitates management of solid waste at generation, storage, collection, transfer and transport, processing, and disposal stages in an environmentally sound manner in accordance with the best principles of public health, economics, engineering, conservation, aesthetics and environmental considerations. Thus, solid waste management includes all administrative, financial, legal, planning, and engineering functions.

The environmentally sound management of solid wastes issue had received the attention of international and national policy-making bodies and citizens. At the international level, the awareness regarding waste began in 1992 with the Rio Conference, where efficient handling of waste was made one of the priorities of Agenda 21.

The Johannesburg World Summit on Sustainable Development in 2002 focused on initiatives to accelerate the shift to sustainable consumption and production, and the reduction of resource degradation, pollution, and waste. Priority is being given to waste minimisation, recycle and reuse, followed by the safe disposal of waste to minimise pollution.

The Government of India has taken many initiatives and implemented new technologies and methods by giving loans for setting up composting plants to encourage proper management of solid waste since the 1960s. The municipal solid waste management (MSWM) problem was compounded with rapid urbanisation. Due to increased public awareness of MSWM, a public litigation was filed in the Supreme Court, which resulted in the Municipal Solid Wastes (Management and Handling) Rules, 2000. Government, for the first time, now has included private organisations in providing this public service.

New methods of storage, collection, transportation, processing and disposal are being explored and implemented. It is necessary to evaluate the current process at this stage to understand if the methods being implemented are suitable for the Indian scenario and to identify the lacunae in the adopted methods. This requires an auditing of all functional elements of MSWM, considering the environmental constraints. An environmental audit of MSWM in Bangalore city was undertaken apart from evaluating the Indian MSWM scenario to understand the shortcomings.

Environmental auditing first began with the principle of 'polluters pay', to prevent liabilities towards the government. The companies voluntarily carried out audits of their operations and processes to prove that their products are environment friendly, with the increasing awareness of the public about environmental protection. Waste audits are undertaken for a variety of reasons, which is to:

1. Ensure regulatory compliance.
2. Compare actual practices to best practice guidelines.
3. Develop baseline generation data.
4. Identify waste minimisation opportunities.
5. Establish sustainable development indicators or bench marks.

In general, there are three different approaches for conducting a solid waste audit, namely:

1. The back end approach, which measures the material generated by the entire facility, i.e. no attempt is made to assess the manner in which the wastes and recyclables are generated within the facility.
2. The activities approach, which tracks the waste and recyclables as they are generated throughout the facility, by performing waste audits within each activity area, e.g. an office, warehouse or cafeteria.
3. The input/output approach, which tracks the material input and output associated with each activity area.

Environmental audit was introduced in India to minimise generation of wastes and pollution. In this regard, a gazette notification was issued by the Ministry of Environment and Forests on March 13th, 1992 and later amended on April 22nd, 1993. This applies to an industry, operation or process requiring consent to operate under Section 25 of the Water (Prevention and Control of Pollution) Act, 1974 or under Section 21 of the Air (Prevention and Control of Pollution) Act, 1981 (14 of 1981) or both, or authorisation under the Environmental Protection Act, 1986 (29 of 1986). The notification requires that an Environmental Statement for the financial year ending on 31st March be submitted to the concerned State Pollution Control Board, on or before 30th September of the same year.

The improvement of solid waste management is one of the greatest challenges faced by the Indian government. The government and the local municipal authorities have taken many initiatives towards the improvement of the current situation. The private sector has been included in the management of the MSW recently. To understand the level of success in the initiatives, it is necessary to carry out an audit. An audit will identify and bring out the lacunae and the loopholes in the current system with respect to the compliance with environmental regulations, occupational health, resource management, pollution prevention systems and occupational health and safety. This could be one of the best ways to increase awareness about the most suitable approaches to MSWM, the issues likely to be faced and the alternative measures that can be adopted, considering the local scenario.

Developed countries have provided technical assistance in SWM to developing countries focusing on SWM as a technical problem with the assumption that the solid waste problem can be solved with mechanisation.

The 'blind technology transfer' of machinery from developed countries to developing countries and its subsequent failure has brought attention to the need for appropriate technology to suit the conditions in developing countries (type of waste, composition, treatment, etc.). Composition of the waste provides a description of the constituents of the waste; this varies widely from place to place as is evident from Table 31.1. The most striking difference that can be seen is the difference in organic content which is much higher in the low income countries than the high income countries, while the paper and plastic content is much higher in high income countries than low income countries. This shows the difference in consumption pattern, cultural and educational differences. In higher income countries disposable material, magazines and packaged food are used in higher quantities; this results in the waste having higher calorific value, lower specific density and lower moisture content. In the case of lower income countries, the usage of fresh vegetables to packaged food is much higher and mostly materials that are reusable are used. This results in a waste composition that has high moisture content, high specific weight and low calorific value.

Table 31.1. Relative composition of household waste in low, medium and high-income countries.

	Parameter	Low-income countries	Medium-income countries	High-income countries
Contents physical	Organic (putrecible), %	40–85	20–65	20–30
and chemical properties	Paper, %	1–10	15–30	15–40
	Plastics, %	1–5	2–6	2–10
	Metal, %	1–5	1–5	3–13
	Glass, %	1–10	1–10	4–10
	Rubber, leather, etc. %	1–5	1–5	2–10
	Other, %	15–60	15–50	2–10
	Moisture content, %	40–80	40–60	5–20
	Specific weight, kg/m^3	250–500	170–330	100–170
	Calorific value, kcal/kg	800–1100	1000–1300	1500–2700

MUNICIPAL SOLID WASTE MANAGEMENT (MSWM) AUDIT

Auditing has become an increasingly popular tool to assess the environmental policies, quality of implementation, compliance with national law and regulation, etc. Auditing has also been widely used in India, especially in industries. The most popular audits that are carried out in India are energy audits followed by environmental management systems audits of which a waste minimisation audit is an integral part. Audits on MSWM in India are however, very rare. In western countries however, audits on urban waste management have increasingly been carried out with respect to performance, compliance, risk, monitoring, existence of waste policy, quality of implementation, etc. Most of the countries have established an auditing institution to carry out the above given assessments.

The Estonian Government had carried out an audit to assess the necessary conditions for successful implementation of the waste policy. Reports and questionnaires were used for the audit and it was observed that the management had serious shortcomings, such as insufficient finance, in comparison to the goal, lack of organisation in the management no national waste management plan and poor monitoring. The Audit Institution of Costa Rica had carried out an audit on SWM in two municipalities, with multiple focus such as pollution prevention system, management system and site audit. The audit was carried out by going through the reports, questionnaire interviews, and site surveys. The various aspects that were looked included compliance with national law and regulation, occupational health and safety, operational risk, pollution prevention and resource management. The audit identified that the ministries were not integrated, resulting in repetition of many working plans. The other findings were, insufficient public awareness programs, lack of new methodologies and technologies, insufficient financial support and improper monitoring. The management and control of the dump was investigated with regard to national health legislation and technical regulations. Checklists and site surveys were used as tools for this audit. From this audit, it was observed that there was no urban cleanliness plan charted out by the municipal authority, serious violations of the legislations, no proper monitoring by the supervision agency and delayed closure of the dump.

This section presents an audit of the MSWM in Bangalore city. This would help to disseminate the innovative practices that have been adopted for managing municipal solid waste. The study explores the role of various stakeholders in MSWM, the current practices, the role of each entity, the shortcomings of the current practices and issues to be addressed to improve the condition. Auditing of MSWM involved the following objectives:

1. To review the existing MSWM practices.
2. To audit the MSWM practices, considering the case of Bangalore city.

METHODS

The approach to the case study was mainly qualitative. Information was gathered using a variety of methods to gain a better understanding of the situation, issues, perspectives and priorities. Data collection methods included document/literature review, semistructured interviews, checklists and observation. Different types of audits were carried out to achieve various objectives:

1. Compliance audit: To check if the current waste management process is being carried out as per the legislation.
2. Operational risk audit in combination with pollution prevention audit: To check the frequency with which an environmental damage occurs and what the consequence of it is. The measures that have been taken against these possible environmental damages were verified.
3. Resource management audit: To check the optimal utilisation of water, energy and material resources.
4. Occupational risk audit: To verify the measures of occupational safety.

A CASE STUDY—ENVIRONMENTAL AUDIT OF MSWM IN BANGALORE CITY

The city of Bangalore (12.97°N and 77.56°E), the state capital of Karnataka is located on the southern part of the Deccan Plateau at the border of two other South Indian states, Tamil Nadu and Andhra Pradesh. At an elevation of 900 m, it is known for its mild, salubrious climate. Since the 1980s, Bangalore has enjoyed the reputation of being one of the fastest growing cities in Asia. The Bangalore metropolitan area covers an area of 223 sq km, and is the fifth largest city in India. However, with a burgeoning population and the increasing necessities of the Information Technology (IT) sector, the local authorities are not able to provide the necessary services like solid waste management, water supply, road maintenance, etc. to a satisfactory level. The authorities however have taken initiatives and measures to achieve compliance with regulations and reduce complaints from citizens, especially in the MSWM sector. The case study would help to identify techniques suitable for the present scenario, the lacunae or the loopholes in the adopted methods and the possible alternatives.

The Bangalore City Corporation (BCC), which has 200 wards within its municipal jurisdiction, has a population of 65 lacs accounting for 80 per cent of the total population of Bangalore Urban Agglomeration. This high growth rate can be attributed not only to the extension of the municipal limits of Bangalore City but also to the ever-increasing population.

The amount of waste generated in Bangalore city varies from 2700 MT/day to 3000 MT/day and the composition of waste is given in Table 31.2. The Bangalore Metropolitan Area is, on the whole, divided into 30 ranges and 100 Revenue wards under the jurisdiction of Bangalore Mahanagara Palike (BMP). BMP is responsible for the SWM policy, setting up targets and objectives.

Table 31.2. Physical characteristics of Bangalore MSW.

Organic waste (%)	60
Dust (%)	5
Paper (%)	12
Plastic (%)	14
Glass (%)	4
Metal (%)	1
Biomedical waste (%)	1
Card board (%)	1
Rubber (%)	1
Miscellaneous (%)	1

A list of necessities was listed in a checklist with regard to the specific target and the presence and absence of each was marked for MSWM auditing. Site survey was done in seven representative sample wards (Shivajinagar, Malleswaram, Koramangala, Indian Institute of Science campus (IISc), Hindustan Machine Tools colony (HMT), Airport Road and Chikpet). A checklist was prepared prior to the visit to check the presence or absence of techniques used, safety measures adopted, compliance with regulatory measures, and the pollution prevention system adopted. Interviews with health Inspectors, workers and lorry drivers were done at the ward level. Discussions with range health officers, zonal health officers, the chief health officer and the special commissioner helped in understanding the structure and management of the system, which helped to understand the objectives, strategies, success, failure of strategies and the issues faced while implementing strategies. The site surveys and ward level interviews helped to verify the process and to identify the lacunae in each functional element. Site visits to the Karnataka Compost Development Authority, Terra Firma Biotechnologies, Betahalli dump yard, K.R. Puram dump yard and the quarry site in Bomanhalli were done during the study to understand waste processing and disposal. The current MSWM is explained in detail in Fig. 31.1.

Collection

The most common method of collection in Bangalore city is door-to-door collection, followed by community bin collection. In 2003, the door-to-door collection method was implemented in 60 health wards. As per the BMP, all wards in the city are supposed to have door-to-door collection and all the community bins have been removed. However, during the site survey, it was observed that many of the wards still have community bins that are in a very dilapidated state. A large quantity of organic waste is generated from 25 commercial vegetable markets. This waste is collected using separate trucks every morning and evening.

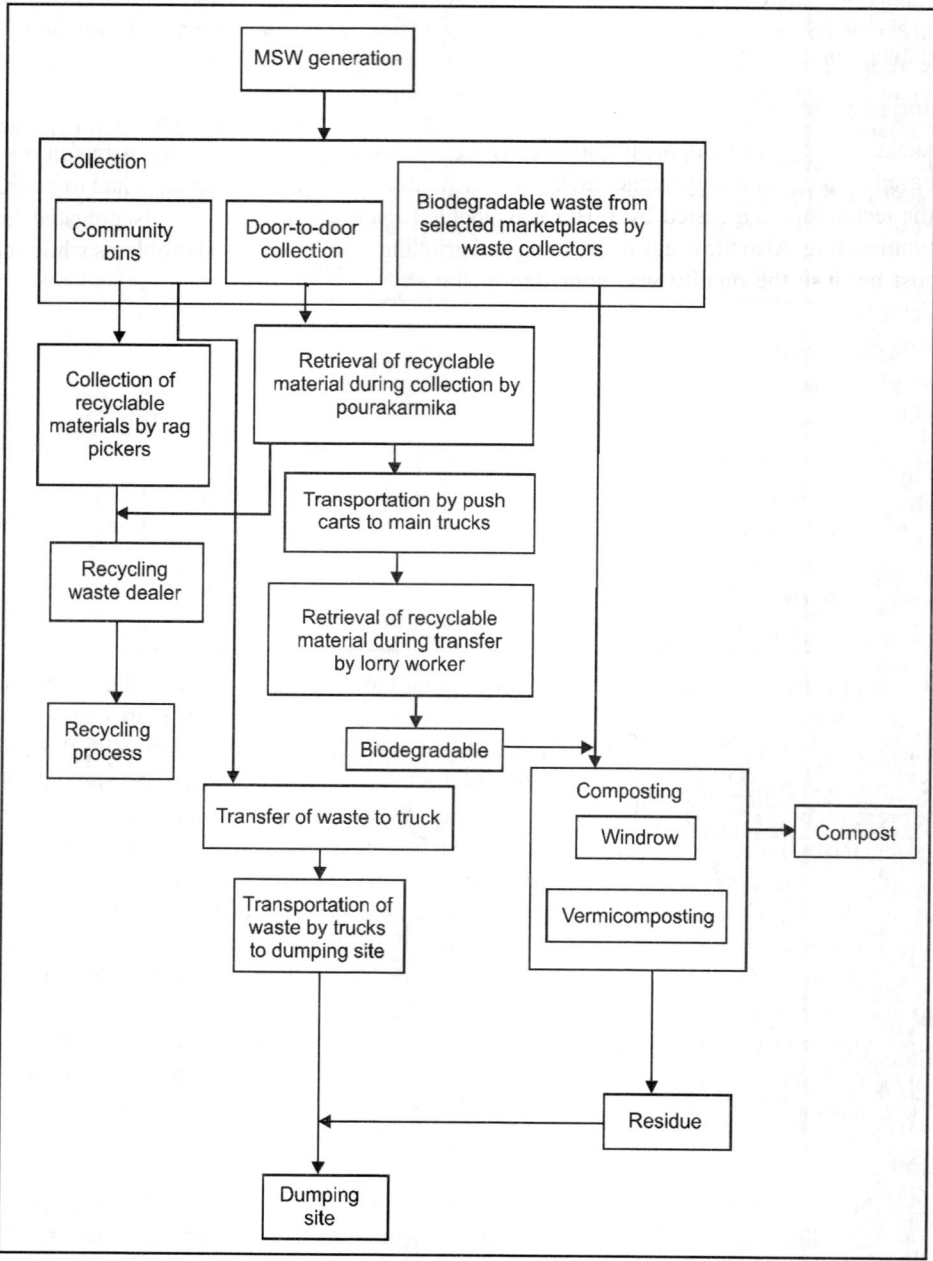

Fig. 31.1. Current MSWM practice in Bangalore city.

The waste collected in pushcarts from lanes is transferred to a truck at a meeting point called a synchronisation point. The truck arrives at the designated point at a specified time and place. The waste is transported to the disposal site by means of a large capacity tipper truck, and in a few wards by a small capacity tipper truck or dumper placers. The truck is covered with a mesh and a polythene sheet to prevent scattering. Currently, Bangalore city has no transfer stations for intermediate storage of waste and intermediate segregation of waste.

Processing of Wastes

The Karnataka Compost Development Corporation (KCDC) was one of the 15 composting units set up in 1975, based on the technology suggested by WHO. Within a year, 10 of these units had to be closed, because the technology suggested by WHO was unable to successfully handle unsegregated Indian waste for composting. Also the usage of crushing and grinding machines caused problems while selling the compost because the quality was poor due to the existence of glass splinters and other non-biodegradable material. In the 1970s, KCDC processed 50–60 tons of mixed waste per day. By 2002 the capacity was expanded to 150 tons/day. Currently the units process 250 tons/day of mixed waste, plus 50 tons/day of market waste, which is collected using vehicles owned by KCDC. City waste is disposed off at Betahalli (Mavallipuram) dump yard situated 18 km northwest of Bangalore city. The waste is brought in by the municipal and contract lorries. This waste is dumped in the yard in the form of a heap. There are three JCB's (Front End Loaders) in the dump yard for waste levelling. The waste is sprayed with effective micro-organisms (EM) solution, covered with a 10 cm layer of debris and sprayed with water after levelling. The solution used for spraying is prepared by mixing 4 litres of EM solution with 8 kg of molasses or jaggery and 150 litres of water. After mixing it is allowed to stand for 7–8 days, after which the pH reduces to 3.4. The EM stock solution consists of actinomycetes, photosynthetic bacteria, and yeast, Lactic acid bacteria (*Lactobacillus* sp., *Streptococcus* sp., *Streptomyces* sp., *Rhodopseudomonas* sp., *Saccharomyces* sp., *Propionibacterium* sp.), which speed up the degradation process and reduce the volume, the flies and the odour. The observations made on the site are:

1. A large number of rag pickers collect recyclable waste from the landfill and pay a small amount to have access to the waste.
2. There is a recycling dealer in the dump yard who buys the recyclable material from the rag pickers and there is one dealer on the way to the dump yard who buys the recyclable waste from the lorry driver.
3. The levelling of MSW after dumping is not carried out efficiently due to fewer number of front-end loaders.
4. There is emission of methane gas from the dump yard, due to which the waste can be easily set on fire.
5. There is always a queue of at least 5–10 lorries waiting to unload; this is due to the lack of number of front-end loaders to level the MSW.
6. This dump yard has no fencing, weigh bridge or no proper approach roads.

The stakeholders and their responsibilities:

The MSWM system and the relationship among the stakeholders are depicted in Fig. 31.2. Various stakeholders are:

1. Ministry of Environment and Forests is responsible for all of the environmental policies at the national level, including the management of waste. The Ministry has an overview of all the activities of the MSWM sector and makes sure that it is performed well.

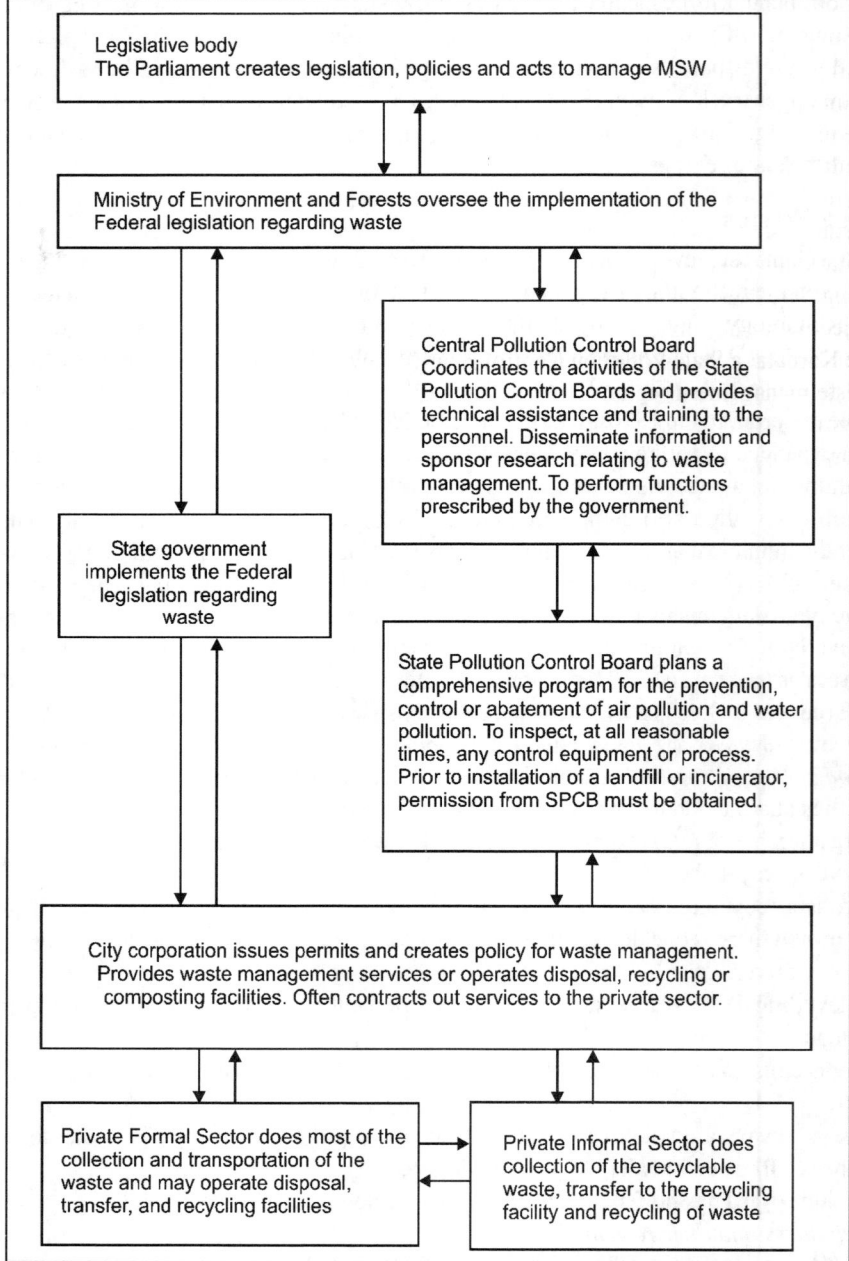

Fig. 31.2. Municipal solid waste management (MSWM) system in India.

2. Central Pollution Control Board keeps a check on all the activities that have potential to pollute the environment, which includes the monitoring of the MSWM in the country. It has divisions in each state that report to CPCB on the environmentally hazardous activities in the state, the

actions taken to check them and the improvements made by the industries and public towards a cleaner environment.

3. Karnataka Pollution Control Board keeps a check on all the activities that have the potential to pollute the environment, which includes the monitoring of the MSWM in the state. It reviews the Environmental Impact Assessment carried out by the agencies prior to the construction of a landfill site, installation of an incinerator or any other processing plant. It carries out public participation meetings to make the public aware of the proposed project and its benefits. Public participation is especially important so that once the project is started there should not be any agitation against the project.

4. Bangalore Mahanagara Palike is responsible for management policy, setting up the targets and objectives. They are responsible for managing the solid waste in the city and are answerable to the Karnataka State Pollution Control Board. They also have the authority to privatise the solid waste management sector.

5. Organisational Structure of the Health Department. The hierarchy of the Health Department in charge of SWM is a pyramidal structure headed by the Chief Health Officer. For effective administration, the city has been divided into three zones namely east, west and south. A Zonal Health Officer administers each zone. There are two Deputy Health Officers to assist him. Each zone consists of ten ranges headed by a Medical Officer of Health. Each Medical Officer of Health is assisted by Senior, Junior Health Inspectors and Sanitary Daffedars. The field worker who is employed in the sanitation work is known as Pourakarmika.

6. NGO, Swabhimana, waste-wise, Swachha Bangalore, Shuchi Mitras, etc. are some non-governmental organisations (NGOs) that support the MSWM. Their functions are stated below:
   (a) They carry out public grievance meetings to identify the problem spots and convey these complaints to the authorities.
   (b) They collaborate with authorities to carry out door-to-door collection of segregated waste.
   (c) They identify public volunteers to monitor the SWM in their respective areas.
   (d) Few NGOs have also set up decentralised composting plants in residential areas and for this they also carry out door-to-door collection and educate the public to segregate the waste prior to disposal.
   (e) They carry out public meetings in schools, colleges, public places, etc. to educate the public about segregation of waste, non-littering, etc.

7. Private formal sector: Currently, out of the 294 health wards in Bangalore city, 182 wards have been given out on a private contract. This includes the functions of collection of waste, transfer of waste to trucks, transport of waste to the specified dump yard. The dump yards that are currently being used are all owned by private entities. They have the responsibility of disposing of the waste by alternative layering of waste and soil, spraying it with EM solution and water. Processing of wastes is done by:
   (a) Karnataka Compost Development Corporation (KCDC), which is a government-aided organisation. This carries out the function of composting (windrow and vermicomposting).
   (b) Terra Firma Biotechnologies, which is a private organisation that carries out vermicomposting.
   (c) Ramky Consultants, which is a private consultancy proposing to setup a sanitary landfill site in Bangalore.
   (d) Srinivas Gayathri Resource Recovery, which is a private consultancy proposing to setup a waste to energy plant and a sanitary landfill site in Bangalore.

(e) Private informal sector: The informal sector in the city is very large and plays a very vital role in the MSWM. It comprises the rag pickers who retrieve recyclable waste from the community bins and landfills, the people who buy recyclable waste from households usually called as 'batli wallas', the middlemen who buy waste from the rag pickers and 'batli wallas' and sell it to either bigger dealers or to recycling factories. Municipal workers like the pourakarmika collect waste from the households and retrieve the recyclable waste; even the lorry workers retrieve the recyclable waste before transferring the waste into the lorry. The waste retrieved by them is sold to the informal sector.

(f) Donor agencies: Development corporation of Norway (DCN), Deutsch Gesellschaft fur Technische Zusammenarbeit (GTZ) and World Health Organisation (WHO) are a few of the international organisations that have sponsored projects in Bangalore. WHO has sponsored large scale composting plants all over India and DCN has sponsored decentralised plants all over Bangalore.

(g) Service users comprise the entire public in the city, including the tourists visiting the city.

RESULTS AND DISCUSSION

The techniques and the shortcomings of the techniques adopted have been identified in all sampled wards and Table 31.3 lists the Malleswaram ward. Door-to-door collection is adopted in Malleswaram area, which has resulted in efficient collection of waste and reduction of littering, foul odour and unaesthetic appearance of bins. However, in commercial areas, due to the absence of community bins, sudden waste, generated at odd hours, is disposed in the street. A few waste heaps can be found on the roadsides in commercial areas. All the trucks that are used for transportation of waste have meshes that prevent littering of waste, but 40 per cent of the trucks have partial polythene cover and 20 per cent have no polythene cover and this results in scattering of waste and foul odour during transport.

Table 31.3. MSWM in Malleswaram.

Function	Shortcoming	Suggestion
Storage		
The waste is stored in households and in shops until it is collected by the door-to-door collector	–	–
Collection		
Door-to-door method adopted in the whole ward, for residential and commercial areas	Seventy per cent of drums are not painted as per the regulations of green for biodegradable, white for recyclable and black for mixed	Painting of drums at regular intervals to make it more convenient to workers
	Segregation not carried out by worker nor householder, though separate bins are provided	Workers accept only segregated waste from households
The recyclable waste is retrieved by the worker and sells it separately to the informal sector	PET bottles and thin plastic bags are not retrieved. The soiled recyclable material cannot be retrieved	–

(Contd ...)

Function	Shortcoming	Suggestion
	Waste heaps found near commercial areas	Placement of large community bins in commercial areas (in commercial area there is a possibility of sudden generation of a large quantity of waste that cannot be stored in the shop till the next day).
		Small litter bins should be provided for the pedestrians in commercial areas and bus stands
Sweeping	The dirt is pushed into the drains which blocks the drains	The workers educated on the affects of blocked drains and regular inspection of drains
	Workers do not use the gloves and footwear that are provided for protection	Mandatory usage of the protection gear provided
Transfer and transport		
The waste collected in pushcarts from narrow lanes and meet at a synchronisation point at a specified time. The waste is transferred from the pushcart to the truck	The waste even if segregated by the workers and stored in separate drums, the waste gets mixed during transfer from pushcarts to lorry. This is because there is no facility in the lorry for separate storage of waste	A small capacity truck and a large capacity truck can be assigned for the collection of dry and wet waste respectively. A better option is to have a partition in a single truck for the collection of segregated waste
The lorry worker retrieves recyclable material during transfer of waste from push cart to lorry	PET bottles and thin plastic bags are not retrieved. The soiled recyclable material cannot be retrieved	Only segregated waste should be accepted to be filled into the lorry
BMP truck—3 large capacity tipper	Mesh covering – 5 trucks, no polythene covering – 1, partial polythene covering – 2 trucks, complete polythene covering – 2 trucks	Trucks completely covered with polythene to prevent scattering of waste and foul odour
Contract truck –1 large capacity tipper	There is leakage of wet waste from truck during transportation	Provision of proper enclosure
Trip truck – 1 large capacity tipper	Foul odour emitted from the waste during transportation	Regular inspections
	The waste is not segregated at an intermediate level and is directly transported to the disposal site	Transfer stations to be provided where waste can be further segregated and higher efficiency for transportation can be achieved by increasing the number of trips made by each truck
	Long distance from ward to dump site, hence only one trip a day is made by each truck	
	Manual transfer of waste	Mechanical loading collection vehicles or proper equipment for transfer of waste
Process	No processing carried out prior to disposal	Recycling of the recyclable material retrieved from waste

(Contd ...)

Function	Shortcoming	Suggestion
		Composting
		High quantity of yard waste generated in the ward and also high quantity of organic waste generated from the market and households
Disposal		
Dump yard in Betahalli	Foul odour, flies and bird menace	Usage of higher quantity of EM solution
	Stray dog nuisance	
	Waste burnt emitting toxic fumes and causing air pollution	Waste burning should be prohibited and strict action should be taken if still continued
	Waste is dumped in heaps causing scattering	Usage of front end loaders for levelling and use soil cover
	Soil contamination	
	The lorry workers and drivers are exposed to diseases	Provision of masks and safety gear
Rag pickers retrieve the recyclable material from the landfill	High exposure to diseases	Provision of masks and safety gear
		Closure of dumpsite and replacement with sanitary landfill

The recycling process is carried out by the informal sector that has resulted in high efficiency of recovery of recyclable material. There is no other process carried out, leading to the entire waste being disposed. There is a large quantity of organic waste that is produced in this ward, including organic waste generated in a market. The waste is disposed off in the Betahalli dump yard, causing foul odour, scattering, leachate formation, and air pollution from burning and methane emission from decomposing organic matter.

In the sampled wards of Bangalore, the waste is stored in open or closed community bins. Out of the community bins present, the average percentage of bins covered is 49 per cent. The collection of the waste is carried out by the community bin method and the door-to-door method. It is essential to have community bins along with the door-to-door collection in commercial areas to avoid littering. The percentage of area covered by community bin in commercial areas is 17.5 per cent.

The door-to-door collection method has been implemented in all areas of the city as it is a suitable method for collection from residential areas and also suitable for collection of segregated waste. In Bangalore door-to-door collection has been implemented in 94 per cent of the residential areas. However, only 3 per cent of the waste is segregated at source. There are currently no transfer stations in Bangalore and all the waste is directly transported to the disposal site. This is very expensive and the efficiency of the trucks is not utilised to the maximum. As per the regulations, all trucks should have mesh and polythene covering.

However, only 96 per cent of the trucks have mesh covering and 41.43 per cent of the trucks have polythene covering.

The quantity of waste processed is very small. The informal sector in the city manages the recycling sector, 18 per cent of the total waste generated is recycled by this sector. The other process method adopted in the city is composting. 3.14 per cent of the waste is reduced through composting. The final quantity of waste sent to the dump yard and quarry (open dump) is 60.71 and 21.14 per cent, respectively. The compliance audit through checklist was attempted and indicate that the regulations being followed by the authorities and private companies responsible for the MSWM. Functional unit-wise compliance of regulations are given below:

1. Storage: From the audit it was observed that the placement of bins has not been done keeping in mind the population density and the quantity of waste generated. There is a lack of community bins in a few of the commercial areas. Due to the high generation of waste in commercial areas, the waste is not always stored on site, but is disposed on the roadsides, causing unaesthetic appearances. Well-designed community bins have to be placed in commercial areas, depending on the quantity of waste generated. The maintenance of the present bins is poor and has resulted in rusted bins having sharp edges. This can prove to be dangerous to the collection staff and also to the users. The staff must be provided with well fitting gloves for safety. Community bins should be provided with a partition for separate collection of waste and proper colouring and labelling on the bins. To improve the separation of waste at source and throughout the MSWM process, adequate staffing, supervision, procedures, training, posters, verbal reminders, reporting, meetings and equipments are required.

2. Collection: Adopting the door-to-door collection method has proved to have many advantages. The complaints from residents due to unaesthetic bins near their houses have stopped, the number of stray dogs and stray cattle has reduced and the no bin system has also improved the waste handling by people or residents. This method is also better suited for collection of segregated waste. However, the door-to-door collection method has its own considerations. In commercial areas, due to the higher quantity of waste generation, the shopkeepers find it difficult to store the waste on site and hence this waste ends up on the street. Though separate drums have been provided for collection of segregated waste, neither the household nor the pourakarmika carry out segregation. This is due to the poor awareness and the general attitude of public and pourakarmikas. The number of awareness programs and training programs carried out by the authorities need to be increased and should be at a regular frequency. It has to be kept in mind that such practices are not easy to instal and will take many months, or even years, to implement. Here again, adequate staffing, supervision, procedures, training, posters, verbal reminders, reporting, meetings and equipment are required to make it possible. The participation of NGOs in such programs can prove to be very helpful to the authorities in making this a success. During door-to-door collection the pourakarmika manually segregates the waste. It is very important that this is carried out with proper protection. The staff should be provided with gloves, footwear, apron, masks and goggles for safety, as they are constantly exposed to waste every day.

3. Transfer and transport: The innovative idea of synchronisation that has been adopted by the municipality to transfer waste from pushcarts to trucks has proved to be successful. This has reduced the spillage, no space is occupied for intermediate storage and collection happens on time as the workers and trucks have to meet at a specified time and location for the transfer. The

transfer of small drums is also much easier and safer than the transfer of waste from large community bins. The trucks that are currently used do not have provision for separate collection of waste. This results in the mixing of waste even if the waste is collected separately. Trucks can either be provided with partition or two trucks can be provided—one truck for the collection of organic and mixed waste and another truck for collection of recyclable waste. The truck for recyclable waste can have a frequency of once in three days as the quantity of recyclable waste generated is less when compared to organic waste. Transfer of waste is carried out manually so it is very important to have proper safety gear like gloves, apron, masks and goggles during transfer. The vehicles used for the transportation of waste should be in a good condition. Most of the trucks have a mesh covering and about 50–60 per cent also have polythene covering. However, there is no proper enclosure provided to prevent the wet waste from leaking on to the road. It is very essential that all trucks have mesh and polythene covering with a proper enclosure to prevent scattering of waste, foul odour and leakage while travelling on crowded roads.

4. Treatment process: The only treatment option that is provided for Bangalore city is composting. This is carried out only for 400 MT/day while the total amount of waste generated is about 2300 MT/day. There have been proposals for setting up three integrated waste management sites that have composting and sanitary landfills. This action needs to be hastened to prevent the excessive damage being caused by open dumping of large quantities of waste every day. Other treatment options also should be considered like decentralised anaerobic digesters near markets. This will not only produce biogas but also reduce the transportation cost of waste to landfill sites. Waste to energy plants like production of refuse derived fuels and incineration plants can be set up to use waste from commercial areas once the source segregation process is set in place.

5. Disposal: In the current MSWM system, the function that has been totally ignored is that of final disposal of MSW. The current method of disposal adopted, as explained earlier, is extremely hazardous to the environment and can cause irreversible damage to the surrounding areas. Unauthorised open dumping of waste is also carried out near crowded slum areas. This is extremely hazardous to the people living around that area. The identification and closure of such dumps should be given the topmost priority. The setting up of the proposed sanitary landfill sites with integrated composting plants should be hastened.

Some important factors that need to be considered for the overall improvement of the waste management system are:

1. Data management: To improve data management there should be commitment to improving reliability of the data on waste from the staff and authorities. Greater confidence in data will help in monitoring the efficiency of the collection, transportation, process and disposal options. Geoinformatics would help in monitoring the unauthorised activities, by monitoring the number of trucks and trips made by trucks to the specified disposal site.

2. Training and education: Environmental education is a way of increasing understanding of problems, cooperation among stakeholders, environmental entrepreneurship and environmental performance. The training should be a regular feature of MSWM, with hands on training on sorting and collection. After training there should be follow up of the practices.

3. Health and safety programs: It has been a common observation that in Bangalore, maintenance staffs do not use the protection gear that is provided to them. Regular health and safety programs

are required to educate the staff on the ill effects of manual handling of waste, walking barefoot in dump yards and continuous exposure to waste. Regular health check ups should be carried out to monitor the health of the workers.

4. Involvement of the community: Community involvement in waste management monitoring programs like that of Suchi Mitra should be encouraged and more people should be involved in such activities. This increases the environmental awareness of the participants and other people. This is one of the fastest and most effective ways to make the public understand the importance of activities like sorting.

5. Integration of waste pickers: NGOs should organise waste pickers, and, instead of the waste pickers retrieving waste at the dump yard which is extremely hazardous to their health, safer methods of retrieving waste from the source by the waste pickers should be developed. Additionally, the waste pickers should be paid to retrieve waste from process plants and dump yards, instead of them paying to access the waste. Ways of improving the working conditions of the waste pickers and providing safety gear for them should be developed.

6. Planning: The waste management that is carried out currently comprises more low cost measures in order to comply with regulation and avoid public agitation and complaints. There is no environmental management planning that is taken into consideration. Improper planning before setting up the sanitary landfill sites has led to increased public agitation and legal complications that have delayed the projects for a very long period. Although an informal approach to problem solving may have worked reasonably well while the program was relatively small, a more systematic and proactive approach to management is required when the complexity of the program increases. This would help to ensure that requirements are handled in a consistent and professional way and problems are addressed promptly and effectively. This would also ensure that the staff has clear objectives and goals while carrying out their activities.

7. Monitoring: Monitoring during collection, transfer, process and disposal needs to become an integral part of the waste management system. The municipal authority not only has to monitor their own staff's activities but also the activities carried out by the private organisations. The State pollution control board has to carry out regular inspections of the dump yards and stop open dumping as it causes serious air and water pollution problems.

8. Public participation: Currently the main hindrance to the implementation of the sanitary landfill sites is due to lack of information dissemination to the public. It is very essential that before any project is implemented, a public participation meeting be held to make the public aware of the technology used in sanitary landfill and the impacts.

CONCLUSION

The audit has brought out the key issues that need immediate attention and minor lacunae that pose major hindrances in the further process of the system. In the storage function, only 49 per cent of the present bins are covered. In collection, 17.5 per cent of the commercial areas have community bins and 94 per cent of the residential areas have adopted the door-to-door method. With these methods of collection, only 3 per cent of waste segregation has been achieved. There are no transfer stations present and out of the trucks present, only 41.43 per cent have polythene covering. Recycling is carried out mainly by the informal sector achieving a high level of efficiency. 3.14 per cent of waste reduction is achieved through composting and 60.71 per cent of the waste is disposed in dump yards and 21.14 per cent is disposed in open quarry sites.

Waste disposal needs immediate attention and strict monitoring. The setting up of sanitary landfill sites has to speed-up and this needs to be given top priority. The number of treatment process plants has to be increased to manage total quantity of waste generated. Many new techniques have been implemented for storage, collection, transfer and transportation. These techniques have brought about many positive changes and have increased the efficiency of the MSWM system. However, segregation of waste at each step is not being carried out. The segregation of waste during storage, collection and transportation has to be set in place for the efficient running of the process plants. Proper training and education needs to be provided to the workers and public awareness programs should be conducted regularly. The occupational and health and safety measures taken by the authorities are not sufficient. Health and safety programs have to be conducted regularly to check the health condition of the workers in the various areas of MSWM and they should be educated on the health hazards related to their work and the importance of wearing the safety gear.

Chapter 32

Integrated Waste Management

INTRODUCTION

This chapter discusses the integration of treatment and disposal options to introduce the concept of 'integrated waste management'. The different approaches to integrated waste management are described.

INTEGRATED WASTE MANAGEMENT

The treatment and disposal of waste has developed from its early beginnings of mere dumping to a sophisticated range of options including reuse, recycling, incineration with energy recovery, advance landfill design and engineering and a range of alternative technologies, including pyrolysis, gasification, composting and anaerobic digestion. The further development of the industry is towards integration of the various options to produce an environmentally and economically sustainable waste management system.

Integrated waste management has been defined as the integration of waste streams, collection and treatment methods, environmental benefit, economic optimisation, and societal acceptability into a practical system for any region. Integrated waste management implies the use of a range of different treatment and disposal options, including the areas covered in this book, i.e. waste reduction, reuse and recycling, landfill, incineration and alternative options such as pyrolysis, gasification, composting and anaerobic digestion. However, integration also implies that no one option of treatment and disposal is better than another and each option has a role to play, but that the overall waste management system is the best environmentally and economically sustainable one for a particular region (Fig. 32.1).

Environmental sustainability means that the options and integration of those options should produce a waste management system that reduces the overall environmental impacts of waste management, including energy consumption, pollution of land, air and water and loss of amenity. Economic sustainability means that the overall cost of the waste management system should operate at a cost level acceptable to all areas of the community, including householders, businesses, institutions and government. In assessing the most environmentally and economically sustainable system, the local existing waste management infrastructure, such as availability of landfill sites, existing incinerators, the types of waste to be managed, waste tonnages generated, etc. should all be considered.

Figure 32.1 shows that at the centre of an integrated waste management system is the collection and sorting of the waste, since this influences the treatment and disposal options of the waste, for example, recycling, composting, use for energy recovery, etc. Material recycling enables the usable materials of the waste to be removed, e.g. paper glass, metals, etc. in a materials recycling facility. The residual waste may then be processed as refuse derived fuel or combusted in an incinerator to recover energy. The waste may be landfilled to produce landfill gas and energy recovered from the combustion of the derived gas. Biological treatment of the waste via anaerobic digestion to produce a combustible gas or treatment to produce compost, may also be an option. In most cases the treatment options require landfill as a final disposal route for the residual product. An integrated waste management system would include one or all of the above options.

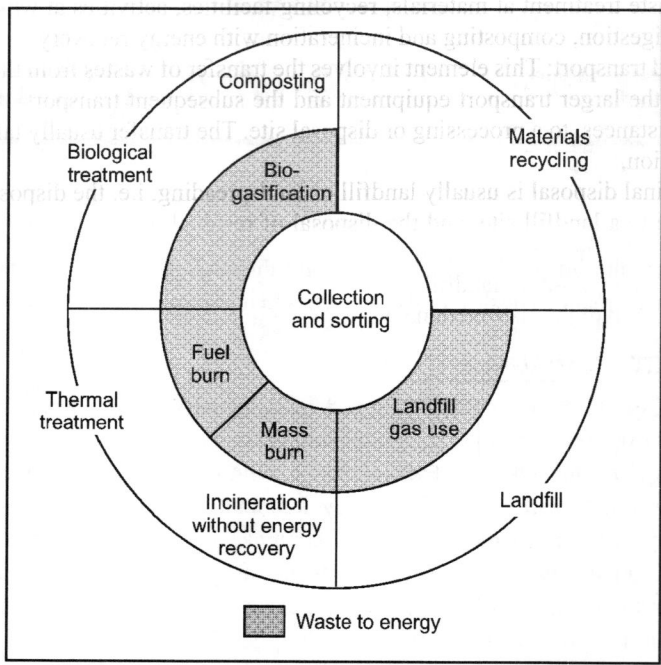

Fig. 32.1. Elements of an integrated waste management system.

Integrated waste management any also be interpreted as integration in terms of the management of wastes from different sources such, as commercial, household and industrial or else in terms of different materials, such as metals, paper and putrescible wastes, or of waste from different product areas, such as packaging waste, white goods, etc. In a truly integrated waste management system wastes such as demolition products, sewage sludge, hazardous, agricultural, industrial and household wastes, would all be included in the waste management system. However, such diverse wastes are often covered by different authorities, are subject to different legislation and arise in different amounts, and are therefore more difficult to integrate than for example 'municipal solid waste'.

Tchobanoglous define integrated waste management in terms of the integration of six functional elements:

1. Waste generation: Assessment of waste generation and evaluation of waste reduction.
2. Waste handling and separation, storage and processing at the source: involves the activities associated with the management of wastes until they are placed in storage containers for collection. This may include source separation of household waste into recyclable and non-recyclable materials. Provision for suitable storage for the wastes, which may encompass a wide variety of different types, is also part of this element. Processing includes such processes as compaction or composting of putrescible materials.

3. Collection: This element of the waste management system covers the collection and transport of the waste to the location where the collection vehicle is emptied. This location may be for example, a materials recycling facility, waste transfer station or landfill disposal site.
4. Separation, processing and transformation of solid waste: The recovery of separated materials, the separation and processing of waste components and transformation of wastes are elements which occur primarily in locations away from the source of waste generation. This category includes waste treatment at materials, recycling facilities, activities at waste transfer stations, anaerobic digestion, composting and incineration with energy recovery.
5. Transfer and transport: This element involves the transfer of wastes from the smaller collection vehicles to the larger transport equipment and the subsequent transport of the waste, usually over long distances, to a processing or disposal site. The transfer usually takes place at a waste transfer station.
6. Disposal: Final disposal is usually landfill or landspreading, i.e. the disposal of waste directly from source to a landfill site, and the disposal of residual materials from materials recycling facilities, residue from waste incineration, residues from composting or anaerobic digestion, etc. to the final disposal in landfill.

The interrelationships of the six functional elements of an integrated solid waste management system are shown in Fig. 32.2.

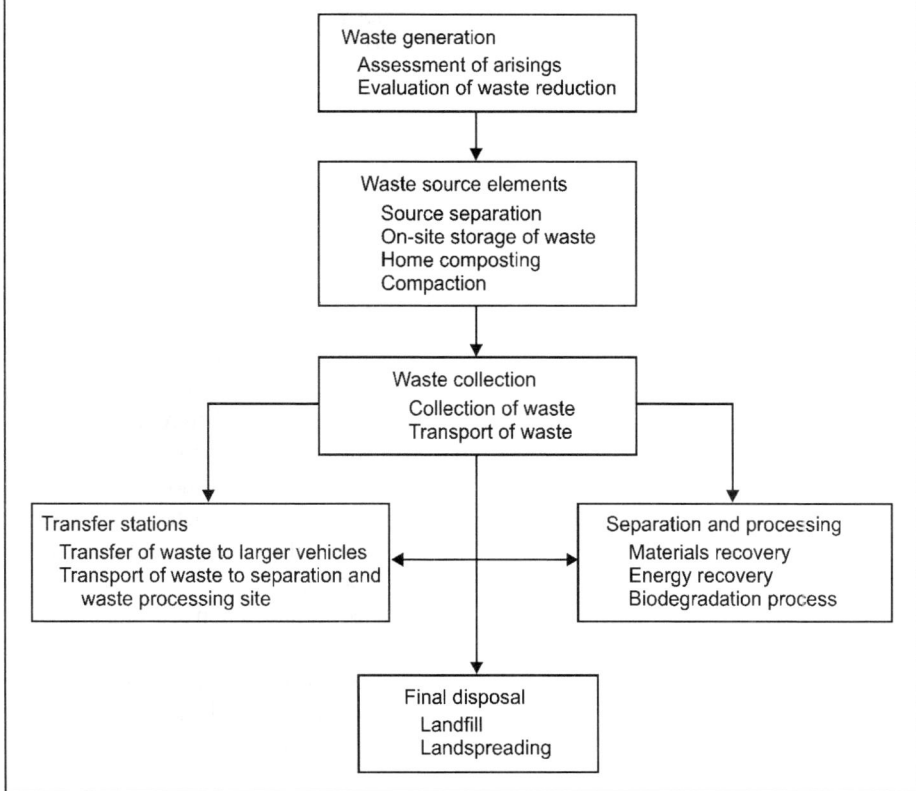

Fig. 32.2. Schematic diagram of the interrelationship between the functional elements in a solid waste management.

Integrated waste management as described by Tchobanoglous involves evaluation of the use of the functional elements and the effectiveness and economy of all the interfaces and connections, to produce an integrated waste management system. They define integrated waste management as the selection and application of suitable techniques, technologies and management programs, to achieve specific waste management objective and goal. This chapter also discusses the concept of ISWM with respect to three perspectives, viz.: lifecycle, waste generation and waste management.

LIFECYCLE-BASED INTEGRATED SOLID WASTE MANAGEMENT

The first concept of ISWM is based on lifecycle assessment of a product from its production and consumption point of view (Fig. 32.3). The reduction in consumption, and utilisation of discarded products within the production system as a substitute for new resources, can lead to reduced end-of-cycle waste generation; thus, less efforts and resources would be required for the final disposal of the waste.

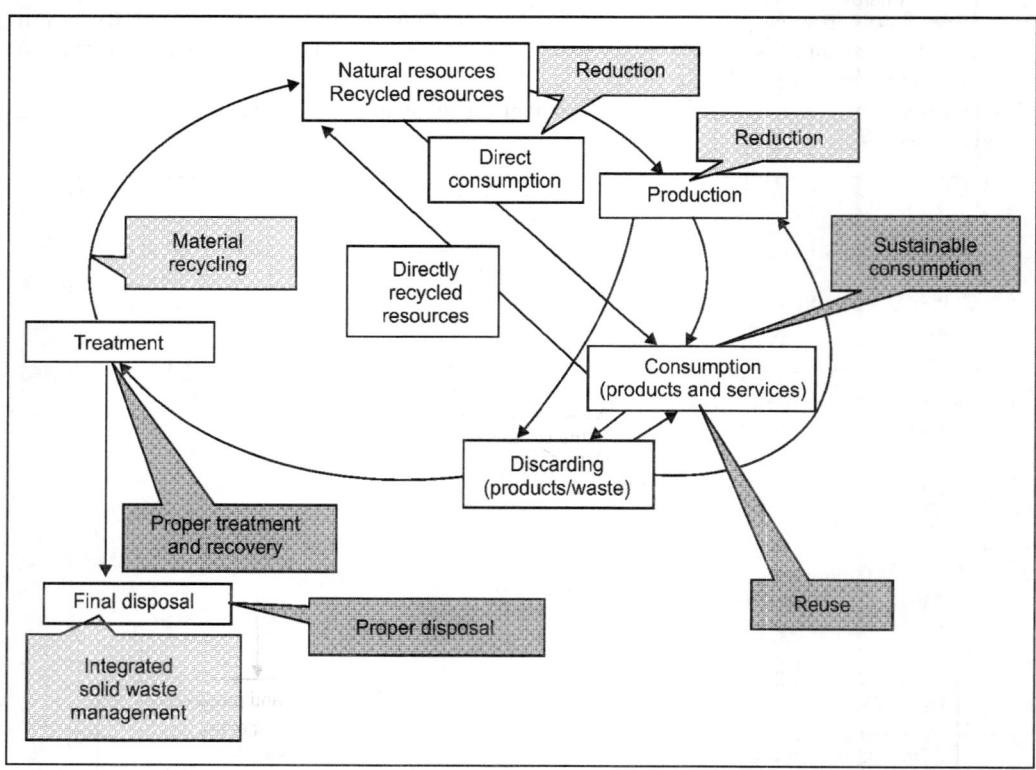

Fig. 32.3. Lifecycle-based ISWM.

GENERATION-BASED INTEGRATED SOLID WASTE MANAGEMENT

The second concept of ISWM is based on its generation from different sources including domestic, commercial, industrial and agriculture. This waste could be further classified as hazardous and non-hazardous waste (Fig. 32.4). The former has to be segregated at source and treated for disposal in accordance with the strict regulations. 3R approach (reduce, reuse and recycle) is applicable both at source as well as at the different levels of solid waste management chain including collection, transportation, treatment and disposal.

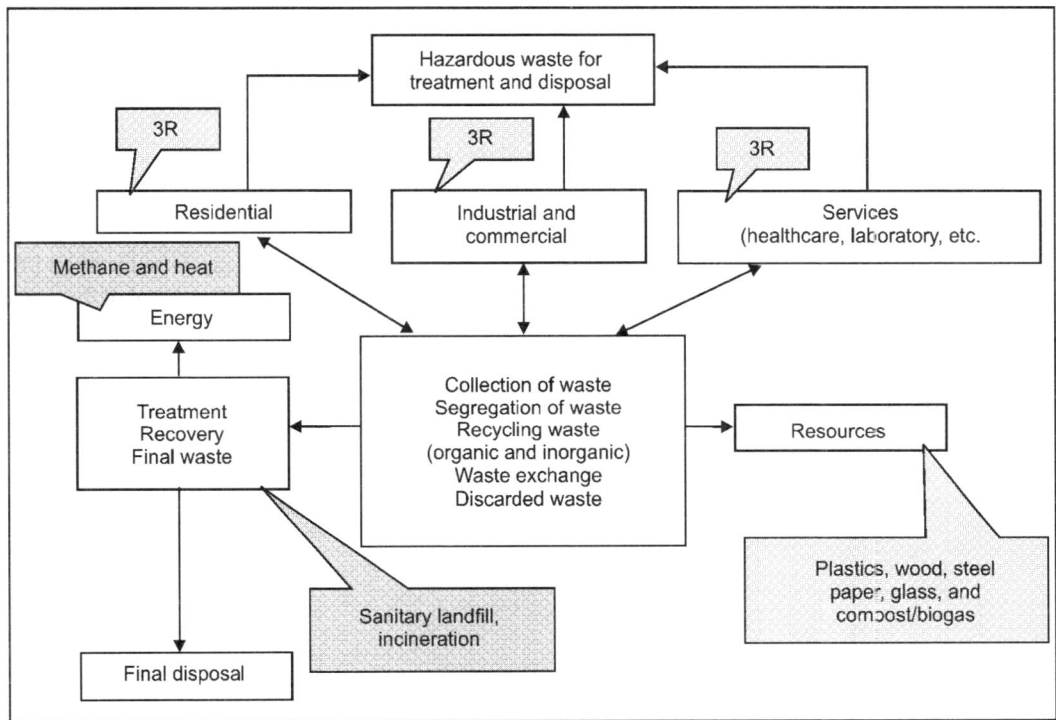

Fig. 32.4. Generation-based ISWM.

MANAGEMENT-BASED INTEGRATED SOLID WASTE MANAGEMENT

The third concept of ISWM is based on its management (Fig. 32.5) which includes regulations and laws, institutions, financial mechanisms, technology and infrastructure, and role of various stakeholders in the solid waste management chain.

ORGANISATION OF GUIDELINES

This is the second set of guidelines that focuses on the assessment of solid waste management systems. The first set, which is available separately, focuses on the quantification and characterisation of solid waste streams from different sources.

The assessment of the management systems may follow the following roadmap:

1. Coverage: In case of SWM, before starting to assess the management systems, it is important to define them. There may be more than one management systems to address solid waste from different sources and/or different types of solid wastes. SWM systems include regulations and laws, institutions, financial mechanisms, technology and infrastructure and stakeholder participation in solid waste management chain.

2. Assessment of individual management systems: If there is more than one management system to handle solid waste, either from different generators or different types of solid waste (hazardous

and nonhazardous), then the individual systems should be analysed separately and the data so obtained could be compared to see their similarities and differences. For general purpose, the guidelines classify solid waste management in three systems, viz.: municipal solid waste, industrial solid waste, and hazardous solid waste. However, there might be fewer or more systems available in a particular city or country. To analyse solid waste management systems, the data and information is required to be collected on the following aspects:

(a) Policies.
(b) Institutions.
(c) Financing mechanisms.
(d) Technology.
(e) Stakeholder participation.

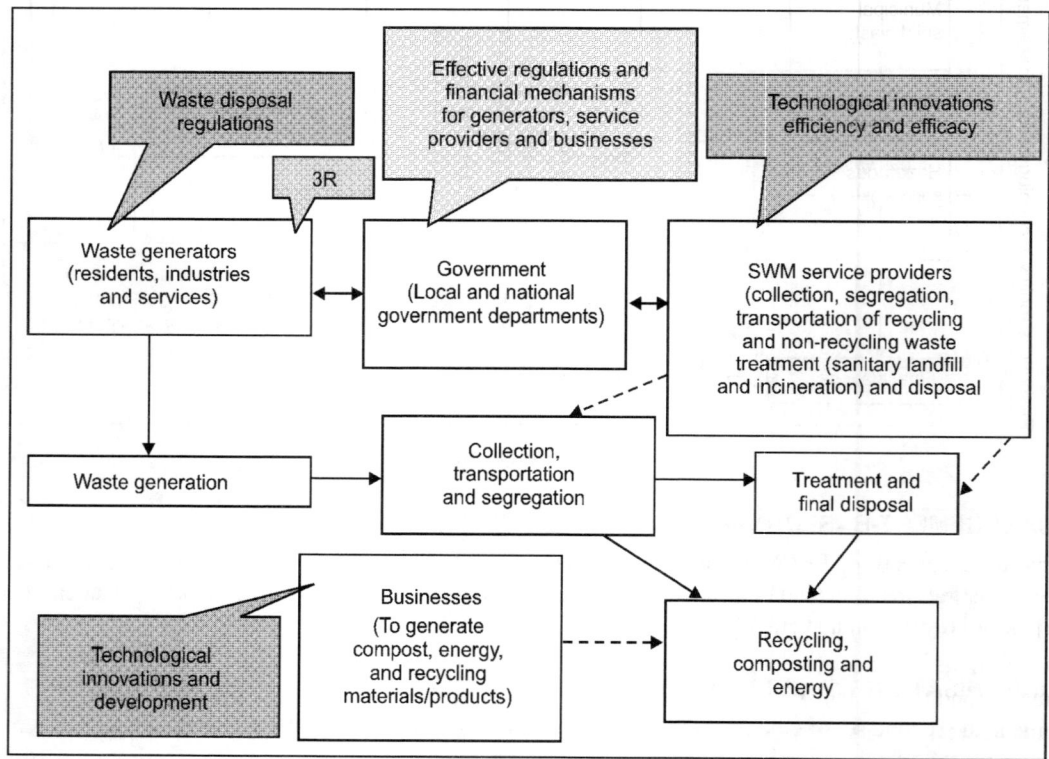

Fig. 32.5. Management-based ISWM.

ASSESSMENT OF SOLID WASTE MANAGEMENT (SWM)

Solid waste management may vary from country to country and city to city. In most countries, the local governments are responsible for municipal solid waste management; however the other two viz.: industrial and hazardous solid waste is the responsibility of the national government. In some places, the local governments with different departments, manages all the three kinds of wastes individually. If there is only one institution or department, responsible for all types of SWM, then it may be considered as a single management system. However, if there is more than one institution responsible for different types of solid waste or waste generated by different sources, then it is considered as separate management systems. Therefore, it is recommended to collect the data and information separately, for different types of systems even if there is some overlapping in terms of regulations and laws, financial mechanisms, technology and infrastructure and stakeholder participation (Fig. 32.6).

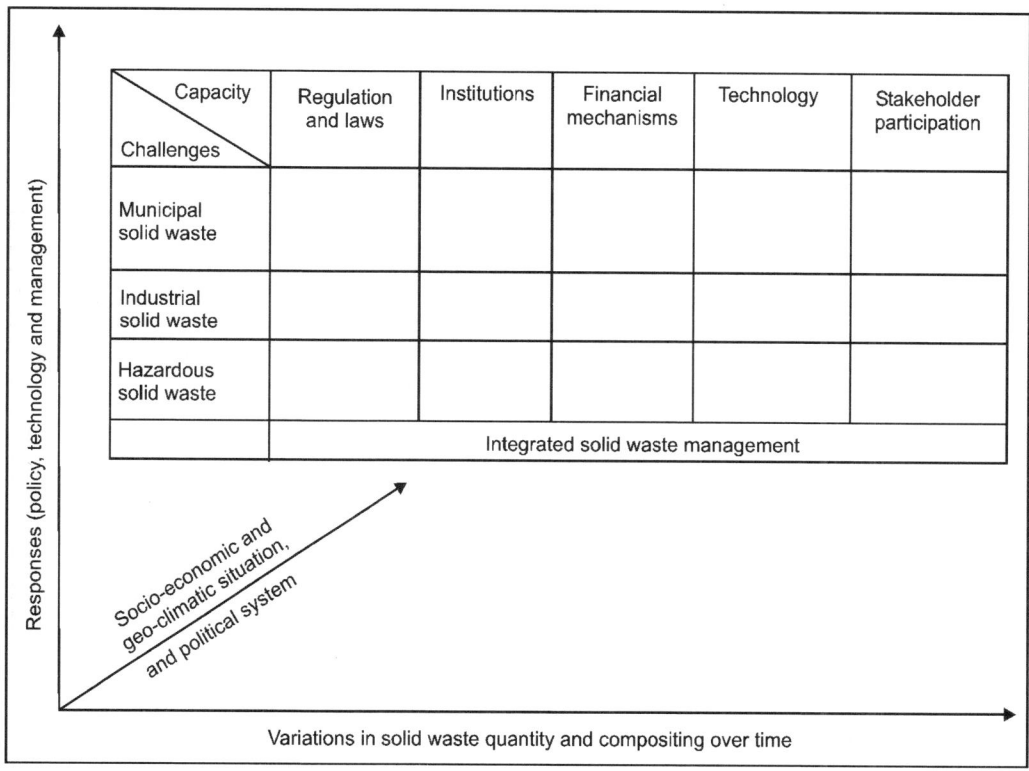

Fig. 32.6. Solid waste management system.

It may also be noted that management systems may be evolved over a period of time depending on the variations in solid waste, political and administrative systems, socio-economic situation, and geo-climatic conditions. Hence, it is useful to capture the evolving process with respect to laws, institutions, financial mechanisms, technology and infrastructure and stakeholder participation.

Importance of Data Collection and Analysis

Development of integrated solid waste management (ISWM) plan demands the assessment of current solid waste management systems apart from their quantification and characterisation which would further be useful for:

1. Analysing the availability, enforcement and impact of regulations and economic tools.
2. Assessing the institutional framework, resources and jurisdictions for current institutions.
3. Analysing the efficiency and effectiveness of collection, treatment and disposal system including technologies.

4. Understanding the role of different stakeholders at different levels of solid waste management chain.
5. Identifying the challenges and opportunities to improve SWM.

Flowchart for Data Collection and Analysis

It may be helpful to create a flowchart for data collection and analysis for all the stages of SWM (Fig. 32.7) to avoid duplication of efforts and to assign a clear role to the team members who are responsible for the same.

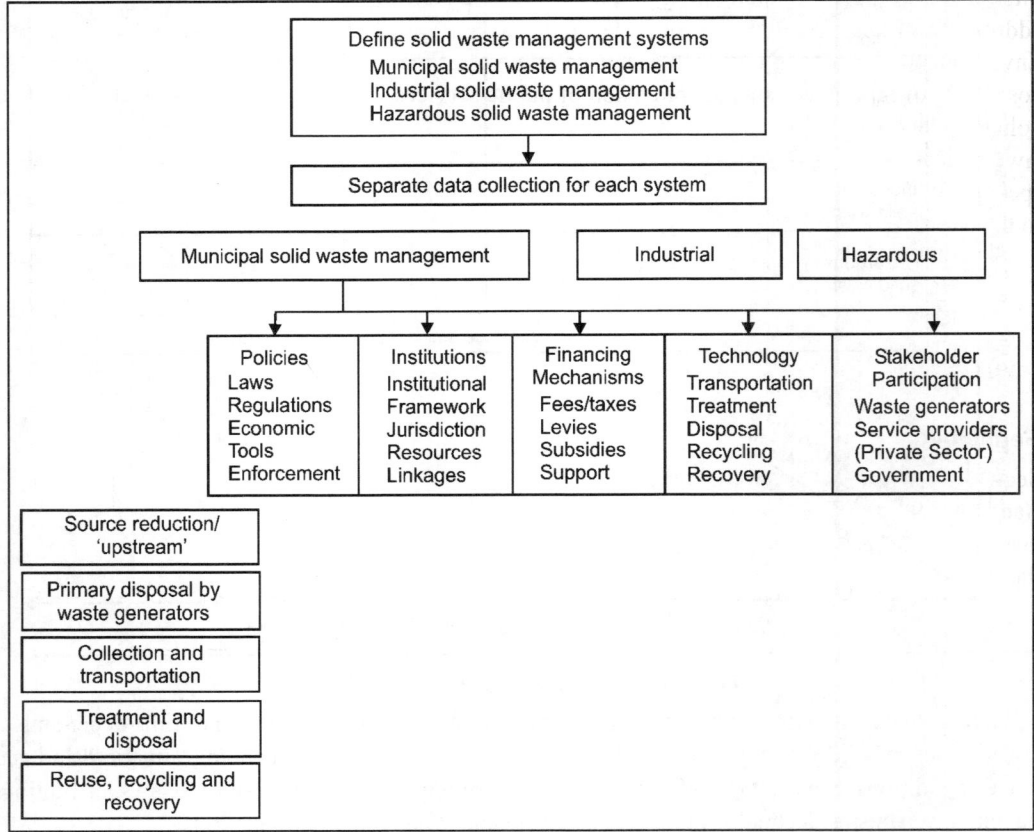

Fig. 32.7. Flowchart for data collection and analysis.

POLICIES

A wide range of policies could be available at international, national, and local level. At international level, various multilateral and bilateral treaties and agreements, including Basel Convention, are available. National policies may have more than one perspective: they may help to improve SWM with respect to local conditions and/or they may assist to comply with international treaties and agreements. Furthermore, local policies could have an importance as in many countries, SWM is a local issue dealt by local governments. The aim of these guidelines is to collect existing national and local policies.

Policies are translated into regulatory and economic instruments for their implementation. The former, also known as command and control, specify the standards or limits to be followed and the latter, also known as market-based instruments, provide incentives and disincentives. There may also be some voluntary instruments agreed by the stakeholders. Regulatory, economic and voluntary instruments may be available either in general or for every stage of solid waste management chain.

Data Collection

Laws and acts

Firstly, the laws and acts pertaining to SWM should be collected. There could be direct laws or acts addressing either overall SWM or a particular aspect of its chain, say, recycling and recovery. Environmental protection laws and acts usually cover SWM at national and local level along with a possibility of other laws, such as provision of public services, covering all or some aspects of it. The policies addressing various economic sectors, such as industries and agriculture, and also the specific laws for healthcare facilities, construction and demolition activities may, directly or indirectly, contain specific clauses on SWM. Laws or acts on SWM may or may not cover hazardous waste management. In that case, some of the separate laws and acts addressing hazardous waste management could be:
1. Environmental protection law/act.
2. Hazardous waste management law/act.
3. Recycling or resource recovery law/act.
4. Clean air act—incineration, landfill gases.
5. Public services act—solid waste management.

Regulations

Secondly, information on all the relevant regulations should be collected which may include various standards covering every stage of existing solid waste management chain. The standards may also be available for technology and infrastructure, for example construction and operation of landfills and incinerators. Some examples of regulations or standards could be:
1. Regulations on production and consumption—upstream measures.
2. Regulations on segregation of recyclable and non-recyclable waste.
3. Regulations on electronics waste.
4. Regulations pertaining to extended producer's responsibility.
5. Regulations on handling of hazardous waste.
6. Regulations on collection and transportation of industrial waste.
7. Regulations on construction and operation of landfills.
8. Regulations on construction and operation of incinerators.
9. Regulations on construction and operation of composting plant.

Economic instruments

Thirdly, the information should be collected on all the relevant economic instruments addressing one or more aspects of solid waste management chain. Financial disincentives (in the form of charges, levy, fine and penalty for waste generators) and Economic incentives (such as subsidies or payback for recycling) could be common economic instruments. Some examples could be:
1. Levy on use of fresh resources in industrial production.
2. Subsidies for recycling in industrial production.

3. Volume-based solid waste fee on non-recyclable waste.
4. Penalties on hazardous waste.
5. Subsidies for resource recovery, including power-generation at landfill.

Enforcement

Enforcement becomes the most crucial aspect of policies for SWM as they could only make a difference if these are properly enforced at all levels. Therefore, an assessment of the level enforcement is vital. But it can become a challenging task as the criteria or benchmarks to ascertain the level of enforcement may not be available and the opinion on the enforcement levels may differ within the different stakeholders. Hence, the opinions from all the major stakeholders should be sought to get a comparatively appropriate assessment.

Datasheet

Based on the collected information on laws, regulations, economic instruments and enforcement, a data sheet should be prepared with the relevant documents annexed.

INSTITUTIONS

Traditionally solid waste management was the responsibility of local governments. However, with the increasing rate of solid waste from diversified, unconventional sources (like industries and laboratories respectively), and awareness and regulations (for recycling and recovery, hazardous waste management and source reduction by intervening at production and consumption level), various institutions got involved into one or more aspects of solid waste management chain. This transition from public to private institutions for undertaking various public utilities and services demanded governments to establish strong regulatory institutions to make sure that the service providers deliver effective and efficient services. There may be more than one institution involved at the same level or for the same type of activity, for example, informal and formal sector for recycling or public and private sector for collection and transportation of municipal waste.

We need to collect detailed information on all the institutions, currently responsible at any level of the solid waste management chain to identify their role or mandate, institutional framework, human resources and sources for financing their activities.

FINANCING MECHANISMS

In many countries, SWM being a local issue, all the financial activities like its annual budget, subsidies from national government, and international cooperation were taken care by the local governments. However, the demands for huge investments, to bring improvements in many aspects of the solid waste management chain, started rising with a rapid increase in waste generation rates and awareness for effective and efficient SWM practices to protect public health and environment. This further paved way for a transition for which the governments started adopting various financing modes and some of the widely practiced ones are as follows:

1. User charges: In many countries, user charges are being introduced. They are still low for municipal sectors but for commercial and industrial sector, the charges could be high to meet the costs in accordance with the polluter's pay principle. However, these charges also motivate waste generators to reduce the waste. Volume-based charges for municipal waste are quite common in some countries.

2. Penalty, fine and levy: This form of direct income is also becoming an important financing tool for governments to finance SWM. The terminology and rate of the penalty/fine/levy may vary from country to country.

3. Environmental bonds: In some countries, these bonds are floated by local governments as a major source to arrange funds for environmental infrastructure and services including SWM and other developmental activities.

4. Environmental fund: Some countries set a revolving fund to assist local governments in meeting their financing needs for environmental infrastructure and services. This fund is financed through various modes including national bonds, annual budget, loans from international financing institutions and international cooperation.

5. Direct loans: Local governments may take direct loans either from domestic or international financing institutions.

6. International cooperation: There is an increasing trend of a direct multilateral and bilateral cooperation with local governments. International agencies are providing support to local governments to improve the local environment. Various bilateral initiatives, including sister cities, are also helping local governments to seek assistance for financing their development projects including SWM.

7. National subsidies: This is still a major source for many local governments to finance environmental infrastructure and services.

8. Annual budget: Local governments allocate substantial portion of their development budget to finance SWM. This is usually cross-subsidised from the profit-making avenues of local governments.

9. Private sector participation (PSP): There is an increasing trend of private sector participation in solid waste management chain. The activities under SWM (collection, transportation, treatment, disposal, recycling and recovery) can be easily separated from each other enabling various organisations to involve in one or more aspects of the chain. There are quite a few established forms of PSP based on the level of investment and ownership as shown in Fig. 32.8. In many countries, the primary collection system, also known as door-to-door collection, has been under community-based private sector where households pay monthly fee to the service provider. Infrastructure, such as landfills and incinerators, are being awarded to private companies on BOT (build-operate-transfer) basis. Franchise is another common way for PSP, where the private sector has the right to collect waste within the agreed location and sell recyclable waste.

Datasheet

The information should be collected on the financing mechanisms for all the activities under solid waste management chain as shown in Table 32.1. There may be more than one organisation and financing mechanism involved for one activity.

TECHNOLOGY

Solid waste management chain requires intensive use of environmentally sound technology (ESTs) for its activities which could be as simple as containers for primary collection to as complicated as incinerators for disposal of hazardous waste. The possible technological interventions within SWM chain are as follows:

1. Primary collection and transfer stations: This may include the waste collection bins for segregated municipal waste and special containers for hazardous waste. Material, construction, labelling

and storage of the collection containers are important. Similarly construction and location of transfer station is also crucial to avoid adverse effects due to odour, breeding of vectors such as flies and mosquitoes, and entry of birds or cats and dogs. The transfer stations should be located and constructed in such a way that it is convenient for small carts to unload solid waste and for bigger vehicles to collect and transport that waste.

2. Transportation: This covers all types of vehicles under operation to transport solid waste from its generation point to the transfer station; and from there to the treatment and disposal site. All the vehicles in operation should be listed out including manually driven small carts, mechanically driven sophisticated transportation vehicles and special vehicles for special wastes—hazardous, bulky and recyclable wastes.

3. Treatment: This includes separation of different types of waste, hence, the technology equipped at this level may enable separation of various types of materials for recovery and recycling, equipment for shredding and treatment of final disposable waste. In some countries, incineration is covered at this level and ash from it is sent to landfill for final disposal. Incineration is a high-tech process— the negative impacts of which could be worse for both public health as well as the environment.

4. Final disposal: Though sanitary landfill is the most common technology around the world, the conventional and environmentally unfriendly methods including open-burning, open-dumping and non-sanitary landfill can still be evidenced. However, in most countries these are officially banned allowing only sanitary landfill for final disposal. Sanitary landfills can be operated with fully aerobic, semi-aerobic and anaerobic methods. The technologies may also vary in accordance with the type of final disposable waste, for example, some landfills may be used for co-disposal of special wastes. The landfills for hazardous wastes could be more complicated and are known as secure landfill. The location of landfill is also an important factor considering the transportation costs and its impacts on the urban environment.

5. Recycling and recovery: This includes various types of activities like recycling of reusable materials (e.g. plastic and glass containers), recycling of materials for industrial production (e.g. paper and iron), converting waste into energy (e.g. burning tyres in cement kiln to produce heat), and converting waste into a resource (e.g. composting and landfill gas). Hence technology can determine the level and sophistication of recycling and recovery activities.

Table 32.1. Financial mechanisms for solid waste management chain (Data sheet).

Type of service	Organisation	Financing mode		
		Direct revenue	Local government/ national government/ international cooperation	Private sector (mention type of PSP)
Municipal solid waste management				
Collection				
Transportation				
Treatment				
Disposal				
Recycling				
Industrial solid waste management				

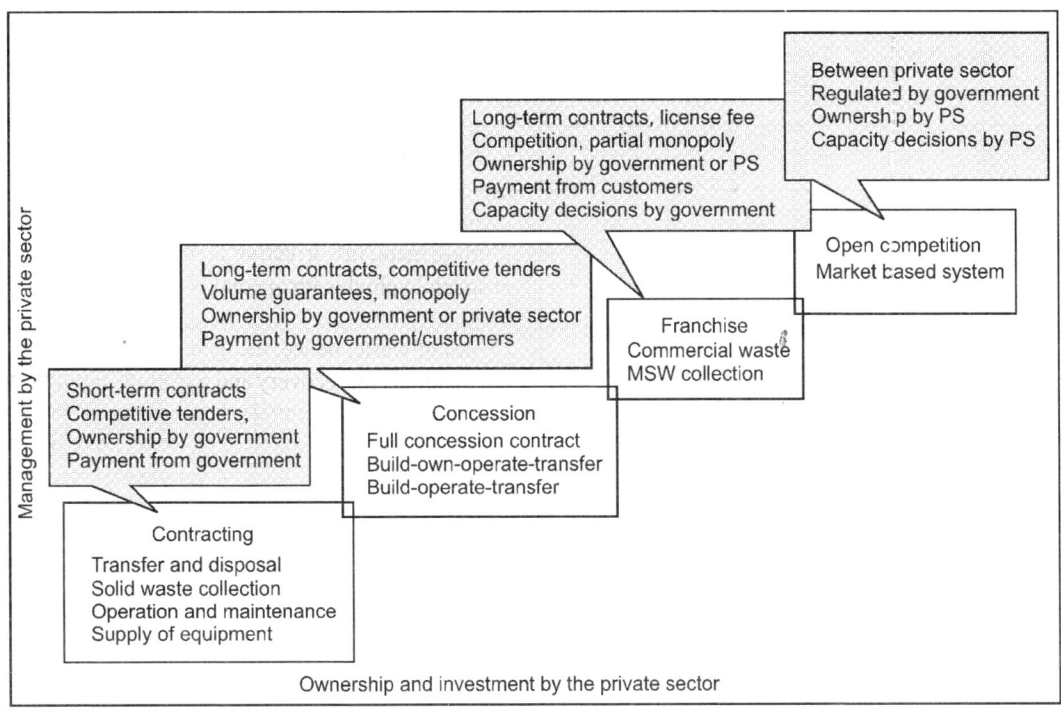

Fig. 32.8. Management and ownerships in various forms of private sector participation.

STAKEHOLDER PARTICIPATION

Stakeholder participation is becoming an essential part of SWM. Major stakeholders include waste generators, regulators, service providers such as organisations involved in waste collection and disposal, and organisations involved in recycling and recovery. Each stakeholder has a specific, clear and active role to improve the efficacy and efficiency of SWM by active participation and continuous interaction. Waste generators — traditionally considered as passive partners — have a major responsibility to reduce, segregate, and properly discard the waste as per the regulations. A close cooperation would be required between waste generators and waste collectors to increase the coverage and effectiveness of the waste collection system, proper disposal of waste, and recycling and recovery of materials. Furthermore, with rapid changes in quantity and composition of solid waste, regulatory organisations or governments have to be in continuous dialogue with the stakeholders to introduce appropriate regulations which can help bring the required improvements in SWM system.

Datasheet

The information on stakeholder participation would be required at two levels. Firstly, to motivate stakeholders to participate with the help of materials, campaigns, meetings, and other political and social interactions and secondly on the ways and means for stakeholder participation, for example, stakeholder representation in decision-making bodies such as a regulatory body and monitoring committee. Stakeholder participation in the decisions to set the level and type of service, such as door-to-door collection or location of transfer station, could also indicate the level of participation. Table 32.2 may be helpful in obtaining and filling in this information.

Table 32.2. Process and level of stakeholder participation in SWM (Data sheet).

Type of service	Major stakeholders	Measures to improve stakeholder participation	Level of stakeholder participation
Municipal solid waste management			
Industrial solid waste management			
Waste management at healthcare facilities			

SECTION X

Case Studies

Chapter 33

Case Studies

INTRODUCTION

This chapter discusses various case studies related to sustainable waste management-Nagpur (Maharashtra); solidwaste management in Jalandhar City (Punjab); environment impact assessment of municipal solidwaste landfill in Jordan; and solidwaste management in Mangalore (Karnataka).

CASE STUDY 1: SUSTAINABLE WASTE MANAGEMENT—NAGPUR (INDIA)

There has been a significant increase in municipal solidwaste (MSW) generation in India in the last few decades. This is largely because of rapid population growth and economic development in the country. The per capita of MSW generated daily in India ranges from about 100 gm in small towns to 500 gm in large towns. The increased MSW generation can be ascribed to our changing lifestyles, food habits and change in living standards. In India the amount of waste generated per capita is estimated to increase at a rate of 1–1.33 per cent annually. It is estimated that the total waste quantity generated in by the year 2047 would be approximately about 260 million tons per year, more than five times the present level of about 55 million tons. The enormous increase in solidwaste generation will have significant impacts in terms of the land required for waste disposal. It is estimated that if the waste is not disposed off in a more systematic manner, more than 1400 sq. km of land which is equivalent to the size of city of Delhi would be required in the country by the year 2047 for its disposal.

In our country municipal corporations are primarily responsible for solidwaste management. But with the growing population and urbanisation municipal bodies are facing financial crunch and can no longer cope with the demands. The limited revenues earmarked for the municipalities make them ill equipped to provide for high cost involved in the collection, storage, treatment and proper disposal of waste. Municipalities are only able to provide secondary collection of waste, means they only collect waste from municipal bins or depots. A substantial part of the municipal solidwaste generated remains unattended and grows in the heaps at poorly maintained collection centres.

Open dumping of garbage facilitates breeding of disease vectors such as flies, mosquitoes, cockroaches, rats and other pests. At present the standard of solidwaste management is far from being satisfactory. The environmental and health hazards caused by the unsanitary conditions in the cities were epitomised by the episode of Plague in Surat in 1994. That triggered public interest litigation in the Supreme Court of India. Based on the recommendations of the committee set up by the apex court in that Public Interest Litigation (PIL), the Government of India has framed Municipal Solidwaste (Management and Handling) Rules, 2000, under the Environmental Protection Act, 1986.

The Municipal Solidwaste (Management and Handling) Rules, 2000 are as follows:

1. Collection of municipal solidwastes: Organising doorstep collection of municipal solidwaste from houses, hotels, restaurants, office complexes and commercial areas.
2. Segregation of municipal solidwastes: Municipal authority shall organise awareness programs for segregating the waste at source as dry and wet waste and promote recycling or reuse of segregated materials.
3. Storage of municipal solidwaste: Municipal authorities shall establish and maintain storage facilities such that wastes stored are not exposed to open atmosphere and shall be aesthetically acceptable and user friendly and it should have easy to operate design for handling, transfer and transportation of waste.
4. Transportation of municipal solidwastes: Vehicles used for transportation of waste shall be covered and waste should not be visible to public, nor exposed to open environment and shall be so designed that multiple handling of wastes prior to final disposal is avoided.
5. Processing of municipal solidwastes: Municipal authorities shall adopt suitable technology or combination of such technologies to make use of wastes so as to minimise burden on landfill.
6. Disposal of municipal solidwaste: Landfilling shall be restricted to nonbiodegradable, inert waste and other waste that are not suitable either for recycling or for biological processing. Landfilling of mixed waste shall be avoided unless the same is found unsuitable for waste processing.

Swachta Doot Aplya Dari: A Scientific and Innovative Approach for Municipal Solidwaste Management

As can be seen from the above guidelines, collection and segregation of municipal solidwaste is a primary requirement for implementation of MSW Rules, 2000. Primary collection of garbage is important to prevent littering of waste on the streets. As per the MSW guidelines, waste has to be collected in segregated form so that it can be recycled to the extent possible by adoption of suitable technology. This recycling will minimise the burden on landfills.

Though doorstep collection of segregated waste is important for municipal solidwaste management, it is not carried out by many of the municipal bodies in the country as they are lacking in financial resources or the expertise to comply with those rules and they often make little effort to revise outdated and deficient waste management systems. As the authorities were hardly able to provide cost-efficient service to citizens, one possibility was to outsource solidwaste management by putting in charge professional private organisations like Centre for Development Communication (CDC). The key concept is a daily door-to-door collection of segregated domestic waste, but the model includes all aspects of solidwaste management from waste generation to waste processing (e.g. recycling and vermicomposting) and the final disposal. The end consumer is both main contributor and main beneficiary, as he should segregate the waste instead of littering it and, in turn, profits from the cleanliness of the city and creation of a new awareness that CDC work is generating. Presently the Swachta Doot project is being successfully being implemented in several cities of India.

The Swachta Doot Project is a major solidwaste management program that includes the following aspects:

1. Daily door-to-door garbage collection.
2. Waste segregation.
3. Garbage lifting and transportation.
4. Employment generation.
5. Awareness building.

Daily door-to-door garbage collection

Daily door-to-door garbage collection is the core of the CDC model. It is most essential for complying with the norms prescribed by the Municipal Solidwastes (Management and Handling) Rules, 2000 as well as the Supreme Court Guidelines for solidwaste management.

Swachta doot

Rag pickers and private sweepers who were previously working in the same sector and spent their life at foul-smelling and most unhygienic places rummaging through debris with bare hands and getting an uncertain and irregular low payment for this dirty work are brought into organised sector by CDC, and now called as Swachta Doots.

Training

Swachta Doots undergo a special training that equips them with the abilities necessary for their job:
1. Handling the waste in a proper and hygienic manner.
2. Polite and helpful behaviour towards local residents.
3. Discipline, sincerity commitment to their work.

The properly trained workers collect the waste from households and shops seven days a week and 365 days a year. This service is provided in the morning time (between 6.00 am to 1.00 pm). They wear colourful work clothes (uniforms) so that residents and shopkeepers can easily identify them. Training and neat public appearance helps the worker to be better accepted by the community.

The garbage is directly transported and unloaded to local containers (transfer stations) using specially designed vehicles. Those containers are brought to landfill sites outside the city by municipal corporation workforce. There is a close cooperation between CDC and the municipal bodies so that waste is not stored longer than necessary in residential areas.

This way of domestic waste management is in compliance of MSW guidelines and guarantees that:
1. Waste is handled only once.
2. It is exposed nowhere.
3. There is no need of burning the garbage or dumping it in streets, drains and open places.

Equipment

CDC has developed a microplan to adopt waste collection to the special conditions that prevail in different areas. For example, in slum areas different type of waste is produced than in posh colonies and the size of the streets is varying. To implement this microplan, CDC uses different types of vehicles.

Swachta Doots collect the garbage with specially designed mechanical tricycle rickshaws and multi bucket wheelbarrows. They have several advantages:
1. Workers can access even very narrow roads (for example in slum areas).
2. Segregated waste collection.
3. The waste can be directly unloaded in the container. It does not have to be touched. So, hygienic conditions for workers are improved.

CDC follows a dynamic model of selection of equipment and modifies the equipment according to the requirements of a particular locality, town or a city.

Evolution of primary collection equipment

1. Open body tricycle rickshaws: The open body tricycle rickshaw is the most basic primary collection equipment and its capacity is about 50 kg. Its benefits include low cost and since it

collects the mixed waste, it takes less time for waste collection from individual households. This type of equipment is no more in use as it collects the waste in mixed form, which is in contravention of MSW Rules. Further the collected waste is often dumped onto the ground or road and attracts rag pickers to sort out the recyclable part of the waste, exposing them to serious health risks. Additionally, since the waste is transported open to atmosphere, it is non-aesthetic and it is also observed that waste keeps spilling from the sides of the rickshaw.

2. Container tricycle rickshaws: In order to overcome some of the problems faced in the above model, CDC evolved a 6-container tricycle rickshaw for primary collection of waste. The major advantages of this type of equipment is that it allows collection of waste in segregated form and prevents multiple handling as the waste can be directly unloaded to the secondary collection vehicle. Due to these reasons, this type of equipment complies with the MSW guidelines. Its cost is a little more than the open body tricycle rickshaw and can carry 75 kgs of waste in one trip. The only limitation with this type of equipment is that it led to frequent thefts of plastic containers requiring their replacement. This unnecessarily adds to the costs.

Types of containers	
Metal containers	Plastic containers
Heavy, corrosive and expensive	Light weight and durable and noncorrosive

3. Closed body container with 2 compartments: In some of the areas, a new type of tricycle rickshaw has also been utilised viz. closed body container with 2 compartments, which aids in segregated collection of waste, and since it's a closed body type the waste is not exposed to the atmosphere and therefore has aesthetic appearance. This type of equipment can carry 150 kgs of waste and costs about Rs. 8000/-, which is a little more than 6 container type tricycle rickshaw. The only limitation with this equipment is that since the waste cannot be unloaded to the secondary collection vehicle, it leads to multiple handling, which is in contravention to MSW guidelines. This type of equipment is highly suitable for smaller cities where the waste can be directly transferred after collection to the landfill site.

4. Closed tipping bodies with two containers: To overcome the problem of multiple waste handling, a further improvement is in the design of rickshaw has been incorporated, that includes hydraulic lifting of the rear side of the rickshaw unto the height of secondary collection vehicle so that the waste can be directly disposed off. This type of equipment is most expensive and requires more maintenance as compared to the earlier ones.

5. Mechanised auto rickshaws with closed container body with compartments: In some cases CDC has deployed mechanised equipment in the form of auto rickshaws for primary waste collection. The benefits of this type of equipment include high coverage area and can handle more waste. Further, since it has a closed container body, it is aesthetic in appearance and can directly unload the waste to final disposal point. This type of equipment is most optimum for smaller cities and towns as the overall economics is favourable as compared to manually driven rickshaws. The only demerit with this equipment as compared to the earlier models is that the initial capital cost is higher.

6. Mechanised operations with medium utility vehicles: In line with the above-mentioned mechanised operations, CDC has also introduced medium utility vehicles for primary waste collection in some of its projects. The advantage of this type of system is that it can handle more

waste than auto rickshaws and is most ideal for bin-free cities. It fully complies with the MSW guidelines and is aesthetic in appearance, as waste is not exposed in the open atmosphere. The demerit is that it is more expensive.

Door-to-door garbage collection—the monitoring system

Supervisors and zonal in-charges inspect the field everyday. They regularly get in touch with households and shops to check for feedback, complaints and suggestions so that a satisfying service can be maintained. The CDC customer care service, telephone number is available from 6.00 am till 8.00 pm. In most cases complaints are redressed within 30–60 minutes during working hours or the next morning if problems arise outside working hours.

Waste segregation

Waste is not all the same. It has different characteristics according to which it can be divided accordingly:

1. Recyclable, e.g. glass, paper, plastic.
2. Organic, e.g. food leftovers, garden waste.
3. Toxic, e.g. tin, batteries.
4. Reusable, e.g. plastic bottles, polythene bags.

While recyclable waste is dry in nature, the organic kind is wet and 100 per cent biodegradable. Hence, bacterial action is faster in the latter. If waste is segregated, it is easier to handle, does not cause much pollution and can be reused, recycled or decomposed.

The CDC model of waste management is based on the principle of segregating waste and treating it according to its characteristics. Waste should be segregated at the place or source of origin. In order to realise this concept CDC, implements the following approach:

1. Educating the community about waste characteristics and the consequences of inappropriate waste dumping.
2. Collecting the waste in a segregated manner every day.
3. Using specially designed multichambered rickshaws for garbage collection.

Garbage lifting and transportation

Since June 2004, CDC is also lifting the local containers and transporting them to landfill sites in few cities. This enables CDC to co-ordinate the different processes of primary waste collection and transportation. Containers are lifted before they overflow and waste is not stored longer than necessary in residential areas. Furthermore, superfluous containers can be removed for better public convenience.

Employment generation

As CDC is not a profit-oriented organisation, it is committed to improve quality of life, especially for the deprived section of the society. For this reason, most grassroots workers have been recruited from slum areas. CDC started its work initially with few workers. This number increased every day and now CDC could create livelihood for about 6000 persons in the Swachta Doot Project. The services have grown in various cities and presently CDC is catering to a population of nearly 6 million.

Table 33.1 highlights some of the current projects of CDC in various cities including employment generated approximate quantity of waste collected, utilised and finally transported as inert material to landfill sites.

Table 33.1. The current projects of CDC in various cities.

City	Households covered	Road sweeping (length in km)	Employment generation	Waste collected (T/day)	Organic matter (T/day)	Recyclable content[a] (T/day)	Inert and inorganic (T/day)
Jaipur	50,000	150	300	100	60	10	30
Nagpur	4,50,000	200	2500	900	540	90	270
Ahmedabad	30,000	35	125	60	36	6	18
Gandhinagar	60,000	0	100	120	72	12	36
Surat	1,00,000	0	100	200	120	20	60
Nanded	10,000	0	150	20	12	2	6
Delhi	2,00,000	450	629	126	N/A	N/A	N/A
Bilaspur	10,000	0	30	10	N/A	N/A	N/A
Udaipur	3000	40	30	N/A	N/A	N/A	N/A
Bharatpur	10,000	60	94	15	N/A	N/A	N/A
Hanumangarh	17,000	60	90	N/A	N/A	N/A	N/A
Beawar	10,000	85	100	16	N/A	N/A	N/A
Jaisalmer	3000	30	32	N/A	N/A	N/A	N/A
Total	9,53,000	1110	4280	1551	840	140	420

Average 5 members per household	N/A indicates non-availability of data

[a]Recyclable contents include paper, glass, plastic and metal.

Introduction of door-to-door collection service has improved the financial condition of Swachta Doots who now receive regular payment as compared to the earlier situation when they could earn low (12–15/per day) and irregular incomes. They are now enjoying dignified working conditions at CDC, as their profession is viewed with greater respect amongst society members compared to their earlier work. As both genders are treated equally the number of women is almost as high as the number of male grassroots workers. Women are given the chance to make an own contribution to the living standard of their families. Additionally unemployed or underemployed educated youth profit from these job opportunities as they are recruited as Managers and Supervisors.

Awareness building

It is CDC's conviction that the cleanliness of city is a collective good. It can only be achieved with the participation of all concerned. Therefore, CDC encourages and motivates people to keep their surroundings clean. They are provided education regarding sanitation and garbage disposal through various means of communication such as:

1. Posters, folders, booklets, leaflets.
2. Exhibitions.
3. Wall paintings.
4. Living society meetings.
5. Debates and painting competitions in schools.
6. Regular talks with citizens.

Benefits of swachta doot project

Social

1. Improving social standard of Swachta Doots by providing training and financial stability.
2. The community is made aware of the consequences of unscientific waste throwing and can participate actively.
3. As the citizens are also involved in the project they develop a sense of belongingness.
4. People appreciate the service and consider it as necessary and essential. This makes the project self-sustainable.

Economic

1. City's image as a 'green and clean' city can boost local economy especially in tourism branch.
2. Creates new avenues of employment.
3. Composting of organic matter and recycling of paper, glass, plastics and metals yield productive outcomes and reduces burden on landfill site.

Public health and life quality

1. Waste is handled in a hygienic and scientific manner, so no pollution is caused at any stage.
2. Garbage on the roads is tremendously reduced.
3. Drains are no longer clogged with garbage—no smell, no breeding site for malaria spreading mosquitoes, no meeting place for pigs and other stray animals.
4. Quality of life improves as the whole city looks clean and aesthetic.

This shows that the model is not only a convincing theoretical concept but also a successful intervention in the field of municipal solidwaste management.

Methodology

1. The idea of door-to-door (D2D) collection of garbage was not new for Nagpur Municipal Corporation (NMC). This work was started in 1996 in some wards of the city covering about 30 per cent of the total population. But after the Supreme Court directive regarding 100 per cent D2D garbage collection, it was implemented throughout the city.
2. Requirements for implementing the directive included equipments like cycle rickshaws, ghanta gadi and manpower.
3. In the city of Nagpur there are different types of residential areas like skyscrapers, slums, independent houses, bungalows, government colonies, etc. These categories have been grouped and suitable volunteers have been deployed for garbage collection.
4. The volunteers were trained and oriented about waste disposal.
5. Every volunteer, i.e. 'Swachta Doot', cover about 200–300 households everyday depending upon the category.
6. All the volunteers have been provided with uniforms and safety kit, which includes hand gloves, face mask, cap, etc.
7. To oversee project implementation, one supervisor and coordinator has been appointed for every zone.

Key highlights

1. First project in India involving D2D collection of garbage from 100 per cent households on all 365 days.

2. The only project on solidwaste management in India to recruit physically challenged persons. At present there are 37 such persons working on this project and thereby earning livelihood for themselves and their families.
3. No dependence on external funding.
4. Community involvement on a large scale.
5. Well-defined roles for NMC, NGO and the community under the scheme.

Strategy followed

All concerned stakeholders were consulted before finalising the implementation plan which is given below:

1. The work of D2D collection of waste by 'volunteer' begins at 6.00 am daily.
2. Every household in the given group is attended daily.
3. The doot goes around the demarcated households and announces his arrival by blowing a whistle.
4. When the cycle rickshaw is completely filled with waste, it is unloaded in the nearby community dustbins. After unloading, the volunteer covers the remaining households.
5. During every unloading, the recyclable waste is separated by the volunteer.
6. At no point of time, he does manual handling of the waste collected.
7. Monitoring indicators have been set in consultation with the NGO, which have further ensured prompt implementation of the project, e.g. households covered, timely complainant redressal, regular and surprise field visits, community feedback, etc. are monitored regularly.

Results

Benefits to NMC

1. Successfully implemented the Supreme Court guidelines.
2. Savings worth Rs. 5 crores in terms of lower costs for providing D2D garbage collection service to the citizens.

Benefits to CDC (NGO)

1. Other municipal corporations in India are adopting Swatchta Doot model. CDC has assisted in preparation of policy in Rajasthan.
2. CDC's budget for the financial year 2004–2005 was Rs. 53 lakhs. After being appointed as implementer for this project, its financial credibility has escalated to Rs. 15 crores. Moreover, we have been appointed to implement similar project in other cities of India.
3. Financial institutions like Kotak Mahindra, ICICI, Tata Finance, etc. have come forward to sponsor equipments required for D2D collection. To date credit worth Rs. 1 crore has been availed from these institutions.
4. NMC has further reposed its confidence in CDC activities and has handed over the responsibility of secondary transportation of Municipal Solidwaste also to a joint venture involving CDC.
5. Improvement in sanitation.

Benefits to citizens

1. Regular D2D collection of garbage and active participation in the zero garbage drive.
2. Better and prompt service at minimum costs.

Stakeholders/partners

'Swachta Doot Aplya Dari' is a joint initiative undertaken by various stakeholders/partners to maintain good hygiene and jointly create a cleaner city. In the following section we would like to highlight those who have made a significant contribution towards the initiative. These include:

1. Nagpur Municipal Corporation (NMC): Keeping the city clean has taken on a whole new meaning in the orange city, i.e. Nagpur, the second capital of Maharashtra. Being innovative, clean and green has had a significant bearing on the city's future competitiveness and attractiveness as a business and travelling hub. Firm determination and hard administrative measures have contributed towards the success of the efforts of NMC. The innovative steps taken by NMC in MSW handling and disposal have led to visible changes in the city. Nagpur is recognised as one of the cleanest cities in the country.

2. Civic cops (Nuisance detectors): The civic cops supervise the work of Swachta doots i.e. the volunteers and addresses the grievances of the citizens. Furthermore, they collect the previous day's regarding information and report it to the higher authorities at the zonal office in writing, which was collected at the main office. Supervision by higher officials at the zonal office and appreciation of their efforts keeps these civic cops motivated and on their toes.

3. Nongovernmental Organisations (NGOs) Center for Development Communication (CDC): CDC had the most Herculean task, i.e. implementation, wherein it had to involve people in implementing the program, apart from recruiting volunteers, training and orientation, convincing the masses, etc. 'Show and tell' observation of a functioning D2D primary collection system was used to train the volunteers. As it is rightly said 'public support comes when people see tangible results and benefit from such change'. The residents not only co-operated but also applauded the joint efforts undertaken by CDC, who have been responsible for making Nagpur a 'Green and Clean' city.

4. Swachata Doot (volunteer): The volunteers or the Swachta Doots were the vital links and key contributors for making Nagpur a 'Green and Clean' city. These volunteers some of who were earlier know as 'Rag Pickers' would previously collect and sort the recyclable dry waste like paper, plastics, metals, rags, etc. for their daily living. Commencement of 'Swachta Doot Aplya Dari' scheme has had positive impacts on the lives of these rag pickers and substitute sanitation workers resulting in livelihood creation for the members of the most downtrodden segment of society giving them dignity of work as 'Swachta Doots' having uniform and protective.

5. Citizens: The citizens have contributed to the scheme by wholeheartedly adopting the idea of waste disposal to volunteers only and making voluntary token monetary contribution towards the welfare of the volunteers.

CASE STUDY 2: SOLIDWASTE MANAGEMENT IN JALANDHAR CITY (PUNJAB)

Solidwaste (SW) can be defined as the material that no longer has any value to the person who is responsible for it and is not intended to be discharged through a pipe. It does not normally include human excreta. It is generated by domestic, commercial, industrial, healthcare, agricultural and mineral extraction activities and accumulates in streets and public places. The words 'garbage', 'trash', 'refuses' and 'rubbish' are used to refer to some forms of SW.

Status of Solidwaste Management in Jalandhar

In this comprehensive investigation, Jalandhar city was chosen as a model region of the Punjab province. Besides a dwelled suburban area, the population of this region is around 20,00,000. The city is located almost 375 km from Delhi and about 90 km from Amritsar.

The sources of SW are residential, commercial institutional, constructional and demolition, municipal service, industrial, agriculture and dead animals.

A total of 300–400 tons of garbage is collected and disposed off daily in the city from the above-mentioned sources.

Physical characteristics based on wet basis are given in Table 33.2 and average chemical characteristics are given in Table 33.3.

Table 33.2. Physical characteristics based on wet basis (%).

Parameters	Average	Range
Metal (ferrous)	0.01	0.00–1.00
Metal (nonferrous)	0.13	0.00–0.04
Earthware stone, etc.	5.17	0.50–15.00
Glass/ceramics	0.57	0.00–2.10
Fine earth	24.15	19.92–25.72
Paper/cardboard	3.43	0.2–10.80
Wooden matter	0.09	0.00–0.30
Rags	3.95	0.10–9.80
Rubber, leather	1.31	0.00–4.00
Plastics	7.42	3.20–14.5
Moisture content	44.53	6.90–68.10

Table 33.3. Average chemical characteristics.

Parameters	Average
pH	8.04
Calorific value	1616.18
Compostable matter	53.75
Nonvolatile matter (%)	70.05
Volatile matter (%)	29.95
Nitrogen (%)	2.51
Phosphorous (%)	0.31
Potassium (%)	0.49
Organic carbon (%)	10.38
Hydrogen (%)	1.20
C/N	4.13

The waste is 85 per cent nonhazardous, 10 per cent infectious and 5 per cent noninfectious. It is 54 per cent compostable, 64 per cent combustible, 24 per cent inert and 11 per cent recyclable.

Status and Analysis of Functional Elements

The functional elements of SW management in this city include waste generation, collection, transportation, processing and final disposal.

Waste generation

For efficient SW management, study of the generation rate is very important. The generation rate varies from activity to activity and the exact generation rate must be determined after averaging over the time factor and including several of the waste producers. The average waste generated in Jalandhar city is 350 TPD. In all, there are about 320 garbage bins placed in city. Garbage is finally disposed off in an open dump at Suchi-Pind village on Hoshiarpur road and Wariana on Kapurthala road. Garbage-lifting vehicles involved in this operation are given in Table 33.4.

Table 33.4. Garbage-lifting vehicles.

Type of vehicle	Number
Truck tipper	7
Tractor trailer	5
Refuse collector	2
Dumper placer	5
Tricycle	300
Wheel barrow	200

Sources of waste generation is given in Table 33.5.

Table 33.5. Sources of waste generation.

Name	Content	Sources
Garbage	From cooking food and domestic work contents	Households and hotels, etc.
Rubbish	Markets refuse rags, cloth and leather	Stores and markets, etc.
Ashes	Residue form	Fire
Bulk waste	Large auto parts, tyres, etc.	Service station
Street refuses	Dust and dirt	Street sweepings and litter, etc.
Special waste	Hazardous waste	Hospitals and industry

Onsite handling, processing and storage

In Jalandhar city, most of the habitable/residential areas have limited storage spaces. In these areas, the waste is of mostly of a biodegradable nature. In some places, open dumping of the garbage is noticed, which causes health hazards as well as fly nuisance.

Handling

It refers to the activities associated with managing SW until they are placed in the containers used for their storage before collection or return to drop off and recycling centres. There are various problems that could be related to handling and storage of SW are seen, like there are a number of places where there are dumps of SW and thus these points if unattended create small nuisance and health hazards. Stray animals like pigs, dogs and cows further aggravate the problem of spreading and littering of SW as they are seen at the site of handling and storage of SW in the study area.

Solidwaste segregation

SW is not segregated; rag pickers collect SW from the streets, bins and deposit sites. Storage spaces are not often adequate. People drop the SW outside the bins.

Collection

SW is collected from the bins from every point and collection from residential areas is carried out daily as the organic matter decomposes rapidly due to a hot climate.

The World Health Organisation recommends that collection of SW should be carried out twice a week from bins in Jalandhar city. Collection is only performed when the bins are filled up. Hand driver cart pullers collect the SW from door-to-door. These cart pullers segregate the plastic bags, polythene and metal, which is then sold to the kabariwalas. By this, nondegradable solids are separated from organic substances. This collection system is also economically feasible.

Transportation

Transportation means 'transfer' of SW from the storage place to the dumping ground. For this purpose, vehicles are dependent on the physical layout of the roads and the cost of manpower available, maintenance provisions, truck tippers, tractor trailer, etc. that are used for final transportation of SW to the site. About 350 TPD waste is generated on a daily basis.

Recycling Process and Recovery Recycling Process

The Municipal corporation, Jalandhar, has signed a MoU with the Punjab Grow More Fertilisers Ltd. A plant has been set up at the village Wariana Basti Bawa Khel. It was earlier the dumping site of MCJ. By this process, organised SW is converted into manure by the process of composting. This site has 14 acres of land having a capacity of 600 TPD. About 100 TPD SW is transferred to the Wariana site. Nearly 250 TPD is dumped at the Suchi-Pind site.

In order to utilise the entire SW produced in the city of Jalandhar for making organic manure, the following two requirements must be met:
1. More land is required at the dumping yard with a provision that it can used for at least 30 years.
2. Proper marketing by government so that the organic manure produced should be sold out. As 300 TPD SW is produced per day, 150 TPD of organic manure is thus generated. Therefore, marketing 150 TPD organic manure should be planned.

Waste Disposal Options

A number of waste disposal options are available in the form of repetitive disposal technology. Table 33.6 lists the leading ones along with the equipments required for them.

Table 33.6. Repetitive disposal technology and equipments required.

Technology	Major equipment
Sanitary landfills	Reactors/digesters, synthetic liners, pumps
Fuel pelletisation	Pelletisation equipment
Composting	Mechanical composters

Sanitary landfilling

SWs are placed in the sanitary landfill system (trenches, pits) in alternate layers of 80 cm thick refuse and then covered with an earth fill of 20 cm thickness. After 2–3 years, the SW volume shrinks by

25–30 per cent and the land can be used for parks, roads or as land for small buildings with normal compaction. A landfill site can take 500 bags of refuse per cubic metre of trench space available. Care should be taken while locating the site for dumping refuse as the land should be selected after taking into account that it can be used for 25–30 years. Landfilling depends on the availability of land area, soil conditions, grand water table, topography, distance from the residential area and ultimate usage of site after reclamation.

The landfill operation is a biological method of waste treatment. SW can be stabilised by dividing it into five distinct phases with the overall process. In the first phase, aerobic bacteria deplete the available oxygen as a result of aerobic respiration and the temperature increases. In the second phase, anaerobic conditions become established and hydrogen and carbon dioxide is evolved. In the third phase, methane is librated and in the fourth phase, methonogic activities become stabilised. In the fifth phase, the system returns to aerobic conditions within the landfill. The duration of each phase varies with the environmental conditions.

Biomethane technology

This process is used for the production of methane from the SW. In this, first of all separation and size reduction of the SW is carried out. After this, moisture and nutrients are added. The pH is adjusted to about 6.7 and temperature of the slurry is increased to 55°–60°C. The slurry is mixed well for about 7–10 days. After this, storage of the gas is carried out.

Incineration

Incineration involves the burning of the SW at very high temperatures. In this method, the volume of the SW is reduced up to 90 per cent. The unburnt SW, which is left, is about 25 per cent of the original waste.

Composting

This method is an aerobic method of decomposition of the SW. Many types of micro-organisms already present in the waste stabilise the SW. The organisms include bacteria, which predominates at all stages, fungi, which appear after the first week and actinomycetes, which assist during the final stages. Mesophillic bacteria present oxidise the organised matter in the refuse to CO_2 and release it as heat. The temperature increases up to 45°C. Thermophillic bacteria take over and continue the decomposition. The temperature further increases to 60°C. After this, the SW is turned. After about 3 weeks, the composites are stabilised.

Waste sanitisation treatment method

SW is first of all treated with biological inoculum at the collection point. The celrich substrate DF-BC-01 (manufacturer USA) is a mixture of biological enzymes and herbal extracts that is spread over the SW. The material is nonhazardous and nontoxic. The SW becomes free of hazardous pathogens, which eliminates the foul smell from the SW. Dumping points get hygienically upgraded. The SW becomes free of flies, insects and other disease-carrying vectors. This provides better working conditions and reduces the chances of smoke, fire and explosion hazards at the dump yards as the production of methane is reduced by this.

Advantage of the Technology

1. Area required in this process is very less.
2. Corporation gets an annual lease rent as well as royalty to meet the collection cost partially.

3. Clean refuse is generated, which can be used for landfilling.
4. Polythene and plastic material that is segregated can be recycled.

Use of Plastics and Their Recycling

Plastics, due to its advantages like its durability, lightness, and ease to be moulded, is used everywhere, for example:

1. In domestic purposes: As carry bags, pet bottles, trash bags, containers.
2. In air, road, rail travel: As cold drink or mineral water bottles, plastic plates, cups.
3. In hospitals: As glucose or other IV fluid bottles, disposable syringes and injections, catheters, wine bags, gloves.
4. In shops and hotels: As packing items, plastic bags and disposable utensils.

Plastic contains a certain component called dioxin that is highly toxic and carcinogenic. After burning, especially PVC, it releases this dioxin and also furan into the environment. To avoid health hazards, plastics should be recycled. Plastic recycling can be performed by the 'green recycling process'. A sketch of the pilot used for this purpose is given in Fig. 33.1.

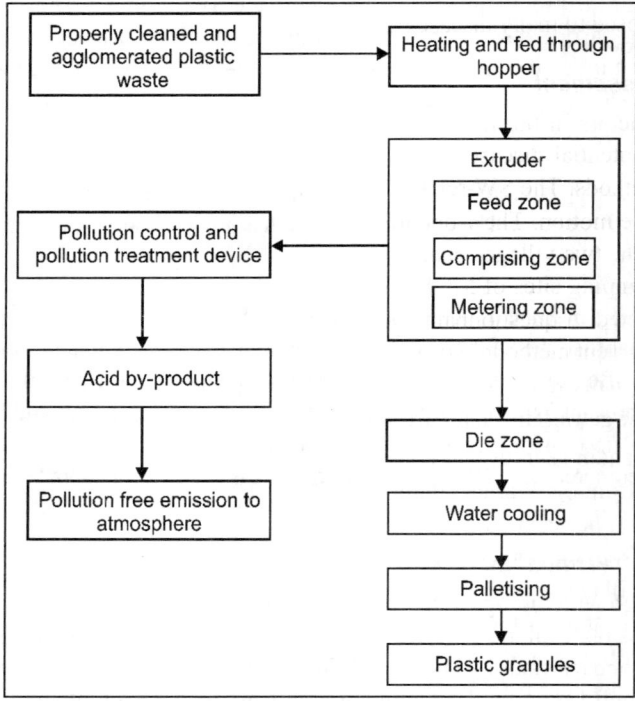

Fig. 33.1. Green recycling process.

Health Impact of Solidwaste

As SW is a major part of environment pollution, it is responsible for spreading many harmful and infectious diseases. The increase in SW is due to an increase in the population. As the population increases, the demand for food and other essentials also increases such that waste is also increased. Some people throw this waste into streets, roads and at other public places, which attracts flies, insects, rats, etc. which helps in spreading the diseases. Unattended waste is normally wet and has a bad odour due to decomposition. This type of waste leads to epidemics in various parts of the country.

1. Domestic waste decomposes and spreads diseases.
2. Waste from agriculture and industries can also cause serious health diseases because these wastes may include some chemicals, pesticides, metals, etc.
3. Uncollected SW may also affect water bodies. When water gets infected, it causes many water-borne diseases to the surrounding community. From this SW everyone gets affected, e.g. collectors of this waste material and persons working in various shops or factories.
4. Like in an industry where some chemicals are used, certain chemicals, if released untreated, e.g. cyanides, mercury and polychlorinated biphenyls, are highly toxic and their exposure can cause severe disease or also death.
5. Plastic is also very harmful to the health. The unhygienic use and disposal of plastic causes very toxic effects as some coloured plastics contain heavymetals that are very toxic, e.g. copper, lead mercury, lead, chromium, cobalt, selenium and cadmium.
6. As we know that there is manual collection of SW from door to door, persons collecting this SW are exposed to many diseases and infections.

Health impact assessment

A number of risk factors in the form of foetal diseases are associated with the malpractices of SW management. The potential diseases that were identified in the area of MCJ may occur due to the prevailing SWM methods. The SW-related vector-borne diseases identified are malaria, dengue, Kala Azar, fever and loose motion. The worst affected areas that are closer to the dumping sites were also selected. These are the two villages, namely Suchi village and Wariana village. All these locations are very close to the dumping sites of MCJ. One more area that is away from the dumping sites, 'Pucca Bagh' was also selected. A questionnaire was prepared keeping in mind all the environmental indices and the impact assessment methodology and established strategies. It was circulated among the residents of these villages originally in the local language 'Punjabi' to collect feedback from them. The modes of survey carried out for getting feedback from these villages is as under:

1. Wariana: A random survey was conducted for 150 persons in the Wariana village. This village is about 5 km from the Wariana dumping site and composting plant. In this village, 3.5 per cent had malaria, 0 per cent from dengue, 0 per cent from Kala Azar, 90 per cent from fever and 95 per cent from loose motions.
2. Suchi-Pind village: A random survey was conducted for 150 persons in the Suchi-Pind village. This village is about 1 km from the Suchi-Pind dumping site. In this village, 1.2 per cent had suffered from malaria, 0 per cent from dengue, 0 per cent from Kala Azar, 92 per cent from fever and 94 per cent from loose motions.
3. Pucca Bagh: A random survey was conducted for 150 persons in the Pucca Bagh area. This area is located in centre of Jalandhar city where proper SW management is performed. SW is collected properly from each house. In Pucca Bagh, 0.9 per cent people suffered from malaria, 0 per cent from dengue, 0 per cent from Kala Azar, 50 per cent from fever and 52 per cent from loose motions.

The received feedback data were compiled and analysed using the established statistical methods. The findings related to the vector-borne diseases caused by SW are summarised in Tables 33.7 and 33.8.

Table 33.8 provides the percentage extent of problems caused by the agents and vectors due to malpractices of SW management adopted in these areas.

Table 33.7. Results of the survey conducted for solidwaste-related vectorborne diseases.

Vectorborne diseases	Wariana (n = 150)	Suchi-Pind (n = 150)	Pucca bagh (n = 150)
Malaria	5 (3.5)	2 (1.2)	1 (0.9)
Dengue	0 (0.0)	0 (0.0)	0 (0.0)
Kala Azar	0 (0.0)	0 (0.0)	0 (0.0)
Fever	135 (90)	138 (92)	75 (50)
Loose motion	142 (95)	141 (94)	78 (52)

Figures in paranthesis are in percentage.

Table 33.8. Household perception of the impact of malpractices of SWM.

Household perception	Wariana (n = 150)	Suchi-Pind (n = 150)	Pucca bagh (n = 150)
Solidwaste pollution	16.0	8.0	10.0
No problem	–	–	–
Problem	92.0	90.0	52.0
Major problem	2.0	3.0	2.0

Figures are in percentage.

In order to ascertain that the higher degree of infections in the villages of Suchi-Pind and Wariana are only because of the impact of SW, a survey was also conducted for the smoking and drinking habits of the residents of the three locations, i.e. Suchi-Pind, Wariana and Pucca Bagh.

It is clear from the data that occurrence of dengue and Kala Azar is not observed in any of the locations; however, other diseases like malaria and loose motions have been reported. A large number of residents, up to 90–95 per cent, were found suffering from fever and loose motions more than once every year. This is indicative of a strong to moderate health impact on the resident population due to the SW being dumped in their vicinity. There is no established correlation of occurrence of these infections due to the smoking and drinking habits of the residents of these areas. Thus, the sole cause of these illnesses lies in the faulty disposal of the wastes closer to these locations.

How can We Prevent these Diseases

1. Proper disposal: A proper method should be used to dispose SW because unattended waste is responsible for spreading these infectious diseases.
2. By avoiding water pollution: SW collection and disposal should be away from a water supply or a big water body because, as we disposed earlier, that there are many diseases that can spread by water pollution, e.g. cholera, typhoid, diarrhoea, etc. Hence, village waste should not be dropped near wells or taps, etc. The wells should be covered.
3. All the eatables should be covered so that flies cannot infect them. Boiled water should be used for drinking and other purpose in the pitches.
4. Early diagnosis and treatment: There should be different surveillance programs arranged in urban and rural areas, e.g. the malarial surveillance program. Also, to detect the symptoms of a patient, various diagnostic tests should be carried out as soon as possible so that proper treatment can be started.

5. Isolation of the patient having a contagious disease: Isolation of the patient is necessary to avoid further spread of the disease and to avoid major problems like epidemics. Vaccinations of some contract calls should be performed according to the severity of the disease.

Discussion

The problem related to SW management and its heath impact is investigated in two phases. In first phase, the prevailing SW management practices in the MCJ have been evaluated vis-*a*-vis the standard SWM methods, and suggestions have been put forward keeping in mind the ground realities and system limitations. In the second phase, the health impact assessment has been performed in the affected areas using the survey technique.

Recommendations to modernise the waste management practices prevailing in MCJ

It has been observed during the course of the study that the prevailing SWM practices and boundary conditions are different for urban and rural regions of MCJ and hence the modernisation strategies for the two are also different accordingly.

Recommendations for urban areas

The problem of preliminary storage of SW is increasing as the population is growing rapidly. The old methods of SW management are not correct. SW should be disposed off after separation into various parts.
1. Biodegradable waste.
2. Recyclable waste (i.e. plastic, metal, glass, leather), etc.

People should be made aware of collecting the waste in different bins, i.e. organic and recyclable.
1. Waste should be divided into two parts: Biodegradable and nonbiodegradable. Waste from the hospital should be collected daily and disposed off to the incineration plant.
2. Chemical, pesticides, batteries and other domestic hazardous/toxic waste material should be collected in different bins.
3. Each area should be given to a particular sweeper for the door-to-door collection of waste. In densely populated areas, about 400 m of road length is given to a particular sweeper. In less-populated areas, about 600 m of road length should be given to a sweeper. Sweepers should collect the SW at one fixed time.

Collection of recyclable/nonbiodegradable waste by NGOs should be motivated to collect the recyclable waste, such as polythene, plastic, glass, metals, so that these can be recycled. Particular residential areas should be given on lease depending on the work of the NGOs.

Public Awareness

SW generally comes from the residential and commercial areas, for example houses, vegetable markets, hotels, marriage palaces, hospitals, institutions, etc. The public should be made aware by arranging awareness camps that the waste should not be spread on streets, roads, nalis, etc. People should be made aware of the fact that if the waste is properly disposed off from the house then the environmental will not get polluted. Many severe diseases can spread by improper disposal of SW. There should be environmental engineers and public health engineers for the SW management in addition to Health Officers related to community medicines. Qualified engineers will work to overcome the drawbacks of this system. There should be trained collectors who know all the details that required for collecting SW from door to door and from streets, roads, etc.

There should be literacy classes in which they learn how the SW can lead to various problems and diseases and how these problems are reduced. There should be health check up camps of community waste collectors. The number of engineers and other staff members should be adequate according to the population of the area, provision of bins, containers, rickshaw and trolleys and trucks. Their number should be sufficient and the government should take care that the number of these equipments and material of the containers should be okay. Proper finances and system to system coordination is an important factor.

Health Education

Health education is major part of the control program of these diseases. Environmental awareness, i.e. impact of various environmental factors on human beings, is yet another important factor that must be addressed. The health worker and doctors should tell people about the common diseases against which care should be taken. Preventive measures should be told to people and they should be told to covers all the eatables.

Boiled water should be used for drinking purposes. Hand should be washed before eating anything. Ammunition schedule should be followed by the people. People should be made aware of population control. Basic contraception methods should be told to the people. Doctors and environmentalists should arrange awareness camps.

CASE STUDY 3: ENVIRONMENTAL IMPACT ASSESSMENT OF MUNICIPAL SOLID WASTE LANDFILLS—JORDAN

Landfilling has been used for many years as the most common mean for solidwaste disposal generated by different communities. Despite the intensive efforts that are directed to the recycling and recovery of solidwastes, landfills remain and will remain an integral part of most solidwaste management. Solidwaste disposed in landfill usually subjected to series of complex biochemical and physical processes, which lead to the production of both liquid and gaseous emissions. As water percolates through the solidwaste matrix, leachate produced which contains soluble components and degradation products from the refuse. Greenhouse gases such as methane and carbon dioxide are generated during the stabilisation of solidwaste organic fraction. Volatile components of the solidwaste tend to be emitted into the atmosphere with the evolved landfill gases. The contamination of groundwater by landfill leachate has been reported by several researchers. Lee and Jones reported the potential adverse impacts of municipal solidwaste (MSW) landfills on those who own or use properties near such facilities. Hirshfeld reported that the property values near MSW landfills are adversely impacted by the landfill for distances of a mile or more from the area where waste deposition occurs. Adverse environmental impacts, public health and socio-economic issues associated with MSW landfills have led to issuance of stricter regulations and increased public opposition to the sitting of such facilities (not in my backyard syndrome). As a result, the sitting of a new landfill has become one of the most difficult tasks faced by most communities involved in MSW management.

It is important to convince the decision-makers and the public that the selected site is environmentally friendly and its adverse impacts will be minimal. One tool to achieve this objective is to conduct EIA study of the selected site, so as to eliminate and/or minimise the adverse impacts of the site. The objective of this chapter is to present the major environmental issues considered in the EIA study conducted for the Al Ghabawi Landfill for the capital City of Jordan (Amman) and to highlight the mitigation techniques recommended to minimise the adverse impacts.

Amman is the capital city of Jordan with population of about 38,01,589 in the year 2008. Solidwaste management in the city is the responsibility of the Municipality of Greater Amman (MOGA). Until recently, MOGA used to dispose of the solidwaste generated by the city at Ruseifeh landfill, which is located at the eastern edge of the city. Due to the location of the landfill in a populated area, and to the unsanitary landfilling process at the site, several adverse environmental impacts were created by this landfill. As a result, MOGA decided to close this landfill and move to a new site. After an overview of several potential sites, Al Ghabawi site was selected to construct the new landfill.

Methodology

The environmental assessment procedure for Al Ghabawi landfill site was conducted in such a way to highlight the key environmental issues and to achieve proper and accurate prediction of potential impacts. To achieve the above objective, the study conducted according to the following sequence:

Data collection

Data on the site location, capacity, topography, geo-hydrology, climatology and other information related to the area that will be served by the landfill were collected. These data were obtained from MOGA's documents, walkover survey to the site, and pictorial recording of certain site features and operations. Data collected were presented in the environmental setting chapter.

Scope of the study

In order to determine the nature, significance and the extent of the project impacts, a scoping session was conducted. The session was attended by many stakeholders, including representatives specialists from regulatory agencies, NGO's, local communities, proponent, academic and research institutions, media and other stakeholder.

Impacts identification

Based on the output from the brainstorming at the scoping session, the significance of each impact was evaluated, and the study focused on the most significant ones. The impact magnitude was evaluated by direct measurement, such as, noise from landfill operation and transportation or using modelling techniques based on the available data such as in the case of groundwater and air pollution or based on the visual inspection and judgement.

Mitigation and monitoring measures

After the magnitude of the significant impacts was evaluated. The proper mitigation, management and monitoring techniques were recommended.

Post closure monitoring plan

After reaching the full capacity of the site, the landfilling process should be stopped and the site must be closed. The closure and post closure monitoring plan is recommended for the site.

Site Characteristics

Al Ghabawi is located in the eastern area of Amman, at 23 km east from Central Amman and at 17 km from Ruseifeh landfill site. The total surface area of the site is about 20,00,000 m². The selected site is part of Azraq Basin, which is one of the four largest desert basins in Jordan.

The area used to be as a military land, and characterised by an arid nature without any residential, historical, cultural land uses. The site was connected by an access dual carriageway for a distance of about 23 km. The topography of Al Ghabawi is semi-flat with general slope of 1.4 per cent from south-east down to northwest. The elevation of the site boundaries ranging from 800–810 m above mean sea level.

The site area itself is characterised by a desert nature and used to be as a military training area. About 2.5 km to the east from the landfill there is a cow farm. Along the access highway, there are some residential communities. The nearest one to the site is Al-Manakher, which is about 7 km from the site. In addition, there are some agricultural activities on both sides of the access road.

As for the seismic activities, the new site at Al Ghabawi and the surrounding area is located, which has 0–5 factor of intensity, and considered as a region with moderate potential of earthquakes occurrences. However, historically there were no earthquakes over the site and surrounding area.

Climate and meteorology

The study area is characterised by an arid climate, which manifests itself by hot dry summers, warm winters, and little rainfall. The region is affected mainly by desert climate with the following characteristics:

1. The colder months of the year are January and February, with temperature about 8°C, while the hottest months are July and August, with average temperature 32°C.
2. The average precipitation in the rainy months (January to May) is about 60 mm/month, while the average precipitation is less than 1 mm for the period from June to September.
3. The lowest evaporation is during January (about 80 mm/month) while the highest is during July (about 334 mm).

Geology and soil

The geology comprises the formations and structures existing in the area. To evaluate the geological and hydrogeological characteristics of the site, a geological survey was conducted on site in 2008. Five boreholes were implemented. The vertical receptivity distribution detected 4 zones as follows:

1. Top soil with thickness ranging from 0.5 to 1 metre.
2. Muwaqqar-chalk Marl (B3) with thickness from 80–32 metres.
3. Dry zone of Al-Hisa- Amman and Wadi As Sir formation (A7/B2) with thickness of 108 metres.
4. Water saturated zone with aquifer system.

From hydrogeological point of view, the Muwaqqar Chalk Marl formation is considered as aquaclude. All the test realised on the site ranged the values of hydraulic conductivity between 10^{-3}–10^{-7} mm/s. The variation in hydraulic conductivity values is due to the presence of chert lenses within the Muwaqqar-chalk Marl formation. The presence of fine and very find materials mixed with sand or silty also affecting the value of permeability.

Groundwater

The water saturation zone aquifer system is one of the main water aquifers in Jordan and it represents the aquifer on the site and surrounding area. The depth of water saturation zone is about 248 m, which is the depth of the groundwater well available at the site.

The flow direction in this aquifer is east-southeast with a hydraulic gradient of 0.007. The thickness of the water in this aquifer is approximately 40 metres. The analysis of water samples withdrawn from the aquifer by the Water Authority of Jordan, revealed that the water quality is not suitable for drinking purposes as per the Jordanian standards of drinking water.

Surface water

Surface water is limited in Al-Ghabawy area to flash storms occurring during winter. This surface water is not exploited as most of them are either evaporated or percolated into ground. The drainage system over the site has two directions:

1. To the north–north west towards Zarqa river through Wadi—Janna'a in the north and Wadi Al-Hajar in the north west of the site.
2. To the north east toward Zarqa river through Wadi Al-Ghabawi which is laying to the east and turns to the north to Zarqa river.

Biodiversity (Flora and Fauna)

The soil in the region is poor with low annual rainfall, this will results in poor vegetative cover. The vegetation is mostly composed of fleshy and drought tolerant plants that can resists the conditions of hot and dry climate. Vegetation is restricted to the places close to the wadi system where enough soil moisture exists. Most of the fauna species in the eastern desert in general are of Saharo-Arabian origin. Between the soft silt dunes, many varied species have evolved. The majority of the Herpeto-Fauna (Reptiles) of Jordan are found in the Eastern Desert, where a wide range of a wide range of microhabitats exists due to its large area and unpredictable harsh climatic conditions.

The present study was prepared one month after the start of landfilling at the site. During the visits to the site, there were no birds noticed in the landfill area. However, as the filling process will proceed, it is expected that the site will attract several types of birds that will come to the site to look for food.

Environmental Consequences and Mitigation Measures

Most waste management projects aim at improving the environmental conditions. Nevertheless, a series of potential environmental impacts should be considered, as well as prospective mitigative measures that can reduce negative impacts of waste handling on health and environment. Thus, assessing the possible environmental impacts of the landfill site can help identify which activities, if any, are likely to give rise to significant adverse impacts. These impacts then can be eliminated or minimised through proper mitigation and management techniques.

The output of the data collection and scoping session revealed that, the key areas of potential environmental impacts of the Al Ghabawi landfill site are as follows:

1. Ecological impacts.
2. Hydrology and water quality impacts.
3. Public health and safety impacts.
4. Air quality impacts.
5. Socio-economic impacts.
6. Noise impacts.
7. Off-site and on-site traffic impacts.
8. Visual and landscape impacts.
9. Archeological and cultural sites.

This section will focus on air and water quality impacts. Impacts were classified as either positive or negative. The potential significance of impacts was then evaluated through brainstorming with a core group of specialists, and through evaluation of baseline data on the surrounding environment, and the specifics of the project. Impacts were classified as to their potential significance. The classification system is based on degrees of impact significance as follows: (i) high, (ii) moderate, (iii) minimal, and (iv) none.

Air quality impacts

Gas and odour emissions

Once solidwaste dumped into the landfill it will be subjected to series of reactions. Initially the waste is aerobically and the main reaction products are carbon dioxide gas and water. This stage takes several days to week. With the progress of degradation, the oxygen is depleted and the degradation converted into anaerobic. As waste decompose in the landfill, landfill gases will be generated due to the anaerobic degradation of the organic fraction of the waste. Gas will start to be given off within few weeks of the waste deposition and will continue to be emitted even after the site closure. The main components of the landfill gas are methane and carbon dioxide. Both of methane and carbon dioxide are greenhouse gases (GHG's), which contribute to global warming phenomena. In addition, the methane gas is a potentially flammable and explosive gas for a concentration of 15 per cent of the air volume. Furthermore, some of the gases that can be produced as a result of anaerobic degradation are hydrogen sulphide and ammonia. These gases are mainly causing a bad odour in the vicinity of the landfills and it has been reported in studies that the offensive odour can reach for a distance of one mile or more from the landfill site.

At Al Ghabawi site, the odour potential at the site and the surrounding area was measured by nose smelling at various distances from the working face on the site at various intervals during the day. Several persons were asked to stay for a while close to the working face and then move gradually into different directions from the filling place. At every 20 m distance from the working face each person stayed for a while and recorded his feeling regarding the odour. The measurement revealed that, there is a moderate to strong odour in the vicinity of the filling work face, while it is moderate at the edges of the filling cell and it is minimal at the site corners. Outside the site boundaries, no odour could be smelt.

The odour problem, so far, limited to the site boundaries, this is because the filling is just started one month ago. It is expected that odour potential will be increased with time, as more waste will be received on the site and the anaerobic degradation will advance. However, the site location and the design features will contribute to keep the odour level minimal:

1. The nearest populated area to the site is about 7 km on the wind side (not down wind), thus the probability of being affected by odours is very low.
2. The site design includes a gas collection and recovery system, which will lead to prevent the odour and release of GHG's to the atmosphere.

Leachate reduction

The problems associated with leachate may be minimised by limiting the amount of water getting into the solidwaste matrix. This can be achieved into a number of simple design and operational measures:

1. Ensuring surface water does not enter the landfilled areas or areas prepared for future landfilling by construction intercepting ditches between the working areas and surrounding unused parts of the site.
2. Ensuring water does not accumulate in the working area where waste is being landfilled.
3. Keeping the open areas at the tipping face as small as practicable.
4. Applying soil cover to the wastes at the end of each working day.
5. Progressively completing and grading areas of the site with a capping layer, as they reach their final design heights.

Surface water diversion is an important issue, since it will not only reduce the leachate quantities, but it also removes flooding by surface water which can destabilise waste slopes, resulting in slip failures (Table 33.9).

Table 33.9. Leachate calculation for Al-Ghabawi landfill site.

	Leachate flow M³/hr	Leachate 1000 (m³)	Waste water 1000 (m³)	Deposited waste 1000 (m³)	Evaporation 1000 (m³)	Surface runoff 1000 (m³)	Direct infiltration 1000 (m³)	Rainfall 1000 (m³)
Jan	24.92	17.94	13.97	139.7	116.9	3.1	3.968	124
Feb	25.19	18.14	13.97	139.7	122.78	3.255	4.1664	130.2
Mar	23.49	16.92	13.97	139.7	86.76	2.3	2.944	92
Apr	19.94	14.36	13.97	139.7	11.32	0.3	0.384	12
May	19.64	14.14	13.97	139.7	5.09	0.135	0.1728	5.4
June	19.41	13.98	13.97	139.7	0.189	0.005	0.0064	0.2
Jul	19.4	13.97	13.97	139.7	0	0	0	0
Aug	19.4	13.97	13.97	139.7	0	0	0	0
Sep	19.42	13.98	13.97	139.7	0.377	0.01	0.0128	0.4
Oct	20.14	14.5	13.97	139.7	15.65	0.415	0.5312	16.6
Nov	21.62	15.56	13.97	139.7	46.96	1.245	1.594	49.8
Dec	23.67	17.04	13.97	139.7	90.528	2.4	3.072	96
Annual	21.35	184.5	–	–	496.58	13.165	16.85	526.6

CASE STUDY 4: SOLID WASTE MANAGEMENT—MANGALORE CITY

Mangalore is an important city in Karnataka and is situated on the west coast. After integration the city is developing fast in all directions viz. in the field of education, industry and commerce.

Mangalore is headquarters of Dakshina-Kannada District, largest urban coastal center of Karnataka and the fourth largest city in the State. The city is an administrative, commercial, educational, and industrial center. Mangalore city has population of 5,79,306 (year 2009).

Mangalore City Corporation presided by a Mayor. The hierachical structure for handling solidwaste management (SWM) is shown in Fig. 33.2.

Health Department plays a key role in Mangalore City Corporation managing basic services for the citizens for instance sanitation, public health services, managing the entire solidwaste management as per MSW Rules 2000 and issuing birth and death certificates.

Solidwaste management is one the most important service which is handled by the Health Department. The most pressing problem faced is rapid urbanisation and changing lifestyles have led to the generation of huge amount of garbage and wastes in the urban areas, so much so, over the past few years; just handling this Municipal Solidwaste has assumed the proportion of major organisational, financial and environmental challenges.

Despite Municipal Solidwaste Management being major task of the local government, typical accounting for a sizeable portion of the municipal budget, yet the Urban Local Bodies is unable to provide effective services. Previously the waste was disposed in an unscientific manner, with crude open dumping in low lying areas being the prevalent practice followed by most Urban Local Bodies. The results of these are foul smell, breeding of flies and other pests and generation of liquid runoffs (Leachate), which pose a serious threat to the underground water reserves. The area coming under the jurisdiction of Mangalore City Corporation produces an average of 220 TPD of wastes, with a daily collection frequency of 200 TPD. The waste collected has a composition of 60 per cent of organic, 25 per cent of inorganic, 5 per cent of combustible and 10 per cent of recyclable wastes.

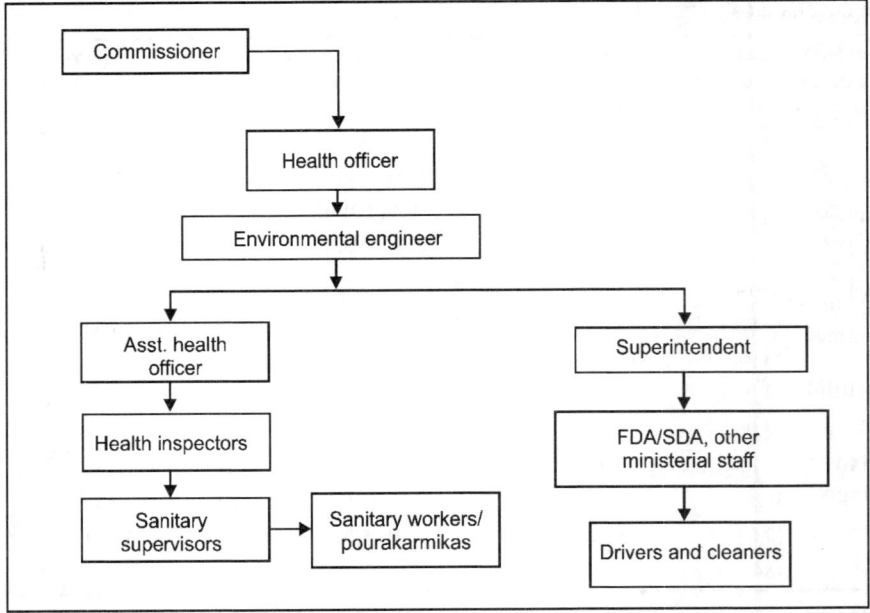

Fig. 33.2. Administrative hierarchical structure for handling SWM.

As per Municipal Solidwaste (Management and Handling) Rules, 2000, including all administrative, financial, legal planning and engineering functions involved in the whole spectrum of solutions to problems of solidwastes thrust upon the community by its inhabitants. The major components of solidwaste management are:

1. Segregation at the source.
2. Primary (door to door) collection.
3. Secondary storage.
4. Transportation.
5. Treatment and processing.
6. Disposal.

Segregation

1. Segregation of wastes into wet, dry/recyclables and household hazardous waste.
2. Familiarising people about the solidwaste management system adopted.
3. Training program for retrievers regarding importance of segregation, proper handling of waste and its hazards due to improper handling.
4. Littering of waste to be banned.

Primary Collection

1. Programmatic street sweeping A, B and C type roads.
2. MSW is not mixed with hospital and industrial waste.
3. No burning of waste in Mangalore.
4. Around 40 per cent of door to door waste collection system is carried out in Mangalore.
5. Recycling biodegradable waste.
6. Levying fines on one who doesn't follow the process of waste segregation.

7. House owners should be levied fine for throwing the garbage in open space.
8. Conducting awareness campaigns for schools, colleges, institutes and for citizens on door to door garbage collection.
9. Primary collection is been outsourced with the regular sanitation works into six zones.

Secondary Storage

1. Closed metal secondary storage containers are provided.
2. Closed bins systems for secondary storage are provided.
3. Manual handling of waste is minimised.
4. Reducing the secondary storage location by transferring garbage directly from door to door collection system to the transportation vehicle.

Transportation

1. Covered transportation vehicles are provided.
2. Avoiding multiple handling of waste.
3. Regular day wise clearance frequency is maintained.
4. Twin container dumper placers, compactors (Back Loaders), side loaders being used for transportation of waste to the processing site.
5. Rs. 214.00 lakhs was the capital investments made on the procurement of the transportation vehicles and containers for Municipal solidwaste transportation and reduce manual handling under Karnataka Urban Development and Coastal Environmental Management Projects (KUDCEMP).

Details on Processing and Disposal Site

Mangalore City Corporation processing and disposal site is having an extent of 37.32 acres of land for the disposal of solidwaste located on top of a hill. The site is divided into two portions by a road passing through with 26.69 acres on the north side and 10.63 acres on the south side. To the south side of the existing landfill 25.4 acres of land have been acquired for the construction of new Sanitary landfill site. The landfill area is adequate for about 25 years of life. The landfill is constructed in phases to enable progressive development. The development of the landfill over the life would be done in 4 phases. Phase I for 6 years time frame, phase II, III and IV for 3 years, 6 years and for 10 years time frame. To the north side, compost yard for a capacity of 120 TPD of waste is being treated and 68 vermin-composting pits for a capacity of 25 TPD is been constructed. The construction of the Municipal Solidwaste Processing and Sanitary landfill site was been constructed under Karnataka Urban Development and Coastal Environmental Management Projects (KUDCEMP) funded through Asian Development Bank loans. Rs. 1587.00 lakhs was the capital investment made on the construction of processing and sanitary landfill site.

Treatment and Processing

1. Aerobic composting and vermin-composting is provided.
2. Provision of 120 TPD of waste is aerobically composted and 25 TPD of waste is vermin-composted.
3. Rejects from compost plant will be transported to sanitary landfill site.
4. Presently Mangalore City Corporation is handling the operation and maintenance of aerobic compost plant.
5. Compost plant is generating 4.0 to 5.0 tons of compost/manure through windrow method of aerobic composting.

6. Operation of vermin-composting is not yet started.
7. Provision is made for running the aerobic compost plant in 2 shifts.
8. Arrangements have been made for outsourcing the entire treatment and processing unit.

Disposal/Landfilling

1. Sanitary landfill site is provided in a 6 acre land as I phase.
2. Rejects from the compost plant will be landfilled.
3. Daily soil top cover of 10 cm will be provided.
4. Presently the sanitary landfill site constructed is not being used.
5. Previous dumped landform will be covered with soil and will be provided with green cover.

Ongoing Projects

1. The entire sanitation of the Mangalore will be divided into 2 zones (North and South) from door to door collection, street sweeping, vegetation cutting, drain cleaning, garbage transportation will be out-sourced along with the operation and maintenance of the compost plant and sanitary landfill site.
2. Establishment of a new modernised abattoir is under process.
3. Additional construction of processing concrete yard is under process.

Information, Education and Communication (IEC)

1. Door to door campaigning for educating the public on segregation and door to door garbage collection.
2. Educated school and college students through IEC activities on SWM.
3. Conducted school and college level competitions for students with cash prizes.
4. Conducted ward level meetings to educate the publics on MSW Rules, 2000.
5. Conducted street jathas to create awareness among public.
6. Conducted several plastic raids in the city with the Pollution Control Board personnel to reduce the sale of plastic below 20 microns.

Biomedical Waste Management

1. The biomedical waste (Management and Handling) Rules notified in July 1998 under the Environment (Protection) Act, 1986, make mandatory for all healthcare facilities irrespective of their size to treat biomedical waste generated by them. In order to comply with the provisions of the Rules, some of the healthcare facilities have installed their own treatment facilities and others are availing services of Common Biomedical Waste Treatment Facilities (CBWTF).
2. Common biomedical waste treatment facilities (CBWTF) has been set by M/s. Medicare Incin Private Limited.
3. Mangalore City Corporation has issued notices to all the Hospitals/Clinics/Nursing homes to treat the biomedical waste as per the BMW rules 1998 and keep informing regularly to Hospitals/Clinics/Nursing Homes not mix the biomedical waste with municipal solidwaste.
4. M/s. Medicare Incin Private Limited has installed 7500 kg/day capacity incineration plant but Karnataka state pollution control board has given authorisation for a capacity of 2500 kg/day.
5. M/s. Medicare Incin Private Limited has provided their own vehicles for collection of biomedical waste from the Hospitals/Clinics/Nursing Homes/Laboratories of total 6200 beds throughout Dakshina Kannada. Father Muller's Hospital and Yenepoya Medical College have installed their own treatment facilities.

Glossary

Abiotic	:	Characterised by absence of life; abiotic materials include non-living environmental media (e.g. water, soils, sediments); abiotic characteristics include such factors as light, temperature, pH, humidity, and other physical and chemical influences.
Absorbed dose	:	The amount of a substance penetrating the exchange boundaries of an organism after contact. Absorbed dose for the inhalation and ingestion routes of exposure is calculated from the intake and the absorption efficiency. Absorbed dose for dermal contact depends on the surface area exposed and absorption efficiency.
Absorption	:	Absorption is the passage of one substance into or through another.
Absorption efficiency	:	A measure of the proportion of a substance that a living organism absorbs across exchange boundaries (e.g. gastrointestinal tract).
Activated sludge	:	The product that results when primary municipal waste-water is mixed with bacteria-laden sludge and then agitated and aerated to promote biological treatment, speeding the breakdown of organic matter in raw municipal waste undergoing secondary waste-water treatment.
Adsorption	:	Adsorption is the adhesion of molecules of gas, liquid or dissolved solids to a surface. The term also refers to a method of treating wastes in which activated carbon is used to remove organic compounds from waste-water.
Adverse ecological effects	:	Changes that are considered undesirable because they alter valued structural or functional characteristics of ecosystems or their components. An evaluation of adversity may consider the type, intensity, and scale of the effect as well as the potential for recovery.
Anaerobic	:	A living system or process that occurs in or is not destroyed by the absence of oxygen.
Anaerobic decomposition	:	A type of decomposition that does not use oxygen. The second phase of decomposition that typically occurs in landfilled wastes. The end products are mainly methane (CH_4) and carbon dioxide (CO_2) gases. Anaerobic decomposition creates odour problems.
Aquifer	:	An aquifer is an underground rock formation composed of such materials as sand, soil or gravel that can store groundwater and supply it to wells and springs.
Biodegradable material	:	Materials that can be broken down by micro-organisms into simple, stable compounds such as carbon dioxide and water. Most organic materials, such as food scraps and paper, are biodegradable.
Bioaccumulation	:	General term describing a process by which chemicals are taken up by an organism either directly from exposure to a contaminated medium or by consumption of food containing the chemical.

521

Bioassessment	:	A general term referring to environmental evaluations involving living organisms; can include bioassays, community analyses, etc.
Bioassay	:	Test used to evaluate the relative potency of a chemical by comparing its effect on living organisms with the effect of a standard preparation on the same type of organism. Bioassay and toxicity tests are not the same—see toxicity test. Bioassays often are run on a series of dilutions of whole effluents.
Bioavailability	:	The degree to which a material in environmental media can be assimilated by an organism.
Bioccumulation factor (BAF)	:	The ratio of the concentration of a contaminant in an organism to the concentration in the ambient environment at steady state, where the organism can take in the contaminant through ingestion with its food as well as through direct contact.
Bioconcentration	:	A process by which there is a net accumulation of a chemical directly from an exposure medium into an organism.
Biodegrade	:	Decompose into more elementary compounds by the action of living organisms, usually referring to micro-organisms such as bacteria.
Biodegradability	:	Biodegradability is the capability of a substance to break down into simpler substances, especially into innocuous products, by the actions of living organisms (that is, micro-organisms).
Biomagnification	:	Result of the process of bioaccumulation and biotransfer by which tissue concentrations of chemicals in organisms at one trophic level exceed tissue concentrations in organisms at the next lower trophic level in a food chain.
Biomarker	:	Biochemical, physiological, and histological changes in organisms that can be used to estimate either exposure to chemicals or the effects of exposure to chemicals.
Biomonitoring	:	Use of living organisms as 'sensors' in environmental quality surveillance to detect changes in environmental conditions that might threaten living organisms in the environment.
Characterisation of ecological effects	:	A portion of the analysis phase of ecological risk assessment that evaluates the ability of a stressor to cause adverse effects under a particular set of circumstances.
Contaminant	:	A contaminant is any physical, chemical, biological or radiological substance or matter present in any media at concentrations that may pose a threat to human health or the environment.
Contaminant of (ecological) concern	:	A substance detected at a hazardous waste site that has the potential to affect ecological receptors adversely due to its concentration, distribution, and mode of toxicity.
Contaminants of potential concern	:	Chemicals that are potentially site-related and whose data are of sufficient quality for use in a quantitative risk assessment.
Corrosivity	:	Corrosive wastes include those that are acidic and capable of corroding metal such as tanks, containers, drums, and barrels.
Critical exposure pathway	:	An exposure pathway which either provides the highest exposure levels or is the primary pathway of exposure to an identified receptor of concern.
Cumulative distribution function (CDF)	:	Cumulative distribution functions are particularly useful for describing the likelihood that a variable will fall within different ranges of x. $F(x)$ (i.e. the value of y at x in a CDF plot) is the probability that a variable will have a value less than or equal to x.

Cumulative ecological risk assessment	:	A process that involves consideration of the aggregate ecological risk to the target entity caused by the accumulation of risk from multiple stressors.
Degradation	:	Conversion of an organic compound to one containing a smaller number of carbon atoms.
Disposal	:	Disposal is the final placement or destruction of toxic, radioactive or other wastes; surplus or banned pesticides or other chemicals; polluted soils; and drums containing hazardous materials from removal actions or accidental release. Disposal may be accomplished through the use of approved secure landfills, surface impoundments, land farming, deep well injection or ocean dumping.
Ecological component	:	Any part of an ecosystem, including individuals, populations, communities, and the ecosystem itself.
Ecosystem	:	The biotic community and abiotic environment within a specified location and time, including the chemical, physical, and biological relationships among the biotic and abiotic components.
Ecotoxicity	:	The study of toxic effects on nonhuman organisms, populations or communities.
Environmental audit	:	An environmental audit usually refers to a review or investigation that determines whether an operating facility is in compliance with relevant environmental regulations. The audit may include checks for possession of required permits, operation within permit limits, proper reporting, and record keeping. The typical result is a corrective action or compliance plan for the facility.
Environmental risk	:	Environmental risk is the chance that human health or the environment will suffer harm as the result of the presence of environmental hazards.
Ex situ	:	The term *ex situ* or 'moved from its original place', means excavated or removed.
Fate	:	Disposition of a material in various environmental compartments (e.g. soil or sediment, water, air, biota) as a result of transport, transformation, and degradation.
Filtration	:	Filtration is a treatment process that removes solid matter from water by passing the water through a porous medium, such as sand or a manufactured filter.
Fly-ash	:	Small, solid particles of ash and soot generated when coal, oil or waste materials are burned. Fly-ash is suspended in the flue gas after combustion and is removed by pollution control equipment.
Habitat	:	Place where a plant or animal lives, often characterised by a dominant plant form and physical characteristics.
Hazard	:	The likelihood that a substance will cause an injury or adverse effect under specified conditions.
Hazard identification	:	The process of determining whether exposure to a stressor can cause an increase in the incidence or severity of a particular adverse effect, and whether an adverse effect is likely to occur.
Hazard index	:	The sum of more than one hazard quotient for multiple substances and/or multiple exposure pathways. The hazard index (HI) is calculated separately for chronic, subchronic, and shorter-duration exposures.
Hazard quotient	:	The ratio of an exposure level to a substance to a toxicity value selected for the risk assessment for that substance (e.g. LOAEL or NOAEL).

Heavy metal	:	The term heavy metal refers to a group of toxic metals including arsenic, chromium, copper, lead, mercury, silver, and zinc. Heavy Metals often are present at industrial sites at which operations have included battery recycling and metal plating.
Ignitability	:	Ignitable wastes can create fires under certain conditions. Examples include liquids, such as solvents that readily catch fire, and friction-sensitive substances.
In situ	:	The term *in situ*, 'in its original place', or 'on-site', means unexcavated and unmoved. *In situ* soil flushing and natural attenuation are examples of *in situ* treatment methods by which contaminated sites are treated without digging up or removing the contaminants.
Landfill gas	:	A mixture of primarily methane and carbon dioxide that is generated in landfills by anaerobic decomposition of organic wastes.
Leachate	:	A leachate is a contaminated liquid that results when water collects contaminants as it trickles through wastes, agricultural pesticides, or fertilisers. Leaching may occur in farming areas and landfills and may be a means of the entry of hazardous substances into soil, surface water or groundwater.
Leachate collection system	:	A network of pipes or geotextiles/geonets placed at low areas of the landfill liner to collect leachate from a landfill for storage and treatment. Flow of leachate along the liner is facilitated by the use of a soil drainage blanket or geonet.
Mechanical separation	:	The separation of waste into components using mechanical means, such as cylcones, trommels, and screens.
Migration pathway	:	A migration pathway is a potential path or route of contaminants from the source of contamination to contact with human populations or the environment. Migration pathways include air, surface water, groundwater, and land surface. The existence and identification of all potential migration pathways must be considered during assessment and characterisation of a waste site.
Moisture content	:	The fraction or percentage of a substance or soil that is water.
Municipal solid waste (MSW)	:	MSW means household waste, commercial solid waste, nonhazardous sludge, conditionally exempt small quantity hazardous waste and industrial solid waste.
Pathogens	:	Disease-causing agents, especially micro-organisms such as bacteria, viruses, and fungi.
Permeability	:	Permeability is a characteristic that represents a qualitative description of the relative ease with which rock, soil or sediment will transmit a fluid (liquid or gas).
Pesticide	:	A pesticide is a substance or mixture of substances intended to prevent or mitigate infestation by or destroy or repel, any pest. Pesticides can accumulate in the food chain and or contaminate the environment if misused.
Plume	:	A plume is a visible or measurable emission or discharge of a contaminant from a given point of origin into any medium. The term also is used to refer to measurable and potentially harmful radiation leaking from a damaged reactor.
Point source	:	A point source is a stationary location or fixed facility from which pollutants are discharged or emitted or any single, identifiable discharge point of pollution, such as a pipe, ditch or smokestack.
Radioactive waste	:	Radioactive waste is any waste that emits energy as rays, waves or streams of energetic particles. Sources of such wastes include nuclear reactors, research institutions, and hospitals.

Refuse-derived fuel (RDF)	:	Product of a mixed waste processing system in which certain recyclable and noncombustible materials are removed, with the remaining combustible material converted for use as a fuel to create energy.
Risk assessment	:	Qualitative or quantitative evaluation of the risk posed to human health and/or the environment by the actual or potential presence or release of hazardous substances, pollutants or contaminants.
Salvaging	:	At landfills or material recovery facilities, salvaging is the controlled separation of recyclable and reusable materials. Controlled means that the separation is monitored by operators.
Scavenging	:	At a landfill or material recovery facility, scavenging is the uncontrolled separation of recyclable and reusable materials. Uncontrolled means that the operator does not monitor the removal of materials, and in many cases prohibits it. Material scavenging of recyclables may also occur at the curb or at drop-off centres.
Scrap	:	Discarded or rejected industrial waste material often suitable for recycling.
Sensitivity	:	In relation to toxic substances, organisms that are more sensitive exhibit adverse (toxic) effects at lower exposure levels than organisms that are less sensitive.
Shredder	:	A mechanical device used to break waste materials into smaller pieces by tearing and impact action. Shredding solid waste is done to minimise its volume and make it more readily combustible.
Sludge	:	Sludge is a semisolid residue from air or water treatment processes. Residues from treatment of metal wastes and the mixture of waste and soil at the bottom of a waste lagoon are examples of sludge, which can be a hazardous waste.
Solid waste	:	Any garbage, or refuse, sludge from a waste-water treatment plant, water supply treatment plant, or air pollution control facility and other discarded material, including solid, liquid, semisolid or contained gaseous material resulting from industrial, commercial, mining, and does not include solid or dissolved materials in domestic sewage.
Source reduction	:	The design, manufacture, acquisition, and reuse of materials so as to minimise the quantity and/or toxicity of waste produced. Source reduction prevents waste either by redesigning products or by otherwise changing societal patterns of consumption, use and waste generation.
Surface water	:	Surface water is all water naturally open to the atmosphere, such as rivers, lakes, reservoirs, streams, and seas.
Toxic mechanism of action	:	The mechanism by which chemicals produce their toxic effects, i.e. the mechanism by which a chemical alters normal cellular biochemistry and physiology. Mechanisms can include: interference with normal receptor-ligand interactions, interference with membrane functions, interference with cellular energy production, and binding to biomolecules.
Toxicity test	:	The means by which the toxicity of a chemical or other test material is determined. A toxicity test is used to measure the degree of response produced by exposure to a specific level of stimulus (or concentration of chemical) compared with an unexposed control.

Toxicity value	:	A numerical expression of a substance's exposure-response relationship that is used in risk assessments.
Underground storage tank (UST)	:	A UST is a tank and any underground piping connected to the tank that is used to contain gasoline or other petroleum products or chemical solutions and that is placed in such a manner that at least 10 per cent of its combined volume is underground.
Volatile organic compound (VOC)	:	A VOC is one of a group of carbon-containing compounds that evaporate readily at room temperature. Examples of VOCs include trichloroethane; trichloroethylene; and BTEX. These contaminants typically are generated from metal degreasing, printed circuit board cleaning, gasoline, and wood preserving processes.
Waste reduction	:	Waste reduction is a broad term encompassing all composting—that result in reduction of waste going to a combustion facility or landfill.
Waste-to-energy system (WTE)	:	A method of converting MSW into a usable form of energy, usually through combustion.
Windrow	:	A large, elongated pile of composting material, which has a large exposed surface area to encourage passive aeration and drying.

References

Allen, D.M., *Environmental Engineering*, Marcel Dekker Inc., New York.

Arceivala, K.W., and Lloyd, R., *Waste Minimisation*, Academic Press, London.

Baker, J.L., *Handbook of Industrial Pollution*, Applied Science Publishers, London.

Brown, R.K., *Treatment of Water and Solid Wastes*, Springfield, New York.

Bryan, G.C., *Health Hazards and Human Health*, Academic Press, London.

Budyko, M.K., *Chemical and Biological Methods of Water Pollution Studies*, Academic Press, London.

Cattabeni, M., Cavallaro, A. and Galli, G., *Fundamental of Ecology*, S.P. Medical and Scientific Books, New York.

Connell, D.N. and Miller, G.J., *Chemistry of Water Pollution*, John Wiley & Sons, New York.

Coolingwood, R.S., *Ecological Aspects of Used Water Treatment*, John Wiley & Sons, New York.

Curds, C.R. and Hawkes, H.A., *Basic Hazardous Waste Management*, Academic Press, London.

Dix, H.M., *Waste-Water Engineering, Treatment and Disposal*, John Wiley & Sons, New York.

Dugan, P.R., *Environmental Engineering and Sanitation*, Plenum Publishing Corporation, London.

Goldberg, E.D., *Hazardous Waste Management*, Gordon and Breach, Science Publishers, New York.

James, A. and Evison L., *Treatment of Industrial Wastes*, John Wiley & Sons, New York.

Kathern, R.L., *Pollution Prevention and Waste Minimisation*, Harwood Academic Publishers, New York.

Lehr, J.H., Tyler, E.G., Wayne, A.P. and Jack, D., *Handbook of Solid Waste Management*, McGraw-Hill, New York.

Lenihan, J. and Fletcher, W.W., *Introduction to Environmental Science and Technology*, Blackie, Glasgow and London.

Liptak, Bela G., *Industrial Pollution—Origins, Characteristics and Treatment*, Chilton Book Company, Radnor, Pennsylvania.

Lowman, F.G., *Biochemical Ecology of Waste-Water*, National Academy of Sciences, Washington DC.

Masters, G.M., *Toxic Metal in the Environment*, John Wiley & Sons, New York.

McCaull, J. and Crossland, J., *Pollution Prevention*, Harcourt Brace Jovanovich, New York.

Neff, J.S., *Treatment of Industrial Wastes*, Applied Science Publishers, London.

Nemerow, N.C., *Industrial Waste Treatment*, Addison-Wesley Publishing Company, Philippines.

Odum, E.P., *Integrated Solid Waste Management*, John Wiley & Sons, New York.

Palmer, C.L., *Environmental Technology*, Castle Housing Publications Ltd., London.

Reid, G.C., *Sewage and Sludge Treatment*, Reinhold Publishing Corporation, New York.

Webster, J., *Handbook of Solid Wastes*, Cambridge University Press, Cambridge.

Index